Exploratory and Robust Data Analysis

Exploratory and Robust Data Analysis: A Modern Applied Statistics Guide Using SPSS and R is an essential resource for students, researchers, and professionals seeking a comprehensive yet practical approach to modern statistical analysis. This book bridges traditional statistical methods with contemporary techniques, emphasizing exploratory and robust data analysis while integrating powerful computational tools such as R and SPSS.

Designed for intermediate-level courses and research applications, the book begins with fundamental concepts in exploratory data analysis, graphical methods, and confirmatory statistical procedures. It then introduces robust statistical methods, including M-estimators, high breakdown estimators, bootstrap techniques, and Monte Carlo simulations, equipping readers with tools to handle complex and real-world data scenarios. Key topics include regression analysis, multiple linear models, nonparametric regression, and generalized linear models, ensuring broad applicability across disciplines.

What sets this book apart is its emphasis on theoretical foundations and hands-on applications. Annotated computer sessions guide readers through statistical analysis, enabling them to apply techniques effectively while understanding their theoretical underpinnings. This book fosters an analytical mindset that encourages critical thinking and data-driven decision-making by combining classical statistical procedures with modern computational methods.

With real-world datasets, practical exercises, and detailed software integration, this book is an indispensable guide for those looking to master data analysis in an era where statistical rigor and computational efficiency are paramount.

Exploratory and Robust Data Analysis

A Modern Applied Statistics Guide Using SPSS and R

Abdel-Salam G. Abdel-Salam
and Jeffrey B. Birch

CRC Press
Taylor & Francis Group
Boca Raton London New York

CRC Press is an imprint of the
Taylor & Francis Group, an **informa** business
A CHAPMAN & HALL BOOK

Designed cover image: Shutterstock_ 2448232829

First edition published 2026
by CRC Press
2385 NW Executive Center Drive, Suite 320, Boca Raton FL 33431

and by CRC Press
4 Park Square, Milton Park, Abingdon, Oxon, OX14 4RN

CRC Press is an imprint of Taylor & Francis Group, LLC

ISBN: 978-1-032-93192-0 (hbk)
ISBN: 978-1-032-93308-5 (pbk)
ISBN: 978-1-003-56532-1 (ebk)

DOI: 10.1201/9781003565321

Typeset in Palatino
by Apex CoVantage, LLC

Dedication to

My wife, sweet children, parents, all my family and colleagues.
Abdel-Salam Gomaa

All my family, students, friends, and colleagues.
Jeffrey Birch

Contents

Preface

The field of statistics has undergone significant transformations over the past few decades, especially with AI nowadays. A notable example is the evolution of introductory statistics courses. Twenty or thirty years ago, these courses typically began with an introduction to probability theory, covering topics such as permutations and combinations. They emphasized mathematical and statistical concepts, including the expectation and variance of a random variable. Inferential techniques, such as hypothesis testing and confidence intervals, were illustrated using simple, often artificial, examples. The datasets used in textbooks were typically small, consisting of ten to twenty observations, with minimal emphasis on real-world data analysis.

In contrast, contemporary introductory statistics courses prioritize data analysis from the outset. Students are taught that proper analysis begins with graphical exploration. Basic plotting techniques, such as dot plots and box plots, guide future analyses. Modern courses introduce a broader range of statistical methods and emphasize comparative analysis. Textbooks now incorporate real data from practical problems, and computers play a central role in enabling complex statistical analyses efficiently and affordably.

This book aligns with this modern approach, offering an intermediate-level applied statistics course that integrates contemporary statistical philosophy and procedures. It teaches students to analyze data using various graphical and statistical methods, providing both theoretical justification and practical application. As a result, the book offers a deeper theoretical foundation than typical statistical methods textbooks.

For the student

The book presents statistical concepts through the complexity of data structures. The initial chapters (1–4) focus on the single-variable or single-sample scenario (the "location" model):

- **Chapter 1** introduces basic plotting techniques and the philosophy of exploratory data analysis, including data transformation.
- **Chapter 2** discusses confirmatory data analysis (inferential methods), covering classical statistical procedures such as least squares, weighted least squares, maximum likelihood estimation, and the delta method.
- **Chapter 3** introduces robust data analysis, emphasizing M-estimation principles (Huber, 1964).

- **Chapter 4** focuses on computing M-estimators and high-breakdown estimators (least median of squares, least trimmed squares). It also covers computational methods such as bootstrapping and Monte Carlo simulations, which are applied to the location model.

These foundational chapters set the stage for subsequent topics. The core concepts—data examination, transformation, application of classical, nonparametric, and robust methods, and comparative analysis—are then extended to other data structures. Bootstrapping and Monte Carlo methods are further applied across different statistical settings.

Later chapters cover:

- **Chapter 5**: The two-sample and k-sample problems.
- **Chapter 6**: Simple linear regression, including nonparametric regression methods such as kernel regression and local polynomial regression, with robust variations.
- **Chapter 7**: Multiple linear regression, incorporating robust and nonparametric techniques.
- **Chapter 8**: Generalized linear models for non-normal data, including binomial and Poisson distributions.

Throughout the text, the emphasis is on **the practical application** of each technique using **R** and **SPSS software. Where appropriate, and when it can be done conveniently, statistical theory is enhanced by mathematics.**

<u>There are several unique features of this text:</u>

1) This book teaches students to use basic graphical methods to examine data and **use their observations in subsequent analysis steps**.
2) The text exposes the students to **modern methods of robust data analysis**. It uses intermediate-level theory to motivate, explain, and provide theoretical background for the methods. Students **learn to use the methods by seeing them applied to real data sets** through annotated computer sessions.
3) The **R and SPSS statistical software** are emphasized throughout the text. While both are excellent packages, the R system is emphasized. Their availability on both the PC and Mac platforms, the ease with which R or SPSS can be taught to the beginning student, and the affordability of R and SPSS all led to this decision. The software packages are used in several ways.

First, basic R and SPSS commands are used when appropriate.

Second, the students learn to write R code to enable new methods to be performed. For example, the book includes codes to perform M-estimation in the linear model, bounded-influence estimation, kernel and local linear regression, and many others. All codes and macros will be available on the website that accompanies this text.

Third, each SPSS and R session is heavily annotated to convey the ideas and thoughts as the data analysis unfolds on the screen. This is the most unique feature of the text. The student **is interactively guided through the analysis.** For example, in a multiple regression problem, the student begins the analysis by plotting the data. This leads to the formulation of a model and a variety of comparative analyses, such as the effect of data transformations on removing outliers or stabilizing variance or a robust analysis to deal with high influence points. The results of each form of analysis are compared through the examination of the residuals, with a heavy emphasis on plotting. Finally, conclusions are given based on the observations that were previously discussed.

Fourth, the text is designed to encourage the students to use the computer, through R and SPSS, to address "what if" questions. For example, in the classical location model, the observations are usually assumed to be independent from a normal distribution with a constant variance. Under these assumptions, the students are taught in any basic statistic course that the sample mean is the optimal estimator of the location parameter, and the t-test is the most powerful test for the location parameter. **What if the observations are from a normal distribution with nonconstant variance? What if the observations are from some non-normal distribution with constant or nonconstant variance?** What happens to the performance of the sample mean, or what happens to the **power of the t-test**? Statistical software is extremely useful for examining these kinds of questions due to the simplicity with which Monte Carlo simulations may be performed. ***The text asks many of these "what if" questions*** and shows the students how to address them on R and SPSS with fully annotated solutions. After reading this text, students should thoroughly understand how to use the Monte Carlo method to address other interesting issues.

The book emphasizes **applications with enough theory** to provide the necessary substance and motivation for each new concept. The book is not a regression textbook, even though three chapters of this text deal heavily with regression. Nor is the text a book on numerical methods in current use in statistics, even though a considerable amount of attention is given to various numerical methods. Consequently, this text is not a mathematical statistics text, a methods text, a regression text, or a numerical methods text, but a combination of all these topics, always emphasizing **the applications.**

The exercises are chosen to guide the student through the techniques presented in the preceding chapter. Some exercises ask the students to derive some concept using algebra, others ask students to analyze a data set, mimicking the techniques discussed in the section, and others ask the students to address various "what if" questions. The latter form of exercise is useful since it asks the student to design and carry out the Monte Carlo experiment.

Most of the **data sets presented in the text are real data** taken from articles, theses, dissertations, and consulting projects. All data presented in the text are also available on the book's website.

To the instructor:

This book is a product of years of teaching statistics at **Virginia Tech, Qatar University, and Oklahoma State University**. Most students in these courses have prior exposure to applied statistics, probability, inferential statistics, and experimental design, often with a background in regression analysis. However, even advanced students will find new material in this text.

The following table suggests the percentage of course time allocated to each chapter in a **15-week semester**:

Percentage of Time Devoted to Each Chapter (15-week course).

Chapter	Percentage of Time (in %)
1	8%
2	8%
3	15%
4	15%
5	6%
6	22%
7	21%
8	5%

Early chapters can be covered more quickly, as students may already be familiar with some concepts. More time should be devoted to regression (Chapters 6 and 7). If students lack prior exposure to regression, instructors may need to adjust the pacing.

The book is also suitable for advanced undergraduates who have completed a year of statistical methods and a calculus-based mathematical statistics course. Theoretical sections in **Chapters 3, 7, and 8** require more sophistication; instructors may choose to cover these at a slower pace or supplement them with additional explanations in courses where students are less experienced with matrix algebra.

Importance of This Book to Researchers

This book serves as an essential resource for researchers by:

- Covering both **classical** and **modern** statistical methods, including robust techniques and nonparametric regression.
- Emphasizing **real data analysis**, ensuring applicability to real-world research problems.

- Introducing **computer-intensive methods** such as bootstrap and Monte Carlo simulations to evaluate estimators and tests in complex scenarios.
- Highlighting **robust techniques** for handling data that deviate from classical assumptions (e.g., non-normality, heteroscedasticity, and outliers).
- Providing **detailed, annotated R and SPSS sessions**, making statistical methods accessible for researchers unfamiliar with certain techniques.
- Encouraging an **exploratory mindset**, fostering deeper insights through comparative analyses and visualization.
- Balancing **practical application with theoretical rigor** ensures researchers understand both the implementation and underlying principles of statistical methods.
- Including **"what if" scenarios**, guiding researchers in critically assessing assumptions and evaluating robustness.
- The **accompanying website** offers supplementary materials, including code and datasets, for hands-on practice and experimentation.

Finally

Exploratory and Robust Data Analysis: A Modern Applied Statistics Guide Using SPSS and R is a valuable resource for researchers. Its comprehensive coverage, practical focus, emphasis on robust methods, and integration of exploratory and confirmatory data analysis make it an essential guide for conducting rigorous and insightful statistical research. This book stands out because it emphasizes real-world data analysis, providing practical guidance and hands-on experience with real data sets. Researchers are equipped with the necessary tools, methods, and mindset to conduct rigorous and innovative statistical analyses, ultimately enhancing the quality and impact of their work. The book's detailed coverage of modern statistical techniques, practical application examples, and integration of robust methods make it a must-have resource for any researcher aiming to produce high-quality and impactful statistical research.

The Authors

Abdel-Salam G. Abdel-Salam

Dr. Abdel-Salam is an Associate Professor of Statistics at Qatar University, holding prestigious titles as a Chartered Statistician (CStat) from the Royal Statistical Society (UK) and a Certified Professional Statistician (PStat) from the American Statistical Association (USA). He earned BS and MS degrees in Statistics from Cairo University (2004), an MS (2006) and Ph.D. (2009) from Virginia Tech (USA), and an MBA from Qatar University (2023). His academic career includes teaching at Cairo University, Virginia Tech, and Oklahoma State University, alongside corporate experience as Assistant Vice President in Risk Management at JPMorgan Chase & Co. With numerous publications and international presentations, his expertise spans statistical modeling, quality control, and economic capital models. His achievements have been recognized with awards such as the Teaching Excellence Award (Department of Statistics, Virginia Tech), the Freud International Award, the Mary G. Natrella Scholarship (ASA & ASQ), and the Outstanding Faculty Service Award (Qatar University, 2018).

Dr. Abdel-Salam's research interests cover statistical process control, regression analysis, multivariate analysis, AI applications, and public health surveillance. Since 2005, he has been an active statistical consultant, contributing to the Department of Statistics' Laboratory for Interdisciplinary Statistical Analysis and leading the Statistical Consulting Unit at Qatar University since 2013. He provides expert guidance to researchers across various fields, translating complex data challenges into actionable insights. With a strong passion for interdisciplinary collaboration, he continues to advance research in applied statistics and artificial intelligence.

Jeffrey B. Birch

Prof. Birch is a distinguished statistician and academic known for contributing to robust statistical methodologies and data analysis. He earned his Ph.D. in Biostatistics from the University of Washington in 1978 and has held a longstanding academic career at Virginia Polytechnic Institute and State University (Virginia Tech). Birch served as the Director of Graduate Programs in Statistics from 2001 to 2016 and has been a Professor in the Department of Statistics since 1999. His research spans areas such as robust regression, nonparametric statistics, and statistical applications in various scientific fields. Throughout his career, he has received multiple teaching excellence awards and has been an active member of professional societies, including the American Statistical Association and Sigma Xi.

Acknowledgments

We are forever thankful for the encouragement and love from our parents, wives and families, colleagues, and students, who inspired and motivated us to complete this book.

Abdel-Salam Gomaa
Jeffrey Birch

Introduction

Confirmatory and exploratory data analyses are two methodologies essential for the successful statistical evaluation of data. Confirmatory data analysis (CDA) involves the application of standard statistical inferential methodologies, such as confidence interval techniques and hypothesis testing, to confirm statements. On the other hand, exploratory data analysis (EDA) primarily deals with the initial study and evaluation of data, essentially exploring the data to identify interesting features such as trends, shapes, tendencies, and suggestions of model characteristics, as well as unusual features like outliers. Proper use of EDA techniques allows the user to formulate models that may be more appropriate for the situation under study than other models that are often used simply because they have become "standard models" taught in most textbooks.

The relationship between EDA and CDA can be illustrated by considering the following outline of a typical experimental procedure in which statistics is used.

Step 1: Formulate the research question.

Step 2: Design and conduct the experiment.

Step 3. Gather the data.

Step 4. Formulate models.

Step 5. Statistically evaluate the models

Step 6. Make decisions regarding the research question.

CDA emphasizes all aspects of the model formulation process and the inferential decision-making process. Consequently, it is used primarily in steps 5 through 6.

EDA examines the data and helps formulate the models. Thus, it is mainly used in steps 4 and 6 of our six-step process.

Consider the following example illustrated using the statistical modeling process for the location model.

In the location model, we observe n responses of the random variable y: $y_1, y_2, \ldots y_n$. We naturally formulate the model:

$$y_i = \mu + \varepsilon_i$$

where $\varepsilon_i \sim N(0, \sigma^2)$. That is, the assumption of normality and constant variance are automatically made. At least, this is the approach often taken in elementary and intermediate statistical texts. Under these assumptions, our statistical theory directs our next steps. The estimation of μ is easily obtained as $\hat{\mu} = \bar{y}$, this is the least squares (LS) and the maximum likelihood estimate (MLE). Estimation of σ^2 is obtained as $\hat{\sigma}^2 = \frac{\sum (y_i - \bar{y})^2}{n-1}$.

Inference on μ is accomplished by the use of the standard t distribution. For example, to test $H_0 : \mu = \mu_0$ versus $H_1 : \mu \neq \mu_0$. We use the t-test as

$$t = \frac{\bar{y} - \mu_0}{\hat{\sigma} / \sqrt{n}} \sim t'(n-1, \lambda),$$

where $t'(n-1, \lambda)$ denotes a non-central t distribution with n-1 degrees of freedom and non-centrality parameter λ, where

$$\lambda = \frac{\mu - \mu_0}{\sigma / \sqrt{n}}.$$

The model may be evaluated by examination of the residuals,

$$r_i = y_i - \hat{y}_i = y_i - \hat{\mu} = y_i - \bar{y}.$$

Further inference on μ consists of a $(1-\alpha)100\%$ confidence interval for μ of the form $\left(\bar{y} \pm t_{1-\frac{\alpha}{2},\, n-1}\left(\frac{\hat{\sigma}}{\sqrt{n}}\right)\right)$ where $t_{1-\frac{\alpha}{2}, n-1}$ is the $\left(1-\frac{\alpha}{2}\right)$ percent point of the t distribution with $n-1$ degrees of freedom.

We note that once the model has been formulated, all steps in our procedure are determined from statistical theory, using optimal procedures under the given assumptions. However, if the model provides an incorrect representation of the true situation, then the procedures are no longer optimal, and other methods may be significantly better. For example, if the normality assumption is not true, then the t-test procedure may be seriously compromised, as the most powerful test of μ and other procedures may be preferred. Additionally, confidence intervals based on the t distribution and of the form given above may be misleading, neither accurate in terms of true confidence level nor precise interval estimates of μ.

To illustrate some of these points, consider the following data consisting of $n = 22$ observations representing the percentage of wastage per week in cloth at a mill producing cloth for Levi Strauss in Albuquerque, New Mexico (Koopmans, 1987). This data is part of a larger data set, which will be examined in more detail in Chapter 4. It is desirable to have small values of wastage, with values close to zero preferred. Negative values result when workers can make the product requiring less material than specified by the computerized layouts of patterns. A production system under proper control should provide values of wastage whose average is close to zero with small variability. The 22 values of wastage are:

0.12, 1.01, 0.15, -0.20, -0.30, -0.07, 0.32, 0.27, -0.32, -0.17, 0.24, 0.03, 0.35, -0.08, 1.94, 0.28, 1.30, 4.27, 0.14, 0.30, 0.24, 0.13

SPSS could be used to perform the descriptive statistics as follows:

Steps

Go to *SPSS – Analyze – Descriptive Statistics – Descriptives*

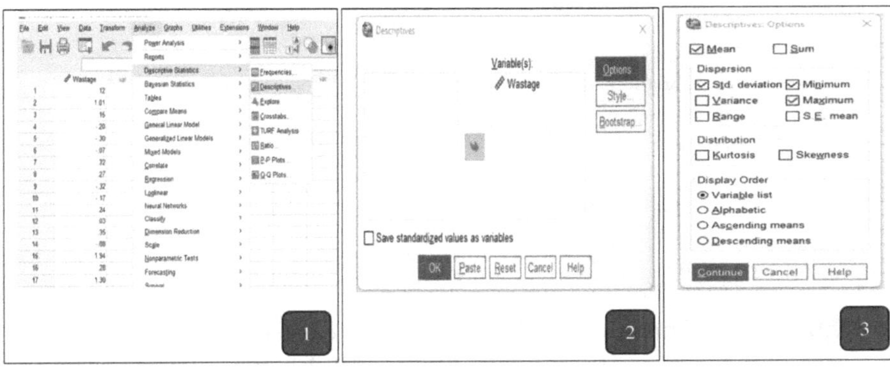

Then, the following table will be obtained:

Descriptive Statistics

	N	Minimum	Maximum	Mean	Std. Deviation
Wastage	22	-.32	4.27	.4523	1.00320
Valid N (listwise)	22				

Go to *SPSS – Analyze – Descriptive Statistics – Explore*

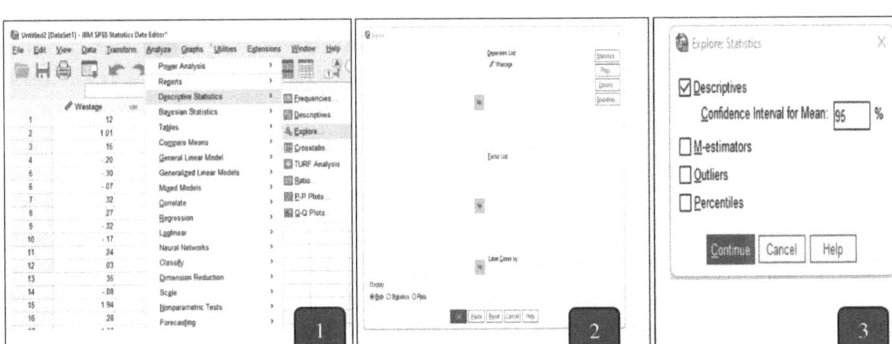

The following output will appear:

Descriptives

			Statistic	Std. Error
Wastage	Mean		.4523	.21388
	95% Confidence Interval for Mean	Lower Bound	.0075	
		Upper Bound	.8971	
	5% Trimmed Mean		.2947	
	Median		.1950	
	Variance		1.006	
	Std. Deviation		1.00320	
	Minimum		-.32	
	Maximum		4.27	
	Range		4.59	
	Interquartile Range		.40	
	Skewness		2.998	.491
	Kurtosis		10.240	.953

Without careful thought or a desire to evaluate the assumptions, the careless user of statistics might attempt to quantify the average percentage of wastage per week at this mill (μ) using the sample mean, calculated as $\bar{y} = 0.452$, and by stating a 95% confidence interval for μ as (0.0075, 0.897). This interval suggests that possible values could be as small as 0.0075 or as large as 0.897, with 95% confidence. However, a more thorough examination of this data, to be performed in Chapter 1, reveals that the normality assumption is violated and that the sample mean and the resulting confidence interval are misleading estimates. There are better ways to summarize this data, as we shall later observe in Chapter 4.

EDA examines the data before the model formulation stage to determine its characteristics, such as patterns, trends, and unusual observations. These characteristics, plus our knowledge of the subject matter, lead us to formulate models that make sense from a subject matter viewpoint and are supported by the data. EDA is also used at the model evaluation stage to determine the adequacy of the model. Typically, this involves the study of the residuals. Finally, EDA leads us to formulate more flexible models that are less sensitive to assumption violations. This is a major part of the topic of robust statistics.

For example, we might formulate the location model as before, with the errors following a distribution f where f is a member of the family of heavy-tailed distributions rather than the normal distribution. The optimal methods of the former model are no longer optimal under the latter model. In fact, an interesting question is what procedures are optimal under these new conditions. It turns out that optimal techniques do not exist, so we concentrate on **"robust"** procedures that are nearly optimal throughout the members of this family of distributions.

The following chapters will present and illustrate various EDA and CDA techniques using the *SPSS* and *R* statistical software systems. We will also present and study some modern, robust procedures, as well as other methods that do not require normal random variables. Heavy-tailed distributions represent one possible non-normal situation where heavy-tailed distributions are frequently used to model outliers. Other sources of non-normal data are those situations where the random variables are discrete, such as when counting data or $0-1$ data is observed. In both situations, the classical statistical models used for normal data are altered, in the first case to accommodate outliers and in the second to accommodate other forms of non-normal data. To this end, we will present three basic model forms: the general linear model, the altered (or robust) general linear model, and the generalized linear model. These models will be presented from an application viewpoint in each case, emphasizing the use and interpretation of the theoretical properties.

1

Exploratory Data Analysis for the Location Model

Summary

In this chapter, several graphical methods for univariate data will be discussed and illustrated in the following: histograms, dot plots, stem-and-leaf displays, box plots, empirical cumulative distribution function (ECDF) plots, symmetry plots, and probability plots. We note that the first three displays utilize the raw data, while the latter four methods require the ordered data. That is, if we let $y_1, y_2, ..., y_n$ denote the raw sample, then $y_{(1)}, y_{(2)}, ..., y_{(n)}$ represents the ordered sample. Here, $y_{(i)}$ is called the *i*th order statistic. To fully understand the behavior of the latter four displays, some of the theory of order statistics will be presented later.

Learning Objectives

By the end of this chapter, students will be able to do the following:

- Understand basic graphical techniques for analyzing univariate data.
- Learn about order statistics and their application in data analysis.
- Acquire skills in using SPSS and R for data visualization.
- Develop the ability to interpret various plots to understand data distribution.
- Analyze data to identify patterns, trends, and anomalies using exploratory tools.

1.1 Some Basic Data Displays

In this section, we will present some of the most basic methods of displaying data. Each technique will be illustrated using the SPSS and/or R statistical software.

DOI: 10.1201/9781003565321-1

1.1.1 Histograms

One of the oldest methods for displaying data is the histogram. Techniques for properly constructing the histogram are given in many textbooks: for example, Ott (1993). The reader is assumed to be familiar with these techniques, and we will not repeat them here. However, several aspects concerning the histogram, pertinent to later discussions, are presented. One problem with the histogram is that the data needs to be grouped to construct it. This often leads to a loss of information in the detail of the data, especially in the "tails" of the distribution.

On SPSS, the histogram is obtained by clicking on "Graphs" and then selecting "Chart Builder." Users can control the interval length used in constructing the histogram.

1.1.2 The Test Score Data

To illustrate the histogram on SPSS and the need for proper interval length determination, consider the following data representing test scores from a recent midterm in one of my statistics courses. The test was taken by 58 students, and 50 was the maximum possible score. The ordered test scores are as follows:

2, 2, 3, 3, 11, 12, 13, 13, 16, 17, 17, 18, 22, 22, 23, 23, 24, 24, 25, 26, 26, 27, 27, 28, 28, 29, 29, 29, 31, 31, 31, 31, 31, 32, 32, 33, 33, 33, 34, 34, 35, 35, 36, 37, 37, 38, 39, 39, 39, 40, 41, 42, 42, 43, 43, 44, 46, 46

On SPSS, the data are entered and saved. To obtain a histogram created without graphics features, type as follows:

1. *Analyze—Descriptive Statistics—Explore.* The outputs are shown in Table 1.1.

It is observed that SPSS has automatically chosen an interval of length 5, with midpoints at 0, 5, 10, and up through 50. Thus, the intervals are [0, 5), [5, 10), through [45, 50). The "frequency" tells us that there are four test scores in the first interval, four in the second interval, and so on. We observe from Figure 1.1 that the data is slightly skewed to the left, toward the smaller scores; has a range "around 0–45"; and is centered "around 30."

The "midpoint" or "cutpoint" option allows us to change the interval length. For example, we can obtain the histogram with intervals of length 25 with intervals of [0,25) and [25,50) by changing the number of intervals to two as in Figure 1.2.

Notice that this histogram is not nearly as informative as the first one, indicating the loss of information in the data by using improper interval length.

TABLE 1.1
Descriptive Statistics

			Statistic	Std. Error
Midterm_Scores	Mean		28.40	1.488
	95% Confidence interval for mean	Lower bound	25.42	
		Upper bound	31.38	
	5% Trimmed mean		28.90	
	Median		31.00	
	Variance		128.419	
	Std. deviation		11.332	
	Minimum		2	
	Maximum		46	
	Range		44	
	Interquartile range		14	
	Skewness		−0.705	0.314
	Kurtosis		−0.001	0.618

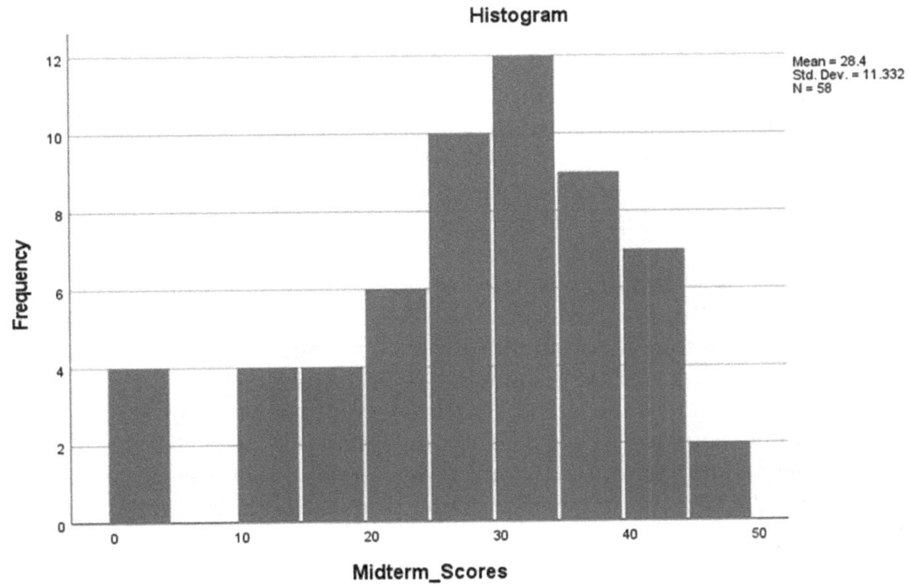

FIGURE 1.1
Histogram for the test scores.

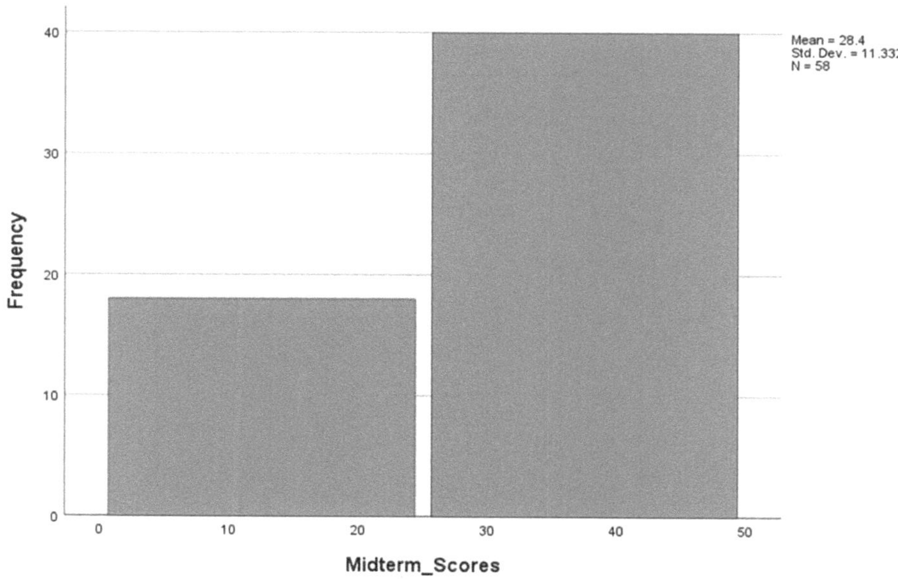

FIGURE 1.2
Histogram for the test scores with two intervals.

The SPSS is able to determine the proper interval length, which works well in most cases. However, several histograms of different interval lengths may be necessary in order to evaluate the data properly. For example, the test score data and an interval length of two results are used in the following histogram, as shown in Figure 1.3.

Four extremely low test scores are now clearly revealed, indicating four students who are having great difficulty with the tested material. Interestingly, other versions of the histogram of this data obscured this important fact. We want our plotting displays to reveal interesting information, not hide it. However, this new histogram also displays several peaks and valleys, indicating the random variation in the data. It is doubtful that this feature is meaningful. Thus, one of the disadvantages of the histogram is that repeated iterations are often required to obtain the best representation of the data.

We'll try one more time, using the cutpoint option and intervals of length 5. It is recommended that the optimal number of intervals be between six and eight. We'll also add a title and remove the border drawn around the plot. While the steps to accomplish this are given, the reader should learn to create graphs.

This histogram (Figure 1.4) gives about the right amount of detail; center, spread, and shape features are clearly revealed. The cluster of four outliers in the lower left is also displayed.

We could use R to obtain a histogram of our data. The test scores have been stored in the data frame named "testscore." We then use the hist command

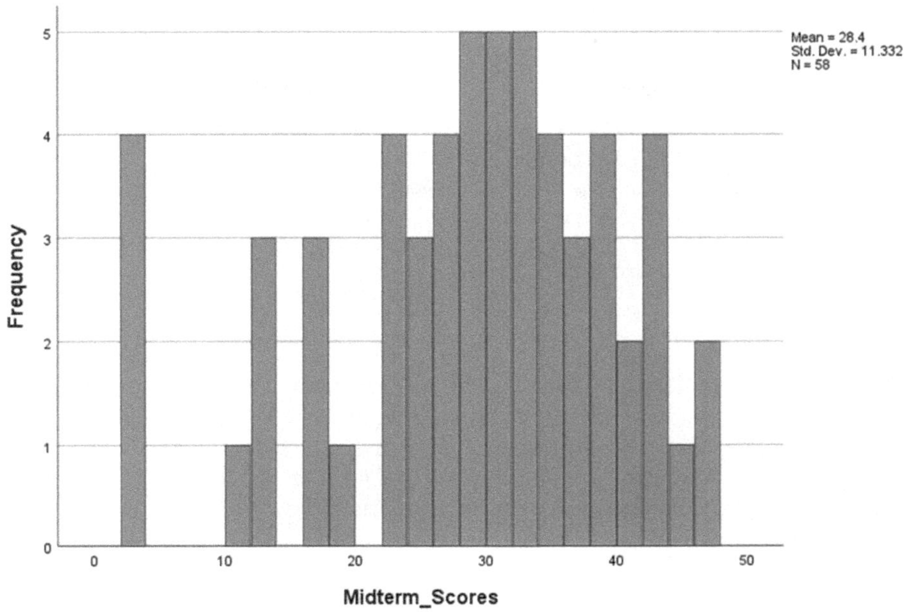

FIGURE 1.3
Histogram for the test scores with length of interval of 2.

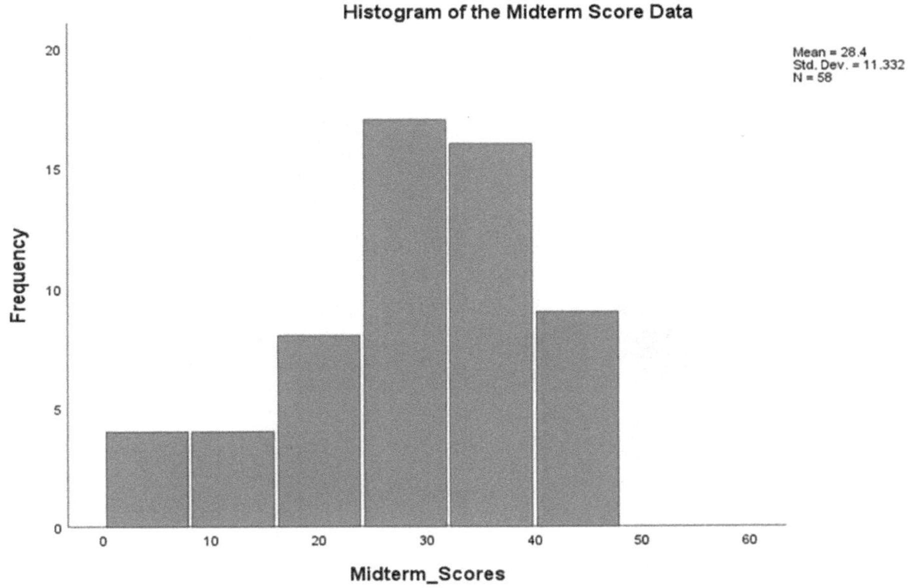

FIGURE 1.4
Histogram for the test scores with six intervals.

to obtain the histogram of the variable, also labeled "testscore." On R, type as follows:

```
# Load the data
data <- read_excel(file_path, sheet = "Sheet1")
# Create the histogram
hist(data$Midterm_Scores, breaks = 5, col = "blue", main =
"Histogram of Midterm Scores", xlab = "Midterm Score",
ylab = "Frequency", border = "black")
# Optional: Add grid lines
grid(nx = NULL, ny = NULL, col = "lightgray", lty =
"dotted")
```

The resulting histogram (Figure 1.5) suffers from the same problem of improper groupings as the original histogram constructed using SPSS.

We can customize our histogram on R by changing the "breaks = 7" as follows.

The resulting histogram appears in Figure 1.6. We notice that it looks very similar to the last histogram we obtained using SPSS.

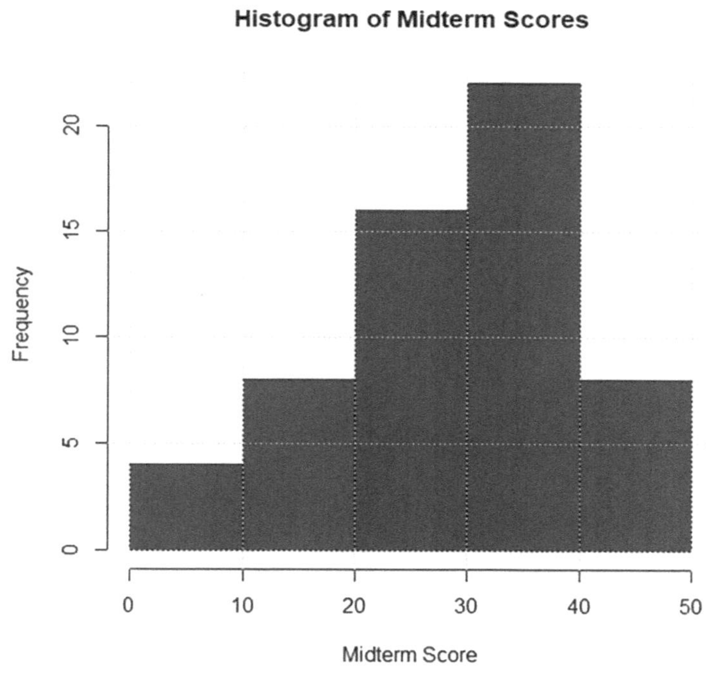

FIGURE 1.5
Histogram for the test scores with five intervals.

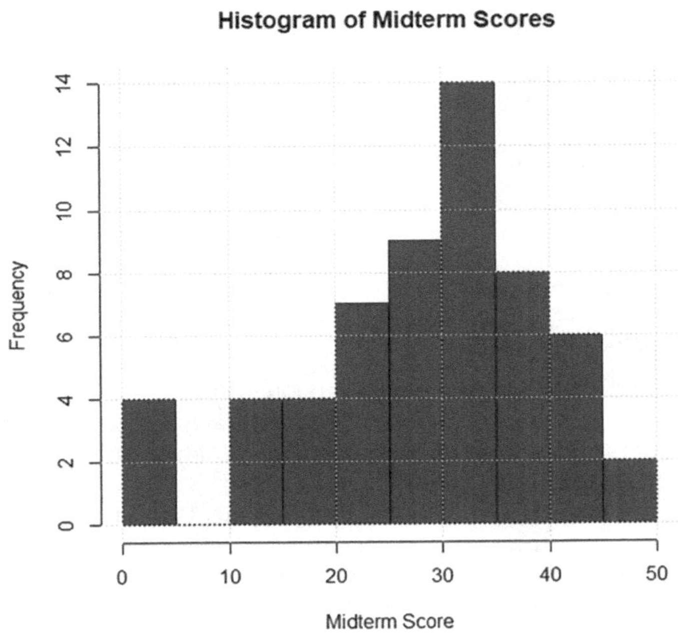

FIGURE 1.6
Histogram for the test scores with seven intervals.

1.1.3 Dot Plots

The dot plot simply places a "dot" on the real line at each occurrence of the variable y. If there are multiple occurrences of the same value, the dots are stacked on top of one another proportionally to the number of occurrences of the value. The dot plot thus plots each individual value, as opposed to the histogram, which groups the data, and gives a very nice representation of the data's center, spread, shapes, and tail behavior characteristics. The dot plot is especially useful for small to modest samples, say with n less than 100 or so.

On SPSS, the dot plot (Figure 1.7) is obtained by clicking on SPSS-Graphs-Scatter/Dotplots, choosing Simple Dot-Define, and then choosing the "X-Axis Variable:" as "Midterm_Scores." The output is shown in Figure 1.7.

Since the data is not grouped when the dot plot is constructed, the plot is not changed by changing the interval length, as is the case with the histogram.

FIGURE 1.7
Dot plot for the midterm score.

1.2 Stem-and-Leaf Displays

Stem-and-leaf displays are a modern version of the histogram. As with each of the displays presented in this chapter, they are designed to graphically reveal the data's characteristics of center, spread, shape, and unusual characteristics such as groupings and outliers.

Consider the test score data again for 58 students in an introductory statistics course. Since the scores range from 2 to 46, only the tens and unit positions will contain information. We'll let the tens position represent the stem and the ones position represent the leaf. The basic stem-and-leaf display would look as here:

Stem-and-Leaf Display of 58 Test Scores		(#)	Rdown	Rup
0	2 2 3 3	(4)	4	58
1	1 2 3 3 6 7 7 8	(8)	12	54
2	2 2 3 3 4 4 5 6 6 7 7 8 8 9 9 9	(16)	28	46
3	1 1 1 1 1 2 2 3 3 3 4 4 5 5 6 7 7 8 9 9 9	(12)	49	30
4	0 1 2 2 3 3 4 6 6	(9)	58	30

where, for example, 4| 2 represents the test score of 42.

The stem-and-leaf plot has five stems and 58 leaves, each leaf representing one digit. The three columns on the right represent the number of leaves in each stem (#), the cumulative ranking from smallest observation to largest (Rdown), and the cumulative ranking from largest to smallest (Rup).

This stem-and-leaf plot visually displays a center of roughly "30 something" and a skewed-left appearance. Its similarity and advantage over the histogram are readily apparent in that the number of leaves in each stem corresponds to the height of the histogram for the appropriate interval. The primary advantage is that the data are also displayed within each stem.

Variations of the basic stem-and-leaf display involve the scale of the stems. A "stretched" stem-and-leaf display can be obtained by stretching the single stem into more than one stem per line. A common example is the two-stem display, shown in the following table for the test data:

Stem-and-Leaf Display of 58 Test Scores	(#)	Rdown	Rup
0 2233	(4)	4	58
0	(0)	4	54
1 1233	(4)	8	54
1 6778	(4)	12	50
2 223344	(6)	18	46
2 5667788999	(10)	28	40
3 111112233344	(12)	40	30
3 556778999	(9)	49	18
4 0122334	(7)	56	9
4 66	(2)	58	2

As is readily apparent, this stem-and-leaf display reveals a cluster of very low scores in the upper portion of the display. These scores were not as easy to identify in the original display. The book *Exploratory Data Analysis* by John Tukey (1977) illustrates many other versions of the stem-and-leaf display.

If the data is contained in column c1, the stem-and-leaf display can be obtained on SPSS by performing the following steps:

- Go to Analyze > Descriptive Statistics > Explore.
- In the Explore dialog, move Midterm_Scores to the Dependent List.
- Click on Plots.
- In the Explore: Plots dialog, check the Stem-and-leaf option.
- Click Continue.

The stem-and-leaf display here is produced by SPSS for the test score data:

```
Midterm_Scores Stem-and-Leaf Plot

 Frequency    Stem &  Leaf
  4.00           0 .  2233
  4.00           0 .
  4.00           1 .  1233
  4.00           1 .  6778
  6.00           2 .  223344
 10.00           2 .  5667788999
 12.00           3 .  111112233344
  9.00           3 .  556778999
  7.00           4 .  0122334
  2.00           4 .  66

 Stem width: 10
 Each leaf: 1 case(s)
```

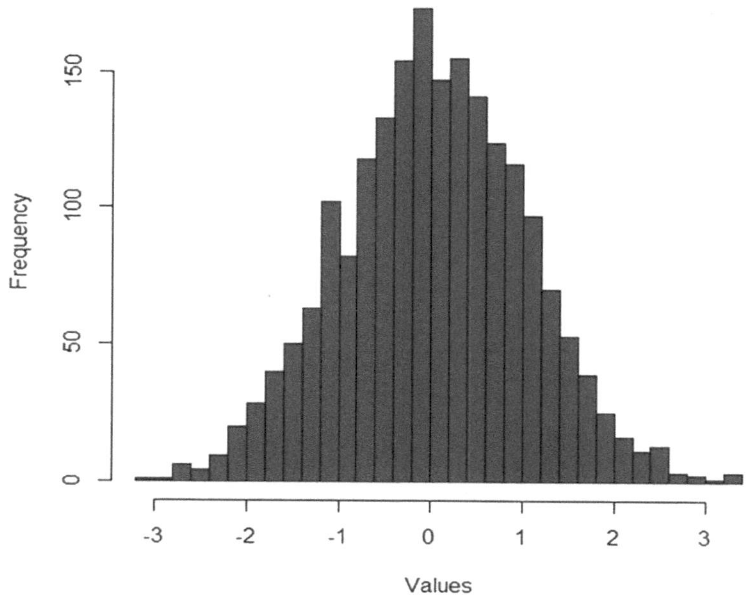

FIGURE 1.8
Histogram of 2,000 N(0, 1) random numbers.

The primary disadvantage of the stem-and-leaf plot is that it does not work well with large datasets. R can be easily used to demonstrate this latter concept. Suppose 2,000 normal random numbers (with a mean of zero and standard deviation of one) are generated by R and displayed by the histogram (Figure 1.8). A sample R session appears as follow:

```
# Set the seed for reproducibility
set.seed(123)
# Generate 2000 N(0,1) random numbers
c1 <- rnorm(2000, mean = 0, sd = 1)
# Plot a histogram
hist(c1, breaks = 30, col = "blue", main = "Histogram of
2000 N(0,1) Random Numbers", xlab = "Values", ylab =
"Frequency", border = "black")
# Plot a dot plot
dotchart(c1, main = "Dotplot of 2000 N(0,1) Random
Numbers",
         xlab = "Values", ylab = "Frequency")
# Create a stem-and-leaf plot
stem(c1, scale = 2)
# Save the results
jpeg("Histogram.jpg")
```

```
hist(c1, breaks = 30, col = "blue", main = "Histogram of
2000 N(0,1) Random Numbers", xlab = "Values", ylab =
"Frequency",
     border = "black") dev.off()
jpeg("Dotplot.jpg")
dotchart(c1, main = "Dotplot of 2000 N(0,1) Random
Numbers",
     xlab = "Values", ylab = "Frequency") dev.off()
# For the stem-and-leaf plot, you can view it in the
console or save the output to a file
sink("Stem_and_Leaf.txt")
stem(c1, scale = 2)sink()
Character Dotplot. Each dot represents 9 points.

                              ..
                             .::::
                          :.:::::::.
                         :::::::::::
                       ::::::::::::::.
                     .:::::::::::::::::.
                  .::::::::::::::::::::....
       .  .  .....::::::::::::::::::::::::::::..... .
-------+---------+---------+---------+---------+---------x
     -3.0      -1.5       0.0       1.5       3.0       4.5

Character Stem-and-Leaf Display

Stem-and-leaf of x          N = 2000
Leaf Unit = 0.10

     1   -3  4
     8   -2  9765555
    32   -2  443332222221111000000000
   108   -1  9999999988888888888888887777777766666666666666666
             6666555555555555+
   281   -1  4444444444444444444444444444443333333333333333333
             3333333333322222+
   600   -0  9999999999999999999999999999999999999999999999999
             9999988888888888+
 (411)   -0  4444444444444444444444444444444444444444444444444
             4444444444444444+
   989    0  0000000000000000000000000000000000000000000000000
             0000000000000000+
   593    0  5555555555555555555555555555555555555555555555555
             5555555555555556+
   323    1  0000000000000000000000000000000000000000000000000
             0111111111111111+
   144    1  5555555555555555555555555666666666666666777777777
             7777888888888888+
    57    2  00000000000011111111111122222222233334444
    16    2  55666666777789
     2    3  04
```

It is observed that while both the histogram and the dot plot can capture the familiar "bell-shape" appearance of the normal curve, the stem-and-leaf display does not.

1.3 The Box Plot

The box plot is an important and extremely useful display. This plot displays the characteristics of the center, spread, and shape and the "tail behavior" of the sample data. The plot is based, in part, on five numbers: the median, the upper and lower quartiles, the extremes, and the minimum and maximum sample observations.

A convenient way to obtain the median and quartiles is to find their location in the following manner. Let $L()$ denote the location of the number of interest in the ordered sample. Let M denote the sample median. Then $L(M) = \frac{n+1}{2}$, implying that M is found at location $L(M)$ if n is odd and M is the average of the two data values found at $L(M) - 0.5$ and $L(M) + 0.5$ if n is even.

Let Q_1 and Q_3 represent the sample quartiles. These can be thought of as estimates of the population quartiles, the 25th and 75th percentiles, respectively. Then $L(Q) = \frac{[L(M)]+1}{2}$, where [] denotes the integer part of $L(M)$. Q_1 is found at the $L(Q)$th smallest ordered observation and Q_3 is found at the $L(Q)$ th largest ordered observation. The same "averaging" rule used to obtain M applies for obtaining Q_1 and Q_3 if $L(Q)$ ends in 0.5.

1.3.1 The Rat Data

Suppose that the variable X represents the time in seconds required for a rat to complete a maze. Ten randomly selected rats, timed under identical conditions, have the following times: 3, 4, 5, 2, 20, 6, 7, 20, 5, 4. The ordered sample is: 2, 3, 4, 4, 5, 5, 6, 7, 20, 20. To find the median, we compute $L(M) = \frac{10+1}{2} = 5.5$. Thus, M is found by averaging the fifth and sixth smallest data points. This results in $M = \frac{5+5}{2} = 5$. To find the quartiles, we compute $L(Q) = \frac{[L(M)]+1}{2} = \frac{[5.5]+1}{2} = 3$. This implies that Q_1 is the third smallest ordered observation, equal to 4, and Q_3 is the third largest ordered observation, equal to 7.

If $n = 11$, instead of 10, then $L(M) = 6$ and $L(Q) = \frac{[6]+1}{2} = 3.5$. Q_1 would be the average of the third and fourth smallest observations, and Q_3 would be the average of the third- and fourth-largest observations.

These values can be displayed conveniently in the following summary, called the "Five-number summary" by Tukey (1977). The general form of the display is as follows:

		n		
M L(M)		**M**		
Q L(Q)	Q_1		Q_3	$Q_3 - Q_1 = \text{IQR}$
E	$x_{(1)}$		$x_{(n)}$	

Here, the "___" means replacing the label in the display with the appropriate sample value. The "E" refers to the extremes of the ordered sample, $x_{(1)}$ and $x_{(n)}$. In this summary, IQR denotes the interquartile range.

For example, the five-number summary for the rat data is as follows:

		$n = 10$		
M 5.5		5		
Q 3	4		7	3 = IQR
E 1	2		20	

It is obvious that the display contains a measure of the center (M), behavior of the middle 50% of the data (Q_1 and Q_3), a measure of spread (IQR), and the shape (distance between the M and Q's and from Q's to E's) of the sample data. These measures of center and spread are resistant statistics that are not unduly influenced by outliers. For example, the sample mean of the rat data is 7.6, compared to the median of 5, which clearly shows that the two sample values of 20 have pulled the mean to the right toward these large observations or outliers.

The IQR is the interquartile range and is a measure of the spread of the data. The value IQR/1.35, sometimes called the pseudo-standard deviation or PSD, is comparable to the sample standard deviation. For the rat data, $S = 6.69$ compared to the PSD of 2.22. Again, this difference in size indicates the lack of resistance of S to outliers in the data, while the PSD is resistant to outliers. Justification for the use of the PSD as a measure of scale derives from the normal curve. If the data is assumed to follow a normal distribution with a mean of α and a standard deviation of σ, then Q_1 and Q_3 are estimates of the 25th and 75th percentiles or $\mu - .6745\sigma$ and $\mu + .6745\sigma$, respectively. It follows that the population interquartile range, or IQR, is computed as $IQR = \mu + .6745\sigma - (\mu - .6745\sigma) = 2 * 0.6745\sigma \approx 1.35\sigma$. Thus, $PSD = IQR / 1.35$ is an estimate for σ.

Construction of the box plot is a simple matter once the five-number summary is obtained. The steps are as follows:

1) Locate M, Q_1, and Q_3 on the "x-axis," draw a box connecting Q_1 to Q_3, and draw a vertical line at M.
2) Compute the "step" = 1.5 * IQR.

3) Determine the "adjacent values":

   The lower adjacent (la) value is the smallest data value between $Q_1 - 1 * \text{step}$ and Q_1.

   The upper adjacent (ua) value is the largest data value between Q_3 and $Q_3 + 1 * \text{step}$.

   Mark the adjacent values with an "|" and connect them to the box with a line.

4) Identify the outliers:

   All observations between one step and two steps away from the quartiles are called mild outliers (mo). All observations greater than two steps away are called extreme outliers (eo). Mark the mos with an "*" and the eos with "o."

   All necessary values for the box plot are displayed in the following tableau:

		step		
adjacent	la		ua	
f	Q_1 − 1*step		Q_3 + 1*step	
list lmo	#lmo		#umo	list umo
F	Q_1 − 2*step		Q_3 + 2*step	
list leo	#leo		#ueo	list uel

For example, #lmo represents the number of lower mild outliers, and #ueo represents the number of upper-extreme outliers. Tukey (1977) refers to "f" as the "inner fence" and "F" as the "outer fence," indicating the boundaries for the mild and extreme outliers, respectively. Figure 1.9 shows the steps for creating the box plot using the SPSS.

SPSS—Analyze—Descriptive Statistics—Frequencies

Then we get the following output:

Statistics

Rat

N	Valid	10
	Missing	0
Percentiles	25	3.75
	50	5.00
	75	10.25

1.3.2 The Rat Data

Recall that $IQR = 3$ for the rat data example presented earlier. Then the step = 1.5 * IQR = 4.5. The value of la is the smallest value between (Q_1 − 1 *

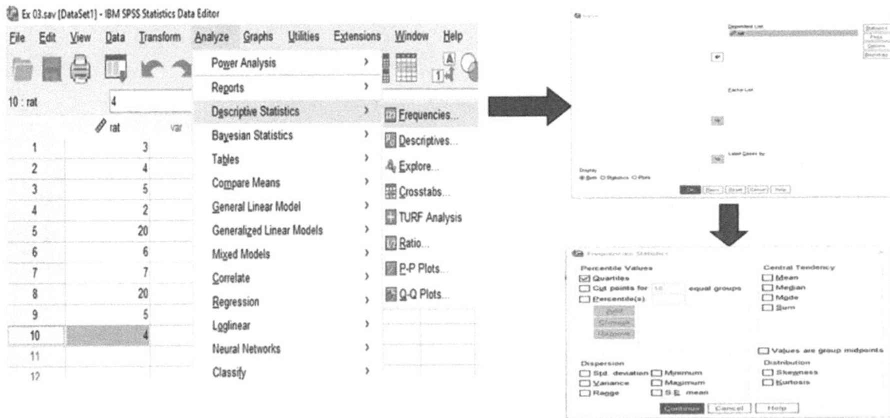

FIGURE 1.9
Steps for creating the box plot using the SPSS.

step, Q_1) = (−0.5, 4), which is two. The value of ua is the largest value between (Q_3, Q_3 + 1 * step) = (7, 11.5), which is seven. The upper F is 7 + 2 (4.5) = 16. Thus, the two observations of 20 are classified as upper-extreme outliers. The lower f is 4 − 4.5 = −0.5, and the lower F is 4 − 2 (4.5) = −5, and since no observations can be below 0, there are no outliers of either type in the lower tail of the data. The tableau for the rat data is as follows:

	4.5		
adjacent	2	7	
f	−0.5	11.5	
F	−5	16	
		2	20, 20

The box plot using the SPSS for this data is displayed in Figure 1.10.

The unconnected whisker at the top of the plot indicates at least one outlier in the data.

As in any graphical analysis, the proper interpretation of the box plot results from comparing the plot for a given dataset to some particular reference plot or "ideal box plot." Often, the assumption of normality is required for the optimality of our statistical procedures, and for this reason, the ideal box plot is usually taken as that for the normal curve. Figure 1.11 gives the normal density and the box plot corresponding to a very large sample from a normal population.

A comparison of the box plot for the rat data to the ideal box plot acts as an informal test of normality. It appears obvious that the rat data is far from

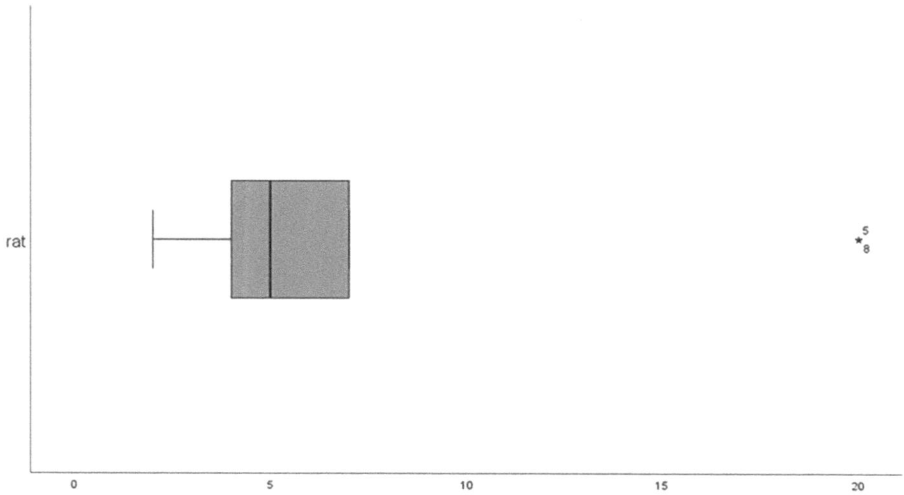

FIGURE 1.10
Box plot for the rate data.

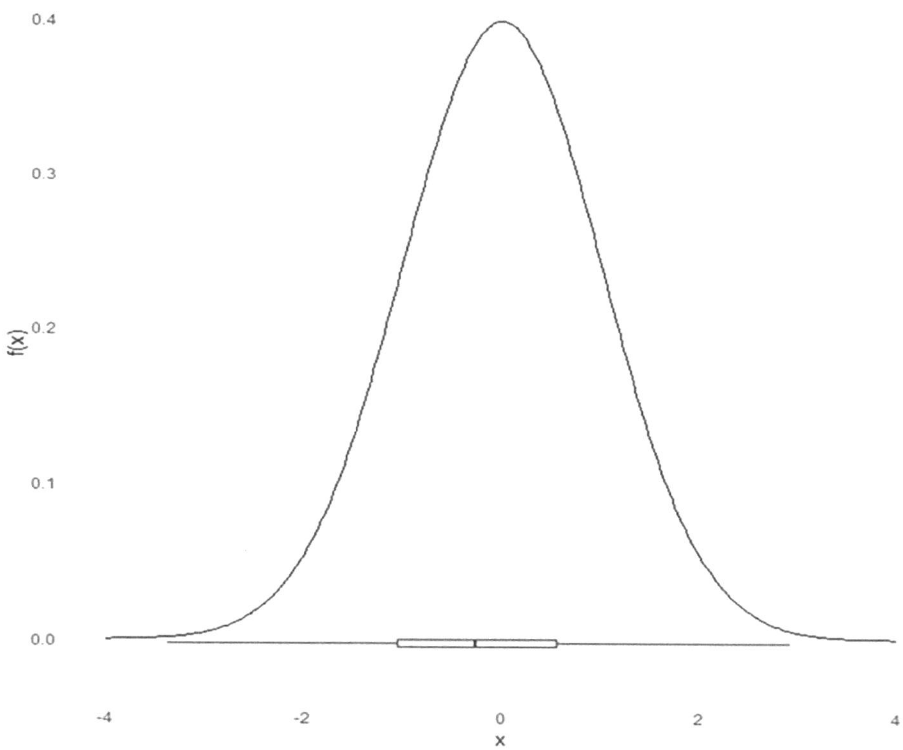

FIGURE 1.11
A box plot for a very large sample from an ideal normal population.

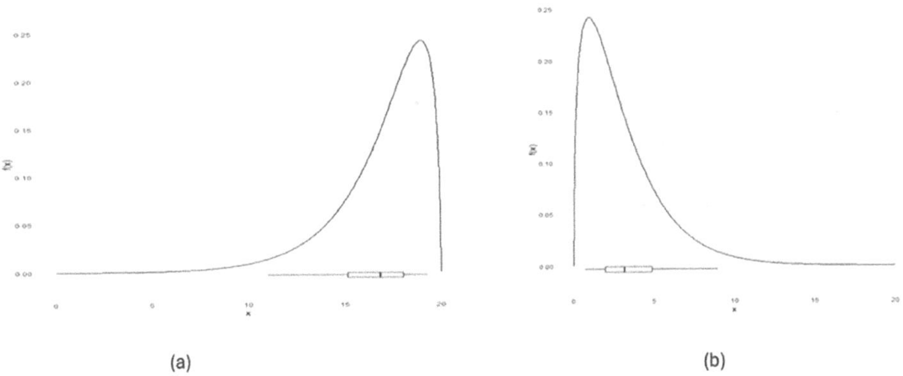

FIGURE 1.12
(a) A skewed-left distribution-chi-square with three degrees of freedom; (b) A skewed-right distribution-chi-square with three degrees of freedom.

normal. At this point, a word of caution must be made as to the validity of using the box plot as a "test" of the hypothesis on the distribution of a set of data, normal or otherwise. Small samples from any distribution often exhibit characteristics opposed to those of the true underlying distribution. As we shall see in Chapter 2, using the box plot as a test for normality can often lead to erroneous conclusions.

The box plot readily reveals other common distributional shapes. Figure 1.12 gives two other examples of common distributional shapes and their corresponding box plots.

One extremely useful feature of the box plot is that it facilitates the comparison between several groups. Side-by-side box plots help reveal similarities and differences that exist between the groups. In Figure 1.13, side-by-side box plots are given for five groups representing the automobile hydrocarbon emissions by various years of manufacture (Koopmans, 1987). Note the differences in center and spread and the number and severity of the outliers.

The emissions data is displayed in the side-by-side box plots in Figure 1.13. The response variable, "emission," is the amount of hydrocarbon emissions measured in randomly sampled automobiles from varying years of manufacture. The labels in the display of 1 through 5 represent the year groups "pre-1963," "1963–1967," "1968–1969," "1970–1971," and "1972–1974." Of special interest is the noticeable drop in hydrocarbon emissions in groups 3–5 (representing year groups 1968 and newer), coincident with the use of federally mandated emission control devices. Also noticeable are the outliers and different IQRs across the five groups. Further confirmatory analysis of these data should account for these features.

The reader is encouraged to use the graph menu of SPSS to obtain this plot and all other SPSS plots.

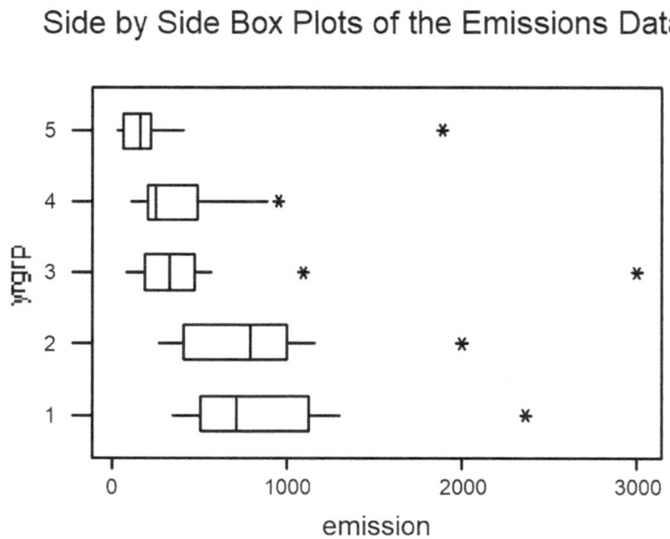

FIGURE 1.13
Side-by-side box plots for five groups.

1.4 The Empirical Cumulative Distribution Function Plot

The ECDF plot is obtained by plotting the ith ordered data point, $y_{(i)}$, the ith order statistic, versus i/n, for $i = 1,\ldots,n$. Thus, the ECDF is always a step function with jumps of $1/n$ at each sample value. However, if some sample values are identical ("ties"), then the jump size is equal to "the number of ties at that sample value"/n.

It can be easily created using R's programming capabilities. The following provides a step-by-step guide to achieve this.

- Before plotting the ECDF, you need to install and load the ggplot2 package, which provides powerful tools for data visualization in R.

```
install.packages("ggplot2") # Install ggplot2 package
library(ggplot2) # Load ggplot2 library
```

- Define a function for the ECDF plot to create and plot the ECDF. The function will sort the data, compute the ECDF values, and use ggplot2 to create the plot.

```
# Function to create and plot ECDF
plot_ecdf <- function(data) {
  # Sort the data
```

```
sorted_data <- sort(data)
# Create the ECDF values
n <- length(sorted_data)
ecdf_values <- (1:n) / n
# Create a data frame for plotting
ecdf_df <- data.frame( x = c(sorted_data,
max(sorted_data) + 1), ecdf = c(ecdf_values, 1))
# Plot the ECDF
ggplot(ecdf_df, aes(x = x, y = ecdf)) + geom_
step() + labs( title = "Empirical Cumulative Dis-
tribution Function Plot", x = "X", y = "ECDF" )
+ theme_minimal()}
```

- To plot the ECDF for your data, simply call the plot_ecdf function with your data as the argument.

```
# Example data
test_scores <- c(45, 50, 55, 60, 65, 70, 75, 80, 85,
90, 95, 100)
# Plot ECDF for the example data
plot_ecdf(test_scores)
```

This approach in R has several advantages over traditional methods:

- The function can be customized for various types of data and plot styles.
- Unlike certain methods that may overwrite data, this function creates a new data frame for plotting.
- Using ggplot2, you can easily enhance and customize the plot with additional features like themes, labels, and colors.

This example serves to illustrate several important features of the ECDF plot. It is noted that the raw data is plotted through the order statistics, and thus the data is not grouped, as in histograms or stem-and-leaf plots. Outliers and clusters of observations are clearly revealed by "flat spots" in the display. The slope of the ECDF plot (Figure 1.14) reveals where the data occurs infrequently (small slopes) and where the data is frequent (large slopes).

The slope of the ECDF plot reveals where the data occurs infrequently, indicated by small slopes, and where the data is frequent, indicated by large slopes. This is seen in the plot in Figure 1.15, where the steepest part of the curve is around 30, the center of the middle portion of the data. The slope is at its smallest value around 10 and 45, where the tails of the data occur.

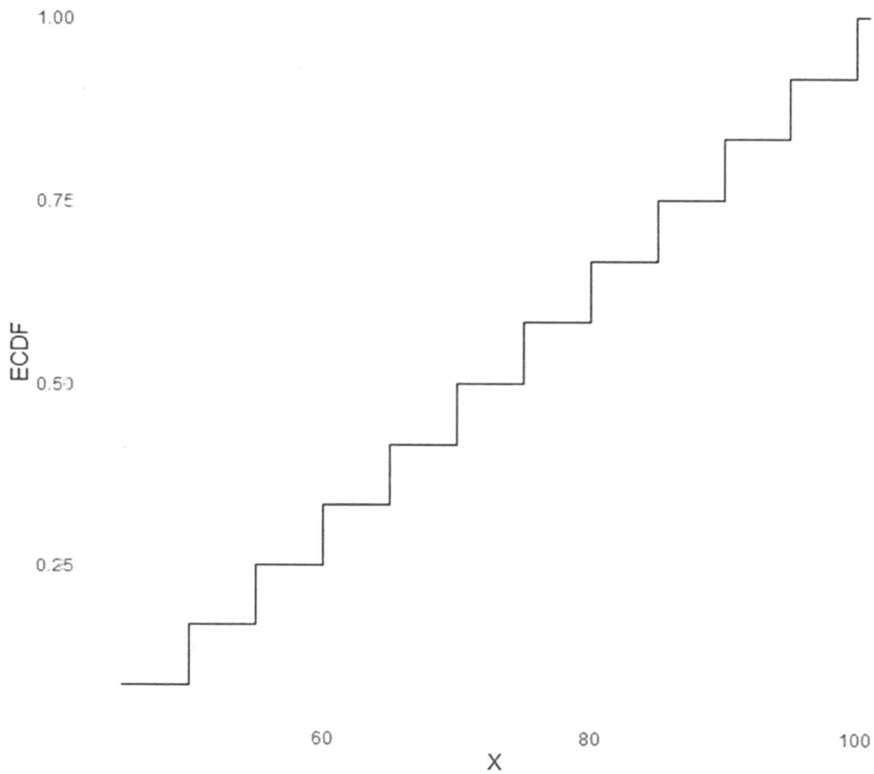

FIGURE 1.14
Empirical cumulative distribution function plot.

One can easily estimate the percentiles of the sample data from the ECDF plot. For example, the sample 50th percentile, or M, is found to be extending a horizontal line from 0.5 on the "y-axis" of the plot to the data and then dropping a vertical line down to the "x-axis." The intersection of this vertical line and the x-axis is an estimate of the 50th percentile. The process is illustrated in the ECDF plot (Figure 1.15), where the 50th percentile appears to be around 32.

If the data can be treated as a "'random sample," then the ECDF can be viewed as an estimator of the true cumulative distribution function (CDF), F_Y. Let Y denote the random variable; then the CDF of Y is $F_Y(y) = \text{Prob}(Y \leq y)$. The ECDF is defined as follows:

$$F_n(y) = 0 \text{ if } y < y_{(1)} \text{ and } F_n(y) = \frac{(\# y\text{'s} \leq y)}{n} \text{ for } y \geq y_{(1)}.$$

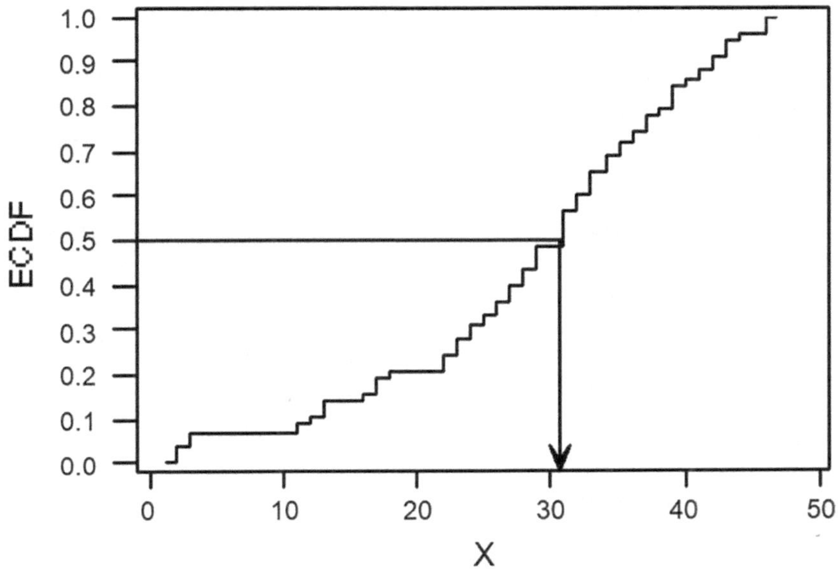

FIGURE 1.15
Empirical cumulative distribution function plot with 50th percentile.

The ECDF, $F_n(y)$, can be viewed as the "nonparametric maximum likelihood estimate" of $F_Y(Y)$ and has the property that as n gets large, that $F_n(y)$ converges to $F_Y(Y)$. This result is the famous Glavenko-Cantelli lemma (Rao, 1965).

1.5 Other Summary Statistics Based on Letter Values

In the section on box plots, the summary statistics of M and Q_1 and Q_3 were introduced as the median and the 25th and 75th percentiles, respectively. Tukey (1977) termed such statistics as examples of "letter values" and gave a methodology for computing other letter values as estimates of appropriate percentiles. For example, the median, M, splits the sample data into two groups of equal size. The quartiles, the Q's, split the upper and lower 50% of the data into four groups of equal size. The upper and lower 25% of the data can again be split into two equal groups, each representing 12.5% (or one-eighth) of the data. If we let "E" represent the "eighth," then the location of the eights is found in a manner similar to that of the quartiles. That is, $L(E) = \{[L(Q)] + 1\}/2$.

Thus, the lower eighth, E_l, is found at the $L(E)$th smallest ordered observation, and the upper eighth (the estimate of the 87.5 percentile), E_u, is found at the $L(E)$th largest observation. As in the case of the quartiles, if $L(E)$ ends in 0.5, the eighths are found by averaging the appropriate data values.

Additional letter values can now easily be expressed by this process of splitting the lower and upper tails into two pieces of equal size. The letter "S" can be used to represent the "sixteenths" and found at locations $L(S) = \{[L(E)] + 1\}/2$ from the left and right of the ordered data. More letter values can be similarly defined, though it is often not useful to go beyond E or S.

Letter values can be obtained on R by using the following code. For example, other letter values for the test score data can be obtained by typing as follows:

```
# Function to compute letter values
compute_letter_values <- function(data) {
  # Sort the data
  sorted_data <- sort(data)
  n <- length(sorted_data)
    # Compute Median (M)
  M <- median(sorted_data)
    # Compute Quartiles (Q1 and Q3)
  Q1 <- quantile(sorted_data, 0.25)
  Q3 <- quantile(sorted_data, 0.75)
    # Compute Eighths (E1 and E7, representing 12.5% and
    87.5% percentiles)
  E1 <- quantile(sorted_data, 0.125)
  E7 <- quantile(sorted_data, 0.875)
    # Compute Sixteenths (S1 and S15, representing 6.25%
    and 93.75% percentiles)
  S1 <- quantile(sorted_data, 0.0625)
  S15 <- quantile(sorted_data, 0.9375)
    # Create a data frame with the results
letter_values <- data.frame(
    Depth = c(n, n/2, n/8, n/16),
    Lower = c(M, Q1, E1, S1),
    Upper = c(M, Q3, E7, S15),
    Mid = c(M, (Q1 + Q3) / 2, (E1 + E7) / 2, (S1 + S15) / 2),
    Spread = c(NA, Q3 - Q1, E7 - E1, S15 - S1) )
# Assign letter values names
rownames(letter_values) <- c("M", "Q", "E", "S")
    return(letter_values)}

# Example data
test_scores <- c(45, 50, 55, 60, 65, 70, 75, 80, 85, 90,
95, 100)
# Compute letter values for the example data
```

```
letter_values <- compute_letter_values(test_scores)
# Print the letter values
print(letter_values)

   Depth   Lower   Upper  Mid Spread
M 12.00 72.5000 72.5000 72.5     NA
Q  6.00 58.7500 86.2500 72.5 27.500
E  1.50 51.8750 93.1250 72.5 41.250
S  0.75 48.4375 96.5625 72.5 48.125
```

1.6 The Ideal Box Plot

Box plots are a common method recommended by many elementary statistics texts to evaluate the presence of outliers in the data with respect to a given model. The question remains as to the validity of such a procedure. Consider the ideal box plot, where it is known that the random variable y follows a normal distribution with a known mean, α, and variance, σ^2. One can imagine a random sample of infinite size from a normal population where the sample statistics are nothing more than the parameters themselves. If a box plot were to be constructed for this data, we would obtain $M = \mu$ and Q_l and Q_u equal to $\mu - .6745\sigma$ and $\mu + .6745\sigma$, respectively. Thus, the probability that a randomly selected y from this population is declared an outlier by the box plot method is, as easily verified, Prob("no outliers") = $\Pr ob(\mu - 2.70\sigma \le y \le \mu + 2.70\sigma) = .993$.

Thus, the Prob("outlier") = 0.007. So, for example, in a sample of size $n = 5$, if we knew the true parameters of the model, we could compute Prob(at least one outlier in five observations) = 1 – Prob("no outliers") = $1 - .993^5 = .034$. This looks like a reasonable number under the circumstances. However, in reality, the parameter values are seldom, if ever, known and must be estimated from the data. In this case, the probability that the box plot will declare an observation as an outlier will change dramatically.

To illustrate this, consider using the box plot technique with estimated values of M and Qs. In our example $n = 5$, so that $M = y_{(3)}$ and Q_l and Q_u are $y_{(2)}$ and $y_{(4)}$, respectively. The probability that in a sample of $n = 5$, the box plot outlier technique will detect an outlier is now

$$P\left(y_{(1)} < y_{(2)} - 1.5 * IQR\right) + P\left(y_{(5)} > y_{(4)} + 1.5 * IQR\right)$$

where $IQR = y_{(4)} - y_{(2)}$. Thus, the actual probability that an outlier will be found is much more complicated than first indicated.

One way to evaluate this probability is to apply the repeated sampling concept by using R to generate random samples, say 500 samples, of size n from a

N(0, 1) distribution (or any other distribution). For each random sample, obtain the box plot. Thus, for each value of n, 500 box plots are obtained. Count the number of box plots that indicate either a mild or extreme outlier. The results may surprise you. The following annotated R session shows how this is done.

```
# Set parameters
n <- 5
num_samples <- 500
# Generate 500 samples of size n from N(0,1)
distribution
set.seed(123) # For reproducibility
samples <- replicate(num_samples, rnorm(n), simplify =
TRUE)
# Check each sample for outliers
outlier_counts <- apply(samples, 2, has_outliers)
# Compute the proportion of samples with at least one
outlier
prob_outlier <- mean(outlier_counts)
# Print the result
print(prob_outlier)
```

Compute the relative frequency of outliers. Then the Prob(outlier) estimate in a sample of $n = 5$ for data from $N(0,1)$ distribution will result in 0.294.

Thus, caution is in order: In using the box plot to "test the assumption of normality," be aware that a very high proportion of normal samples will result in box plots that indicate outliers. This proportion can be as high as 20%–30%.

1.7 Some Comments on the Order Statistics

It has been observed that box plots depend heavily on using order statistics. In fact, Tukey's exploratory data analysis (EDA) methods depend on using letter values to construct the plots and determine the proper data transformations. Some understanding of the theoretical properties of order statistics is essential. For a more detailed discussion of this topic, the reader is referred to Larsen and Marx (1981) and Mood, Graybill, and Boes (1974).

Let the random variable y follow the density (PDF) $f(y)$ with cumulative distribution function (CDF) $F(y)$. Let $y_1, y_2, \ldots, y_n$ be a random sample of size n and $y_{(1)}, y_{(2)}, \ldots, y_{(n)}$ be the n-order statistics. Then $y_{(i)}$ is a random variable whose PDF is given by (Larson and Marx, 1981)

$$y_{(i)} \sim \frac{n!}{(n-i)!(i-1)!} f(y)\left[F(y)\right]^{i-1}\left[1-F(y)\right]^{n-i} = f_{(i)}(y).$$

We note that the PDF of $y_{(i)}$ depends on n, i, and F. Thus, $y_{(i)}$ is a random variable with PDF $f_{(i)}(y)$ and CDF $F_{(i)}(y)=\int_{-\infty}^{y} f_{(i)}(v)dv$. One can compute the expectation, $E(y_{(i)})$, and variance, $Var(y_{(i)})$ of $y_{(i)}$, as well as determine other properties associated with random variables. In particular, let $med(y_{(i)})$ denote the median, or 50th percentile of the distribution of $y_{(i)}$.

Now, by definition, $med(y_{(i)})$ is that value such that $F_{(i)}(med(y_{(i)}))=0.5$. Of course, if the median of y is denoted by $med(y)$, it follows that $F(med(y))=.5$. The relationship between $F(y)$, $F_{(i)}(y)$, $med(y)$, and $med(y_{(i)})$ can be seen in Figure 1.16.

The horizontal line at $P=0.50$, representing the inverse CDF operation, extends across to the first $F_{(i)}(y)$ and then to $F(y)$, as in Figure 1.16. The vertical line intersects the abscissa at $med(y_{(i)})$ and $med(y)$. The second horizontal line is at $\frac{i-\frac{1}{3}}{n+\frac{1}{3}}$, and applying $F^{-1}\left(\frac{i-\frac{1}{3}}{n-\frac{1}{3}}\right)$ intersects the abscissa at approximately $med(y_{(i)})$, as indicated by Blom's conjecture.

It is clear from the figure that the median of both y and $y_{(i)}$ can be obtained by using the appropriate inverse CDF – that is, $F^{-1}(.5)=med(y)$ and $F_{(i)}^{-1}(.5)=med(y_{(i)})$. As can be easily imagined, $F_{(i)}^{-1}$ is a very complicated expression in all but the most elementary situations. However, the need to

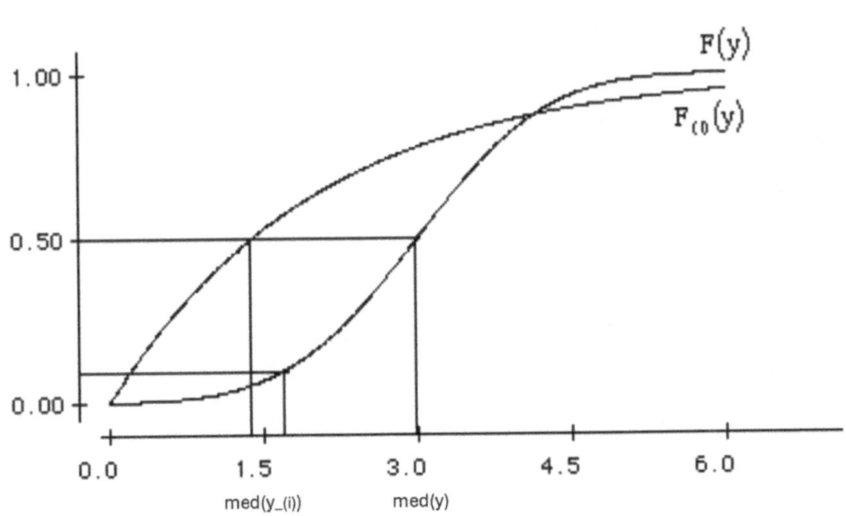

FIGURE 1.16
Plot of the CDFs and $F_{(i)}(y)$.

utilize $F_{(i)}^{-1}$ to determine $med\left(y_{(i)}\right)$ is diminished by a remarkable result attributable to Blom (1958) and referred to as "Blom's conjecture." This conjecture states that

$$med\left(y_{(i)}\right) = F_{(i)}^{-1}(.5) \approx F^{-1}\left(\frac{i-\frac{1}{3}}{n+\frac{1}{3}}\right)$$

for any continuous random variable y. That is, the $med\left(y_{(i)}\right)$ is approximately equal to the $\frac{i-\frac{1}{3}}{n+\frac{1}{3}}$ percentile of y. For example, if n is 10, then $Q_1 = y_{(3)}$ and is viewed as an estimate of the 25th percentile of the distribution of y, even though $3/10 = 30\%$ of the sample data are less than or equal to $y_{(3)}$. Blom's conjecture states that if $y_{(3)}$ is regarded as an estimate of $med\left(y_{(3)}\right)$, then $y_{(3)}$ is an estimate of the $\frac{3-\frac{1}{3}}{10+\frac{1}{3}} = 25.81$ percentile of g.

It should be noted that as $n \to \infty$, the difference between the three percentiles (25%, 30%, and 25.81%) converges to zero, indicating that the letter values are indeed consistent estimates of their associated percentiles. That is, M is a consistent estimate of the 50th percentile, and Q_1 is a consistent estimate of the 25th percentile, and so on.

1.8 Transformations of Data

A technique of great benefit to the data analyst is data transformation. For example, the researcher may find it beneficial to analyze the *log* of the observations rather than the observations themselves. The *log* function is a commonly used transformation. Of particular interest to the data analyst is determining whether a transformation is necessary and, if so, which transformation should be used. The analyst must bear in mind that transformations may help in the initial exploratory stage of the data analysis. For example, interesting features may be revealed in the transformed response, thus guiding the researcher toward some appropriate statistical model. Transformations can also often be used to alter the appearance of the raw data so that, in the transformed space, the data conforms to the specifications of some standard statistical model.

An example of this occurs in the location model, where the response, or random variable of interest, is modeled as

$$y_i = \theta + \varepsilon_i$$

If $\varepsilon_i \sim f$<, where f is symmetric about zero, then we have a reasonable interpretation of θ as the "center of the distribution" of y as $E(y) = med(y)$. In other words, if we are going to summarize the distribution of y using one number, θ is a good choice, the clear center of the data.

If the data is asymmetric, such as displayed in Figure 1.17, it is no longer clear which single number best summarizes the distribution of y since $E(y) \neq med(y)$, and there is no clear interpretation of the "center." Consequently, one of our goals when using a transformation may be to transform the data to achieve symmetry.

Note that $E(y) = 4$ and $med(y) = 3.35$.

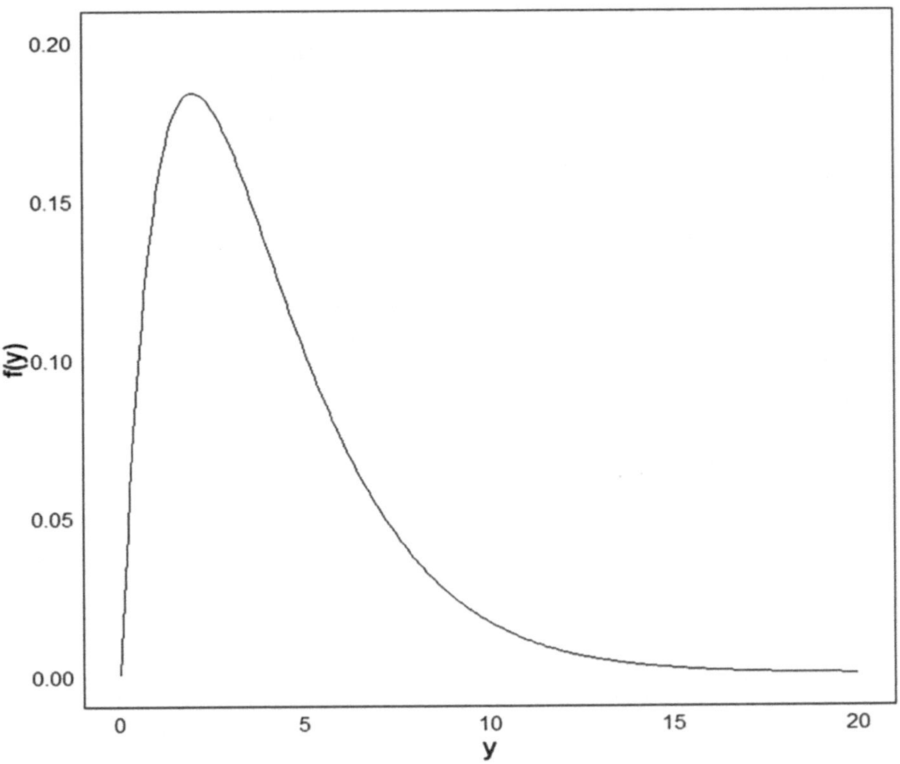

FIGURE 1.17
An example of an asymmetric distribution.

A second reason for transforming data in the location model may be to achieve, in addition to symmetry, the normality assumption of the classic location model – that is, transform y to the new variable, $g(y)$, so that $g(y) \sim N(\theta, \sigma^2)$. Under these circumstances, inferential questions on the unknown parameters can be handled routinely.

One philosophy of transformations then in the location model is that if the original y data do not satisfy the assumptions of the traditional location model, perhaps the transformed data can satisfy the assumptions of this model. In this way, we are able to use a conventional statistical model based on the transformed data rather than some nonconventional model on the raw data.

1.8.1 Types of Transformations

Let y be an original observation and y^* be the transformed value of y, where $y^* = g(y)$ for some function g. If the function $g(y)$ is of the form $g(y) = a + by$, for arbitrary constants a and b, then g is simply the linear transformation. A common example of a linear transformation is the conversion from one unit of measure to another. For example, converting data measured in inches to feet or centimeters requires multiplying the original observations by some appropriate constant. In converting from feet to inches, we would multiply the observations by 12, for example. Or converting from degrees Celsius to degrees Fahrenheit is another common example of a linear transformation. Usually, while convenient for interpretation purposes, linear transformations do nothing to alter the symmetry properties of the distribution of the random variable, y.

Nonlinear transformations, such as the log or square-root transformations, are more beneficial to the statistician. An extremely useful class of nonlinear transformations is the power family of transformations, which can be expressed in two basic forms. The simple form for the power family of transformations is to express y^* as

$$y^* = \begin{cases} y^p & \textit{for } p > 0 \\ \log(y) & \textit{for } p = 0 \\ -y^p & \textit{for } p < 0 \end{cases}$$

Some of the most common values of p and their associated values of y^* are listed in the table below:

p	−3	−2	−1	−1/2	0	1/2	1	2	3
y^*	$-\frac{1}{y^3}$	$-\frac{1}{y^2}$	$-\frac{1}{y}$	$-\frac{1}{\sqrt{y}}$	log(y)	$y^{\frac{1}{2}}$	y	y^2	y^3

Note that the "−" sign is needed if it is desired to preserve the original order of the data. For example, if the third, fifth, and second subjects have the first three ordered observations of 1, 3, and 5, and if the reciprocal transformation, y^{-1}, is used, then the three smallest transformed observations are 1/5, 1/3, and 1, corresponding to the second, fifth, and third subjects, respectively. Using y^{-1} results in the three smallest observations of −1, −1/3, −1/5, preserving the original order of the raw data.

It should also be noted that if it is determined that $p = 0$, suggesting the log transformation, the base of the logarithm is not of great importance. The conversion from the logarithm in one base to another is simply a multiplicative constant, a linear transformation. Thus, the basis of the "log" transformation should be one of the convenience transformations for the researcher. Of course, the most commonly used bases are 10 and e (the natural log), but any other base could be used, if appropriate.

The previous table is called the "ladder of reexpression" by Tukey (1977) and is a very useful device for determining the value of p required to achieve symmetry of the data, as we shall see. A more complex form of the power family is to express y^* as

$$y^* = \begin{cases} \dfrac{y^p - 1}{p} & \textit{for } p \neq 0 \\ \log(y) & \textit{for } p = 0 \end{cases}.$$

This form of power transformation is often called the "Box-Cox" family of transformations. We will first concentrate on using the simple form of the power family of transformations and the determination of p via exploratory data methods. We will return to the complex form later and use more mathematically oriented methods to determine p.

Tukey (1977) introduced a very simple method for determining the proper value of p if the goal is to achieve symmetry in a single batch of data. The method called the method of mid-summaries, examines the sequence of sample mid-summaries of the dataset, where this sequence is defined as follows:

$$\begin{aligned} \text{midM} &= \text{M} \\ \text{midQ} &= (Q_1 + Q_3)/2 \\ \text{midE} &= (E_l + E_u)/2 \\ \text{midS} &= (S_l + S_u)/2. \end{aligned}$$

We usually stop with the midS, but if the dataset is very large, one could employ more mids, utilizing the 32nds, the 64ths, and so on. Usually, these four mids are more than adequate.

One readily observes that if the underlying population distribution is symmetric, then the theoretical mids, those based on the true letter values, should be all equal to the "center" of the symmetric distribution. This will be the population mean or, equivalently, the population median. On the other hand, if the distribution is skewed to the right, then the mids will increase. Conversely, if the distribution is skewed to the left, the sequence should be decreasing. We use this information to guide us toward the proper power transformation. That is, in order to determine the proper value of p, we examine the sample mid-summaries. For a fixed value of p, if the sequence is "increasing," we know that the transformed data, the y^*s, based on the simple form of the power transformations, are skewed to the right. If the sequence is "decreasing," we know that the y^*s are skewed to the left, and if the sequence is "constant," then the y^*s are symmetric. Of course, we realize that we can only deal with the sample data rather than the entire population of data, and consequently, we can only estimate the true mids and that the behavior of the sample mids will reflect random variation. Thus, we are really looking for a sequence of mids that is relatively constant, reflecting a relatively symmetric set of transformed data.

The method of determining the proper p is an iterative technique, a common feature with the EDA methods to be presented in this book. As with all iterative schemes, they must begin at some well-chosen starting value, proceed through some iterative phase, and conclude on the basis of some stopping criteria.

The method of mid-summaries can be described as follows:

1. Select an initial value for p, p_0.
2. Compute the sequence of mids based on the y^*, using the simple form of the power transformations.
3. If the sequence of mids is "constant," quit. Choose $p = p_0$.

If the sequence of mids is "increasing," select a new value of p_0 that is smaller than the last value and return to step 2. While, if the sequence of mids is "decreasing," select a new value of p_0 that is larger than the last value, and return to step 2.

We note that if P_0 is initially set equal to 1, the analysis begins with the raw data. Additionally, one typically must iterate the procedure several times, moving up (by increasing p) or down (by decreasing p) the ladder of reexpression. The amount of movement along the ladder is usually determined by educated "trial and error."

The following examples illustrate the method and use of R to help accomplish the selection process. Recall the test score data introduced in the previous sections. It was noted from the visual displays, such as the dot plot and box plot, that the data appeared to be skewed left. The output from the following R codes also presents the mid-summaries under the heading "mid."

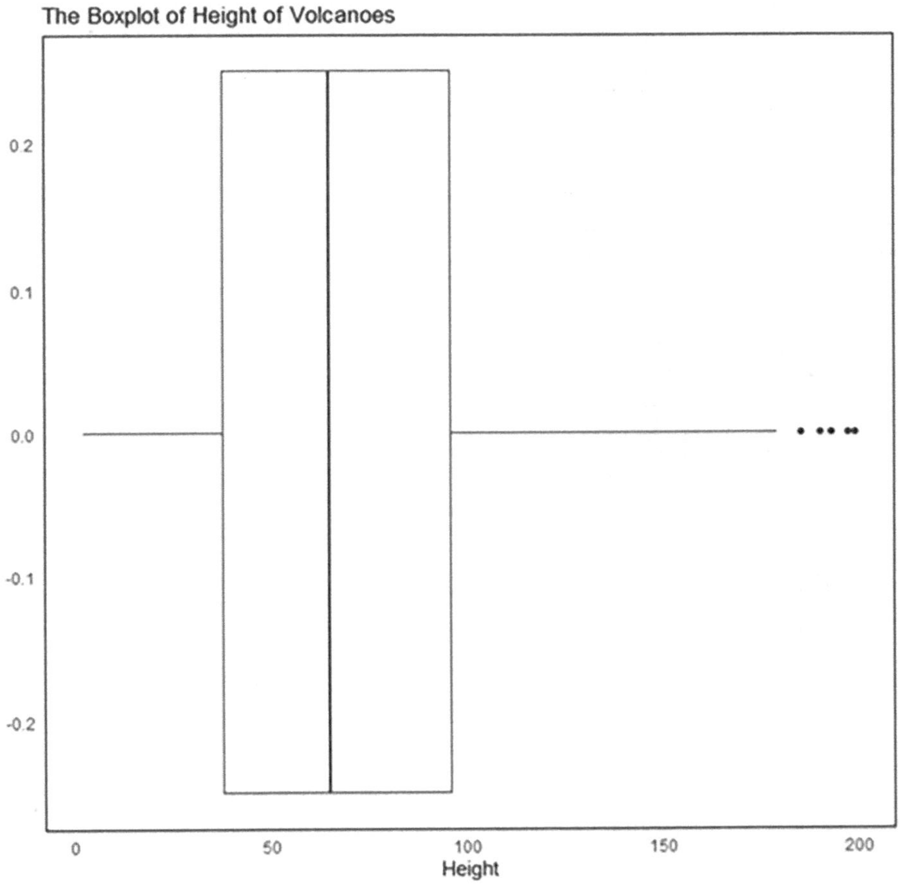

FIGURE 1.18
The box plot for volcano data.

The first four mid-summaries, denoted by M, H, E, and D, are 31, 30, 27, and 25, clearly a decreasing sequence, reinforcing our observation that the data is skewed to the left.

1.8.2 The Volcano Data

Consider the heights of 219 volcanoes, rounded to the nearest 100 feet, found listed in the 1966 *World Almanac*. The R codes to obtain the stem-and-leaf plot, the box plot in Figure 1.18, and the sequence of mid-summaries are listed in the following, along with the results.

```
# Load the data
volcano_data <- read.table("C:\\Users\\VOLCANO.DAT",
header = FALSE)names(volcano_data) <- c("height")
# Generate Stem-and-Leaf Plot and Box Plot for Original
Data
# Stem-and-leaf plot
stem(volcano_data$height)
```

```
Stem-and-leaf of volht N = 219
The decimal point is 1 digit(s) to the right of the |

   0 | 25666799
   1 | 0001366799
   2 | 00112224445566677788999
   3 | 011224455556667899
   4 | 011123333444678899999
   5 | 00112223445566666677799
   6 | 00114455666677789
   7 | 000011123345556788 89
   8 | 122223335679
   9 | 000123344556779
  10 | 0112233445689
  11 | 0112334669
  12 | 11244456
  13 | 03478
  14 | 00
  15 | 667
  16 | 25
  17 | 29
  18 | 5
  19 | 03379
```

The stem-and-leaf plot reveals that this data is strongly skewed to the right.

```
# Box Plot
ggplot(volcano_data, aes(y = height)) + geom_boxplot() +
ggtitle("Boxplot of Height of Volcanoes") +
theme_minimal()
```

The box plot in Figure 1.18 reveals four upper-extreme outliers.

```
# Calculate mid-summaries for original data
> midsummaries_original
  <- calculate_midsummaries(volcano_data$height)
> print(midsummaries_original)

       Depth Lower Upper   Mid Spread
1 219.000000   199     2 100.5   -197
2 109.500000    65    65  65.0      0
```

```
3  54.750000   37    96  66.5    59
4  27.375000   24   121  72.5    97
5  13.687500   16   140  78.0   124
6   6.843750    9   179  94.0   170
7   3.421875    6   193  99.5   187
8   1.710938    5   197 101.0   192
```

The sequence of mid-summaries, especially M through B (where "B" denotes the $\left(\frac{1}{2}\right)^6 = \frac{1}{64}$th percentile), is an increasing sequence, clearly supporting our previous observations that the raw data, with $p_0 = 1$, is skewed to the right. The method of mid-summaries states that a smaller value of p_0 be selected in the power family of transformations if the symmetry of the transformed data is required. Moving down the ladder of reexpression, we'll try $p_0 = 0$, or the log transformation. On R, type:

```
# Log Transformation
volcano_data$log_height <- log(volcano_data$height)
# Stem-and-leaf plot for log-transformed data
stem(volcano_data$log_height)
```

```
Stem-and-leaf of logvht N = 219
The decimal point is 1 digit(s) to the left of the |

   6 | 9
   8 |
  10 |
  12 |
  14 |
  16 | 1999
  18 | 5
  20 |
  22 | 00000
  24 | 06
  26 | 77
  28 | 344
  30 | 0044999888
  32 | 226660033777
  34 | 03377336666888
  36 | 14669111466666888
  38 | 35779999911335557 99
  40 | 1133333344488991166779999
  42 | 00022355556668990222346 6679
  44 | 1111222457900012334455677
  46 | 0122223344568901123345 58
  48 | 00022234790234 4
  50 | 5569159
  52 | 256689
```

The stem-and-leaf display reveals that the log of the height of volcanoes is strongly skewed to the left.

```
> # Calculate mid-summaries for log-transformed data
> midsummaries_log
  <- calculate_midsummaries(volcano_data$log_height)
> print(midsummaries_log)

         Depth    Lower     Upper      Mid       Spread
1 219.000000 5.293305 0.6931472 2.993226 -4.6001576
2 109.500000 4.174387 4.1743873 4.174387  0.0000000
3  54.750000 3.610918 4.5643482 4.087633  0.9534303
4  27.375000 3.178054 4.7957905 3.986922  1.6177367
5  13.687500 2.772589 4.9416424 3.857116  2.1690537
6   6.843750 2.197225 5.1873858 3.692305  2.9901612
7   3.421875 1.791759 5.2626902 3.527225  3.4709307
8   1.710938 1.609438 5.2832037 3.446321  3.6737658
```

The sequence of mid-summaries, M through Z (where "Z" denotes the 1/128th) is a strongly decreasing sequence, supporting our observation that the data is skewed to the left. Obviously, the choice of $P_0 = 0$ was incorrect; we moved too far down the reexpression ladder. Let's try an intermediate value of P_0, say $P_0 = 0.5$.

```
# Step 5: Square-Root Transformation
# Square-root transformation
volcano_data$sqrt_height <- sqrt(volcano_data$height)
# Stem-and-leaf plot for square root-transformed data
stem(volcano_data$sqrt_height)
      Stem-and-leaf of sqrtvht N = 219
      Leaf Unit = 0.10
      The decimal point is at the |

         1 | 4
         2 | 24446
         3 | 0022236
         4 | 001445566777999
         5 | 0011122334445667788 9999
         6 | 000122234445666666668999
         7 | 00000111122233344555555555777788
         8 | 00111111222223444444455567777788889
         9 | 0111111123345555666777778889
        10 | 0001111222344555666 7889
        11 | 000111224567788
        12 | 55578
        13 | 146899
        14 | 01
```

FIGURE 1.19
The box plot of square-root transformation for volcano data.

The stem-and-leaf display reveals that the square-root transformation has resulted in symmetric data. Stem-and-leaf plot for square root–transformed data is given in Figure 1.19.

The box plot in Figure 1.19 is close to the "ideal" box plot.

```
> # Calculate mid-summaries for square root-transformed data
> midsummaries_sqrt
  <- calculate_midsummaries(volcano_data$sqrt_height)
> print(midsummaries_sqrt)
           Depth     Lower     Upper      Mid     Spread
     1 219.000000 14.106736  1.414214 7.760475 -12.692522
     2 109.500000  8.062258  8.062258 8.062258   0.000000
     3  54.750000  6.082763  9.797959 7.940361   3.715196
     4  27.375000  4.898979 11.000000 7.949490   6.101021
     5  13.687500  4.000000 11.832160 7.916080   7.832160
     6   6.843750  3.000000 13.379088 8.189544  10.379088
```

```
7    3.421875  2.449490 13.892444 8.170967   11.442954
8    1.710938  2.236068 14.035669 8.135868   11.799601
```

The sequence of mid-summaries is nearly a constant sequence. Thus, the iteration within the method of mid-summaries ends with $p_0 = 0.5$. This suggests that further analysis with the height of volcanoes data, such as a comparison of heights of volcanoes in North and South America or a comparison of heights of active and inactive volcanoes would require the square-root transformation.

1.9 The Symmetry Plot

In the previous example, we saw that the stem-and-leaf display and the box plot are useful displays for checking the symmetry of a single sample of data. Another plot specifically intended to evaluate the symmetry of data is the "symmetry plot," where $y_{(i)}$ is plotted against $y_{(n-i+1)}$, for $i = 1, \ldots, n$. Theoretically, one could argue that if the underlying distribution of the random variable y is symmetric about θ, then $E\left(y_{(i)}\right)+E\left(y_{(n-i+1)}\right)=2\theta$, for $i=1,\ldots,n$. Consequently, $E\left(y_{(n-i+1)}\right)=2\theta-E\left(y_{(i)}\right)$. Thus, since $y_{(i)}$ is an estimate of $E\left(y_{(i)}\right)$ and $y_{(n-i+1)}$ is an estimate of $E\left(y_{(n-i+1)}\right)$, it follows that a plot of $y_{(n-i+1)}$ versus $y_{(i)}$ should represent random observations about a straight line with a slope of −1.

On the other hand, if the underlying distribution is skewed to the right, then $y_{(n-i+1)}$ is further away from the center than $y_{(i)}$, resulting in a symmetry plot with a concave-up appearance, while a skewed-left distribution will appear as a concave-down symmetry plot. Three symmetry plots for symmetric, skewed-right, and skewed-left distributions are given in Figure 1.20.

A practical application of the symmetry plot is Figure 1.21 using volcano data.

1.9.1 Volcano Data

Again, consider the volcano data from the previous example. The sorted data, log heights, and the square root of heights are saved. The symmetry plot (Figure 1.21) is obtained by simply plotting the data sorted in ascending order against the same data sorted in descending order. On R, type:

```
# Symmetry plot for original data
symmetry_plot(volcano_data$height, "Symmetry Plot of
Original Heights")
```

As expected, this symmetry plot (Figure 1.21 (a)) indicates that volcanoes' heights are skewed to the right. Next, try the log height data.

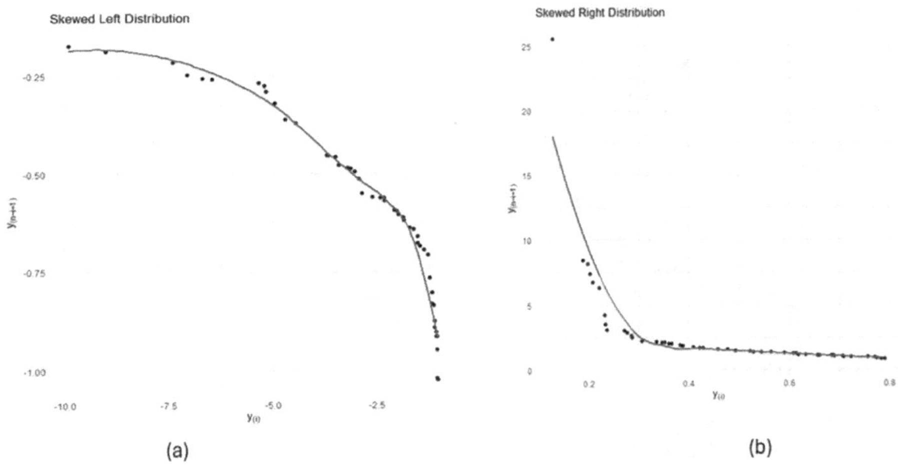

FIGURE 1.20
(a) Symmetry plot of a skewed-left distribution; (b) Symmetry plot of a skewed-right distribution.

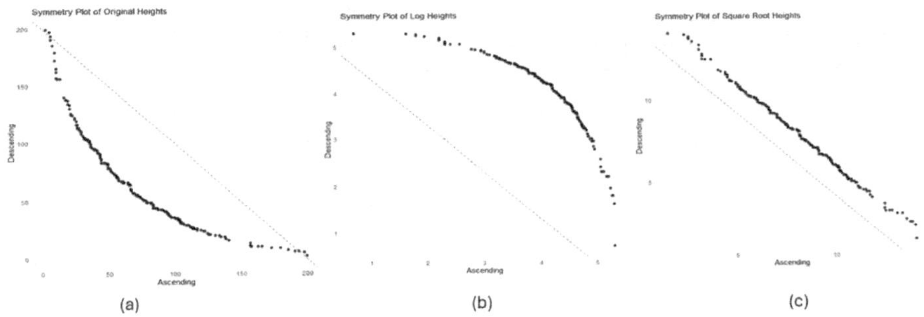

FIGURE 1.21
(a) Symmetry plot of original data; (b) Symmetry plot of log(heights); (c) Symmetry plot of square root of volcano heights.

```
# Symmetry plot for log-transformed data
symmetry_plot(volcano_data$log_height, "Symmetry Plot of
Log Heights")
```

This symmetry plot (Figure 1.21 (b)) indicates that the log of heights is strongly skewed to the left.

```
# Symmetry plot for square root-transformed data
symmetry_plot(volcano_data$sqrt_height, "Symmetry Plot
of Square Root Heights")
```

This symmetry plot of the square root of volcano heights (Figure 1.21 (c)) is nearly a perfect straight line with a slope of –1, indicating nearly symmetric data.

1.10 Probability Plotting

One last plotting technique that we will find useful in the location model, as well as other statistical models, is the probability plot. The most common type of probability plot is the "normal probability plot," which is often discussed in elementary statistics courses. In this section, we will demonstrate that normal probability plots are a special case of the general probability plot. Probability plots can be classified into two basic categories: Q-Q plots and P-P plots. We will emphasize the Q-Q plot here.

1.10.1 Q-Q Plot

The Q-Q plot, or quantile versus quantile plot, is sometimes also called a percentile plot. We define $q_y(p)$ as the quantile of order p for the random variable y. That is, $q_y(p)$ is that value of y such that Prob($y < q_y(p)$) = $F(q_y(p)) = p$, where F is the CDF of the random variable y. Or $q_y(p)$ can be found explicitly as $F^{-1}(p) = q_y(p)$, where F^{-1} is the inverse CDF of the random variable y. We also define $p_y(q)$ as the percent of order q for the random variable y. That is, $p_y(q)$ is that probability and q is that value of y such that Prob($y < q$) = $p_y(q)$. The relationship between $q_y(p)$ and $p_y(q)$ can easily be seen in Figure 1.22. By

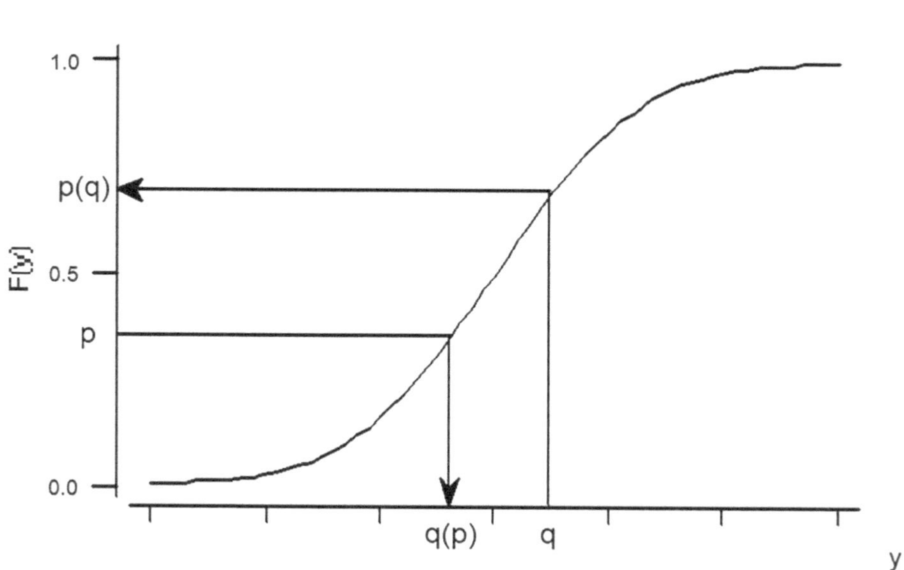

FIGURE 1.22
The relationship between the quantiles and percents.

varying p from 0 to 1 and q from $-\infty$ to ∞, we can obtain all possible values of $q_y(p)$ for any p and $p_y(q)$ for any value of q.

The relationship between the quantile of order p and the percent of order q is established through the CDF and its inverse. For a given value of p, $q_y(p)$ is found using the inverse CDF, F^{-1}. For a given value of q, a value of the random variable y, $p_y(q)$ is found using the CDF, F.

There are three basic types of Q-Q plots: theoretical versus theoretical (t-t), empirical versus theoretical (e-t), and empirical versus empirical (e-e). Our emphasis will be on the first two types.

1.10.2 Theoretical versus Theoretical Q-Q Plots

In a theoretical versus theoretical (t-t) Q-Q plot, the distributions of two random variables are compared by examining their CDFs. One obvious way to compare the CDFs is to overlay the two CDFs on the same plot, as illustrated in Figure 1.23, where the random variable x has a smaller mean and larger variance than the variable y.

The Q-Q plot can provide us with additional information that is not easily seen in Figure 1.23. To compare one CDF with another, plot $q_y(p)$ versus $q_x(p)$ for successive values of p, where $0 \le p \le 1$. This idea is demonstrated in Figure 1.23, where for the given value of p, the inverse CDF is used to obtain

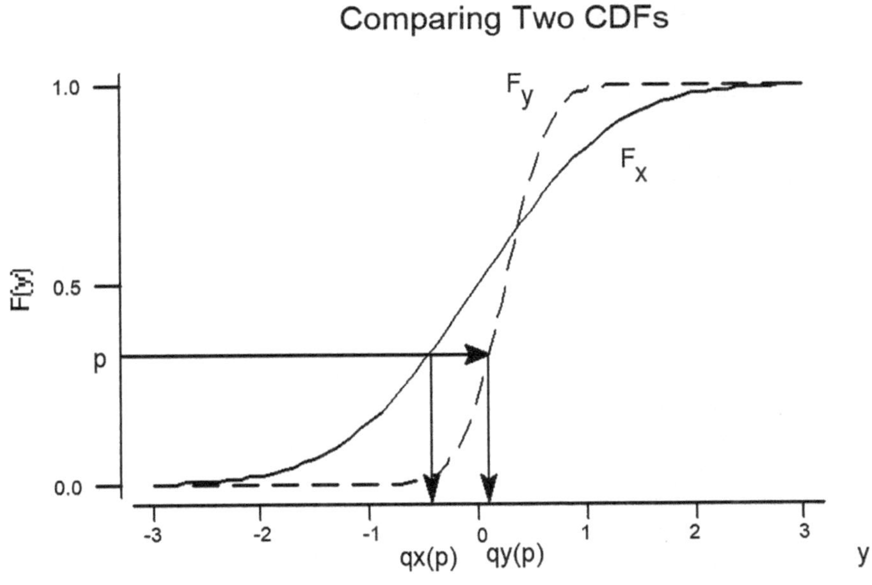

FIGURE 1.23
Comparing two CDFs, F_X and F_Y, for the random variables x and y.

the quantiles for x and y of order p. We will plot $q_x(p)$ on the abscissa of our Q-Q plot and refer to F_x as the "reference distribution." We will use R to demonstrate several of the many possible shapes of Q-Q plots.

Example 1.1: F_x equals F_y.

If the distribution of x is equal to the distribution of y, then $q_x(p)$ and $q_y(p)$ are equal for all values of p. The Q-Q plot will be a straight line with a slope of one and an intercept of zero.

Example 1.2: F_x and F_y of the same distributional form but with different parameters.

Under certain conditions, to be presented shortly, the quantiles of x and the quantiles of y will be linearly related – that is, $q_x(p) = \mathrm{a} + \mathrm{b}q_x(p)$, where a and b relate the parameters of y to those for x. The Q-Q plot will result in a straight line with an intercept of a and a slope of b. The theoretical justification for this result will be presented shortly.

For example, consider the t-t Q-Q plot comparing a N(0, 1) to N(2, 0.25).

The plot is easily obtained in R by typing as follows:

```
# Generate p-values
p <- seq(0.01, 0.99, by = 0.01)
# Quantiles for N(0,1)

q_N01 <- qnorm(p, mean = 0, sd = 1)
# Quantiles for N(2, 0.25)
q_N22.5 <- qnorm(p, mean = 2, sd = sqrt(0.25))
# Q-Q plot for N(0,1) vs N(2,0.25)
qq_plot_N01_N22.5 <- data.frame(q_N01, q_N22.5)
ggplot(qq_plot_N01_N22.5, aes(x = q_N01, y = q_N22.5)) +
geom_point() + geom_abline(intercept = 2, slope = 0.5,
color = "red", linetype = "dashed") + ggtitle("Q-Q Plot:
N(0,1) vs N(2,0.25)") + xlab("N(0,1) Quantiles") +
ylab("N(2,0.25) Quantiles") + theme_minimal() +
theme(panel.grid.major = element_blank(), panel.grid.
minor = element_blank(), panel.border = element_
rect(color = "black", fill = NA, size = 1))
```

Note that the inverse CDF function is used to obtain these quantiles.

Careful inspection of the Q-Q plot in Figure 1.24 reveals a y intercept of two and a slope less than one. These values can be found by regressing $q_y(p)$ versus $q_x(p)$. On R, type as follows:

```
# Perform linear regression
reg_N01_N22.5 <- lm(q_N22.5 ~ q_N01)
summary(reg_N01_N22.5)
```

```
            The regression equation is
               C3 = 2.00 + 0.500 C2
```

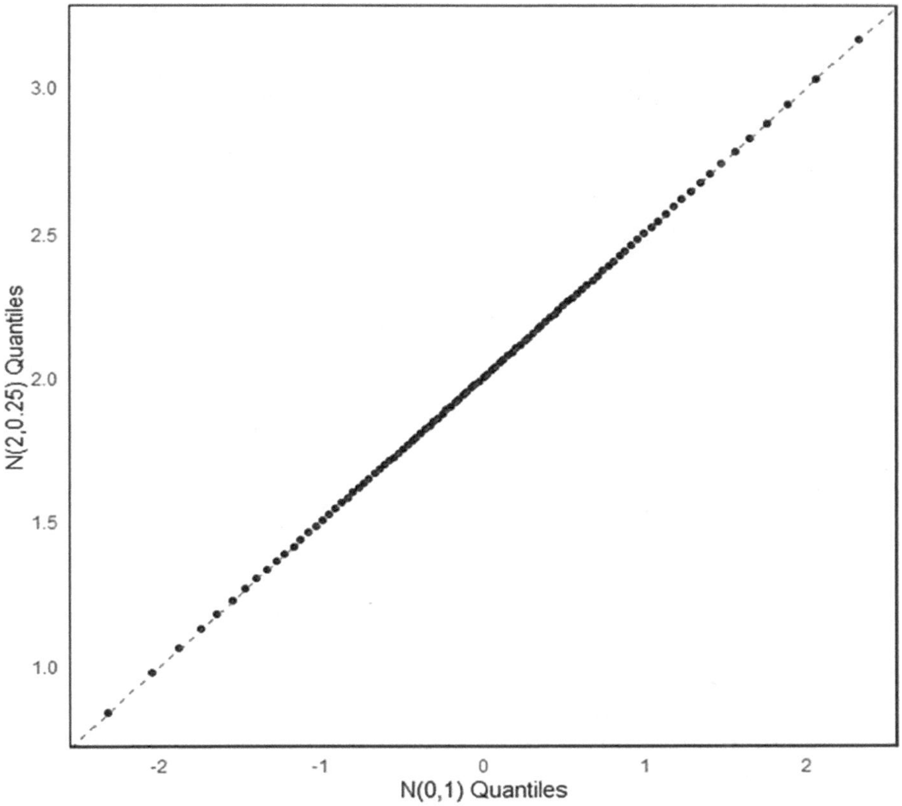

FIGURE 1.24
The Q-Q plot for N(0, 1) versus N(2, 0.25).

We see that the intercept is 2, and the slope is 0.5.

Example 1.3: Comparing N(0, 1) to a heavy-tailed distribution.

In many statistical applications, the assumption of normality is made. As is often the case, the normal distribution is used as the reference distribution to which other distributions are compared. One useful comparison is through the "tail behavior" of the distribution. A heavy-tailed distribution can be loosely defined as one that places more density in the extremes of the distribution than the normal distribution. Consequently, the quantiles for a heavy-tailed distribution are more extreme than the normal distribution. That is, for small values of p, the quantiles are smaller for the heavy-tailed distribution than the normal, and for large values of p, the quantiles are larger for the heavy-tailed distribution.

A simple example is a comparison of N(0, 1) to a *t*-distribution with ν degrees of freedom, denoted by $t(\nu)$. It is well known that the *t*-distributions are heavier tailed than the normal and become less so as the degrees of freedom increase. We all know that at $v = \textit{infinity}$ that the N(0, 1) and $t(\infty)$ are the same distribution. Numerically, if p = 0.01, then the quartile, q(0.01), for the t(2) is –6.97, while the corresponding quartile for the N(0, 1) is only –2.33. The heavy-tailed nature of the *t*-distribution can be seen in the Q-Q plot by a distinctive curved pattern illustrated by comparing a N(0, 1) to t(2) as in Figure 1.25. To obtain the appropriate Q-Q plot on R, type as follows:

```
# Quantiles for t(2)
q_t2 <- qt(p, df = 2)
# Q-Q plot for N(0,1) vs t(2)
qq_plot_N01_t2 <- data.frame(q_N01, q_t2)
ggplot(qq_plot_N01_t2, aes(x = q_N01, y = q_t2)) + geom_
point() +
geom_abline(intercept = 0, slope = 1, color = "red",
linetype = "dashed") + ggtitle("Q-Q Plot: N(0,1) vs
t(2)") + xlab("N(0,1) Quantiles") + ylab("t(2)
Quantiles") + theme_minimal() +
theme(panel.grid.major = element_blank(), panel.grid.
minor = element_blank(), panel.border = element_
rect(color = "black", fill = NA, size = 1))
```

The "reference distribution," the N(0, 1), is plotted on the #abscissa in Figure 1.25. Also plotted is a "reference line" with an intercept of zero and a slope of one. Notice the downward and upward turns at the left and right of the plot. This is the distinctive heavy-tailed feature of Q-Q plots.

Example 1.4: Comparing a N(0, 1) to a short-tailed distribution.

Short-tailed distributions are those whose quartiles are not as "far" (in terms of standard deviation units) from the center of the distribution as they are for the normal distribution. One very common example of a short-tailed distribution is the uniform. For example, a Q-Q plot (Figure 1.26) comparing a N(0, 1) to a uniform distribution on the interval –2.5 to 2.5, denoted by U(–2.5, 2.5) can be obtained on R by typing as follows:

```
# Quantiles for U(-2.5, 2.5)
q_U_2.52.5 <- qunif(p, min = -2.5, max = 2.5)
# Q-Q plot for N(0,1) vs U(-2.5,2.5)
qq_plot_N01_U_2.52.5 <- data.frame(q_N01, q_U_2.52.5)
ggplot(qq_plot_N01_U_2.52.5, aes(x = q_N01, y = q_U_2.52.5)) +
geom_point() + geom_abline(intercept = 0, slope = 1, color =
  "red", linetype = "dashed") + ggtitle("Q-Q Plot: N(0,1) vs
  U(-2.5,2.5)") + xlab("N(0,1) Quantiles") + ylab("U(-2.5,2.5)
  Quantiles") + theme_minimal() + theme(panel.grid.major =
  element_blank(), panel.grid.minor = element_blank(),
panel.border = element_rect(color = "black", fill = NA, size = 1))
```

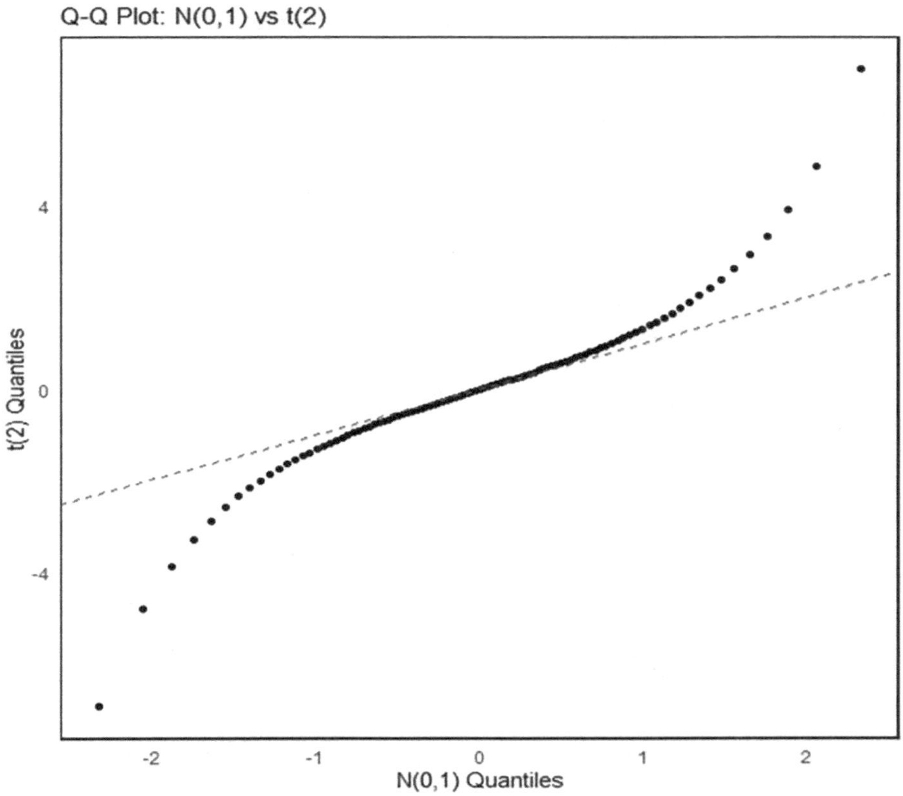

FIGURE 1.25
The Q-Q plot for N(0, 1) versus *t*-distribution with df = 2.

Notice that the upward and downward turns in the plot are now in the left and right, respectively, just the opposite of the Q-Q plot for the heavy-tailed example.

1.10.3 Some Theoretical Justification: A Special Case

Suppose that the random variable of interest, x, is a member of the location-scale family of distributions. That is, the CDF of x, with the location parameter θ_x and scale parameter τ_x, can be written as

$$Pr\,ob(X \leq x) = F_x(x;\theta_x,\tau_x) = F\left(\frac{x-\theta_x}{\tau_x}\right),$$

where F_x is the CDF of x, and F is the CDF of the standardized random variable $z = \frac{x-\theta_x}{\tau_x}$. The CDF F is often called the standard CDF. The previous

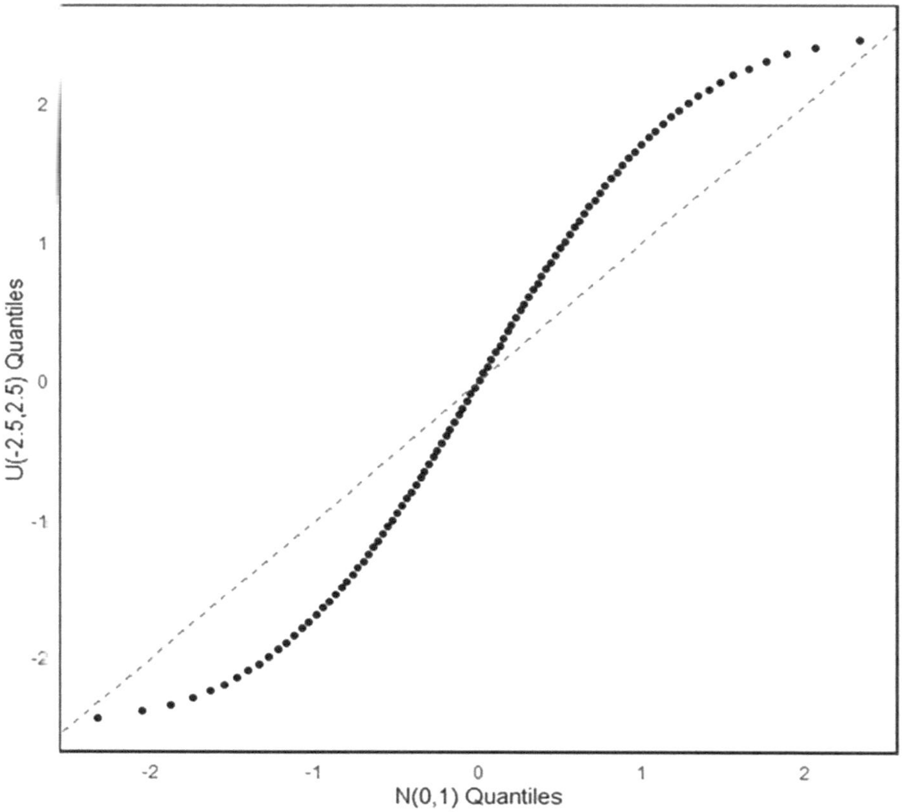

FIGURE 1.26
The Q-Q plot for N(0, 1) versus U(−2.5, 2.5).

expression implies that if x is a member of the location-scale family, any probability statement for x can be computed using the standard CDF.

For example, consider the normal random variable x with mean μ_x and standard deviation σ_x – that is, $x \sim N\left(\mu_x, \sigma_x^2\right)$. Then

$$F_x\left(x; \mu_x, \sigma_x\right) = F\left(\frac{x - \mu_x}{\sigma_x}\right),$$

where F, in this case, is the standard normal, N(0, 1), distribution and $Z = \frac{x - \theta_x}{\tau_x}$ is the standard normal random variable. This implies that, in the normal random variable case, that $\mu_x = \theta_x$, the location parameter, and $\sigma_x = \tau_x$, the scale parameter. Many other examples of distributions are members of the location-scale family. Below are a few of the more commonly used distributions.

Gamma Distribution: The probability density function (PDF), $f(x)$, for the random variable x with a gamma distribution is given by

$$f(x)=\frac{(x-\gamma)^{\eta-1}\, exp\left[-(x-\gamma)/\beta\right]}{\beta^{\eta}\Gamma(\eta)},$$

where $\eta>0, \beta>0$, and $x>\gamma$. It is easily observed that β is the scale parameter, and γ is the location parameter; η is called the shape parameter. Usually, γ is set to zero.

A chi-square distribution with p degrees of freedom, denoted by $\mathrm{X}^2(p)$, results from a gamma distribution if $\eta=p/2$, where p is an integer, and $\beta=2$. A brief summary of some of the useful properties of the chi-square distribution are as follows:

1. If $x \sim \mathrm{X}^2(p)$, then $E(x)=p$ and $Var(x)=2p$.
2. The chi-square distributions are skewed right. The degree of skewness is measured by the skewness parameter, which will be labeled as β_1 and defined as

$$\beta_1=\frac{E(x-E(x))^3}{Var(x)^{\frac{3}{2}}}.$$

It can be verified that β_1 is zero if the distribution is symmetric, a negative number if the distribution is skewed left, and positive if the distribution is skewed right. For the chi-square distribution, one can show that $\beta_1=\sqrt{\frac{8}{p}}$, a positive value that tends to zero as p gets large.

3. If $z_1, z_2, \ldots z_p$ are independent and identically distributed random variables each with a N(0, 1) distribution, then $\sum_{i=1}^{p} z_i^2 \sim \mathrm{X}^2(p)$.
4. As $p \to \infty$, $\mathrm{X}^2(p)$ approaches $N(p, 2p)$.

Laplace or Double Exponential Distribution: The PDF for a random variable with a double exponential distribution is of the form

$$f(x)=.5\lambda\, exp\left(-|x-\theta|/\lambda\right),$$

for $\lambda>0$. We note that the location parameter is θ and the scale parameter is λ. The PDF is plotted in Figure 1.27 for $\theta=0$ and $\lambda=1$.

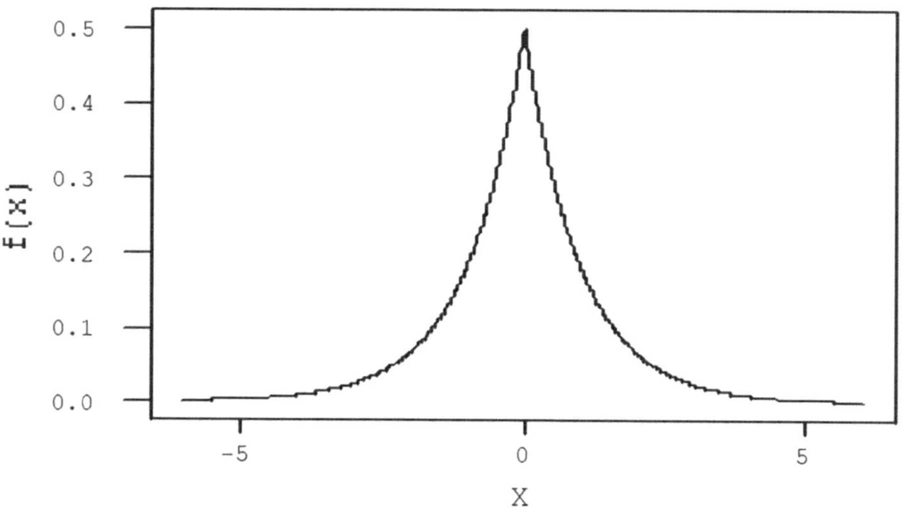

FIGURE 1.27
The double exponential distribution function.

Again, it is apparent that θ and λ are the location and scale parameters, respectively. This distribution is symmetric ($\beta_1 = 0$) and has $E(x) = \theta$ and $Var(x) = 2\lambda^2$.

Logistic Distribution: The PDF for the random variable x with a logistic distribution is

$$f(x) = \frac{exp\left(\frac{x-\alpha}{\beta}\right)}{\left(1 + exp\left(\frac{x-\alpha}{\beta}\right)\right)^2},$$

where $-\infty < \alpha < \infty$ and $\beta > 0$. It is clear that α and β are the location and scale parameters, respectively. The expectation of x is α, and the variance is $\frac{\beta^2\pi^2}{3}$.

Other examples of the location-scale family of distributions may be found in Mood, Graybill, and Boes (1974). Some properties of the location-scale family pertinent to our use are the following:

If x is a location-scale random variable with CDF $F_x(x;\theta_x,\tau_x)$, then

1. The PDF of x is $f_x(x;\theta_x,\tau_x)=\frac{1}{\tau_x}f(z)$, where $z=\frac{x-\theta_x}{\tau_x}$, and f is the standard PDF, $f(z)=\frac{\partial F(z)}{\partial z}$.
2. If the t-t Q-Q plot of the random variable y versus x is a straight line, then y is a member of the same location-scale distribution as x, and the following relationships hold:
   a. $q_y(p)=a+bq_x(p)$, where a and b are the intercept and slope of the straight line in the Q-Q plot.
   b. The location and scale parameters for y are respectively:

$$\theta_y=a+b\theta_x \text{ and } \tau_y=b\tau_x.$$

Since this fact is essential to the proper interpretation of the Q-Q plot, we will prove that property b follows from the property a.

We know that

$$p=Pr\,ob\left(y\le q_y(p)\right)=F_y\left(q_y(p)\right)=F_x\left(q_x(p)\right)$$

$$=F_x\left(\frac{q_y(p)-a}{b}\right)=F\left(\frac{\frac{q_y(p)-a}{b}-\theta_x}{\tau_x}\right)$$

$$=F\left(\frac{q_y(p)-a-b\theta_x}{b\tau_x}\right).$$

Thus, y is a member of the same location-scale distribution of x with location and scale parameters as listed in property b.

One important application of these facts is the t-t normal Q-Q plot, where $x\sim N(0,1)$ and $y\sim N\left(\mu_y,\sigma_y^2\right)$. Then it follows from above that the Q-Q plot will exhibit a straight line with

$$\theta_y=a+b\theta_x=a+b(0)=a=\mu_y$$

and

$$\tau_y=b\tau_x=b(1)=b=\sigma_y$$

(since the normal's location parameter and scale parameter are equal to the mean and standard deviation, respectively). It follows then that $q_y(p)=\mu_y+\sigma_y q_x(p)$. Thus, the intercept of this Q-Q plot is E(y)=μ_y, and the slope of the plot is the standard deviation of y, σ_y. This fact was already demonstrated in Example 1.2, where a Q-Q plot was presented for $N(2,0.25)$ versus $N(0,1)$. The straight line of the Q-Q plot had an intercept of 2, and its slope was 0.5 = sqrt(0.25).

As a second example of a t-t normal Q-Q probability plot, consider the logistic distribution with parameters α and β. The logistic is similar in shape to the normal distribution, except it has slightly heavier tails. R can be conveniently used to study the similarity between these two distributions. For example, to compare a logistic distribution to $N(0,1)$, we first determine that the value of β equal to $\frac{\sqrt{3}}{\pi}$=0.551 and α =0 results in a logistic distribution with a mean of zero and a standard deviation of one, the same values as the N(0, 1). To compare their PDFs, on R type as follows:

```
# Constants
k1 <- sqrt(3) / pi
# Generate data for comparison
x <- seq(-5, 5, by = 0.1)
# Calculate the pdfs
norm_pdf <- dnorm(x, mean = 0, sd = 1)
logistic_pdf <- dlogis(x, location = 0, scale = k1)
# Create data frame for plotting pdfs
pdf_data <- data.frame(x = x, Normal = norm_pdf, Logistic =
logistic_pdf)
# Plot the pdfs
pdf_plot <- ggplot(pdf_data, aes(x = x)) +
geom_line(aes(y = Normal), color = "black", linetype =
"solid") + geom_line(aes(y = Logistic), color = "black",
linetype = "dashed") + ggtitle("PDF Comparison between
Normal(0,1) and Logistic(0, sqrt(3)/pi)") + xlab("x") +
ylab("Density"))
print(pdf_plot)
```

A comparison of the PDFs for N(0, 1) and logistic (0, 0.555) is in Figure 1.28.

It is evident that these two PDFs are very similar, with the logistics being slightly taller in the center and slightly heavier tails. The t-t Q-Q plot should reveal the heavy-tailed shape. On R, type as follows:

```
# Generate probabilities for Q-Q plot
p <- seq(0.01, 0.99, by = 0.01)
# Calculate the quantiles
norm_quantiles <- qnorm(p, mean = 0, sd = 1)
logistic_quantiles <- qlogis(p, location = 0, scale = k1)
# Create data frame for Q-Q plot
qq_data <- data.frame(Normal = norm_quantiles, Logis-
tic = logistic_quantiles)
```

The t-t Q-Q plot for a logistics (0, 0.551) versus a N(0, 1) is given in Figure 1.29.

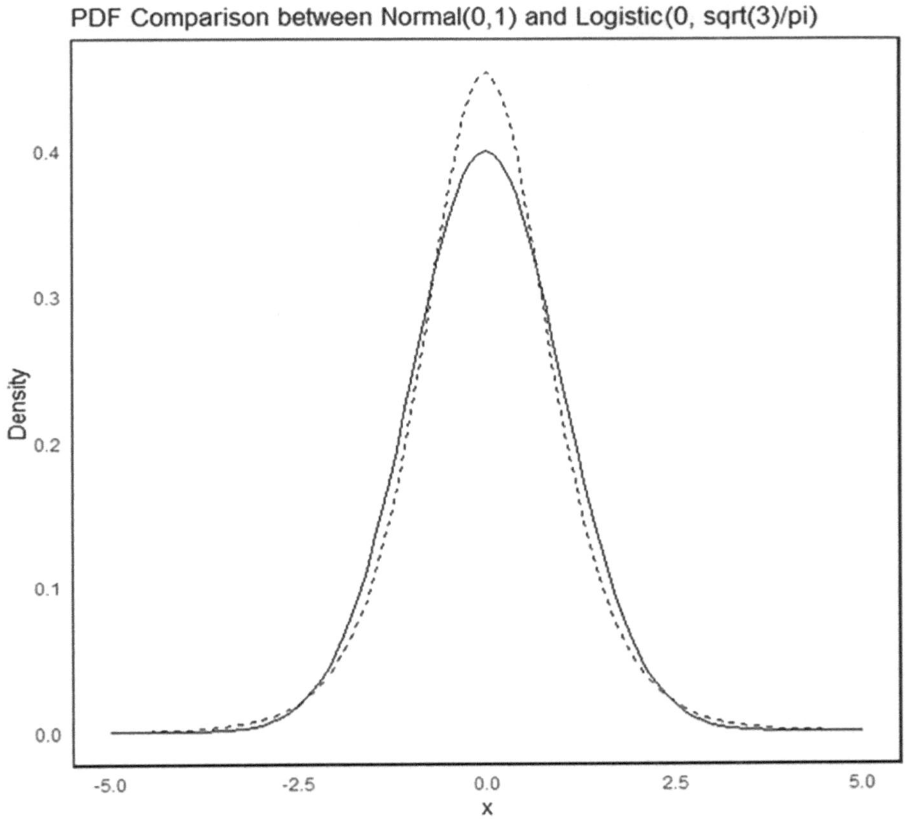

FIGURE 1.28
The probability distribution functions for N(0, 1) and logistic (0, 0.555).

We notice the slight downward and upward shape of the t-t Q-Q plot, indicating the heavy-tailed nature of this logistic distribution.

Since we do not have a straight line, it is inappropriate to determine the location and scale parameters from the intercept and slope of the Q-Q plot. However, the line is nearly straight, so one would expect that the intercept and slope would be very close to the actual location and scale parameters of the logistic distribution. On R, type as follows:

```
# Perform regression to compare the location and scale
parameters
fit <- lm(Logistic ~ Normal, data = qq_data)
summary(fit)
```

The regression equation is

```
logist = 0.00000 + 0.975 N(0.1).
```

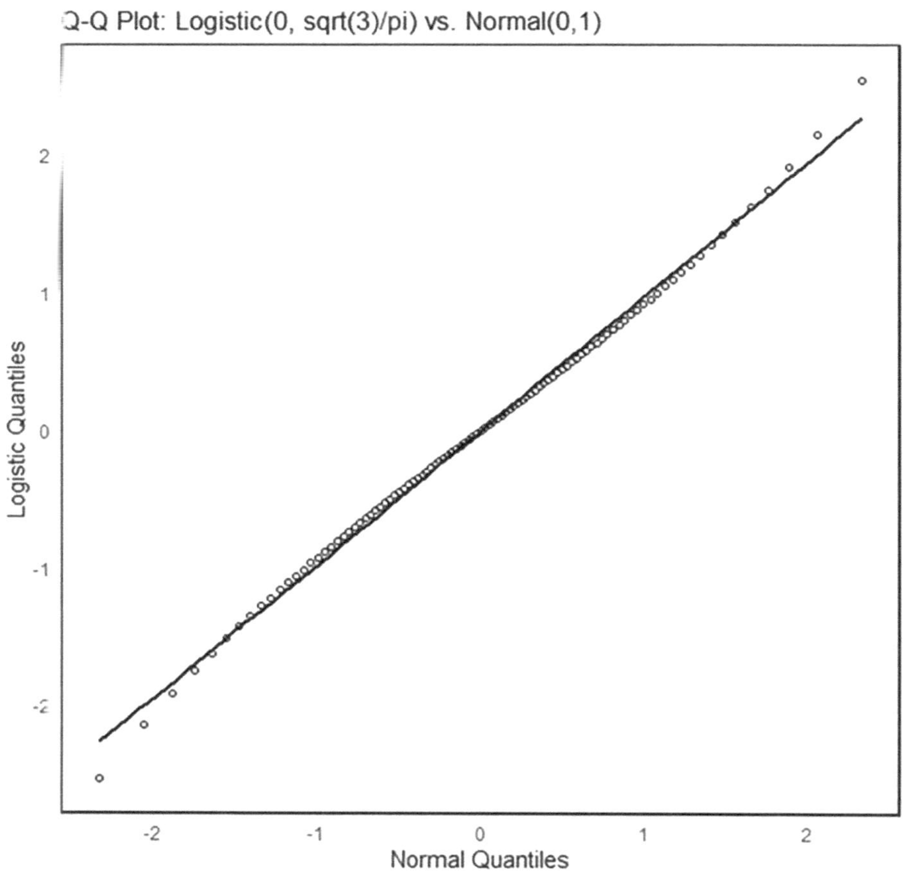

FIGURE 1.29
The T-T Q-Q plot for a logistics (0, 0.551) versus N(0, 1).

We see that the intercept is 0, the correct value of α, and the slope is 0.975, slightly under the true value of β of 1. It is noted that the location parameter, θ_x, and the scale parameter, τ_x, are not necessarily equal to the mean and the standard deviation of x, as they are in the normal case. The double exponential distribution is one example where the scale parameter is not equal to the standard deviation.

Theoretical Q-Q plots represent a convenient starting point in the study of Q-Q plots. The features and patterns observed here carry over to the most important application of Q-Q plots, the empirical versus theoretical (e-t) Q-Q plots.

1.10.4 Empirical versus Theoretical Q-Q Plots

By far, the most common application of Q-Q plots is to compare an observed sample of data to some theoretical reference distribution. For example, one may wish to determine if a given set of data could reasonably represent a sample from a normal distribution. In the empirical versus theoretical Q-Q plot, the sample's empirical cumulative distribution (ECDF) is compared to some appropriately chosen theoretical distribution. The theoretical distribution is called the "reference distribution" and is usually taken to be the standard normal. If this is the case, the e-t Q-Q plot is commonly referred to as a "normal probability plot." Thus, the e-t Q-Q plot is similar to the t-t Q-Q plot, where the CDF of y is replaced with the ECDF for the sample data.

Recall that the ECDF of a sample of size n of the random variable y is defined as

$$F_n\left(y_{(i)}\right) = \frac{i}{n},$$

for $i = 1, \ldots, n$, where $y_{(i)}$ is the ith order statistic. Since the ECDF has jumps of $1/n$ at each of the observed sample points, we consider, in the construction of the Q-Q plot, only those values of $P = i/n$, for $i = 1, \ldots, n$. Thus, for each considered value of p, we plot the sample quantiles versus the theoretical quantiles or $y_{(i)}$ versus $q_x\left(\frac{i}{n}\right)$, for $i = 1, \ldots, n$.

If the e-t Q-Q plot is reasonably straight, then assuming that the data represents a random sample from the reference distribution is reasonable. If the sequence of plotted quantiles is not straight, then the pattern exhibited in the plot indicates the form of the departure of the sample from the reference distribution. The patterns given for the t-t Q-Q plot apply to this plot as well. That is, the sample data may be "heavy tailed" compared to the reference distribution, and the Q-Q plot will reveal a downward and upward pattern in the left and right of the plot. Other characteristics such as short-tailed, skewed-right, and skewed-left are also revealed in a corresponding manner as in the t-t Q-Q plot. The reader should note that small datasets may display a lack of straightness even when they are samples from the reference distribution due to the inherent randomness in data. The "significance" of these departures from straightness will be formally evaluated in Chapter 2.

If the reference distribution is a member of the location-scale family, then a reasonable straight-line pattern in the Q-Q plot suggests that y is also a member of this family with

$$q_y(p) = y_{(i)} \approx a + bq_x\left(\frac{i}{n}\right),$$

where a and b are now the estimated intercept and slope of the straight line. Least squares estimation may be used to obtain a and b. Using the facts established previously for the location-scale family, it follows that a reasonable estimate of the location parameter of y is $\hat{\theta}_y = a + b\theta_x$ and an estimate of the scale parameter of y be given by $\hat{\tau}_y = b\tau_x$. Of course, θ_x and τ_x are specified by the reference distribution.

The most common example is to set the reference distribution equal to the standard normal, where $F(x)$ equals the CDF of a N(0, 1), usually denoted by $\Phi(x)$. There are many variations to this form of "normal probability plot." Some versions are as follows:

$$\text{plot } y_{(i)} \text{ versus } \Phi^{-1}\left(\frac{i-.5}{n}\right), \text{ the "rankits,"}$$

or

$$\text{plot } y_{(i)} \text{ versus } \Phi^{-1}\left(\frac{i-\frac{1}{3}}{n+\frac{1}{3}}\right), \text{ the "normits,"}$$

or

$$\text{plot } y_{(i)} \text{ versus } \Phi^{-1}\left(\frac{i}{n}\right)+5, \text{ the "probits,"}$$

or

$$\text{plot } y_{(i)} \text{ versus } \Phi^{-1}\left(\frac{i-\frac{3}{8}}{n+\frac{1}{4}}\right), \text{ the "normal scores."}$$

There is very little difference between these plots. The normits are useful since, it may be recalled from Blom's conjecture, the ith normit is approximately equal to the median of the ith order statistic in a sample of size n from a $N(0,1)$ distribution. The normal scores are also useful since they provide an approximation to the expected value of the ith order statistic in a sample of size n from a $N(0,1)$ distribution.

Thus, the straightness of the normal probability (using any of the four types) is an informal test of normality, in much the same way as comparing the box plot for the sample data to the "ideal box plot" is also an informal test of normality. Quantifying the "degree of straightness" of the normal Q-Q plot through a test of the hypothesis will constitute one mechanism for formalizing the test of normality. This will be discussed in the Section 1.10.5.

We will now demonstrate the use of the e-t Q-Q using both real and simulated data.

1.10.5 The Volcano Data

Recall the volcano data representing the heights of 219 volcanoes. We observed previously that the raw data was skewed to the right, the log of the data was skewed to the left, and that the square root of heights was symmetric. In the following R session, the sorted raw data is obtained, and the sorted square root of heights is calculated. On R, type as follows:

```
# Create data frames for plotting
data_raw <- data.frame(Height = sorted_raw, NScores =
nscores_raw)
data_log <- data.frame(LogHeight = sorted_log, NScores =
nscores_log)
data_sqrt <- data.frame(SqrtHeight = sorted_sqrt,
NScores = nscores_sqrt)
# Plot raw data with normal scores
plot_raw <- ggplot(data_raw, aes(x = NScores, y =
Height)) + geom_point(shape = 1, color = "black") +
geom_smooth(method = "lm", se = FALSE, color = "blue") +
ggtitle("Q-Q Plot for Raw Volcano Heights") +
xlab("Normal Scores") + ylab("Raw Heights") + theme_min-
imal() + theme(panel.grid.major = element_blank(),
panel.grid.minor = element_blank(), panel.border =
element_rect(fill = NA, color = "black", size = 1))
print(plot_raw)
```

The normal probability plot of volcano heights is shown in Figure 1.30.

Notice the concave upward pattern, which is characteristic of skewed-right data, as in Figure 1.30 (a).

Notice the concave downward tendency, reflecting skewed-left data, as in Figure 1.30 (b). Look at the probability plot for the square root of volcano heights.

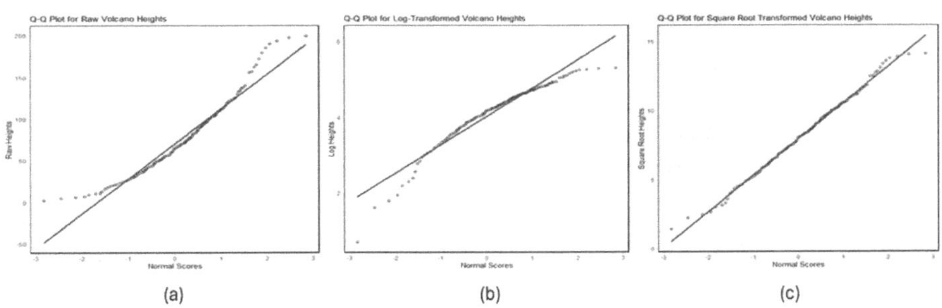

FIGURE 1.30
The Q-Q plot for the (a) Raw volcano data; (b) Log transformation; (c) Square-root transformation.

```
# Plot sqrt-transformed data with normal scores
plot_sqrt <- ggplot(data_sqrt, aes(x = NScores,
y = SqrtHeight)) + geom_point(shape = 1, color = "black") +
geom_smooth(method = "lm", se = FALSE, color = "blue") +
ggtitle("Q-Q Plot for Square Root Transformed Volcano
Heights") + xlab("Normal Scores") + ylab("Square Root
Heights") + theme_minimal() + theme(panel.grid.major =
element_blank(), panel.grid.minor = element_blank(),
panel.border = element_rect(fill = NA, color = "black",
size = 1))
print(plot_sqrt)
```

As expected and seen in Figure 1.30 (c), the square root of heights of volcanoes exhibits a strong linear tendency, with the exception of the upper tail, which is a little shorter than expected if the data were normal.

Nevertheless, the urge to claim that the square-root transformation has achieved symmetry and normality of the heights of volcanoes is very strong. Since the linear trend is strong, one would expect the intercept and slope of the Q-Q plot to provide reasonable estimates of the population mean and population standard deviation of the square roots of volcano heights. On R, type as follows:

```
# Perform regression analysis on square root-transformed
data and normal scores
fit <- lm(SqrtHeight ~ NScores, data = data_sqrt)
summary(fit)
```

The regression equation is

```
sqrtvht = 7.96 + 2.64 nscores

Predictor       Coef      Stdev  t-ratio       p
Constant     7.96004    0.01007   790.46   0.000
nscores      2.64338    0.01017   260.04   0.000

s = 0.1490    R-sq = 99.7%   R-sq(adj) = 99.7%
```

It is seen that the least squares estimate of intercept is 7.96004, and the least squares estimate of slope is 2.64338. These estimates almost equal the sample mean of 7.9604 and sample standard deviation of 2.629.

1.10.6 An Example Using the Crab Data and R

It is required that crab meat must be picked, packed, and put on ice within four hours. This is to prevent levels of bacteria that would be considered unacceptable (>100,000 CFU/g). Researchers tested this claim by examining

18 samples of crab meat placed on ice four hours after picking and measuring the bacteria count (based on aerobic plate count (APC)). The data appears as follow:

200,000 3,400,000 280,000 2,000 13,000 6,000 160,000 37,000 96,000
27,000 59,000 500,000 1,300,000 220,000 110,000 240,000 2,200,000 290,000

The data is part of a larger study conducted in 1997 by Jennifer Smith Asher, a graduate student in food science and technology at Virginia Tech. The sample mean of these 18 observations is 507,800 CFU/g, clearly greater than the unacceptable level. In spite of this, the 95% confidence interval for the population mean of bacteria count after four hours is (55,528.15, 960,027.41), a very wide interval that includes the unacceptable level of 100,000 CFU/g. A *t*-test would fail to conclude that the bacteria count in crab meat is different from the unacceptable level after four hours, contradicting the rule. The problem with this analysis is apparent when one examines the data more closely.

In the command window of R, type as follows:

```
# Crab meat bacteria count data
count <- c(200000, 3400000, 280000, 2000, 13000, 6000,
160000, 37000, 96000, 27000, 59000, 500000, 1300000,
220000, 110000, 240000, 2200000, 290000)
# Calculate the interquartile range (IQR)
iqd <- IQR(count)
# Set the plotting window to display four plots
par(mfrow = c(2, 2))
# Histogram
hist(count, main = "Histogram of Crab Meat Bacteria
Count", xlab = "Bacteria Count (CFU/g)")
# Box Plot
boxplot(count, main = "Boxplot of Crab Meat Bacteria
Count", ylab = "Bacteria Count (CFU/g)")
# Density plot
plot(density(count, width = 2 * iqd), main = "Density
Plot of Crab Meat Bacteria Count", xlab = "Bacteria
Count (CFU/g)")
# Q-Q plot
qqnorm(count, main = "Q-Q Plot of Crab Meat Bacteria
Count")
qqline(count)
```

The plots (histogram, box plot, smoothed histogram, and Q-Q plot) in Figure 1.31 clearly reveal at least four outliers in this dataset. The majority of the data (14 observations) exhibit a strong normal tendency. These outliers are causing the mean count to be shifted toward a larger value and have greatly inflated the standard deviation of these observations, resulting in an

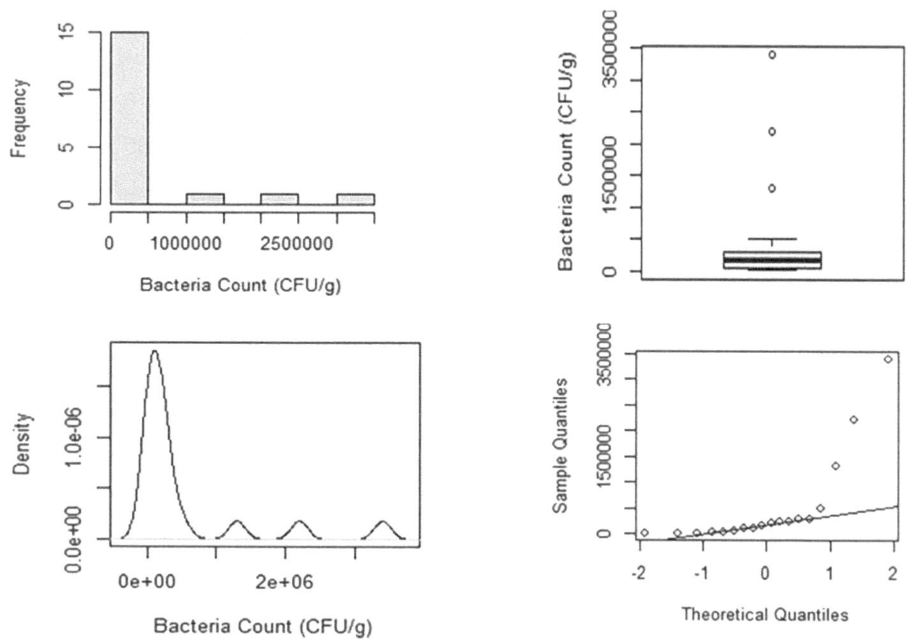

FIGURE 1.31
The histogram, box plot, smoothed histogram, and Q-Q plot.

extremely wide confidence interval. In Chapter 3, we will discuss alternative estimation techniques that will greatly reduce the impact of the outliers in the analysis.

Exercises

1) Use R or SPSS to generate 20 normal random numbers from a N(0, 1) distribution. Use the plotting techniques in this chapter to display your dataset. Comment on the results – that is, which technique reveals the familiar characteristics of the normal distribution.
   Repeat this exercise, now with 2,000 $N(0,1)$ random numbers. Comment on your results.
2) Locate a dataset representing a real-world problem that interests you. For example, the data may be from a project you are involved in, a consulting project conducted by others where your advice has been sought, or an interesting example you have found in a journal article or another textbook. There are even hundreds of great datasets available on the internet. You will be asked to perform the various exploratory and robust procedures learned in this text on your dataset and to write a report containing the results of your analyses.

Your report should be written in proper "report format" using some appropriate word processor. Include sections titled "Introduction," "Methods," "Results," and "Discussion and Conclusions." At the end of each chapter, you will be asked to examine your data using some of the techniques presented in that chapter. Thus, your report will grow longer with each succeeding chapter. At the end of the text, you will be able to present your final conclusions based on a rather extensive analysis of the data. Your dataset should be of moderate size, say a sample size of 30–100. It should contain at least three variables, one of which is a natural "response variable" and two of which are natural "regressor variables." You will want to use this data for univariate, bivariate, and multivariate analyses.

Begin your report by presenting your data. In particular, give a brief description of the research goals of the experiment (Why was this experiment conducted? What "research questions" are this experiment attempting to answer?) Briefly describe the variables measured, and list the data in tabular form.

The Soil Phosphorus Data: The production of higher maize yields has been the central topic of concern among agricultural scientists in the recent past. It has been reported by many scientists that better management of soil fertility has an important role in the maximization of maize yields. Therefore, the application of chemical fertilizers to supplement the nutrition requirement of maize has become a common agronomic practice in maize production. Phosphorus is applied as a chemical fertilizer to maize since it is one of the three major plant nutrients required by maize. Of the amount of phosphorus applied, only a part is absorbed by the crop. The rest is believed to accumulate in the soil. Therefore, the long-term application of phosphorus may lead to the buildup of this element in the soil. It is also opined that the amount of phosphorus accumulated also depends on the fertilizer application rate. Therefore, different plots of land treated with different fertilizers rates may eventually lead to different crop yields.

A long-term experiment was started on a virgin piece of land to test the these hypotheses. Four rates of phosphorus fertilizer – namely, 0, 30, 60, and 120 kilograms per hectare – were applied to the same plots of land that were under maize cultivation over a period of 30 years. Each treatment was replicated seven times. At the 30th year, soils were sampled from each plot and analyzed to determine the phosphorus content of the soil. This analysis used three standard analytical techniques to detect the amount of phosphorus in the soil. The three methods are called Mehlich-1, Bray, and Mehlich-3. These three methods are different because each method used different chemicals to extract the phosphorus. It is known that these chemicals have different ability to extract phosphorus from the soil. In addition, maize plants sampled from each of the treated plots were analyzed to determine the tissue phosphorus content of the plant. The yields of maize were also recorded.

Variables: yield of maize (kg/ha); soil phosphorus by method Mehlich-l (ppm); soil phosphorus by method Bray (ppm); soil phosphorus by method Mehlich-3 (ppm); plant tissue phosphorus (ppm). The objective of the study is to determine the relationship between the yield of maize and soil and the plant phosphorus content of the experiment.

TABLE 1.2

The Soil Phosphorus Data

Fertiliz. (kg/ha)	Mehl-1-P (ppm)	Bray-P (ppm)	Mehl-3-P (ppm)	Tissue-P (ppm)	Yield (kg/ha)
0	3.35	6.88	2.0	0.21	9,216
0	2.10	2.12	0.0	0.16	6,697
0	6.55	5.38	2.5	0.18	8,532
0	5.48	6.47	3.0	0.19	8,845
0	5.60	6.29	3.0	0.19	7,712
0	4.27	4.00	1.0	0.17	8,650
0	3.07	4.12	1.0	0.17	6,697
30	4.12	8.18	4.0	0.18	9,860
30	11.55	11.53	7.0	0.21	11,656
30	11.58	18.38	8.0	0.22	10,680
30	11.23	10.21	6.0	0.22	9,470
30	12.85	15.18	8.0	0.23	13,414
30	14.33	18.85	9.0	0.25	11,559
30	10.83	14.97	6.0	0.22	12,301
60	22.20	35.35	45.0	0.22	13,668
60	13.75	23.53	19.0	0.22	10,622
60	17.95	29.00	23.0	0.23	15,601
60	20.40	33.88	35.0	0.22	12,067
60	24.10	36.82	36.0	0.26	10,094
60	33.50	46.18	48.0	0.27	9,821
60	21.10	28.18	27.0	0.27	13,297
120	58.00	82.71	108.0	0.26	10,290
120	67.80	85.06	107.0	0.23	12,730
120	47.95	65.12	68.0	0.25	14,214
120	64.25	82.47	96.0	0.25	10,114
120	55.75	71.35	83.0	0.28	10,524
120	76.20	128.12	144.0	0.30	13,336
120	56.85	84.88	94.0	0.28	13,062

The data may be found on the website with the name "soil_phosphorus_data.xlsx."

3) Consider the yield variable from the phosphorus data. Perform the following exercises.
   a) Use SPSS to obtain a histogram, dot plot, and stem-and-leaf plot for this variable. Comment on your observations, especially as they relate to the five data characteristics.
   b) Use SPSS to obtain a box plot for this variable. Comment on your observations. Are outliers present in this data? If so, what might be a possible explanation for their presence?
4) Consider the phosphorus data and the tissue-p variable. Repeat parts a) through b) for problem 2.
5) Perform a mini simulation study of the "box plot outlier rule" using the techniques illustrated in Section 1.6. There, $n = 5$ with normal data. You should consider some other question of interest concerning this rule. For example, how does the probability of detecting an outlier using the box plot vary as the sample size, n, increases? Or how does the probability of detecting an outlier using the box plot vary as the distribution of the random variable changes to one with heavier tails, such as a t with decreasing degrees of freedom. The research question that your experiment will address is up to you. One hint: don't be too ambitious and attempt to do too much, but do enough to learn something. Write up your results in report form, with a statement of the problem, the methods used, and a summary of your results. Use tables and graphs whenever possible.
6) This problem concerns the properties of order statistics. Suppose that $y \sim U(0,1)$ and a sample of size 5 is to be taken.
   a) Write out the formulas for the PDFs for each of the five order statistics.
   b) What general class of distributions do all these PDFs belong to (e.g., t, chi-square, gamma)?
   c) Use the result of part b) to find the $med\left(y_{(i)}\right)$, for $i = 1, \ldots, 5$.
   d) Compare the values obtained in part c) to those given by Blom's conjecture.
   e) Use R or SPSS to plot the PDFs, obtained in part a).
7) Suppose that y is a random variable with PDF $f(y)$. Let $y_{.5} = med(y)$. Later in the text, we will prove that if M is the sample median of a random sample of size n of the variable y, that $\text{var}(M) \approx \dfrac{1}{4nf^2(y_{.5})}$.

One useful expression for the approximate standard error of M for data from a Gaussian distribution is $se(M) \approx \frac{1.25 * IQR}{1.35\sqrt{n}}$, where IQR is the sample interquartile range. Justify this standard error formula.

8) Use R to do the following: Generate 30 observations from a chi-square distribution with 1, 5, 10, and 20 degrees of freedom, respectively. This will give you four different datasets. Save your data for later use. Use the method of mid-summaries to find the proper transformation to symmetrize each of these datasets. You should include various displays of the raw and transformed data in your results, such as box plots and symmetry plots. Comment on your results.
9) Use R to examine the behavior of the t-t Q-Q plots. Do this by comparing a t with 1, 5, and 20 degrees of freedom and a lognormal to a N(0, 1). Comment on your results.
10) Consider the response variable of your dataset (from question 2).
    a) Check for symmetry and normality of this variable using the tools presented in class. Use R.
    b) Is a power transformation needed to achieve symmetry? If so, find the proper value of p using the proper technique. Write up your results as continuing analysis of your dataset report.
11) Consider the phosphorus data and the yield variable.
    a) Using R and the ECDF macro, obtain an ECDF plot of this variable. Comment on your observations.
    b) Using R, obtain a symmetry plot of this variable. Comment on your observations.
    c) Using R, obtain an e-t Q-Q plot of this variable. Comment on your observations.
    d) Using the results of parts a) through c) of this problem, is a power transformation needed to achieve symmetry? If so, find the proper value of p using the proper technique.
12) What probability density function for the continuous random variable X has equally spaced letter values? Prove your result.

Snowfall Data: The data in Table 1.3 represents climate data (1961–1990 30-year means) for 22 weather stations in the Southern Appalachian highlands. The data were obtained from the Southeastern Regional Climate Center via the World Wide Web. The goal is to develop a prediction model that can be used to estimate annual snowfall in the Appalachian highlands of North Carolina.

TABLE 1.3

The Snowfall Data

Obs.	Station	Elevation Ave	January Temperature	Annual Precip.	Annual Snowfall
1	Asheville	2,170	36.2	37.92	16.0
2	BanElk[a]	3,750	30.6	50.49	49.0
3	BlkMt[b]	2,395	36.8	49.51	11.5
4	BiRock[c]	3,850	29.3	65.10	37.1
5	Canton	2,662	34.2	41.75	7.8
6	Cullowhe	2,192	36.7	51.03	11.9
7	Franklin	2,113	36.8	52.91	8.4
8	GrandMt[d]	5,300	27.3	62.30	55.1
9	Hendvl[e]	2,200	36.5	55.78	11.5
10	Highland[f]	3,840	33.9	83.76	16.0
11	Jeffersn	3,300	32.1	48.27	24.8
12	Marion	1,425	38.3	56.70	11.3
13	Marshall	1,800	34.8	39.92	13.9
14	PisgFor[g]	2,120	34.7	66.37	10.0
15	NorWilk[h]	1,050	34.4	50.15	13.3
16	MtAiry[i]	1,041	36.0	45.64	11.3
17	Gatlinbr[j]	1,760	35.4	55.81	8.3
18	Galax	2,040	32.0	42.85	21.7
19	Catabo	2,620	32.3	49.72	11.1
20	Brevard	2,200	36.9	66.81	12.2
21	Lenoir	1,300	38.3	49.39	8.7
22	Morgant	1,100	38.3	50.39	9.0

The data may be found on the website with the name "snowfall.xlsx."

[a] ***BanElk*** – Banner Elk.

[b] ***BlkMt*** – Black Mountain.

[c] ***BiRock*** – Big Rock.

[d] ***GrandMt*** – Grandfather Mountain.

[e] ***Hendlv*** – Hendersonville.

[f] ***Highland*** – Highlands.

[g] ***PisgFor*** – Pisgah Forest.

[h] ***NorWilk*** – North Wilkesboro.

[i] ***MtAiry*** – Mount Airy.

[j] ***Gatlinbr*** – Gatlinburg.

13) Consider the **snowfall data** and the "annual snowfall" variable.
    a) Repeat parts a) and b) of question 2.
    b) Repeat parts a) through d) for question 10.

Multiple-Choice Questions

1. Confirmatory data analysis (CDA) primarily deals with which steps in the statistical evaluation process?
   a) Steps 1 through 3
   b) Steps 4 and 6
   c) Steps 5 through 6
   d) All steps
   Answer: c)
2. Exploratory data analysis is mainly used in which steps of the six-step process outlined in the document?
   a) Steps 1 through 3
   b) Steps 4 and 6
   c) Steps 5 through 6
   d) All steps
   Answer: b)
3 In the location model, what is the assumption made about the errors (ε_i)?
   a) $\varepsilon_i \sim N(0, \sigma^2)$
   b) $\varepsilon_i \sim N(\mu, \sigma^2)$
   c) $\varepsilon_i \sim$ Uniform(0, 1)
   d) $\varepsilon_i \sim$ Exponential(λ)
   Answer: a)
4. What is the primary disadvantage of using histograms for data analysis as mentioned in the document?
   a) They are difficult to construct.
   b) They require sophisticated software.
   c) They often lead to a loss of information in the detail of the data.
   d) They are not suitable for large datasets.
   Answer: c)
5. Which assumptions are calculated for the sample mean and confidence interval?
   a) Uniform distribution
   b) Exponential distribution
   c) Normal distribution
   d) Poisson distribution
   Answer: c)

6. Which data visualization technique is the stem-and-leaf display a modern version of?
   a) Box plot
   b) Histogram
   c) Q-Q plot
   d) Dot plot

   Answer: b)
7. Which five numbers are essential for a box plot's construction?
   a) Mean, standard deviation, median, range, variance
   b) Minimum, lower quartile, median, upper quartile, maximum
   c) Mean, mode, median, range, IQR
   d) Minimum, median, mode, upper quartile, maximum

   Answer: b)
8. The empirical cumulative distribution function plot is always a step function with jumps of what size at each sample value?
   a) $1/n$
   b) n
   c) 1
   d) $n/2$

   Answer: a)
9. If the original data does not satisfy the assumptions of the traditional location model, what transformation is suggested?
   a) Linear transformation
   b) Power transformation
   c) Logarithmic transformation
   d) Exponential transformation

   Answer: b)
10. Which of the following is a nonparametric test mentioned in the document?
    a) t-test
    b) ANOVA
    c) Kruskal-Wallis test
    d) Chi-square test

    Answer: c)

2

Combining Exploratory and Confirmatory Data Analysis for the Location Model

Summary

As we have seen, examining the data using various graphical methods is very helpful. Once the data has been examined and studied, the researcher will often want to extend the analysis more formally by establishing and evaluating a statistical model. The confirmatory data analysis (CDA) for this model proceeds with the assessment of the appropriateness of this model to the data. If the model is appropriate, estimation of the parameters of the model along with hypothesis tests on some or all these parameters is usually desired. This process was illustrated for the location model in Chapter 1. If the model is inappropriate, the two basic approaches are to either transform the data so that the model is appropriate for the transformed data, as discussed and illustrated before, or alter the original model. Both these approaches will be presented in more detail in this chapter.

As discussed in Chapter 1, one of the assumptions of the location model is that the error term and, consequently, the random variable Υ is a normal random variable. One of the oldest CDA techniques is to determine if the data obtained by a random sample could reasonably be considered a random sample from a normal population. In the next section, we will discuss some tests commonly referred to as "tests of normality."

Learning Objectives

By the end of this chapter, readers should be able to do the following:

- Integrate exploratory and confirmatory data analysis methods for the location model.
- Apply various statistical tests, including the chi-square goodness-of-fit, to assess whether the data follows a normal distribution.
- Understand how to modify models or transform data to handle non-normality and outliers in location models.
- Use R programming to create histograms, box plots, and probability plots for normality testing.
- Implement and interpret different normality tests, including the Kolmogorov-Smirnov and Lilliefors tests.

DOI: 10.1201/9781003565321-2

2.1 Tests of Normality

Given a set of data obtained by a random sample, it may be desired to test whether it is reasonable to assume that the data represents a sample from a normal population. Thus, when applying the exploratory data analysis (EDA) techniques of Chapter 1 to a given set of data, the researcher is specifically looking for characteristics consistent with the normal curve. For example, one would expect histograms, dot plots, and stem-and-leaf plots to be nearly symmetric with a distinctive bell shape. Box plots should approach the "ideal box plot" with no or few outliers. Box plots, stem-and-leaf plots, and Q-Q plots should show strong straight-line tendencies. The question remains, however, as to "how strong should these characteristics be." How far from a straight line can the data be in a normal probability plot and still reasonably support the assumption that the data is a sample from a normal population. This question is addressed by the tests of normality. As in all statistical hypothesis tests, there are two hypotheses to be tested. The null hypothesis is that the normality assumption of the data is reasonable, while the alternative hypothesis is that the normality assumption is not reasonable.

The R program can be used to illustrate these ideas. Let's generate a random sample of size 25 from a normal distribution with a mean of 50 and a standard deviation of two and examine it with the graphical tools of Chapter 1. On R, type as follows:

```
# Set seed for reproducibility
set.seed(123)
# Generate Random Data from a Normal Distribution
c1 <- rnorm(25, mean = 50, sd = 2)
# Histogram of the Data
hist(c1, main = "Histogram of the 25 Observation from a N(50,
4)", xlab = "Values", col = "lightblue", border = "black")
>dotchart(c1, main = "Dotplot of N(50, 4)", xlab = "Values",
pch = 19, col = "black")
   . ..    .          ..  .  :.  .   .  . ... :  ..:..   .
---+---------+---------+---------+---------+---------+---N(50,4)
 46.8      48.0      49.2      50.4      51.6      52.8
> boxplot(c1, main = "Boxplot of 25 Observations from N(50,
4)", xlab = "Values", col = "lightblue", horizontal = TRUE)


# Stem-and-Leaf Plot
>stem(c1)

Stem-and-leaf of C1 N = 25
Leaf Unit = 0.10
```

```
 1   46 8
 3   47 01
 4   47 7
 4   48
 4   48
 7   49 003
10   49 779
11   50 2
(2)  50 57
12   51 0124
 8   51 58
 6   52 00013
 1   52 8
```

```
# Calculate Normal Scores
c3 <- qnorm((rank(c1) - 0.5) / length(c1))

# Plot of Original Data Against Normal Scores
plot(c1, c3, main = "Probability Plot of 25 Observations from
N(50, 4)", xlab = "Y(i)", ylab = "Normal Scores", pch = 19)
```

Figure 2.1 shows that our random sample does not show us the characteristics we expected. The histogram, box plot, probability plot, dot plot, and stem-and-leaf plots do not show a symmetric, bell-shaped dataset but rather one that appears to be bimodal and slightly skewed to the left. The normal probability plot does not show a strong straight-line tendency but, rather, the distinctive shape of skewed-left data. Clearly, a sample of real normal data does not always result in plots with our expected characteristics. Obviously, the characteristics we see in the dot plot and stem-leaf plot result from the natural sampling variability inherent in the random sample of normal data. Once again, we ask whether the variability we observe in the previous plots could reasonably result from data from a normal population. Of course, the

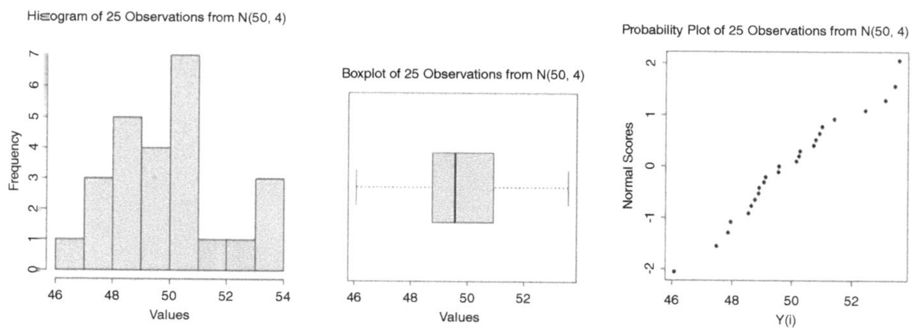

FIGURE 2.1
(a) Histogram, (b) box plot, and (c) probability plots of the 25 observations from N(50, 4).

answer must be "yes" since our data did indeed come from a normal population. The normality tests will account for this sampling variability.

There are many tests of normality. We will discuss only a few, including the oldest (the chi-square test) and those generally considered the "best." The tests of normality that we will discuss fall into four basic categories: those based on comparing the histogram to the normal probability density function (PDF), those based on sample moments, those based on comparing the empirical cumulative distribution function (ECDF) to the normal cumulative distribution function (CDF), and those based on measuring the straightness of the normal probability plot.

2.1.1 The Chi-Square Goodness-of-Fit Test for Normality

Probably the oldest test for comparing a sample of data to a theoretical distribution is the chi-square goodness-of-fit test, introduced by Pearson (1900). This test is not a particularly good one for reasons that will become clear later. However, we will present the test in some detail because it is a very popular procedure. Essentially, the test works by comparing the sample histogram (properly rescaled) to the appropriate theoretical PDF. For example, consider the previous example of 25 random observations generated from a $N(50, 4)$, and the histogram generated using R codes. This histogram certainly did not resemble a symmetric, bell-shaped curve. We can compare this histogram to a normal curve by comparing the observed frequencies in each class of the histogram to the expected or theoretical frequencies for each class. If there is a "close" agreement between the observed and expected frequencies, the normality assumption is a reasonable one; if not, then the assumption should be rejected. The chi-square goodness-of-fit test statistic measures the degree of agreement between these two sets of frequencies.

To construct this test, consider the following setup. The random sample of N observations is first grouped according to the k classes of the histogram. Let O_i denote the observed frequency in the ith class (the height of the bar for the ith class in the histogram is O_i). Since every observation falls into one and only one class, it follows that $\sum_{i=1}^{k} O_i = N$. For each class i and any given theoretical distribution, one can compute the probability that an observation will fall into this class, denoted as P_i. It follows that $\sum_{i=1}^{k} p_i = 1$. The expected or theoretical frequencies, E_i, can then be computed as $E_i = Np_i$, for each i. It follows that $\sum_{i=1}^{k} E_i = N$. The O_i can be compared to the E_i by computing $O_i - E_i$ for each i. If the discrepancies are small, the theoretical distribution provides a reasonable fit to the observed data. To determine if these discrepancies are larger than due to random variation alone, the chi-square goodness-of-fit statistic is computed as

$$\chi^2 = \sum_{i=1}^{k} \frac{(O_i - E_i)^2}{E_i}.$$

The distribution has been shown to approach a chi-square, as N becomes infinite, with $k-p-1$ degrees of freedom, where p is the number of independent parameters estimated from the data, if any. Large values of χ^2 lead to the rejection of the null hypothesis that the data can reasonably be considered as a random sample from the hypothesized distribution, the normal distribution in our case.

Some suggestions for using the chi-square goodness-of-fit tests include the following:

1) Set up the classes. Decide how many classes to set up to cover the range of the sample data. Usually, between 7 and 20 classes are used: 7 for small samples, say 20–50, and 20 or so for larger samples.
2) Use the hypothesized distribution to calculate the expected frequencies, E_i. Often, this requires estimating one or more parameters (the "p" in the formula for degrees of freedom, represents the number of parameters estimated) for the hypothesized distribution. For example, μ and σ for the normal distribution would be estimated by $\bar{X}$, and S, respectively.
3) Theoretical results suggest that small expected frequencies in the tails of the histogram may need to be combined with frequencies in adjacent classes so that all E_i are at least 1.0. Also, it is suggested that no more than 20% of the E_i are less than 5. The number of classes remaining after combining is k.

To illustrate the computations involved with this test, let us consider the random sample of $N=25$ observations generated from a $N(50,4)$ distribution. From the histogram for this data, as in Figure 2.1(a), we see that R has grouped the data into eight classes with equal interval lengths. The P_i are computed using the normal distribution with a mean equal to $\bar{X}=50.359$ and $S=1.765$.

For example, $P_1 = Prob(y \le 47.5) = Prob\left(z \le \dfrac{47.5-50.359}{1.765}\right) = Prob(z \le -1.62)$

$= 0.0526,\ p_2 = Prob(47.5 \le y \le 48.5) = 0.0934 \ldots\ p_{10} = Prob(y \ge 52.5) = 0.1126$

The P_i are computed in this way to ensure that $\sum_{i=1}^{k} p_i = 1$. The expected frequencies, found by computing $N * P_i$ for each i, are 1.32, 2.34, 4.18, 5.46, 5.23, 3.66, and 2.82, respectively. We see immediately that several expected frequencies are less than 5, violating the conditions required for the test. If one were to change the class interval length to three on R instead of one, then three intervals would result, starting at 46.5 and ending at 55.5. The observed frequencies are 7, 17, and 1. It is easy to show that the expected frequencies are 7.84, 14.34, and 2.82, in reasonably close agreement to the observed frequencies. This, perhaps, should be expected since the data are indeed a random sample from a normal distribution. However, one-third of the expected frequencies are now less than 5, again violating the test's requirements. Thus, we see one problem encountered with the chi-square goodness-of-fit test: that it often cannot be applied if the sample size is small.

As a second example, consider the heights of volcanoes presented in Chapter 1. Here, $N = 219$, so we would expect the chi-square procedure to be applicable to this example. We observed that the heights of volcanoes were strongly skewed to the right and that the square-root transformation appeared to symmetrize and normalize this data. As will be demonstrated in the example to follow, for the raw data, using 11 class intervals starting at –10 and ending at 210, with a class interval length of 20, the chi-square goodness-of-fit test is computed as 32.52, with a *p*-value of 0.00003. Note that the 10th and 11th intervals must be combined to result in an expected frequency greater than one. Thus, the assumption of normality is strongly rejected, as expected.

The R program was used to calculate the chi-square goodness-of-fit test using the volcano data. On R, type as follows:

```
# Histogram
hist(volcano_data,
     breaks = 20,
     main = "Histogram of Volcano Heights",
     xlab = "Height",
     ylab = "Frequency",
     col = "lightblue",
     border = "black")
```

From the histogram (Figure 2.2), we observe the class midpoints, the class interval length, and the observed frequencies ("count").

We now calculate the chi-square test statistic as follows:

```
# Chi-square goodness-of-fit test
# Define the class intervals
class_intervals <- seq(-10, 210, by = 20)
observed_frequencies <- hist(volcano_data, breaks =
class_intervals, plot = FALSE)$counts
# Calculate the sample mean and standard deviation
mean_volht <- mean(volcano_data)
sd_volht <- sd(volcano_data)
# Define the expected frequencies under the normal
distribution
prob_intervals <- pnorm(class_intervals, mean = mean_
volht, sd = sd_volht)
prob_intervals_adjusted <- diff(prob_intervals)
prob_intervals_adjusted[1] <- 0 # Adjust the first interval
prob_intervals_adjusted[length(prob_intervals_adjusted)]
<- 1 - prob_intervals[length(prob_intervals) - 1] #
Adjust the last interval
# Combine the last two intervals
```

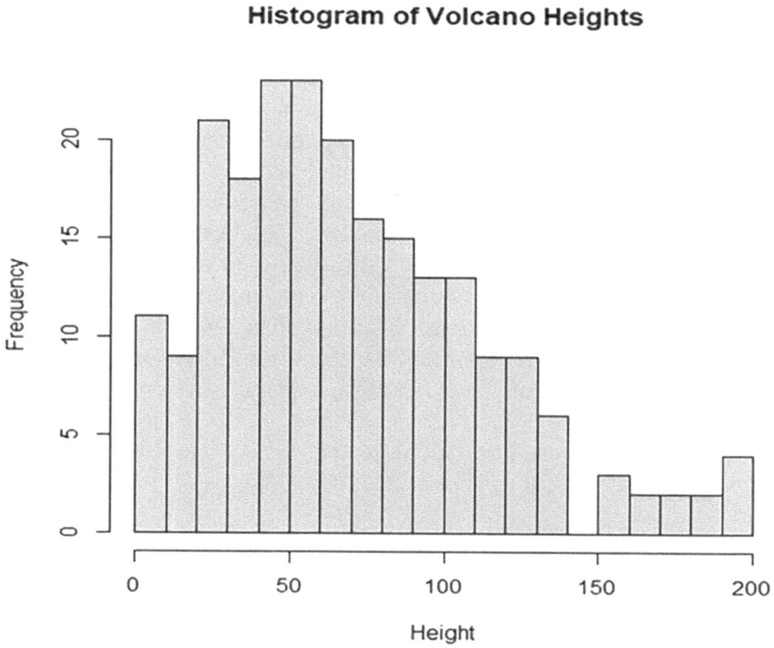

FIGURE 2.2
Histogram for the volcano data.

```
observed_frequencies[length(observed_frequencies) - 1]
<- sum(observed_frequencies[(length(observed_frequen-
cies) - 1):length(observed_frequencies)])
observed_frequencies
<- observed_frequencies[-length(observed_frequencies)]
expected_frequencies <- sum(observed_frequencies) *
prob_intervals_adjusted[-length(prob_intervals_
adjusted)]
# Compute the Chi-square test statistic
chi_square_statistic <- sum((observed_frequencies–
expected_frequencies)^2 / expected_frequencies)
degrees_of_freedom <- length(observed_frequencies) - 3
p_value <- pchisq(chi_square_statistic, df = degrees_of_
freedom, lower.tail = FALSE)
# Print the observed and expected frequencies
cat("Observed Frequencies:", observed_frequencies, "\n")
cat("Expected Frequencies:", expected_frequencies, "\n")
# Print the Chi-square test statistic, degrees of free-
dom, and p-value cat("Chi-square test statistic:", chi_
square_statistic, "\n")
cat("Degrees of Freedom:", degrees_of_freedom, "\n")
cat("p-value:", p_value, "\n")
```

The output is as follows:

ROW	lcb	ucb	Oi	Pi	Ei
1	-10	10	8	0.080683	17.6696
2	10	30	32	0.094064	20.6000
3	30	50	40	0.144198	31.5793
4	50	70	41	0.178769	39.1503
5	70	90	32	0.179237	39.2529
6	90	110	28	0.145335	31.8283
7	110	130	18	0.095303	20.8713
8	130	150	7	0.050539	11.0680
9	150	170	5	0.021672	4.7462
10	170	190	3	0.007515	1.6457
11	190	210	5	0.002687	0.5883

We must combine the last two intervals since the frequencies are too small. Recalculating the chi-square statistic test, we got a chi-square value of 32.5193, seven degrees of freedom, and a p-value of 0.00.

The same approach used before can be followed for the square root of heights to find a chi-square test statistic of 5.17 with a p-value of 0.8791, strong evidence supporting the normality assumption. This result certainly supports our previous observation that the square-root transformation succeeded in normalizing the heights of volcano data.

Some final comments concerning this test are in order. First, this is a commonly used test statistic taught in many books and introductory courses in statistics. The procedure is well known, easy to interpret, and easy to compute. The chi-square goodness-of-fit test is completely general in that a set of data may be compared to any theoretical distribution. The normal distribution is by far the most commonly used. On the negative side, however, the test requires fairly large sample sizes to meet the requirements for the test, as seen in our $N = 25$ dataset example. More importantly, since the test statistic is so general in that it can be used to test any distribution, it is not particularly powerful for testing any specific distribution. For example, there are many tests that are more powerful for testing normality than this one, including the Kolmogorov-Smirnov test, the Shapiro-Wilk test, and the Lilliefors test. Hence, the chi-square goodness-of-fit test for normality is not recommended.

2.1.2 Tests Based on Sample Moments

Another class of tests of normality are based on sample moments, more specifically, the third and fourth moments about the mean. The two measures are the "skewness" statistic and the "kurtosis" statistic. Both are simple to compute and work reasonably well as tests of normality against certain types of alternatives.

The skewness parameter, denoted by β_1, and the kurtosis parameter, denoted by β_2, are defined, respectively, as

$$\beta_1 = \frac{E(X-\mu)^3}{\sigma^3}$$

and

$$\beta_2 = \frac{E(X-\mu)^4}{\sigma^4},$$

where $\mu = E(x)$ and $\sigma^2 = E(x-\mu)^2$, the usual population mean and variance. One can see that β_1 is zero for a symmetric distribution, is a negative value for a skewed-left distribution, and is positive for a skewed-right distribution. Kurtosis is a measure of the "peakedness" of the distribution. One can show that β_2 is 3 for the normal distribution and greater than 3 if the distribution is more peaked than normal. Such distributions are said to be leptokurtic. The double exponential distribution, with $\beta_2 = 6$, and the logistic distribution are examples of leptokurtic distributions. If β_2 is less than 3, the distribution is flatter than normal; an example is the uniform and is referred to as a platykurtic distribution.

Tests of normality based on these two measures amounts to computing the sample estimates of β_1 and β_2 using the method of moments. Thus,

$$\hat{\beta}_1 = \sqrt{n}\frac{\sum_{i=1}^{n}(x_i-\overline{x})^3}{\left(\sum_{i=1}^{n}(x_i-\overline{x})^2\right)^{\frac{3}{2}}}$$

and

$$\hat{\beta}_2 = n\frac{\sum_{i=1}^{n}(x_i-\overline{x})^4}{\left(\sum_{i=1}^{n}(x_i-\overline{x})^2\right)^2}.$$

Values of $\hat{\beta}_1$ around 0 support the normality assumption. Similarly, the values of $\hat{\beta}_2$ around 3 support the normality assumption. It can be shown that the asymptotic distributions of these tests' statistics are $N(0,6/n)$ and $N(3,24/n)$, respectively, when x is normally distributed. However, these statistics' exact finite sample distributions are a function of the specific underlying distribution of x. If x is normally distributed, these distributions have been tabulated and may be found in many sources, including Pearson and Hartley (1958) and Mandansky (1988).

Studies have shown – for example, the simulation study by Shapiro, Wilk, and Chen (1968) – that $\hat{\beta}_1$ provides a reasonably powerful test of normality against skewed alternative distributions while $\hat{\beta}_2$ is a reasonably powerful test against symmetric distributions with a kurtosis parameter different from three. Neither test has a higher power against other types of alternatives than the ones mentioned.

We will illustrate the use of these two tests by computing them on R for the height of volcano data. Recall that the raw data is stored in c1. On R, type as follows:

```
# Calculate the mean of the data
k1 <- mean(volcano_data)
# Compute the necessary sums for skewness and kurtosis
k2 <- sum((volcano_data–k1)^3) # Sum of (c1–mean(c1))^3
k3 <- sum((volcano_data–k1)^4) # Sum of (c1–mean(c1))^4
k4 <- sum((volcano_data–k1)^2) # Sum of (c1–mean(c1))^2
# Calculate the skewness statistic
k5 <- sqrt(length(volcano_data)) * k2 / (k4^(1.5))
# Calculate the kurtosis statistic
k6 <- length(volcano_data) * k3 / (k4^2)
# Print the skewness and kurtosis statistics
cat("Skewness (K5):", k5, "\n")
cat("Kurtosis (K6):", k6, "\n")

Skewness (K5): 0.8344841
Kurtosis (K6): 3.443909
```

The value of $\hat{\beta}_1$ is 0.8345, and the value of $\hat{\beta}_2$ is 3.444. Comparing these values to the tables for $\hat{\beta}_1$ and $\hat{\beta}_2$, found in Pearson and Hartley (1958) and not reproduced here, we see that $\hat{\beta}_1$ exceeds the upper 1% point of 0.403 (at $N = 200$), and $\hat{\beta}_2$ lies within the 5% and 95% values of 2.51 and 3.57, respectively. Thus, the normality assumption is rejected by the skewness test and not rejected by the kurtosis test. This result is not at all surprising since we already know that the height of volcano data is strongly skewed to the right, a situation ideally suited for the skewness test and ill-suited for the kurtosis test. On the other hand, if we compute these two tests for the square root of the height of volcano data, we obtain $\hat{\beta}_1 = 0.0316$ and $\hat{\beta}_2 = 2.6956$. Both these values are well within the "fail to reject" region of their respective distributions as expected.

2.1.3 Tests Based on the Empirical Cumulative Distribution Function

Another class of tests for normality are those based on comparing the shape of the ECDF to the CDF hypothesized by the user. Two tests of this type are the Kolmogorov test and the Cramer-von Mises test. A discussion of both these tests and several others may be found in Conover (1971). We will discuss only the Kolmogorov test.

If we let $F_n(X)$ and $F_x(X)$ denote the ECDF and CDF, respectively, for the random variable x, then the Kolmogorov test is defined as the maximum distance between $F_n(X)$ and $F_x(X)$. That is, the test statistic is computed as

$$k = max_x \left| F_n(X) - F_x(X) \right|.$$

Large values of k lead to rejection of the hypothesized distribution. Any hypothesized distribution may be specified, so the Kolmogorov test is a general test in a manner similar to the chi-square goodness-of-fit test. However, the most commonly used hypothesized distribution is the normal. The sampling distribution of k is "distribution-free" in that it does not depend on the form of the specified distribution, $F_x(X)$. Tables for k may be found in most nonparametric statistics texts, such as Conover (1971).

From a practical viewpoint, one major problem with the Kolmogorov test is that it requires the parameters of the hypothesized distribution to be completely specified – for example, in testing normality, the mean and standard deviation of the normal curve must be specified by the test user. Lilliefors (1967) studied the behavior of a modification of k when $F_x(X)$ is the normal and μ and σ are estimated by $\overline{x}$ and S, respectively. This test statistic is commonly referred to as Lilliefors test and will be denoted by L. Tables for L may be found in Dallal and Wilkinson (1986). Mason and Bell (1986) showed that the upper 5% point for the Lilliefors test can be well approximated by the value $\dfrac{0.895\sqrt{n}}{\left(n - 0.01\sqrt{n} + 0.83\right)}$.

To illustrate the computation of L, consider the test score data from Chapter 1. Recall from our exploratory analysis in Chapter 1 that this data, with 58 observations, was slightly skewed left, bell-shaped in the center, around 32, and contained four outliers. The original data recalled and sorted, ECDF, and normal CDF (based on the sample mean and sample deviation) are calculated using the following R code. The test statistic L is actually computed as

$$L = max_i \left(\max\left(\left| F_n\left(x_{(i)}\right) - F\left(x_{(i)}\right) \right|, \left| F_n\left(x_{(i-1)}\right) - F\left(x_{(i)}\right) \right| \right) \right)$$

and the absolute difference between the ECDF and CDF is calculated as well.

On R, type as follows:

```
# Load the data from the specified file path
# Define a function to calculate the Lilliefors
  statistic
lilliefors_stat <- function(data) {sorted_data
  <- sort(data)
  ecdf_values <- (1:n) / n
  mean_data <- mean(sorted_data)
```

```
    sd_data <- sd(sorted_data)
    normal_cdf <- pnorm(sorted_data, mean = mean_data, sd =
      sd_data)
    abs_diff <- abs(ecdf_values—normal_cdf)
    previous_ecdf <- c(0, head(ecdf_values, -1))
    abs_diff_previous <- abs(previous_ecdf—normal_cdf)
    max(pmax(abs_diff, abs_diff_previous))}
  # Calculate Lilliefors statistic for original and
    square-root transformed data
  k3 <- lilliefors_stat(test_scores)
  sqrt_k3 <- lilliefors_stat(sqrt(test_scores))
  # Calculate the critical value for the Lilliefors test
  k4 <- 0.895 * sqrt(n) / (n - 0.01 * sqrt(n) + 0.83)
  # Output Lilliefors statistics and critical value
  cat("Lilliefors statistic (k3):", k3, "\n")
  cat("Lilliefors statistic for sqrt data (sqrt_k3):",
    sqrt_k3, "\n")
  cat("Critical value (k4):", k4, "\n")
  # Conclusion based on the Lilliefors test
  cat(ifelse(k3 < k4, "Lilliefors test fails to reject
    normality for original data.\n", "Lilliefors test
    rejects normality for original data.\n"))
  cat(ifelse(sqrt_k3 < k4, "Lilliefors test fails to
    reject normality for square-root data.\n",
         "Lilliefors test rejects normality for square-root
  data.\n"))
  # Perform Kolmogorov-Smirnov test for normality
  ks_test <- ks.test(test_scores, "pnorm", mean =
    mean(test_scores), sd = sd(test_scores))
  cat("Kolmogorov-Smirnov test statistic (ks):", ks_
    test$statistic, "\n")
  cat("Kolmogorov-Smirnov test p-value:", ks_test$p.value,
    "\n")
  # Conclusion based on the Kolmogorov-Smirnov test
  cat(ifelse(ks_test$p.value < 0.05,
             "The Kolmogorov-Smirnov test rejects the null
  hypothesis of normality.\n",
       "The Kolmogorov-Smirnov test fails to reject the null
  hypothesis of normality.\n"))
```

This R code will give us the value for *k*3, the Lilliefors statistic, *L*, **K3 = 0.108094.** Let *k*4 = 0.895 * sqrt(*k*1)/(*k*1 – 0.01 * sqrt(*k*1) + 0.83), which is the upper 5% point for the Lilliefors test. Hence, the code gives **K4 = 0.116011.** Since *k*3 is less than *k*4, the Lilliefors test fails to reject normality.

From this output, it is seen that the value of the Lilliefors statistic $L = 0.108$ is much smaller than the upper 5% cutoff value of 0.116011. We cannot reject the assumption of normality. The Lilliefors test can be computed in R using the following code that generates the ECDF with Lilliefors confidence bands and overlays the standard normal CDF.

```
# Sort the data and calculate ECDF
sorted_scores <- sort(test_scores)
ecdf_values <- ecdf(sorted_scores)(sorted_scores)
# Calculate the mean and standard deviation
mean_score <- mean(sorted_scores)
sd_score <- sd(sorted_scores)
# Calculate the normal CDF values
normal_cdf <- pnorm(sorted_scores, mean = mean_score,
sd = sd_score)
# Lilliefors Bands (using the asymptotic approximation)
alpha <- 0.05 # significance level
k4 <- 0.895 * sqrt(n) / (n - 0.01 * sqrt(n) + 0.83) #
Lilliefors critical value
upper_band <- pmin(ecdf_values + k4, 1)
lower_band <- pmax(ecdf_values–k4, 0)
# Create the plot
plot(sorted_scores, ecdf_values, type = "s", col =
  "black", lwd = 2,main = "ECDF with Lilliefors Bands and
  Normal CDF",
      xlab = "Scores", ylab = "Cumulative Probability",
      ylim = c(0, 1))
lines(sorted_scores, normal_cdf, col = "black", lwd = 2,
  lty = 2)
lines(sorted_scores, upper_band, col = "black", lty = 3)
lines(sorted_scores, lower_band, col = "black", lty = 3)
# Add legend
legend("bottomright", legend = c("ECDF", "Normal CDF",
  "Lilliefors Band"),col = c("black", "black", "black"),
  lwd = 2, lty = c(1, 2, 3))
```

Figure 2.3 contains the ECDF for the test score data and the normal curve based on the sample mean and standard deviation. Of course, L represents the maximum distance between these two curves, occurring, in this case, at the test score value of 31. Along with these two curves are the Lilliefors bands, obtained by adding and subtracting the upper 5% point value of 0.1160 from the normal CDF. Thus, if the ECDF crosses either of these curves, the value of L must exceed the 5% cutoff value, and the hypothesis of normality must be rejected at the 5% level. Since the ECDF does not cross these curves in Figure 2.3, we cannot reject the assumption of normality.

As a second example, consider the height of volcano data, which has already been observed to be skewed right.

We observe from Figure 2.4 that the ECDF for the height of volcano data crosses the upper Lilliefors band at a data value slightly more than 50, leading to the rejection of the normality assumption. On the other hand, the square root of the height was determined earlier to be quite symmetric and normal in appearance. So, on R, type as follows:

```
# Calculate the square root of the volcano heights
sqrt_volcano_data <- sqrt(volcano_data)
```

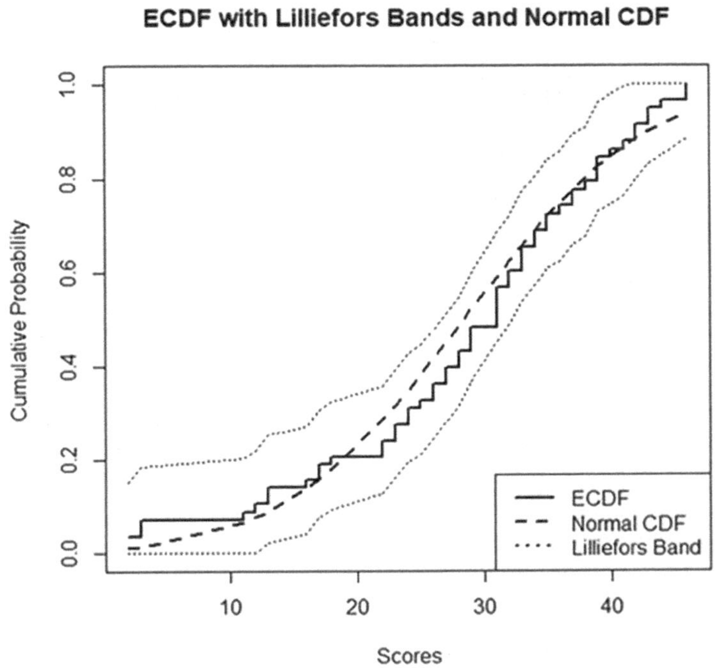

FIGURE 2.3
ECDF with Lilliefors bands and normal CDF.

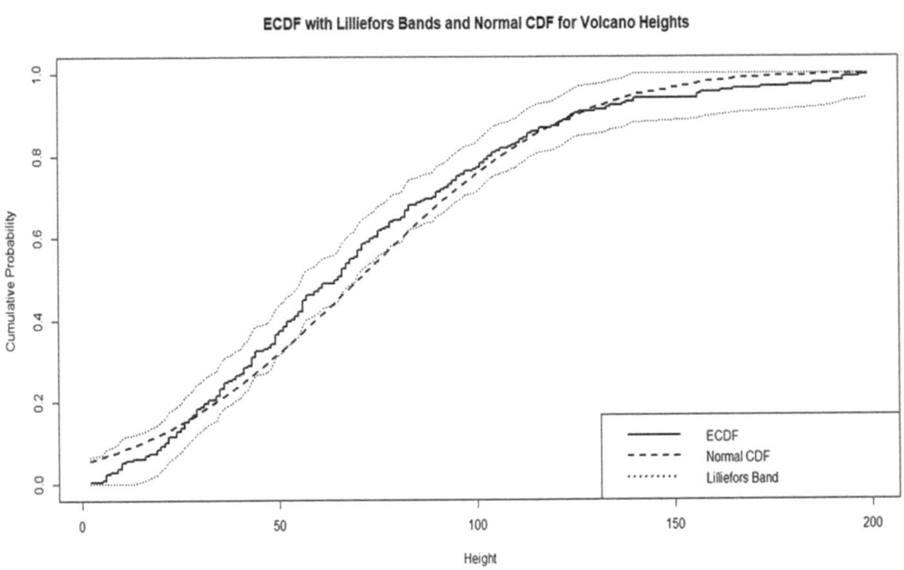

FIGURE 2.4
ECDF with Lilliefors bands and normal CDF for volcano heights.

```
# Number of observations
n <- length(sqrt_volcano_data)
# Sort the data and calculate ECDF
sorted_sqrt_volcano <- sort(sqrt_volcano_data)
ecdf_values <- ecdf(sorted_sqrt_volcano)
(sorted_sqrt_volcano)
# Calculate the mean and standard deviation of the
square-root data
mean_sqrt_volcano <- mean(sorted_sqrt_volcano)
sd_sqrt_volcano <- sd(sorted_sqrt_volcano)
```

Figure 2.5 shows that the ECDF and the normal CDF are nearly identical and that the ECDF does not cross the Lilliefors band. Thus, the normality assumption is reasonable for the square root of heights of volcanoes.

2.1.4 Tests Based on the Normal Probability Plot

The fourth type of normality test concerns the normal probability plot. Recall that the normal probability plot is a specific example of the Q-Q plot where the order statistics, $y_{(i)}$, are plotted against the appropriated percentiles from the standard normal curve. That is, $y_{(i)}$ are plotted against Φ^{-1} ("something"), where "something" refers to $i/n, \frac{i-.5}{n}$, or $\frac{i-.333}{n+.333}$, to name

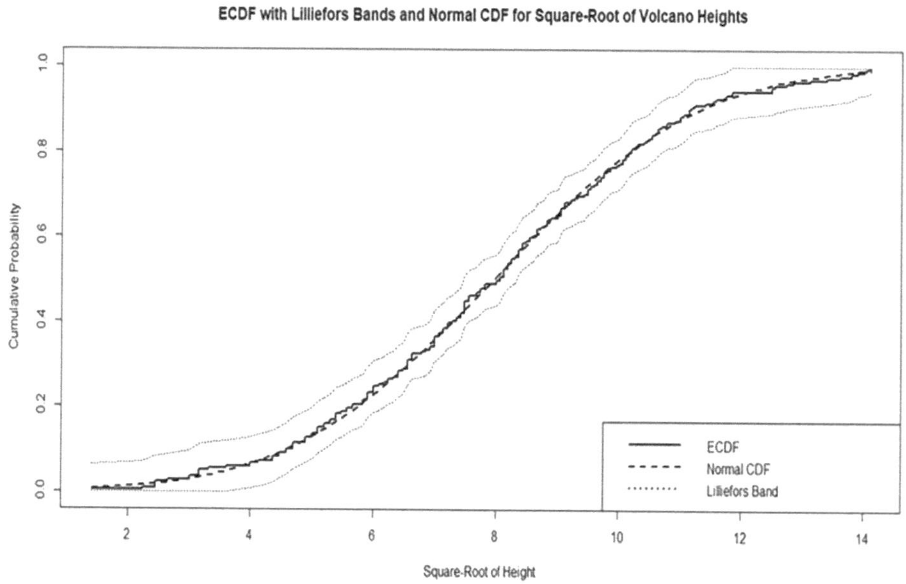

FIGURE 2.5
ECDF with Lilliefors bands and normal CDF for square-root volcano heights.

a few possibilities, where $i = 1,\ldots,n$. We also recall from our discussion on probability plotting that the standard normal (the "reference" distribution) is a member of the location-scale family of distributions and that a straight-line relationship in the normal probability plot suggests that the random sample of data (the y's) are from a normal population where the intercept and slope of the line in the normal probability plot are estimates of μ_y and σ_y, respectively. With these facts in mind, we will consider two tests: the Shapiro-Wilk test and the correlation coefficient test of normality.

2.1.4.1 Shapiro-Wilk Test

Shapiro and Wilk (1968) introduced a statistic specifically designed to test the normality of a random sample of data. In their procedure, the $y_{(i)}$ are plotted against the $a_i, i = 1,\ldots,n$, where the a_i s are functions of $E\left(z_{(i)}\right)$, the expected value of $z_{(i)}$, the ith order statistic from a sample of size n from a $N(0,1)$ distribution. The a_i values are standardized so that

$$\sum_{i=1}^{n} a_i = 0 \text{ and } \sum_{i=1}^{n} a_i^2 = 1.$$

It is easy to show that the least squares estimate of the slope of this normal probability plot is

$$\hat{\sigma}_y = \frac{\sum a_i y_{(i)} - n\bar{a}\,\bar{y}}{\sum a_i^2 - n a^{-2}} = \frac{\sum a_i y_{(i)}}{\sum a_i^2} = \sum a_i y_{(i)}$$

The theory states that this estimate of the population standard deviation is the best linear unbiased estimate (abbreviated as "BLUE") provided that the null hypothesis of normality is true – that is, $E\left(\hat{\sigma}_y\right) = \sigma_y$ under the true null hypothesis of normality.

The strength of the linear association between the $y_{(i)}$s and the a_is can be measured in the standard way by computing the square of the correlation coefficient between them. Of course, this value is just the coefficient of determination, R^2, found by regressing the order statistics (the "response" variable) against the a_is, the expected order statistics from a sample of size n from a $N(0,1)$. Shapiro and Wilk (1968) studied the properties of this statistic, commonly referred to as the Shapiro-Wilk test of normality. The form of this test, denoted by W, is then

$$W = R^2 = \frac{SS_{\text{regression}}}{SS_{\text{total}}} = \frac{(\text{slope})^2\, S_{aa}}{S_{yy}} = \frac{\left(\sum a_i y_{(i)}\right)^2}{(n-1)s^2} = \frac{\hat{\sigma}_y^2}{(n-1)s^2},$$

where s^2 is the usual sample variance of the y's – that is, $s^2 = \frac{\sum(y_i - \bar{y})^2}{n-1}$. We see from the form of the test that W can be interpreted as a comparison of two estimators of σ_y^2, one valid if the null hypothesis is true ($\hat{\sigma}_y^2$) and the other valid regardless of the truth of the null hypothesis (s^2).

Under a true null hypothesis of normality, we expect a large W value, so small W values suggest that the null hypothesis is not true. So the decision rule for this test is to reject the null hypothesis if the observed value of W is less than the cutoff point, usually taken as the lower 5% percentile of the null distribution of W. Simulation studies (see, for example, Shapiro, Wilk, and Chen, 1968) show that W is among the best tests of normality in terms of power against a wide variety of alternative distributions and is highly recommended as a test for normality. From a practical viewpoint, the test requires the use of two tables, one to obtain the a_is for each sample size and the other to obtain the cutoff values. Additionally, the test is tabulated only for n up to 50. For this reason, an adjustment to the Shapiro-Wilk test of normality, discussed below, is recommended.

Blom (1958) suggested that the expected order statistics for a sample of size n from a $N(0,1)$ distribution can be well approximated by the "normal scores"; that is,

$$E\left(z_{(i)}\right) \approx \Phi^{-1}\left(\frac{i-3/8}{n+1/4}\right).$$

Weisberg and Bingham (1975) showed that if a_is are replaced in W by $\Phi^{-1}\left(\frac{i-3/8}{n+1/4}\right)$, the resulting test statistic, denoted by W, has a sampling distribution almost identical to that of W. Now the $\Phi^{-1}\left(\frac{i-3/8}{n+1/4}\right)$ are obtained on R by some easy commands that the statistic W can be computed on R without the use of special tables.

Another equivalent version of W is to compute the correlation coefficient between the order statistics and $\Phi^{-1}\left(\frac{i-3/8}{n+1/4}\right)$ rather than the squared correlation coefficient as a measure of the strength of the linearity of the normal probability plot.

This test, referred to as the "correlation coefficient test of normality," is thus a highly recommended test since it has excellent power against a wide variety of alternative distributions and is simple to compute. Approximate lower 5% points can be obtained for any value of n between 10 and 100 by using the formula

$$p = .8726 + .1115\left(1 - \exp\left(-.0563^{*}n\right)\right),$$

where p is the approximate lower 5% point.

It should be noted that the approximation formula for the lower 5% underestimates the true 5% point for this test statistic for n greater than 100. The approximation is excellent for n between 10 and 100, with a maximum relative error of 0.5%, and considerably less than this for most values of n.

Computation of the correlation coefficient test of normality is a simple task on the R program.

Consider the test score data discussed previously. If the test score data is recalled, then, on R, type as follows:

```
# Calculate the normal scores (nscores)
normal_scores <- qnorm(ppoints(test_scores))
# Calculate the correlation between the original data
and the normal scores
correlation <- cor(test_scores, normal_scores)
# Output the correlation
cat("Correlation of test scores and normal scores
(C10):", correlation, "\n")
```

Correlation of Test Scores and Normal Scores (C10): 0.9752542

We see that the value of the test is 0.975, slightly less than the 0.979 cutoff value found by linear interpolation of the $n = 50$ and $n = 60$ values from the table. We would barely reject the normality assumption.

Also, the P-P plot (Figure 2.6) could be obtained using the following code:

```
# Generate the P-P plot
plot(ppoints(test_scores), pnorm(sort(test_scores),
mean = mean(test_scores), sd = sd(test_scores)), main =
"Q-Q Plot of Test Scores vs. Normal Distribution",
xlab = "Theoretical Probabilities (Normal CDF)", ylab =
"Empirical Probabilities (ECDF)", col = "black", pch = 19)
abline(0, 1, col = "black", lwd = 2, lty = 2) # Line of
equality
```

Reject the assumption of normality if the following test statistic (corrtest = 0.975) is less than the following lower 5% cutoff value percentage 0.979843. Figure 2.1 (c) shows that the data is nearly straight, with exceptions in the lower-left portion of the scatter of points. The value of the test statistic, corrtest, is 0.975, slightly smaller than the lower 5% cutoff value of 0.9798 found by using the approximation formula. Thus, the correlation coefficient test would reject the normality assumption (with a p-value of about 0.05).

The SPSS and R perform the correlation coefficient test for normality (referred to as the "Ryan-Joiner" test or the approximate Shapiro-Wilk test by R and SPSS), as well as several other tests for normality including the "Anderson-Darling" test and the "Kolmorgov-Smirnov" test (our Lilliefors

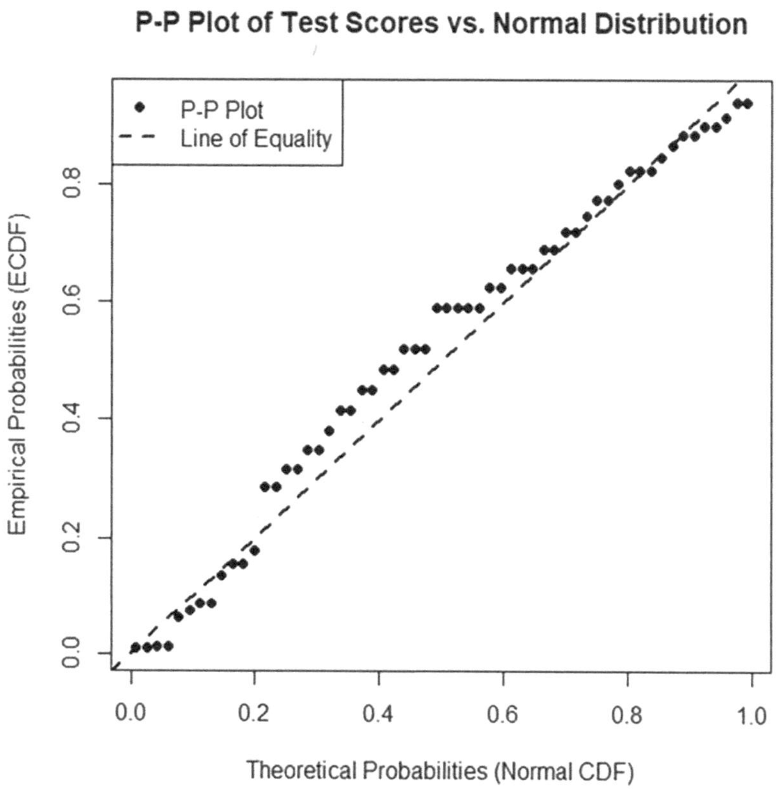

FIGURE 2.6
Normal probability plot for test scores.

test). The p-value is given, indicating that the test statistic is significant at the 0.05 level (p-value = 0.01211883).

2.2 Review of Some Concepts of Statistical Theory

As we begin to study the behavior of statistical models, it is important to emphasize the role of statistical theory in our discussions. It is assumed that the reader has a basic understanding of such concepts of statistical estimation theory and statistical inference as unbiased and consistent estimators, maximum likelihood estimation theory, and hypothesis testing. Each of these concepts, and a few others as well, will be reviewed in this section. Each concept will play an important role in future discussions of robust data analysis. Recall the location model, the primary model of the first two chapters, which can be written as

$$y_i = \theta + \varepsilon_i,$$

for $i = 1, \ldots, n$. The ε_i are independent and identically distributed (i.i.d.) random variables from some symmetric distribution around zero with constant variance σ^2. Usually, the ε_i are considered to be normal. If this is the case, the distribution of the ε_i is $N(0, \sigma^2)$. It follows that $E(y) = \theta$. Our goal in statistical estimation theory is to obtain an estimate of θ, possessing some desirable statistical properties and making inferential statements concerning θ. Let $\hat{\theta}_n$ denote our estimate of θ, based on a sample size n.

2.2.1 Statistical Estimation Theory Concepts

We will now review some of the basic concepts of statistical estimation theory. For further details on the topics discussed in subsections 2.2.2 and 2.2.3, the reader may refer to any one of hundreds of texts on mathematical statistics, such as Mood, Graybill, and Boes (1974) and Larsen and Marx (1981).

2.2.2 Unbiasedness

The estimator, $\hat{\theta}_n$, is said to be unbiased if $E(\hat{\theta}_n) = \theta$ for all n and θ. Unbiasedness is a "repeated sampling" property in that $E(\hat{\theta}_n)$ refers to the average value of $\hat{\theta}_n$ over all possible samples of size n of the random variable y. In practice, the fact that $\hat{\theta}_n$ is unbiased is not particularly comforting when the user has only one sample of size n of the random variable y. Nonetheless, unbiasedness is considered a highly desirable property to be possessed by an estimator. At the very least, an estimator with a large amount of bias would be considered undesirable.

For example, $\bar{y}$ is an unbiased estimator of $E(y) = \theta$, but $n^{-1}\Sigma(y_i - \bar{y})^2$ is a biased estimator of σ^2, while $(n-1)^{-1}\Sigma(y_i - \bar{y})^2$ is unbiased. The reader should verify these statements on their own.

2.2.3 Consistency

The estimator, $\hat{\theta}_n$, is said to be a consistent estimator of θ if for any $\varepsilon > 0$,

$$\text{Prob}\left(\left|\hat{\theta}_n - \theta\right| > \varepsilon\right) \to 0$$

as $n \to \infty$. Consistency is a "large sample" property, saying, in effect, that the sampling distribution of $\hat{\theta}_n$ becomes more tightly concentrated around θ as

n gets large. Thus, as n goes to infinity, the probability approaches one that $\hat{\theta}_n$ is very "close" (within ε) to θ. We say that $\hat{\theta}_n$ converges to θ in probability if $\hat{\theta}_n$ is a consistent estimator of θ. One useful result for showing the consistency of an estimator is the theorem that states that an unbiased estimator of θ is consistent if the variance of that estimator converges to zero as n goes to infinity.

The classic illustration of a consistent estimator is that of $\bar{y}$ as an estimator of θ. It is well known that $E(\bar{y}) = E(y) = \theta$ and that $\text{var}(\bar{y}) = \frac{\text{var}(y)}{n} = \frac{\sigma^2}{n}$. Obviously, $\bar{y}$ is a consistent estimator of θ since $\text{var}(y) \to 0$ as $n \to \infty$. This result holds regardless of the distribution of ε, provided that the $E(\varepsilon) = 0$.

In the case where the distribution of ε is normal so that the distribution of y is $N(\mu, \sigma^2)$, it is easy to show that $(n-1)^{-1}\sum(y_i - \bar{y})^2$ is a consistent estimator of σ^2 This is so since one can show that $(n-1)^{-1}\sum\left(y_i - \bar{y}\right)^2 \sim \sigma^2(n-1)^{-1}\chi^2_{(n-1)}$, where $\chi^2_{(n-1)}$ is a chi-square random variable with $n-1$ degrees of freedom. It follows from the properties of the variance operator that

$$\text{var}\left((n-1)^{-1}\sum(y_i - \bar{y})^2\right) = \frac{2\sigma^4(n-1)}{(n-1)^2} \to 0 \text{ as } n \to \infty.$$

2.2.4 Minimum Variance

In order to discuss the variance properties of an estimator, one needs to define the likelihood function of the random sample. Let the probability function of y be denoted by $f(y|\theta)$, where the "$|\theta$" is used to denote the dependency of the distribution on the location parameter of interest, θ. The likelihood function of the random sample $y_1, \cdots, y_n$ is denoted by $L(\theta|y_1, \ldots, y_n) = L(\theta|\underline{y})$. The likelihood function is simply the joint distribution of $y_1, \cdots, y_n$, viewed as a function of the unknown parameter, θ. If $y_1, \cdots, y_n$ are independent observations, as is assumed in our location model, then it is well known that $L(\theta|\underline{y}) = \prod_{i=1}^{n} f(y_i|\theta)$. To simplify the notation, we will write the likelihood function as L. We will assume that the first and second derivatives of the likelihood function exist and are denoted by $\frac{\partial L}{\partial \theta}$ and $\frac{\partial^2 L}{\partial \theta^2}$, respectively.

The following results hold. Since the integral of the density over y equals one, it follows that

$$\int L d\underline{y} = 1;$$

then it follows that under appropriate conditions,

$$\frac{\partial\left(\int L d\underline{y}\right)}{\partial \theta} = \int \frac{\partial L}{\partial \theta} d\underline{y} = \frac{\partial 1}{\partial \theta} = 0$$

by differentiation of both sides. Now the $\int \frac{\partial L}{\partial \theta} d\underline{y}$ term can be rewritten as

$$\int \frac{\partial L}{\partial \theta} \frac{L}{L} d\underline{y} = \int \left(\frac{\partial L}{\partial \theta} \frac{1}{L} \right) L dy = E\left(\frac{\partial \ln L}{\partial \theta} \right) = 0.$$

Thus, the expected value of the first derivative of the log-likelihood equals zero, a useful result. The second derivative of the log-likelihood function plays a key role in the necessary variance expressions. The second derivative can be found by taking the derivative of the last expression. Hence,

$$\frac{\partial E\left(\frac{\partial \ln L}{\partial \theta} \right)}{\partial \theta} = \int \left[\left(\frac{\partial L}{\partial \theta} \frac{1}{L} \right) \frac{\partial L}{\partial \theta} + L \frac{\partial}{\partial \theta} \left(\frac{\partial L}{\partial \theta} \frac{1}{L} \right) \right] dy = 0.$$

After a little algebra, this expression can be rewritten as

$$\int \left(\left(\frac{\partial L}{\partial \theta} \frac{1}{L} \right)^2 + \frac{\partial^2 \ln L}{\partial \theta^2} \right) L d\underline{y} = 0.$$

Or, expressed in terms of expectations, we have shown that

$$E\left[\left(\frac{\partial \ln L}{\partial \theta} \right)^2 \right] = -E\left(\frac{\partial^2 \ln L}{\partial \theta^2} \right).$$

Additionally, if the *n* observations are i.i.d., the expressions can be written as

$$nE\left[\left(\frac{\partial \ln f(y|\theta)}{\partial \theta} \right)^2 \right] = -nE\left(\frac{\partial^2 \ln f(y|\theta)}{\partial \theta^2} \right) = nI(\theta).$$

where $I(\theta)$ is referred to as the Fisher information number and plays an important role in subsequent material. One important use of this number relates to the minimum variance that an estimator can achieve.

If $\hat{\theta}_n$ is our unbiased estimator, based on a random sample, for the location parameter θ, it can be shown (see, for example, Larsen and Marx, 1981) that the variance of $\hat{\theta}_n$, $\mathrm{var}(\hat{\theta}_n)$, must satisfy the inequality, referred to as the Cramer-Rao inequality,

$$\mathrm{var}(\hat{\theta}_n) \geq n^{-1} I^{-1}(\theta)$$

the inverse of the Fisher information number times n^{-1}. The inverse of $I(\theta)$ times $1/n$ represents the lower bound, the Cramer-Rao lower bound, that the variance of $\hat{\theta}_n$ can achieve. Consequently, if an unbiased estimator achieves its lower bound, we know that no other unbiased estimator can have a smaller variance.

To illustrate some of these concepts, consider the case were $y_1, \ldots, y_n$ is a random sample from a Poisson distribution with probability function $P(y|\lambda) = \frac{e^{-\lambda}\lambda^y}{y!}, y = 0, 1, 2, .., \lambda > 0$. It is clear that the likelihood function is

$$L(\lambda|\underline{y}) = L = \prod P(y_i|\lambda) = \prod \frac{e^{-\lambda}\lambda^{y_i}}{y_i!} = \frac{e^{-n\lambda}\lambda^{\Sigma y_i}}{\prod y_i!}.$$

So

$$\ln L = -n\lambda + \sum y_i \ln\lambda - \sum \ln y_i!;$$

thus,

$$\frac{\partial \ln L}{\partial \lambda} = -n + \frac{\Sigma y_i}{\lambda}.$$

Furthermore, we have

$$\frac{\partial^2 \ln L}{\partial \lambda^2} = -\frac{\Sigma y_i}{\lambda^2}.$$

Since $E(y) = \lambda$, we know that

$$-E\left(\frac{\partial^2 \ln L}{\partial \lambda^2}\right) = \frac{n\lambda}{\lambda^2} = \frac{n}{\lambda} = nI(\lambda).$$

Thus, the Cramer-Rao lower bound for any unbiased estimator of λ is $n^{-1}I^{-1}(\lambda) = \frac{\lambda}{n}$.

Now if the sample mean, $\bar{y}$, is used as the estimator of λ, it follows from the fact that $E(\bar{y}) = \lambda$ that $\bar{y}$ is an unbiased estimator of λ. Also, since $\text{var}(y) = \lambda$, we know the $\text{var}(\bar{y}) = \lambda/n$, achieving the minimum lower bound. Consequently, $\bar{y}$ has minimum variance among all unbiased estimators of λ. Conditions guaranteeing that an estimator's variance achieves its lower bound can be found in texts such as Bickel and Doksum (1977). Extensions of the Cramer-Rao bounds to the multiparameter case can be found in Rao (1965) among others.

2.2.5 Minimum Mean Square Error

To compare unbiased estimators, one need only compare their variances; the estimator with the smaller variance would be preferred over the estimator with the larger variance. However, if one or both of the estimators to be compared are biased, we need a measure that simultaneously accounts for bias and variance. One measure that does this is the mean squared error (MSE) of the estimator, defined as

$$\begin{aligned}\text{mse}\left(\hat{\theta}_n\right) &= E\left(\hat{\theta}_n - \theta\right)^2 = E\left(\hat{\theta}_n - E\left(\hat{\theta}_n\right) + E\left(\hat{\theta}_n\right) - \theta\right)^2 \\ &= \text{var}\left(\hat{\theta}_n\right) + \left(E\left(\hat{\theta}_n\right) - \theta\right)^2 = \text{var}\left(\hat{\theta}_n\right) + \text{bias}\left(\hat{\theta}_n\right)^2.\end{aligned}$$

Thus, the MSE of $\hat{\theta}_n$ is equal to the variance plus the squared bias of the estimator. Two estimators can then be compared by evaluating their MSEs, with the estimator with the smallest MSE being preferred.

One could then consider using that estimator with the minimum MSE over all possible estimates of θ. For example, consider the situation where the distribution of y is $N\left(\mu, \sigma^2\right)$. As discussed earlier, the sample variance estimator of σ^2 is $n^{-1}\Sigma\left(y_i - \bar{y}\right)^2$, a biased estimator, while $(n-1)^{-1}\Sigma\left(y_i - \bar{y}\right)^2$ is an unbiased estimator. Which estimator is preferred?

Let us consider all estimators of the form $\hat{\sigma}^2 = k\Sigma\left(y_i - \bar{y}\right)^2$, where k is some constant, and determine the value of k that results in the minimum MSE of $\hat{\sigma}^2$. We have

$$\begin{aligned}\text{mse}\left(\hat{\sigma}^2\right) &= \text{var}\left(\hat{\sigma}^2\right) + \text{bias}\left(\hat{\sigma}^2\right)^2 \\ &= \text{var}\left(k\sum\left(y_i - \bar{y}\right)^2\right) + \text{bias}\left(k\sum\left(y_i - \bar{y}\right)^2\right).\end{aligned}$$

Now one can show that $k\Sigma\left(y_i - \bar{y}\right)^2 \sim k\sigma^2\chi^2_{n-1}$, a constant times a chi-square random variable with $n - 1$ degrees of freedom. Using the fact that $E\left(\chi^2_v\right) = v$ and $\text{var}\left(\chi^2_v\right) = 2v$, it follows that

$$E\left(k\sum\left(y_i - \bar{y}\right)^2\right) = \sigma^2 k(n-1) \text{ and } \text{var}\left(k\sum\left(y_i - \bar{y}\right)^2\right) = \sigma^4 k^2\left(2(n-1)\right).$$

It follows that

$$\text{mse}\left(\hat{\sigma}^2\right) = \sigma^4\left(2k^2(n-1) + (k(n-1)-1)^2\right) = \sigma^4 f(k),$$

where $f(k)$ is a function of k. We need to find the value of k that minimizes $f(k)$, and, consequently, mse $(\hat{\sigma}^2)$. Taking the derivative of $f(k)$ with respect to k, setting equal to zero, and solving for k, results in $k = 1/(n+1)$. It is interesting to note that for $k = 1/n$, the resulting estimator is the average squared deviation, while $k = 1/(n-1)$ results in the unbiased estimator, the estimator most suggested in texts on statistics, and $k = 1/(n+1)$ results in the minimum MSE. Actually, there is very little difference between these three estimators for all but relatively small n. It is easy to show using $f(k)$ above, that for $k = 1/n$, $\text{mse}(\hat{\sigma}^2) = \frac{2n-1}{n^2}\sigma^4$; if $k = 1/(n-1)$, $\text{mse}(\hat{\sigma}^2) = \frac{2}{n-1}\sigma^4$; and if $k = 1/(n+1)$, we have $\text{mse}(\hat{\sigma}^2) = \frac{2}{n+1}\sigma^4$. A little algebra reveals that the MSE is largest for $k = 1/(n-1)$, followed by the MSE for $k = 1/n$.

To obtain a visual representation of the magnitude of the difference between these MSE values, R can be used to plot the MSE values for differing values of n as follows:

```
# Set the sample sizes, avoiding n=1 for k = 1/(n-1)
sample_sizes <- 2:20 # Start from 2 to avoid division by
zero for k = 1/(n-1)
# Calculate the MSE values for different k values
mse_k1 <- (2 * sample_sizes - 1) / sample_sizes^2 # k =
1/n
mse_k2 <- 2 / (sample_sizes - 1) # k = 1/(n-1)
mse_k3 <- 2 / (sample_sizes + 1) # k = 1/(n+1)
# Plot the MSE values
plot(sample_sizes, mse_k2, type = "b", col = "black",
pch = 16,
      ylim = range(c(mse_k1, mse_k2, mse_k3)), lwd = 2,
      xlab = "Sample Size", ylab = "MSE", main = "Plot of
      MSE for k = 1/n, 1/(n-1), and 1/(n+1)")
lines(sample_sizes, mse_k1, type = "b", col = "black",
pch = 17, lwd = 2)
lines(sample_sizes, mse_k3, type = "b", col = "black",
pch = 18, lwd = 2)
# Add a legend
legend("topright", legend = c("k = 1/(n-1)", "k = 1/n",
"k = 1/(n+1)"), col = c("black", "black", "black"),
lty = 1, pch = c(16, 17, 18), lwd = 2)
```

Figure 2.7 shows that the top plot is $k = 1/(n-1)$, and the middle is $k = 1/n$. It is obvious that by $n = 20$, there is very little difference between these three estimators of σ^2.

2.2.6 Maximum Likelihood Estimation

One important method for obtaining estimates of parameters is the method of maximum likelihood. Knowledge of this technique will provide insight

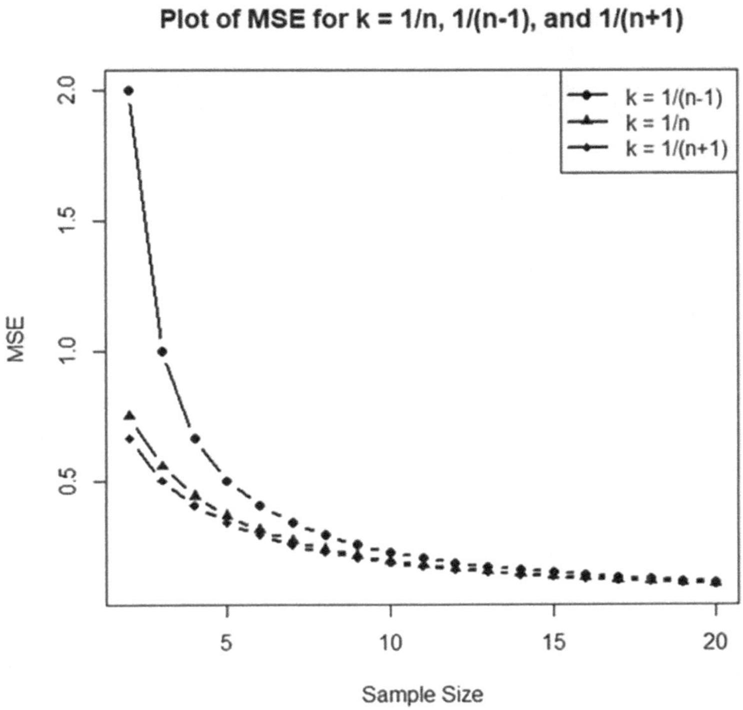

FIGURE 2.7
Plot of MSE for $k = 1/n$, $1/(n-1)$, and $1/(n+1)$.

into several of the methods to be discussed later in the area of robust estimation. Recall that the likelihood function for the random sample $y_1,\ldots,y_n$ of the random variable y, with probability function $f(y|\theta)$, is

$$L(\theta|\underline{y}) = \prod f(y_i|\theta),$$

the joint distribution of the sample. Of course, $L(\theta|\underline{y})$ is a function of the unknown parameter θ. The goal of maximum likelihood estimation is to find the estimator, $\hat{\theta}$, a function of the data that will maximize $L(\theta|\underline{y})$ over all possible values of θ – that is, if $\hat{\theta}$ satisfies

$$\max\text{imum}\, L(\theta|\underline{y}) = L(\hat{\theta}|\underline{y}),$$

then $\hat{\theta}$ is the maximum likelihood estimator. The value of $\hat{\theta}$ for a particular sample, $y_1,\ldots,y_n$, of y is called the maximum likelihood estimate, the MLE. Usually, the MLE is found by taking the derivative of the log of $L(\theta|\underline{y})$, setting the results equal to zero, and solving for $\hat{\theta}$.

Properties of $\hat{\theta}$ can be found in many texts, including Mood, Graybill, and Boes (1974). Several of these properties are of special interest to us: namely that $\hat{\theta}$, under appropriate conditions, is a consistent estimator of θ, though often biased. Additionally, under appropriate conditions, the asymptotic distribution of the MLE can be written as

$$\sqrt{n}\left(\hat{\theta}-\theta\right)\xrightarrow{d}N\left(0,I^{-1}(\theta)\right),$$

an asymptotic normal random variable with expectation equal to θ and variance equal to $I^{-1}(\theta)/n$, achieving the Cramer-Rao lower bound, the minimum variance possible. Thus, the maximum likelihood estimators are asymptotically optimal in terms of minimum variance.

To illustrate a few of these concepts, consider again the case where the random variable y follows a Poisson distribution with an unknown parameter λ. The likelihood function was presented earlier. Taking the derivative and setting equal to zero results in

$$\frac{\partial \ln L}{\partial \lambda}=-n+\frac{\sum y_i}{\lambda}=0.$$

Solving for the maximum likelihood estimator gives us $\hat{\lambda}=\sum y_i/n=\bar{y}$. Since $I(\lambda)=\lambda^{-1}$, it follows that

$$\sqrt{n}\left(\bar{\lambda}-\lambda\right)\xrightarrow{d}N(0,\lambda)$$

expresses the asymptotic distribution of $\hat{\lambda}=\bar{y}$. This result could easily have been anticipated from the well-known central limit theorem, which states that the asymptotic distribution of the sample mean, based on n i.i.d. observations, is normal with a mean of $E(y)=\lambda$ and variance of $\text{var}(y)/n=\lambda/n$, the same result given by MLE theory. The reader should be reminded that these results are asymptotic in nature, and one needs to question their accuracy in those cases where n is small, say less than 20.

Application of the previous results suggests that to test the null hypothesis that λ equals some specified value, say λ_0, the test statistic would be of the form

$$z=\frac{\sqrt{n}\left(\bar{y}-\lambda_0\right)}{\sqrt{\lambda_0}},$$

which has an approximate $N(0,1)$ null distribution. Additionally, approximate $(1-\alpha)\times 100\%$ confidence intervals for λ would be constructed by using the formula

$$\bar{y}\pm z_{1-\frac{\alpha}{2}}\sqrt{\bar{y}/n}.$$

These formulas become more exact as n gets large, around 30, as a rough rule of thumb.

2.3 Least Squares and Weighted Least Squares Estimation in the Location Model

We found two related methodologies very useful in later discussions: least squares (LS) and weighted least squares (WLS) estimation. We will introduce both LS and WLS by examining these forms of estimation for the location model. It is assumed that the reader is already familiar with the concept of least squares, a concept usually presented carefully during the study of regression analysis. Briefly, for the location model of

$$y_i = \theta + \varepsilon_i$$

with the assumption that the ε_i are i.i.d. with $E(\varepsilon_i) = 0$ and $var(\varepsilon_i) = \sigma^2$, for all $i = 1,\ldots,n$, the LS estimator can be defined as the value of θ, $\hat{\theta}_{ls}$, such that the quantity

$$S(\theta) = \sum (y_i - \theta)^2$$

is minimized at $\theta = \hat{\theta}_{ls}$. The LS estimator is found by taking the derivative of $S(\theta)$ with respect to θ, setting the results equal to zero, and solving the equation for the unknown, $\hat{\theta}_{1ls}$. Using $S(\theta)$, we find that the derivative is

$$S'(\theta) = -2\sum (y_i - \theta).$$

Setting $S'(\theta)$ equal to zero and solving, we obtain:

$$S'(\theta) = -2\sum (y_i - \theta) = 0.$$

Or we obtain

$$\sum y_i - n\theta = 0.$$

Solving for $\hat{\theta}_{ls}$ gives

$$\hat{\theta}_{ls} = \frac{\sum y_i}{n} = \overline{y}.$$

We see that the LS estimator in the location model is simply the sample mean, $\overline{y}$. It is easy to show that $\overline{y}$ is unbiased with $E(\overline{y}) = \theta$ and has variance

$\mathrm{var}(\bar{y}) = \sigma^2/n$. It turns out that among all unbiased "linear estimators" that $\bar{y}$ has the minimum variance. A "linear estimator," $\hat{\theta}$, is one that can be written in the form $\hat{\theta} = \sum a_i y_i$, a linear combination of the y's, where $a_1, \ldots, a_n$ are constants. We see that $\hat{\theta}_{ls} = \bar{y}$ is a linear estimator with each a_i equal to $1/n$.

Under the assumptions of our model, – that is, independent error terms with $E(\varepsilon_i) = 0$ and $\mathrm{var}(\varepsilon_i) = \sigma^2$, a well-known result, stated by the Gauss-Markov theorem (see Graybill (1976), for example), is that the LS estimator is optimal in the sense that no other linear estimator of θ can have a smaller variance than the LS estimator. We say that the LS estimator is the best linear unbiased estimator of θ. The acronym BLUE is often used to describe an estimator possessing this property.

Now suppose that associated with each observation, y_i, is a predetermined weight, w_i, for $i = 1, \ldots, n$. Then the WLS estimator of θ is obtained by solving for that value of θ, $\hat{\theta}_{wls}$ that minimizes

$$SW(\theta) = \sum w_i (y_i - \theta)^2,$$

the weighted sum of squares. Proceeding exactly as before by taking the derivative with respect to θ, setting the resulting equation equal to zero, and solving for the unknown leads to

$$\hat{\theta}_{wls} = \frac{\sum w_i y_i}{\sum w_i} = \bar{y}_w,$$

the weighted sample mean of the observations, $y_1, \ldots, y_n$. It is easy to see that this estimator is unbiased and that the variance of this estimator is

$$\mathrm{var}\left(\hat{\theta}_{wls}\right) = \frac{\sum w_i^2 \mathrm{var}(y_i)}{\left(\sum w_i\right)^2},$$

provided that the observations are uncorrelated, as is assumed here.

One interesting application of this result is in the altered location model, where the variance of the ε_i is not constant but rather $\mathrm{var}(\varepsilon_i) = \sigma_i^2$, where the σ_i^2 are not all equal. While this model will soon provide us with several interesting results, it is not a very useful model from a practical viewpoint. Seldom will the researcher find a need for this model. Nonetheless, it is used here to demonstrate several important techniques that will be valuable in subsequent presentations. For this model, $E(y_i) = \theta$ and $\mathrm{Var}(y_i) = \sigma^2$ – that is, the observations are no longer identically distributed. The LS estimator of θ is not BLUE for this situation since the Gauss-Markov theorem cannot be applied: the ε_i are still independent with $E(\varepsilon_i) = 0$, but the variance is no longer constant.

It should be clear that the observations, y_i, that provide the most information concerning the value of the location parameter θ are those with the smaller variances. Thus, if we weigh each observation by the reciprocal of its variance – that is, $w_i = 1/\sigma_i^2$ – larger weights will be given to those observations with smaller variances. This is, in fact, the theoretically optimal choice of weights, resulting in an estimator with minimum variance. Though no proof of this statement will be given here, we can justify it in a simple way. Consider again the altered location model of

$$y_i = \theta + \varepsilon_i,$$

where the $\text{var}(\varepsilon_i) = \sigma_i^2$. Assume for the moment that these variances are known. The method of least squares, though no longer optimal, can still be applied to give $\hat{\theta}_{ls} = \bar{y}$, as before, for the original location model. This estimator is still unbiased but now has $\text{var}(\hat{\theta}_{ls}) = \sum \sigma_i^2 / n^2$, no longer the minimum variance unbiased linear estimator of θ.

Consider multiplying both sides of the location model by $1/\sigma_i$. We get

$$\frac{1}{\sigma_i} y_i = \frac{1}{\sigma_i}\theta + \frac{1}{\sigma_i}\varepsilon_i$$

or

$$y_i^* = \theta_i^* + \varepsilon_i^*,$$

where the ε_i^* are independent with $E(\varepsilon_i^*) = 0$ and $\text{var}(\varepsilon_i^*) = \text{var}\left(\frac{1}{\sigma_i^2}\varepsilon_i\right) = \frac{1}{\sigma_i^2}\sigma_i^2 = 1$. Thus, the ε_i^* have constant variance and, consequently, satisfy the assumptions for the application of the Gauss-Markov theorem. The method of least squares can then be applied to obtain the minimum variance linear unbiased estimator of θ_i^*, which amounts to finding the least squares (LS) estimator of θ since $\theta_i^* = \frac{\theta}{\sigma_i}$, is a function of only one unknown, θ (recall that the σ_i^2 are assumed to be known). The theory of LS tells us to find that value which minimizes

$$\sum (y_i^* - \theta_i^*)^2 = \sum \frac{1}{\sigma_i^2}(y_i - \theta)^2 = \sum w_i (y_i - \theta)^2,$$

which is the previously mentioned WLS problem with $w_i = 1/\sigma_i^2$. The solution is the familiar weighted sample mean:

$$\hat{\theta}_{wls} = \frac{\sum w_i y_i}{\sum w_i} = \bar{y}_w.$$

Using the properties of the variance operator, one can show that

$$\text{var}\left(\hat{\theta}_{\text{wls}}\right) = 1/\sum\left(\frac{1}{\sigma_{\text{i}}^2}\right).$$

Two features of interest concerning this result are as follows. The first is to note that if all variances are equal to a common value – that is, $\sigma_i^2 = \sigma^2$, for all $i = 1,\ldots,n$, then the weighted sample mean reduces to the regular sample mean with $\text{var}\left(\hat{\theta}_{w/s}\right) = \frac{\sigma^2}{\text{n}}$, the minimum value. Second, the Gauss-Markov theorem tells us that in the event that the variances are not all equal, the weighted estimator should have minimum variance among all linear estimators. An interesting comparison is to note that the variance of $\overline{y}$, $\text{var}\left(\hat{\theta}_{\text{ls}}\right) = \sum\sigma_{\text{i}}^2/\text{n}^2$, which also reduces to $\text{var}\left(\hat{\theta}_{\text{ls}}\right) = \sigma^2/\text{n}$ if all the variances are equal, is always greater than the variance for the $\hat{\theta}_{\text{wls}}$. So it is clear that weighing the observations results in a superior estimator than the LS estimator in this case.

In applications, of course, the variances are never known, and consequently, the correct weights used above are also unknown. The natural approach is to estimate the variances whenever possible and use these estimates to obtain estimates of the weights. The WLS technique is then referred to as the estimated weighted least squares (EWLS) method. For example, if multiple observations are available from each of the populations so that the altered location model can be written as

$$\text{y}_{\text{ij}} = \theta + \varepsilon_{\text{ij}},$$

$j = 1,\ldots,n_i$, where n_i is the number of multiple observations in each population, with $\text{i} = 1,\ldots,\text{k}$, where k is the number of populations. In this case, the obvious estimate of the population variance for the ith group, σ_i^2, is

$$\hat{\sigma}_i^2 = \frac{\sum\left(\text{y}_{\text{ij}} - \overline{y}_i\right)^2}{\text{n}_{\text{i}} - 1},$$

the sample variance for the ith group. Then we could use $\hat{\text{w}}_{\text{ij}} = 1/\hat{\sigma}_{\text{i}}^2$ and compute the EWLS estimator as

$$\hat{\theta}_{ewls} = \frac{\sum_{i=1}^{k}\sum_{j=1}^{n_i}\hat{w}_{ij}y_{ij}}{\sum_{i=1}^{k}\sum_{j=1}^{n_i}\hat{w}_{ij}}.$$

While this approach may seem as an obvious solution for the multiple observation case, studies have shown (see, for example, Deaton, Reynolds, and Myers, 1983) EWLS can be problematic if the sample size is small, resulting in imprecise estimates of the variances and, hence, the weights. It has been speculated that if n_i are small, say less than 9, the user is better off using regular least squares. This is an interesting statement that we will evaluate in greater detail.

2.3.1 A Mini R Experiment

R can be used to help illustrate the ideas presented in this section. One can study the behavior of the three estimators, the LS, the WLS, and the EWLS, via rather simple simulation methods. Let us consider the following situation. Suppose we have three populations where the random variable y follows a normal distribution with a mean θ and variances σ_1^2, σ_2^2, and σ_3^2 for populations one through three, respectively. We will consider the situation where the sample sizes from each population are equal. We can then examine the behavior of the three estimators across various combinations of sample sizes and differences in the variances.

For a simple illustration, consider the case where the three variances are 1, 9, and 25, respectively, and each sample size is five. Our goal is to estimate $\theta = 0$. To obtain a random sample for this scenario, on R, type as follows:

```
# Set seed for reproducibility
set.seed(123)
# Generate random samples from three populations
c1 <- rnorm(5, mean = 0, sd = 1) # Population 1: vari-
ance = 1
c2 <- rnorm(5, mean = 0, sd = 3) # Population 2: vari-
ance = 9
c3 <- rnorm(5, mean = 0, sd = 5) # Population 3: vari-
ance = 25
# Stack the samples into a single vector and create a
group identifier
c4 <- c(c1, c2, c3) # Combined observations
c5 <- factor(rep(1:3, each = 5)) # Group identifiers (1,
2, 3)
# Display the data
data <- data.frame(Group = c5, Observations = c4)
print(data)
# Box Plot of the Observations
boxplot(c4 ~ c5, main = "Comparing the Three Samples of
Size Five", ylab = "Population", xlab = "Observations",
names = c("Population 1", "Population 2", "Population
3"), col = c("white", "white", "white"), horizontal =
TRUE)
```

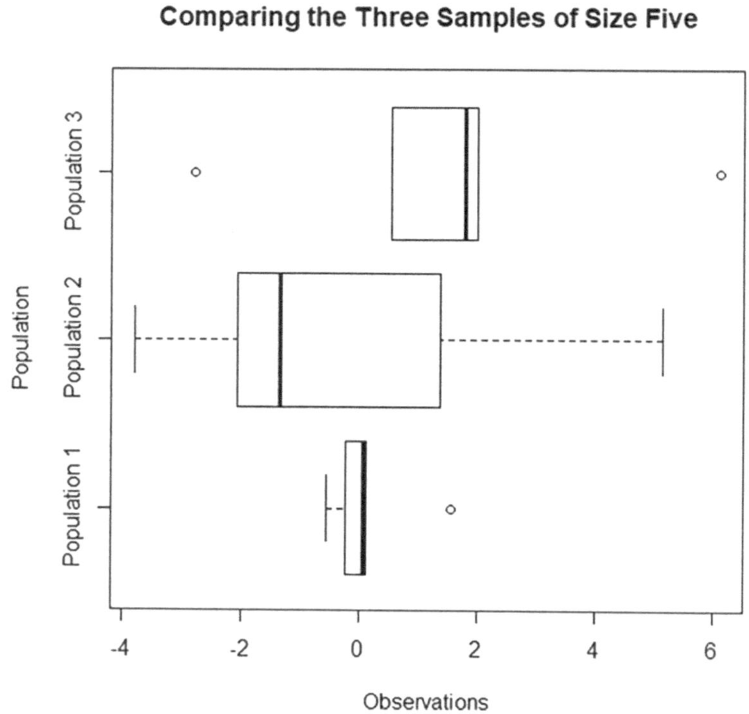

FIGURE 2.8
Side-by-side box plot for three populations.

Figure 2.8 shows that the observations from population 1 are more concentrated around the true mean of zero than the data from the other two populations. To calculate the three estimators, we will use the estimated weights based on the sample variances and the true variances (recall that the true variances must be known in order to obtain the optimal weights). One can get the LS estimate = 0.533374, the WLS estimate = 0.2088223, and the EWLS estimate = 0.2552441. It can be observed that the WLS estimate is much closer to the true value of $\theta = 0$ than either of the other two estimates, with the EWLS estimate being also much closer to zero than the LS estimate. Too much emphasis is then given to the observations far removed from zero, resulting in an inaccurate EWLS estimate.

Of course, the results of a single sample hardly constitute convincing evidence that one method is superior to another. To help remedy this situation, we will study their behavior using the Monte Carlo method. Chapter 3 will present much more of the theory and practice of Monte Carlo methods. For now, think of Monte Carlo studies as using R to simulate the sample distributions of the estimators that we are studying, the LS, the WLS, and the EWLS estimators, in the altered location model.

```
# Monte Carlo Simulation to Study Estimator Behavior
mc <- 100 # Number of repetitions
n1 <- n2 <- n3 <- 5 # Sample sizes
sig1 <- 1 # Population 1 standard deviation
sig2 <- 1 # Population 2 standard deviation
sig3 <- 1 # Population 3 standard deviation
ls_estimates <- numeric(mc)
wls_estimates <- numeric(mc)
ewls_estimates <- numeric(mc)
# Monte Carlo simulation loop
for (i in 1:mc) {# Generate new random samples
  x1 <- rnorm(n1, mean = 0, sd = sig1)
  x2 <- rnorm(n2, mean = 0, sd = sig2)
  x3 <- rnorm(n3, mean = 0, sd = sig3)
  # Combine the samples
  x <- c(x1, x2, x3)
    # Calculate sample variances
ew1 <- var(x1), ew2 <- var(x2), ew3 <- var(x3)
# True variances
w1 <- sig1^2, w2 <- sig2^2, w3 <- sig3^2
# Combine the weights
  estimated_weights <- c(rep(1 / ew1, n1), rep(1 / ew2,
n2), rep(1 / ew3, n3))
  true_weights <- c(rep(1 / w1, n1), rep(1 / w2, n2),
rep(1 / w3, n3))
    # Calculate estimators
  ls_estimates[i] <- mean(x)
  wls_estimates[i] <- sum(true_weights * x) /
  sum(true_weights)
  ewls_estimates[i] <- sum(estimated_weights * x) /
  sum(estimated_weights)}
# Summarize the results
summary_ls <- summary(ls_estimates)
summary_wls <- summary(wls_estimates)
summary_ewls <- summary(ewls_estimates)
cat("LS Estimates Summary:\n", cat("\nWLS Estimates
Summary:\n")
cat("\nEWLS Estimates Summary:\n")
cat("\nVariance of LS estimator:", var(ls_estimates),
"\n")
cat("Variance of WLS estimator:", var(wls_estimates),
"\n")
cat("Variance of EWLS estimator:", var(ewls_estimates),
"\n")
# Visualize the Distribution of Estimators
estimates_data <- data.frame( LS = ls_estimates, WLS =
wls_estimates, EWLS = ewls_estimates )
# Side-by-side box plots of the estimates
boxplot(estimates_data, main = "Side by Side Boxplots
Comparing Three Estimators", ylab = "Estimate", xlab =
```

```
"Estimator", names = c("LS", "WLS", "EWLS"), col =
c("white", "white", "white"), horizontal = TRUE)
# Q-Q plot of LS vs. WLS
qqplot(ls_estimates, wls_estimates, main = "E-E Q-Q Plot
Comparing LS to WLS", xlab = "LS Estimates", ylab = "WLS
Estimates", col = "black") abline(0, 1, col = "black")
# Regression of order statistics (for LS vs WLS)
fit <- lm(ls_estimates ~ wls_estimates)
summary(fit)
```

These random numbers represent a random sample from which the three estimators can be computed. This process can be repeated to obtain another random sample and another set of values for the three estimators. The number of times this process is repeated is called the number of Monte Carlo repetitions; the resulting set of values for each estimator represents the Monte Carlo or simulated estimate of the sampling distribution for that estimator. That is, the values of the estimator obtained via simulation represent the estimator's empirical distribution. As the number of Monte Carlo repetitions increases to a very large number, say 1,000, this ECDF approaches the true CDF of the estimator, its theoretical sampling distribution.

As in the previous case, all three estimators are truly unbiased estimators of θ, which equals zero here, so the values of 0.02 for LS, 0.02 for WLS, and 0.016868 for EWLS differ from zero because of the sampling variability inherent in the Monte Carlo method.

The Monte Carlo estimates of the true variance of each estimator are also very close to the true values. The Monte Carlo results for the LS, WLS, and EWLS estimators are 0.7089349, 0.07089349, and 0.07662655, respectively.

In Monte Carlo studies of this type, the EDA methods of Chapter 1 are useful in evaluating the performance of these estimators. For example, the side-by-side box plots, and the e-e Q-Q plots, as in Figure 2.9, directly compare one estimator with another. The Q-Q plots can be used to evaluate the sampling distribution properties of the estimators. To obtain side-by-side box plots comparing three estimators on R, type as follows:

```
# Side-by-side box plots of the estimates
boxplot(estimates_data, main = "Side by Side Boxplots
Comparing Three Estimators", ylab = "Estimate", xlab =
"Estimator",
   names = c("LS", "WLS", "EWLS"), col = c("white", "white",
"white"), horizontal = TRUE)
```

To obtain Q-Q plot for LS and WLS using R, type as follows:

```
# Q-Q plot of LS vs. WLS
qqplot(ls_estimates, wls_estimates, main = "E-E Q-Q Plot
Comparing LS to WLS", xlab = "LS Estimates", ylab = "WLS
Estimates", col = "black")
abline(0, 1, col = "black")
```

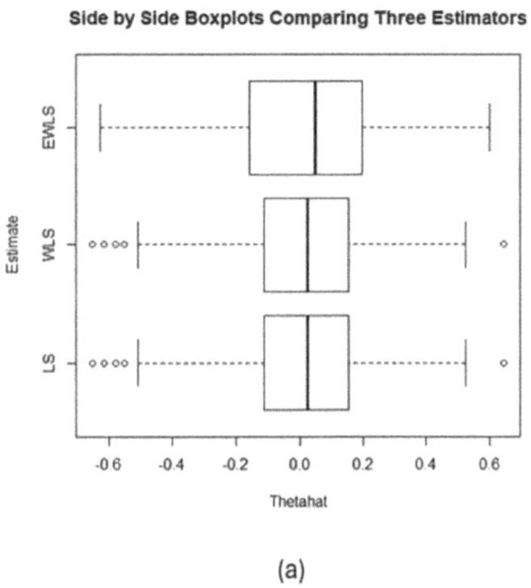

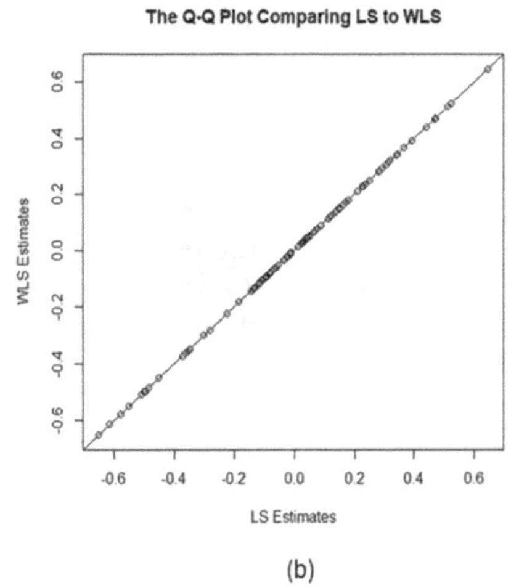

FIGURE 2.9
(a) Side-by-side box plot for the three estimators; (b) The Q-Q plot of LS versus WLS.

The box plot, Figure 2.9 (a), clearly reveals that the LS estimator is certainly more unstable than the other two estimators in this situation. There is little difference between the WLS and the EWLS estimators here.

The Q-Q plot, Figure 2.9 (b), comparing the LS and the WLS estimators, in the case of the equal number of observations for each estimator, is simply the scatterplot of the order statistics for each estimator and reveals nearly a straight-line behavior with several outliers in the tails. The normality assumption for the sampling distribution of the EWLS procedure is reasonable. While this may seem a trivial result, the reader should be reminded that the EWLS procedure is a weighted combination of normal observations, in this case, where the weights are themselves random variables with chi-square distributions (except for a constant). Deriving the exact sampling distribution of the EWLS procedure is by no means a simple process. Yet the Monte Carlo method suggests that whatever this sampling distribution may be, it is very close to a normal distribution.

As one final situation, consider again the sample size of five for each of the three groups and variances of 1, 1, and 9, respectively. Here, WLS should be better than LS, and it is interesting to see if EWLS can continue outperforming the LS estimator.

We see from running the R code for this scenario that WLS, the theoretically optimal estimator, is indeed outperforming the other two estimators, with EWLS having a substantially smaller variance than the LS estimator. A more thorough analysis of this situation would also include our graphical analysis, including the box plots and the various Q-Q plots. We will not proceed any further with this example. Our results indicate that the EWLS procedure appears to be quite promising and raises the point that the condition requiring each n_i to exceed 9 before EWLS is beneficial (Deaton et al., 1983) may be too restrictive. While our small-scale experiment is certainly not detailed enough to allow general conclusions and recommendations, it does serve to illustrate the convenience and usefulness of the Monte Carlo method. It would be a relatively easy matter to increase the number of Monte Carlo repetitions from 100 to 500 and to consider a much broader combination of other parameters of our experiment, including the number of groups, the sample size of each group, and the relationship between the variances of each group. Perhaps the reader can think of other aspects of interest to be considered in an experiment of this type.

2.4 The Approximate Variance of Functions of Random Variables: The Delta Method

Another tool that we will find beneficial in later applications is the use of the Taylor series to find an expression for the approximate variance of a function of one or more random variables. The technique is commonly referred to as the "Delta Method." We will first briefly review the use of the Taylor series to approximate a function of one variable with a polynomial, extending these results to the multivariate case later in the presentation.

2.4.1 The Univariate Case

Let y be the random variable of interest with $\mathrm{E}(y)=\theta$ and $\mathrm{var}(y)=\sigma^2$. Let $g(y)$ be some sufficiently smooth function of y: for example, $g(y)=\ln(y)$ or any other differentiable function. Then a Taylor series expansion of $g(y)$ about some arbitrary point, a is defined as:

$$g(y)=g(a)+(y-a)g'(a)+\frac{(y-a)^2}{2}g''(a)+\frac{(y-a)^3}{3!}g'''(a)+\ldots,$$

where $g'(a)=\left.\frac{\partial g(y)}{\partial y}\right|_{y=a}$ is the first derivative of $g(y)$ with respect to y, evaluated at $y=a$. Similarly, $g''(a)=\left.\frac{\partial^2 g(y)}{\partial y^2}\right|_{y=a}$ is the second derivative of $g(y)$ with respect to y, evaluated at $y=a$, with other terms defined by using the remaining higher-order derivatives.

2.4.2 The Delta Method

We can use the Taylor series expansion to find approximate expressions for the expected value and variance of the function, g, of the random variable y.

Using the Taylor series expansion in this way is referred to as the Delta Method. We let $a = E(y) = \theta$.

To find an expression for $E(g(y))$, we use only the first three terms of the expansion, the second-order Taylor series, as follows:

$$g(y) \approx g(\theta) + (y-\theta)g'(\theta) + \frac{(y-\theta)^2}{2} g''(\theta).$$

Then taking the expectation of both sides, we obtain

$$E(g(y)) \approx g(\theta) + 0 + \frac{\mathrm{var}(y)}{2} g''(\theta).$$

Thus, we obtain the well-known result that $E(g(y)) \neq g(E(y)) = g(\theta)$, unless, of course, all derivatives of the second order and higher of $g(y)$ are zero.

For example, if $g(y) = a + by$, the linear function of y, then $g''(\theta) = 0$ and $E(a+by) = a + b\theta$, demonstrating that the expectation operator is a linear operator. On the other hand, if $g(y) = 1/y$, then $E(g(y)) \neq 1/\theta$, or if $g(y) = y^2$, then $E(g(y)) \neq \theta^2$. It is interesting to note that in the latter example, the application of the approximate expression for $E(g(y))$ results in $E(y^2) = \theta^2 + \mathrm{var}(y)$, an exact result since the square function has derivatives only through the second order, and consequently, a second-order Taylor series expansion is exact for this function. Of course, this result could also have been derived by using the formula for the variance of y as $\mathrm{var}(y) = E(y^2) - E(y)^2$.

To obtain an approximate expression for the variance of a function of the random variable y, we use the first-order Taylor expansion – that is,

$$g(y) \approx g(\theta) + (y-\theta)g'(\theta).$$

Applying the variance operator to both sides of this expression, and noting that all terms that involve θ are constants and not random variables, we obtain the expression

$$\mathrm{var}(g(y)) \approx \mathrm{var}(y)(g'(\theta))^2,$$

a very useful result. If the function of y is $g(y) = a + by$, the linear function, our approximation formula gives $\mathrm{var}(a+by) = b^2\sigma^2$, an exact result since the linear function is exactly represented by the first-order Taylor series. We note that the expression for the $\mathrm{var}(a+by)$ could also have been obtained by using the well-known properties of the variance operator.

The usefulness of the delta method, however, is to obtain an approximate expression for the variance of nonlinear functions of y. For example, consider the random variable y with a Poisson distribution with parameter λ.

It is well known that $E(y)=\lambda=\text{var}(y)$. We note the dependency of the variance of y on the expected value of y. Obviously, as $E(y)$ varies, so does $\text{var}(y)$. Consider the function $g(y)=\sqrt{y}$. Since $g'(y)=\frac{1}{2}y^{-\frac{1}{2}}$, it follows that $g'(\lambda)=\frac{1}{2}\lambda^{-\frac{1}{2}}$, and we have the approximate variance of $\sqrt{y}$ as

$$\text{var}\left(\sqrt{y}\right)\approx\lambda\left(\frac{1}{2\sqrt{\lambda}}\right)^2=\frac{\lambda}{4\lambda}=\frac{1}{4}.$$

The square-root transformation of a Poisson random variable gives a new variable whose variance is no longer dependent on $E(y)$. Here, we say that the square-root transformation is the "variance stabilizing transformation." The use of a transformation to accomplish this task will be discussed in greater detail in Chapter 4.

In the above example, if we had used the ln() transformation, we would obtain $g'(\lambda)=\frac{1}{\lambda}$, and consequently, $\text{var}(\ln(y))\approx\lambda\left(\frac{1}{\lambda}\right)^2=\frac{1}{\lambda}$, a value that continues to be dependent on $E(y)$.

2.4.3 A Mini-Experiment: How Well Does the Delta Method Work?

The delta method, discussed in subsection 2.4.2, raises a few questions that can be nicely addressed using the following R code. We have argued that a second-order Taylor series is sufficient to approximate the expected value of $g(y)$, and a first-order Taylor series provides an adequate approximation to the variance of $g(y)$. In fact, we specifically argued that the first-order Taylor series should not be used for the expected value since that would imply that $E(g(y))=g(E(y))=g(\theta)$. Let us address these issues by evaluating the square-root transformation used for Poisson random variables to stabilize the variance – that is, if $g(y)=\sqrt{y}$, then $\text{var}\left(\sqrt{y}\right)\approx .25$.

On R, we can easily obtain the complete probabilities for a Poisson random variable with any value of the parameter λ. And, since the square-root transformation is a one-to-one transformation in our situation (only positive values of the square root make sense for a Poisson random variable), the probabilities for $g(y)=\sqrt{y}$ are the same as those for y. Thus, R can compute the exact expected value and variance for $\sqrt{y}$, where y is Poisson with any value of λ, and consequently, we can find the $E\left(\sqrt{y}\right)$ and $\text{var}\left(\sqrt{y}\right)$ as follows:

```
# Define the function to compute E(sqrt(Y)) and
Var(sqrt(Y)) for Poisson-distributed Y
compute_sqrt_poisson <- function(lambda) {
  # Set up the range for Y
  y <- 0:100
```

```
  # Compute the Poisson probabilities for given lambda
  prob <- dpois(y, lambda)
    # Apply the square-root transformation
  sqrt_y <- sqrt(y)
    # Compute the expected value E(sqrt(Y))
  E_sqrt_y <- sum(sqrt_y * prob)
  # Compute the variance Var(sqrt(Y))
  var_sqrt_y <- sum(sqrt_y^2 * prob)–E_sqrt_y^2
  return(list(E_sqrt_y = E_sqrt_y, var_sqrt_y =
  var_sqrt_y))}
  # Example: Calculate for lambda = 5
  lambda <- 5
  result <- compute_sqrt_poisson(lambda)
cat("For lambda =", lambda, ":\n")
cat("E(sqrt(Y)) =", result$E_sqrt_y, "\n")
cat("Var(sqrt(Y)) =", result$var_sqrt_y, "\n")
For lambda = 5:
> cat("E(sqrt(Y)) =", result$E_sqrt_y, "\n")
E(sqrt(Y)) = 2.171154
> cat("Var(sqrt(Y)) =", result$var_sqrt_y, "\n")
Var(sqrt(Y)) = 0.2860894
```

Note that the $\text{var}\left(\sqrt{y}\right)$ is 0.286 when $\lambda = 5$, slightly off from the 0.25 value it is expected to be if the delta method were exact. Also, note that R gives us $E\left(\sqrt{y}\right)$ as 2.17, compared to $\sqrt{5} = 2.24$. We see that $E\left(\sqrt{y}\right)$ is not equal to $\sqrt{\lambda}$, for $\lambda = 5$, as expected.

Also, we could iterate this code for different values of λ. In the plot in Figure 2.10, we see the comparison of the exact $E\left(\sqrt{y}\right)$ to the $\sqrt{\lambda} = \sqrt{E(y)}$ for λ ranging from 1 to 40.

In Figure 2.10, the top curve is the $\sqrt{\lambda}$; the bottom curve is $E\left(\sqrt{y}\right)$. There is very little difference between them, less than 2% relative error for λ equal to six or more. Thus, it seems that there is little difference between $E\left(\sqrt{y}\right)$ and $\sqrt{E}(y)$, at least in this situation. However, this is not unexpected since the square-root transformation is quite linear in the range of λ considered here, with a correlation coefficient between λ and $\sqrt{\lambda}$ of 0.977.

A plot of the exact $\text{var}\left(\sqrt{y}\right)$, Figure 2.11, reveals that the delta method provides a reasonable approximation to the exact value, only for λ equal to 20 or more, where the relative error decreases to less than 2%. However, for small values of λ, the approximation is not very good. For example, at $\lambda = 2$, the exact $\text{var}\left(\sqrt{y}\right)$ is 0.39, far from the 0.25 value given by the approximation.

2.4.4 The Bivariate Case

The delta method can easily be extended to the case with more than one random variable. In the next situation, consider the two-variable case. Let x

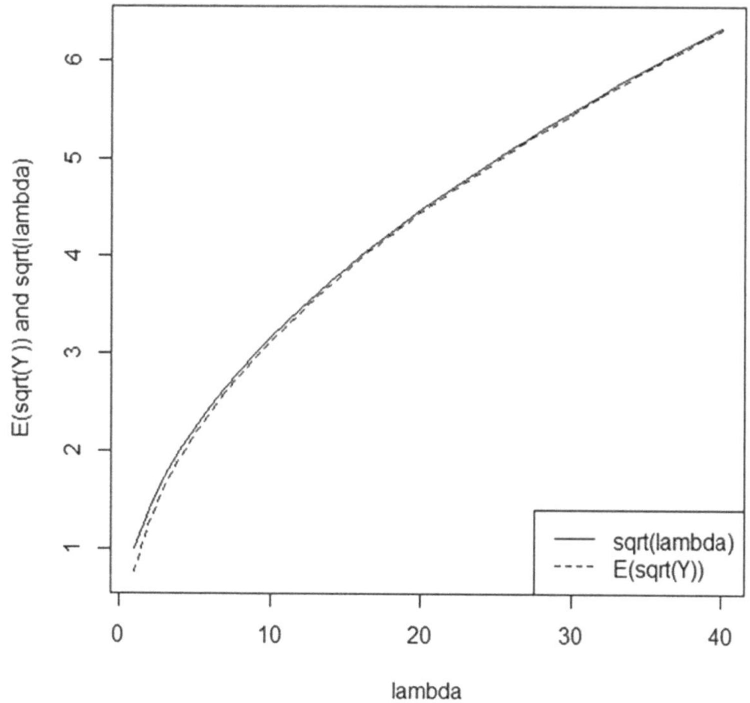

FIGURE 2.10
The comparison of the exact $E\left(\sqrt{y}\right)$ to the $\sqrt{\lambda} = \sqrt{E(y)}$.

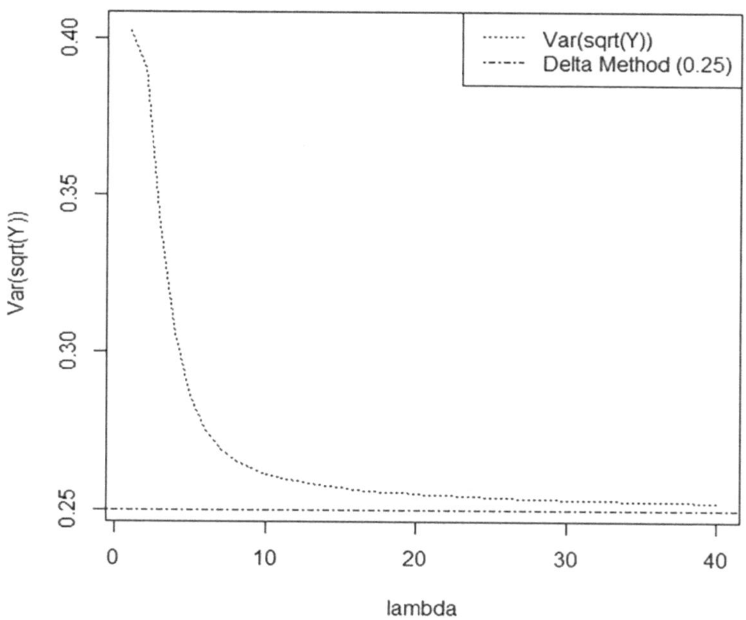

FIGURE 2.11
The exact $\text{var}\left(\sqrt{y}\right)$ versus the delta method.

and y be the two random variables of interest, and $g(x, y)$ be some sufficiently smooth function of them – that is, at least continuous first partial derivatives. Let a and b be two arbitrary points defined on the domain of x and y, respectively. Then we can express $g(x,y)$ in terms of a and b by a straightforward application of a bivariate first-order Taylor series expansion. We obtain

$$g(x,y)=g(a,b)+(x-a)\left.\frac{\partial g(x,y)}{\partial x}\right|_{(a,b)}+(y-b)\left.\frac{\partial g(x,y)}{\partial y}\right|_{(a,b)}+$$

(terms involving higher derivatives of $g(x,y)$).

Without these extra terms and if a and bb are chosen as the expectations of x and y, respectively, we have the following:

$$\mathrm{E}(\mathrm{g}(\mathrm{x},\mathrm{y}))\approx \mathrm{g}(\mathrm{a},\mathrm{b})+0+0=\mathrm{g}(\mathrm{a},\mathrm{b}),$$

and it follows that

$$\mathrm{var}(\mathrm{g}(\mathrm{x},\mathrm{y}))\approx \mathrm{E}\left[(\mathrm{g}(\mathrm{x},\mathrm{y})-\mathrm{g}(\mathrm{a},\mathrm{b}))^2\right].$$

Thus,

$$\begin{aligned}\mathrm{var}(g(x,y)) &\approx E\left[(x-a)^2\left(\left.\frac{\partial g(x,y)}{\partial x}\right|_{(a,b)}\right)^2+(y-b)^2\left(\left.\frac{\partial g(x,y)}{\partial y}\right|_{(a,b)}\right)^2\right.\\ &\qquad\left.+2(x-a)(y-b)\left(\frac{\partial g(x,y)}{\partial x}\right)\left(\frac{\partial g(x,y)}{\partial y}\right)_{(a,b)}\right].\\ &=\mathrm{var}(x)\left(\left.\frac{\partial g(x,y)}{\partial x}\right|_{(a,b)}\right)^2+\mathrm{var}(y)\left(\left.\frac{\partial g(x,y)}{\partial y}\right|_{(a,b)}\right)^2\\ &\qquad+2\mathrm{cov}(x,y)\left(\frac{\partial g(x,y)}{\partial x}\right)\left(\frac{\partial g(x,y)}{\partial y}\right)_{(a,b)}\end{aligned}$$

This expression simplifies to the sum of the variances times the square of the appropriate partial derivative if the variables x and y are uncorrelated. For example, consider the function $\mathrm{g}(\mathrm{x},\mathrm{y})=\mathrm{x}+\mathrm{y}$, the linear function representing the sum of two random variables. The variance operator applied to this function gives the exact result, the well-known expression:

$$\mathrm{var}(x+y)=\mathrm{var}(x)+\mathrm{var}(y)+2\mathrm{cov}(x,y).$$

This is also the same result obtained by the delta method, not surprising since the function is linear, and the first-order Taylor series represents this function exactly.

For a second example, consider the function $g(x,y) = xy$. Then $\frac{\partial g(x,y)}{\partial x} = y$, and $\frac{\partial g(x,y)}{\partial y} = x$. If x and y are also uncorrelated random variables, we have the approximate variance of xy as

$$\text{var}(xy) \approx \text{var}(x)(E(y))^2 + \text{var}(y)(E(x))^2.$$

Consider as a third example the function $g(x,y) = x / y$.

Here, $\frac{\partial g(x,y)}{\partial x} = y^{-1}$ and $\frac{\partial g(x,y)}{\partial y} = -xy^{-2}$, so that we have

$$\text{var}\left(\frac{x}{y}\right) \approx \frac{\text{var}(x)}{(E(y))^2} + \frac{\text{var}(y)(E(x))^2}{(E(y))^4} - 2\frac{\text{cov}(x,y)E(x)}{(E(y))^3}.$$

If we let $m = \frac{E(x)}{E(y)}$, and if x and y are uncorrelated, the expression reduces to

$$\text{var}\left(\frac{x}{y}\right) \approx \frac{1}{(E(y))^2}\left[\text{var}(x) + \text{var}(y)m^2\right].$$

Further extensions can be made to the delta method. Since the following material is slightly more technical than the previous discussions, the reader may skip this presentation and return to it later in the text when the need for the material is more apparent.

2.4.5 The K-Variate Case

One generalization of the delta method results when we consider a function g of $k(k \geq 1)$ random variables, $y_1,\ldots,y_k$ – that is, g() – is a mapping from R^k to R^1. If we let $\underline{y}$ denote the kx1 vector $(y_1,\ldots,y_k)'$ and let $E(y) = \underline{\mu} = (\mu_1,\ldots,\mu_k)'$ denote the expectation vector, then the first-order k variate Taylor series expansion of $g(y)$ is

$$g(\underline{y}) \approx g(\underline{\mu}) + \underline{\partial g}g'(\underline{y} - \underline{\mu}),$$

where $\underline{\partial g'} = (\partial g_1,\ldots,\partial g_k)$, with $\partial g_i = \left.\frac{\partial g(\underline{y})}{\partial y_i}\right|_{\underline{y}=\underline{u}}$ for $i = 1,\ldots,k$, the partial derivative of g with respect to the ith variable, evaluated at the expectation vector. Applying the variance operator to both sides of the expression, we obtain

$$\text{var}\left(g\left(\underline{y}\right)\right) \approx \underline{\partial g}' \, \Sigma \, \underline{\partial g},$$

where Σ is the variance-covariance matrix of the $\underline{y}$ vector.

2.4.6 The Multivariate K-Variate Case

One further generalization results when we consider p functions of $\underline{y}$, – that is, $g_1\left(\underline{y}\right) \ldots, g_p\left(\underline{y}\right)$. From our immediate results, we have a first-order Taylor expression for each of the p functions as

$$g_i\left(\underline{y}\right) \approx g_i\left(\underline{\mu}\right) + \underline{\partial g_i'}\left(\underline{y} - \underline{\mu}\right),$$

for $i = 1,\ldots,p$, where $\underline{\partial g_i'} = \left(\partial g_{i1} \ldots \partial g_{ik}\right)'$ with $\partial g_{ij} = \left.\dfrac{\partial g_i\left(\underline{y}\right)}{\partial y_j}\right|_{\underline{y}=\underline{\mu}}, j = 1,\ldots,k.$

Or rewriting all p expressions simultaneously in vector-matrix notation, we have

$$\underline{g}\left(\underline{y}\right) \approx g\left(\underline{\mu}\right) + \Delta'\left(\underline{y} - \underline{\mu}\right),$$

where $g'\left(\underline{y}\right) = \left(g_1\left(\underline{y}\right), \ldots, g_p\left(\underline{y}\right)\right)'$ and Δ' is a $p * k$ matrix of partial derivatives with the ith row being $\underline{\partial g_i}$. Applying the variance operator to both sides of the expression gives

$$\text{var}\left(\underline{g}\left(\underline{y}\right)\right) \approx \Delta' \Sigma \Delta,$$

the approximate variance-covariance matrix of $\underline{g}\left(\underline{y}\right)$. Note that $\underline{g}\left(\underline{y}\right)$ is a px1 vector and $\text{var}\left(\underline{g}\left(\underline{y}\right)\right)$ is a pxp matrix. For example, consider the functions $g_1\left(\underline{y}\right) = \dfrac{y_1}{y_2}$ and $g_2\left(\underline{y}\right) = y_1 y_2$, two functions of two variables. The approximate variances have already been calculated for both functions. But what is the covariance between g_1 and g_2? The reader can verify that

$$\Delta' = \begin{bmatrix} 1/y_2 & -y_1/y_2^2 \\ y_2 & y_1 \end{bmatrix} \underline{y} = \underline{\mu},$$

and if the variance-covariance matrix of y is

$$\Sigma = \begin{bmatrix} \sigma_1^2 & \sigma_{12} \\ \sigma_{12} & \sigma_2^2 \end{bmatrix},$$

it follows that the approximate variance-covariance matrix for $g\left(\underline{y}\right)$ is

$$\operatorname{var}\left(\underline{g}\left(\underline{y}\right)\right) \approx \begin{bmatrix} \sigma_1^2 / \mu_2^2 - 2\sigma_{12}\mu_1 / \mu_2^3 + \sigma_2^2\mu_1^2 / \mu_2^4 & -\sigma_2^2\mu_1^2 / \mu_2^2 \\ -\sigma_2^2\mu_1^2 / \mu_2^2 & \sigma_1^2\mu_2^2 + 2\sigma_{12}\mu_1\mu_2 + \sigma_2^2\mu_1^2 \end{bmatrix}.$$

Exercises

1) **Growing Season Data:** Many people take pride in their gardens. They plant fruit and vegetables to add to their meals or flowers to add beauty to their yard and home. Regardless of the seed planted, the gardener relies upon a prediction of the length of the growing season. This affects which seeds may be grown or which may be prosperous. Thus, a good prediction helps the gardener achieve his or her goals. The following data represents climate information from various cities in the states of Illinois, Michigan, Minnesota, Missouri, and Wisconsin for the years 1961 through 1990. The variables are season: median length of the growing season (days) for a base temperature of 32^0F; avetemp: annual average temperature; aveprecip: annual average precipitation (inches); degreday: annual growing degree days for a base temperature of 40°F, calculated daily as (average temperature 40); and latitude: latitude of the city (degrees).

The data (Growing Season Data_Ch2.xlsx) can be downloaded from the book's website. Consider the growing season data and the season variable.

a) Repeat parts a) and b) of question 2 of Chapter 1.
b) Repeat parts a) through d) for question 10 of Chapter 1.
c) Formally test normality for your variable by computing the Lilliefors and the "correlation coefficient test of normality" test statistics using R or SPSS. Comment on your results.

2) Consider the phosphorus data and the yield variable. Perform part c) of question 1.
3) Consider the phosphorus data and the tissue-p variable. Perform part c) of question 1.
4) Consider the snowfall data and the "annual snowfall" variable. Then perform part c) of question 1.

TABLE 2.1

Growing Season Data

Obs.	State	City	Season	Avetemp	Aveprecip	Degreday	Latitude
1	MN	Albert Lea	156	53.9	31.71	4,524	43.633
2	MN	International Falls	118	47.7	24.36	3,202	48.567
3	MN	Bemidji	123	48.8	23.99	3,403	47.467
4	MN	Minneapolis	158	54.3	28.32	4,621	44.967
5	MN	St. Cloud	139	52.7	27.43	4,056	45.550
6	MN	Winona	158	56.8	32.57	4,651	44.050
7	WI	Marinette	151	54.2	32.16	4,343	45.067
8	WI	Green Bay	152	53.3	28.83	4,120	44.500
9	WI	La Crosse	160	56.2	30.55	4,829	43.800
10	WI	Eau Claire	145	54.1	31.60	4,300	44.783
11	WI	Milwaukee	172	54.3	32.93	4,430	43.050
12	WI	Madison	141	55.6	30.88	4,408	43.083
13	WI	Washington Island	142	51.4	28.44	3,387	45.300
14	MI	Ann Arbor	178	58.6	32.81	5,068	42.250
15	MI	Battle Creek	151	58.1	35.70	4,766	42.333
16	MI	Marquette	159	50.4	30.67	3,461	46.533
17	MI	Sault Ste. Marie	121	49.6	34.23	3,063	46.483
18	MI	Flint	160	56.3	30.28	4,508	43.000
19	MI	Kalamazoo	169	59.8	37.03	5,206	42.333
20	MO	Carthage	194	68.5	42.49	6,826	37.167
21	MO	St. Louis	202	65.4	37.51	6,725	38.650
22	MO	Farmington	167	65.8	41.00	6,218	37.767
23	MO	Camdenton	193	68.6	43.53	7,089	38.017
24	MO	Lebanon	177	67.2	41.37	6,766	37.667
25	IL	Chicago	217	59.9	37.22	5,590	41.817
26	IL	Peoria	179	60.4	36.28	4,948	42.267
27	II.	Rockford	157	57.6	36.28	4,948	42.267
28	IL	Springfield	182	62.3	35.25	6,026	39.767
29	IL	Urbana	185	61.1	39.71	5,717	40.167
30	II.	Dixon	168	58.7	35.58	5,176	41.833

5) Consider the Poisson example discussed in Section 2.3. One formula for the $(1-\alpha)\times 100\%$ confidence interval for λ is given in that section. Another formula is $\bar{y} \pm z_{1-\frac{\alpha}{2}} \frac{s}{\sqrt{n}}$, where s is the usual sample variance. Which formula gives the best confidence interval performance? Run a mini-experiment to find out. Generate 500 repeated samples for n equal to 5, 20, and 30 with λ fixed at five. Compute the 95% confidence intervals for all 500 repeated samples for both

methods. Determine the proportion of 95% intervals that contain $\lambda = 5$, for each value of n. Compare the two methods to each other and to the nominal 95%. Use tables or graphs where appropriate to best present your findings. What have you learned? Can you reach any general conclusion? What other simulations might you run to study this question further? You are not asked to perform any further analyses.

6) Consider the typical regression setting where n pairs of numbers are observed in the form $(y_i, x_i), i = 1,\dots,n$. The variable y is random, representing the number of bubbles found in plates of glass. The variable x is considered as fixed, representing the temperature of the glass mixture before it is rolled into a glass. One model for this problem is to use Poisson regression. Here, in a simplified version of Poisson regression, consider the following model: $y_i \sim \text{Poisson}(\lambda_i)$, where $\lambda_i = \beta x_i$. That is, the parameter λ_i, the density parameter, is considered as a linear function of x (without an intercept, the "no intercept" model). Answer the following questions.
   a) Write out an expression for the likelihood of $y_1,\dots,y_n$ using β and x_i.
   b) Find the MLE of β, $\hat{\beta}$.
   c) What is the asymptotic distribution of $\hat{\beta}$? Be sure to explicitly state the Fisher information number.
   d) Give a formula for an approximate 95% confidence interval for β. Give the most simplified version possible.
   e) Use the delta method to find an approximate expression for the variance of $g(\hat{\beta}) = e^{x_i \hat{\beta}}$.
7) Consider the random variable y with a chi-squared distribution with v degrees of freedom. Use the first-order Taylor series to find expressions for the $E(\ln(y))$ and $\text{Var}(\ln(y))$.
8) Suppose that $\hat{\beta}_0$ and $\hat{\beta}_1$ are two unbiased estimators of β_0 and β_1 with variances σ_0^2 and σ_1^2, respectively, and covariance σ_{12}.
   a. Use the formula in Section 2.4 to find an approximate expression for the variance of $F(\hat{\beta}_0 + x_i\hat{\beta}_1)$, where x_i is a constant. Use the fact that $f(u) = \dfrac{dF(u)}{du}$.
   b. Let $\underline{\hat{\beta}} = \begin{pmatrix} \hat{\beta}_0 \\ \hat{\beta}_1 \end{pmatrix}$ and $\Sigma =$ the variance-covariance matrix of $\underline{\hat{\beta}}$. Let the vector $\underline{x}_i' = (1 x_i)$. Use the formula in Section 2.4 to rewrite your variance of $F(\hat{\beta}_0 + x_i\hat{\beta}_1)$ formula from part (a) in terms of Σ and $\underline{x}_i$.

9) Consider the functions $g_1(\underline{y}) = \frac{y_1}{y_2}$ and $g_2(\underline{y}) = y_1 y_2$, two functions of two variables. What is the covariance between g_1 and g_2?

10) Consider the case where $y_1, \ldots, y_n$ are i.i.d. from a $N(\mu, \sigma^2)$ and the estimator $\hat{\sigma}^2 = k\sum(y_i - \bar{y})^2$.

    a. Show that for $k = 1/n$, that $mse(\hat{\sigma}^2) = \frac{2n-1}{n^2}\sigma^4$; for $k = 1/(n-1)$, that $mse(\hat{\sigma}^2) = \frac{2}{n-1}\sigma^4$; and for $k = 1/(n+1)$, that $mse(\hat{\sigma}^2) = \frac{2}{n+1}\sigma^4$.

    b. Show that the MSE is the largest for $k = 1/(n-1)$, followed by the MSE for $k = 1/n$.

11) Suppose that the random variable y is a Bernoulli random variable with $Prob(y = 0) = P$. It is well known that $E(y) = P$ and $Var(y) = P(1-P)$. Suppose we have a random sample of size n of $y, y_1, \cdots, y_n$.

    a. Show that the MLE of P is $\hat{P} = \frac{\sum_{i=1}^{n} y_i}{n}$, the sample proportion.

    b. What is the asymptotic distribution of $\hat{P}$? Be sure to explicitly state the Fisher information number.

    c. Use the delta method to find an expression for the approximate variance of the function $\log\left(\frac{\hat{P}}{1-\hat{P}}\right)$, called the logit of $\hat{P}$.

12) Suppose we have a random sample of size n of the random variable y where $E(y) = \theta$ and the $Var(y) = \sigma^2$. Recall that the coefficient of variation, a commonly used statistic, is defined as $CV = \frac{s}{y} x100\%$, where s is the sample standard deviation and $\bar{y}$ is the sample mean. Use the methods discussed in this chapter to give an expression for the standard deviation of cv, $sd(cv)$.

Multiple-Choice Questions

1) What is the main purpose of combining exploratory data analysis and confirmatory data analysis in the location model?

   a) To simplify the data analysis process
   b) To visually examine the data without any statistical testing
   c) To extend the data analysis by establishing and evaluating a statistical model
   d) To apply transformations without checking assumptions

   Answer: c)

2) In the context of the location model, what is the purpose of the Q-Q plot?
   a) To compare the sample histogram to the theoretical probability density function
   b) To visually assess the straight-line tendency in the data
   c) To calculate the correlation coefficient
   d) To test the goodness of fit for a Poisson distribution

   Answer: b)

3) What is the chi-square goodness-of-fit test primarily used for in the context of normality tests?
   a) To compare two different distributions
   b) To compare the observed and expected frequencies in a sample
   c) To calculate the variance of a normal distribution
   d) To test the correlation between two variables

   Answer: b)

4) Which transformations are considered a "variance stabilizing transformation" for Poisson random variables?
   a) Logarithmic transformation
   b) Square-root transformation
   c) Reciprocal transformation
   d) Exponential transformation

   Answer: b)

5) What does the correlation coefficient test of normality measure?
   a) The strength of the linear relationship between variables
   b) The skewness of the data
   c) The degree of normality in the data
   d) The variance of the residuals

   Answer: a)

6) In the weighted least squares method, what is the optimal choice of weights?
   a) Equal weights for all observations
   b) Weights are inversely proportional to the variances of the observations
   c) Weights based on the sample mean
   d) Weights based on the standard deviation

   Answer: b)

7) What is the approximate variance of the square root of a Poisson random variable, as given by the delta method?
   a) ½
   b) ¼
   c) ⅓
   d) 1

   Answer: b)

8) What is the purpose of the Kolmogorov-Smirnov test in normality testing?
   a) To compare the sample data against a normal distribution
   b) To calculate the mean and standard deviation of the data
   c) To transform the data for better visualization
   d) To assess the linearity of a probability plot

   Answer: a)

9) Which test statistic is used in the Lilliefors test for normality?
   a) Mean
   b) Median
   c) Variance
   d) Maximum deviation of the empirical CDF from the normal CDF

   Answer: d)

10) What does a significant result in the correlation coefficient test of normality indicate?
   a) The data follows a normal distribution
   b) The data does not follow a normal distribution
   c) The data has no outliers
   d) The data is skewed

   Answer: b)

3

Robust Estimation in the Location Model

Summary

In this chapter, we continue discussing and evaluating statistical procedures applied to the location model. This chapter is concerned with outliers' effect on estimating and testing the location parameter. In Chapter 1, we have seen many ways of examining data and how we can alter data by transformation to conform to the location model. We also discussed methods for testing the normality of the data in the location model. We will now consider other methods by which we can estimate the parameter θ, besides using $\overline{y}$ or maximum likelihood estimation. We will address such questions as how we can evaluate the ability of an estimation technique to "withstand" the outliers. What is meant by a robust estimator and a robust test, and how can the robustness of an estimation procedure or a testing procedure be measured? What are some new forms of robust estimators, how good are they, and are they ever better than using $\overline{y}$? All these questions will be addressed in this chapter.

Learning Objectives

By the end of this chapter, readers should be able to do the following:

- Explain the impact of outliers on location models, and define robust estimators that mitigate their influence.
- Describe how parameters and estimators relate to cumulative distribution functions (CDFs) and empirical cumulative distribution functions (ECDFs), and apply this concept to robust estimation.
- Define the influence curve, and analyze how it measures an estimator's sensitivity to data changes.
- Compare different finite sample influence curves, such as the empirical influence curve (EIC) and sensitivity curve, to assess estimator robustness.
- Use metrics like gross-error sensitivity and breakdown points to evaluate the robustness of various estimators.

3.1 Parameters and Estimators as Functionals

A functional is defined as a "function of a function." In our case, the latter function is the CDF, F, or the ECDF, F_n. It will be demonstrated in this section

DOI: 10.1201/9781003565321-3

that the functions of the CDF turn out to be parameters of interest, and the functions of the ECDF turn out to be the estimators of these parameters. We will need the integral notation of $\int dF(x)$ to represent the "integral" of either a discrete or a continuous random variable. That is,

$$\int dF(x) = \begin{cases} \int f(x)dx & x \text{ continuous} \\ \sum f(x_i) & x \text{ discrete.} \end{cases}$$

Justification for this notation results from $dF(x)$ as the "differential operator." In the continuous case, consider the function $F(x)$ as continuous with the derivative defined as

$$\frac{dF(x)}{dx} = \lim_{\varepsilon \to 0} \frac{F(x+\varepsilon) - F(x)}{\varepsilon} = f(x),$$

where $f(x)$ is a continuous function and is simply the derivative of $F(x)$ with respect to x. If $F(x)$ is the CDF, then $f(x)$ is the probability density function (PDF).

On the other hand, if x is discrete, then $F(x)$ is a step function with "jumps" of $f(x_i)$ at each distinct value, x_i, of the random variable x. An example of such a discrete CDF is shown in Figure 3.1.

The frequencies f2, and f6, indicated in the CDF in Figure 3.1, represent the probabilities that $x = 2$ and $x = 6, f(2)$, and $f(6)$, respectively. The

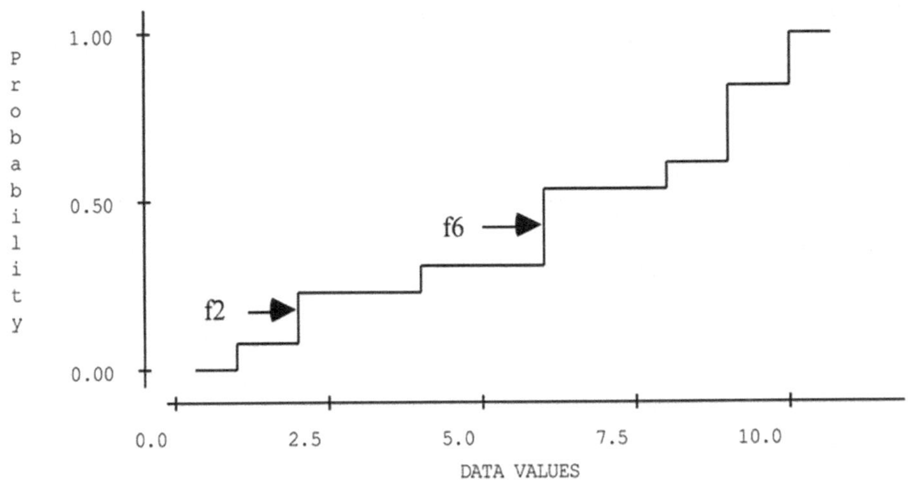

FIGURE 3.1
The cumulative distribution function (CDF).

differential operator, $dF(x)$, applied to the discrete CDF, results in $f(x_i)$ at each distinct value of x assigned a nonzero probability and is zero for all other values of x. This is because the difference in $F(x+\varepsilon)$ and $F(x)$ is zero for all values of x, except at each x_i, the values of x where the jumps of size $f(x_i)$ occur. Thus,

$$dF(x) = \begin{cases} f(x_i) & x = x_i \\ 0 & \text{otherwise.} \end{cases}$$

Using the functional notation, some examples of parameters as functionals are:

1. The expectation of $x: T_1(F) = \int x dF(x) = \mu = \begin{cases} \int xf(x)dx & x \text{ continuous} \\ \sum x_i f(x_i) & x \text{ discrete.} \end{cases}$
2. The median of x: $T_2(F) = F^{-1}\left(\frac{1}{2}\right) = x_{.5}$.
3. The quantiles of x of order p: $T_3(F) = F^{-1}(p) = q_x(p)$.
4. The variance of $x: T_4(F) = \int (x - T_1(F))^2 dF(x) = \sigma^2$.

There are many other examples, some of which we shall see shortly.

To express estimators as functionals, suppose we have a sample $x_1, \ldots, x_n$ with the CDF F and the ECDF, $F_n(x)$, defined in Chapter 1 as: $F_n(x) = \frac{\#(x_i \le x)}{n}$. This implies that $F_n(x) = 1/n$ at $x = x_i$. Then often, $T(F_n)$ estimates $T(F)$. If so, $T(F_n)$ is called the "empirical substitution principle" T(F) estimator. The first example that we will consider is the estimator of the expected value of x,

$$T_1(F_n) = \int x dF_n(x).$$

Recall that since $F_n(x)$ is discrete, the differential $dF_n(x)$ picks up the jumps of $1/n$ at the observed sample points. That is, as x moves from $-\infty$ to ∞ in the integral, the value of x is weighted by $dF_n(x) = 0$ at all x other than the observed sample values. At the values $x_1, \ldots, x_n$, the weights are $1/n$. Thus,

$$T_1(F_n) = \frac{\sum_{i=1}^{n} x_i}{n} = \overline{x}.$$

It has previously been demonstrated that $\overline{x}_0$ is an unbiased and consistent estimator of μ.

As a second example, consider the estimator, written as a functional, of $T_4(F_n)=\sigma^2$. Replacing $F(x)$ by $F_n(x)$ in the definition of T_4 results in

$$T_4(F_n)=\int (x-T_1(F_n))^2\,dF_n(x)=\frac{\sum(x_i-\bar{x})^2}{n}=s^2.$$

It can be shown that s^2 is a biased and consistent estimator of σ^2.

As a final example, consider the empirical substitution principle estimator of the population median, $T_2(F)=F^{-1}\left(\frac{1}{2}\right)$. This example is not as straightforward as the first two illustrations. This is due to the dependency on the sample median on whether *n* is odd or even. We know that the sample median is the single middle value of the ordered data if the sample size is odd and the average of the two middle ordered values if the sample size is even. Expressing this idea as a function of the inverse CDF is accomplished by writing

$$T_2(F_n)=M=\begin{cases} F_n^{-1}\left(\dfrac{n+1}{2n}\right) & \text{if n} \quad \text{odd} \\ \dfrac{1}{2}F_n^{-1}\left(\dfrac{1}{2}\right)+\dfrac{1}{2}F_n^{-1}\left(\dfrac{1}{2}+\dfrac{1}{n}\right) & \text{if n} \quad \text{even.} \end{cases}$$

Since this expression can be theoretically inconvenient, it is usually replaced by $T_2(F_n)=F_n^{-1}\left(\frac{1}{2}\right)$, for large *n*. The fact that the ECDF converges uniformly to the CDF as *n* goes to infinity (the "GlivenkoCantelli lemma") implies that estimators, based on the empirical substitution principle, converge in probability, to the parameter – that is, $T(F_n)\xrightarrow{p}T(F)$. In each of our three examples, we see that this is indeed the case as $\bar{x}\xrightarrow{p}\mu$, $s^2\xrightarrow{p}\sigma^2$, and $M\xrightarrow{p}x_{.5}$.

The consistency property of our estimators provides a useful interpretation of the parameter: the parameter can be thought of as an estimator based on a very large sample. Hence, studying certain properties of the estimator, at least when based on large samples, can be accomplished by studying those properties of the parameter. This concept will be explored in the next section.

3.2 The Influence Curve

We know that if we have a function of *x*, say $g(x)$, then we can learn a great deal about the behavior of $g(x)$ by examining the derivative of $g(x)$, $dg(x)/dx$.

In a similar fashion, if we desire to learn about the behavior of a functional, $T(F)$, we can do so by studying its "derivative." This "derivative" is called the influence curve. Before defining the influence curve of a function, we need to know the following notations and concepts. Let $\theta = T(F)$ be the parameter of interest. Let $\tilde{F}(x) = (1-\varepsilon)F(x) + \varepsilon\delta_{x_0}(x)$, where the indicator function $\delta_{x_0}(x)$ places the point mass of one at the point x_0 – that is,

$$\delta_{x_0}(x) = \begin{cases} 0 & x < x_0 \\ 1 & x \geq x_0. \end{cases}$$

It is useful to think of $\tilde{F}(x)$ as an altered CDF that places special emphasis on the point x_0. It is easy to show that $\tilde{F}(x)$ is itself a CDF, satisfying the properties required for CDFs. A plot of $\tilde{F}(x)$ appears in Figure 3.2.

The smooth curve in Figure 3.2 is the CDF $F(x)$, and the curve with the jump is $\tilde{F}(x)$. The jump of size ε occurs at x_0.

Now, let $\tilde{\theta} = T(\tilde{F})$ be the parameter of interest for the altered CDF. For example, if $\theta = T_1(F) = \mu = \int x dF(x)$, then

$$\begin{aligned} &\tilde{\theta} = T_1(\tilde{F}) = \tilde{\mu} = \int x d\tilde{F}(x) = \\ &(1-\varepsilon)\int x dF(x) + \varepsilon \int x d\delta_{x_0}(x) = \\ &(1-\varepsilon)\mu + \varepsilon x_0 = \mu + \varepsilon(x_0 - \mu). \end{aligned}$$

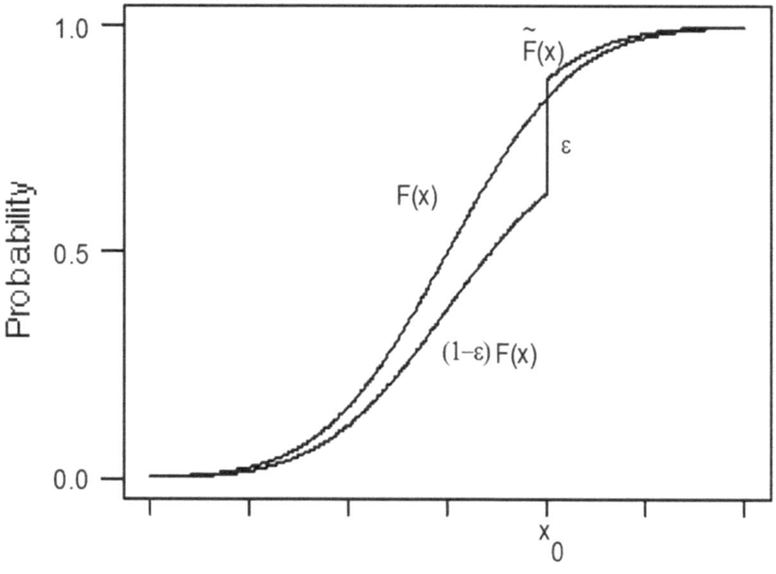

FIGURE 3.2
A comparison of the CDF and the altered CDF.

Thus, the mean of the altered distribution, μ, reflects the mass of size ε at the point x_0.

The influence curve, IC, of the parameter θ, as a function of x_0, is defined as

$$\mathrm{IC}(x_0,\theta)=\lim_{\varepsilon\to 0}\frac{T(\tilde{F})-T(F)}{\varepsilon}=$$

$$\lim_{\varepsilon\to 0}\frac{\tilde{\theta}-\theta}{\varepsilon}=\left[\frac{dT(\tilde{F})}{d\varepsilon}\right]_{\varepsilon=0},$$

a "derivative" of the functional $T(\tilde{F})$. One interprets the influence curve as measuring the "influence" of x_0 on the parameter θ. That is, it measures the change in θ, in the limit, as the point x_0 is given special emphasis through the weight ε.

Some examples will help to illustrate the usefulness of the influence curve. Consider the parameter $\theta=T_1(F)=\int x dF(x)=\mu$ and $\tilde{\theta}=\tilde{\mu}=\mu+\varepsilon(x_0-\mu)$. Then the influence curve of μ at x_0 is defined as

$$\mathrm{IC}(x_0;\mu)=\lim_{\varepsilon\to 0}\frac{\tilde{\mu}-\mu}{\varepsilon}=\lim_{\varepsilon\to 0}\frac{\mu+\varepsilon(x_0-\mu)-\mu}{\varepsilon}=x_0-\mu.$$

We note that $\mathrm{IC}(x_0;\mu)$ is linear and unbounded in x_0, with a slope of one. A plot of $\mathrm{IC}(x_0;\mu)$ versus x_0 (Figure 3.3), where the linear and unbounded nature of the influence curve are clear.

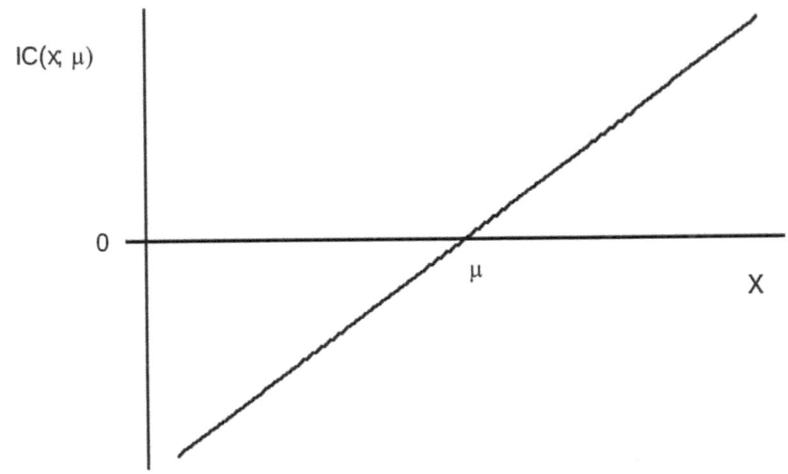

FIGURE 3.3
The influence curve for the mean.

The influence curve for μ could be computed directly as a derivative of $T(\tilde{F})$ with respect to ε as

$$IC(x_0;\mu)=\left[\frac{dT_1(\tilde{F})}{d\varepsilon}\right]_{\varepsilon=0}=\left[\frac{d\tilde{\mu}}{d\varepsilon}\right]_{\varepsilon=0}=\left[\frac{d(\mu+\varepsilon(x_0-\mu))}{d\varepsilon}\right]_{\varepsilon=0}=x_0-\mu,$$

giving the same results as above.

As a second example, consider the variance as a functional $T_4(F)=\int(x-T_1(F))^2\,dF(x)=\sigma^2$. A little algebra is required to find the expression for $\tilde{\sigma}^2$ as

$$\tilde{\sigma}^2=\int(x-\tilde{\mu})^2dF(x)=\int\left[(x-\mu)+\varepsilon(\mu-x_0)\right]^2dF(x)=$$
$$\int\left[(x-\mu)^2+2\varepsilon(x-\mu)(\mu-x_0)+\varepsilon^2(\mu-x_0)^2\right]d\left[(1-\varepsilon)F(x)+\varepsilon\delta_{x_0}(x)\right]=$$
$$(1-\varepsilon)\sigma^2+0+\varepsilon(x_0-\mu)^2+\text{terms of order }\varepsilon^2\text{ or higher.}$$

Thus, $\tilde{\sigma}^2-\sigma^2=\varepsilon\left[(x_0-\mu)^2-\sigma^2+\right]$ terms of ε^2 or higher. Consequently, $IC(x_0;\sigma^2)=\lim_{\varepsilon\to 0}\frac{\tilde{\sigma}^2-\sigma^2}{\varepsilon}=(x_0-\mu)^2-\sigma^2$, an unbounded, quadratic function in x_0, that equals zero at $\mu\pm\sigma$ and is minimized at μ. A plot of this influence curve is presented in Figure 3.4.

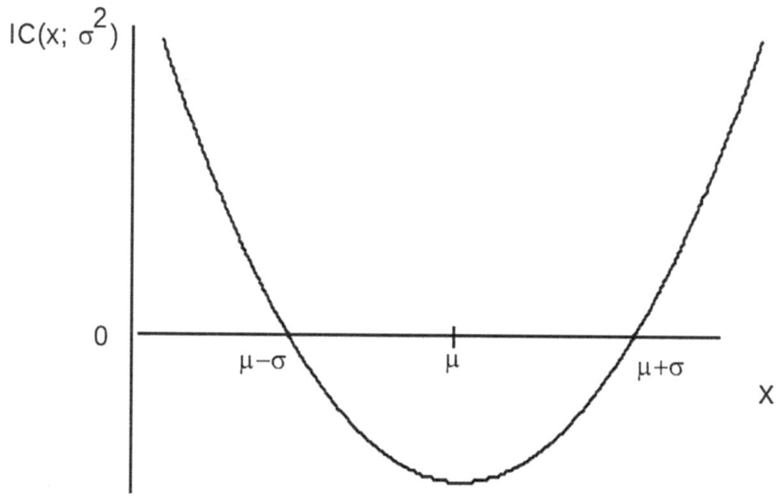

FIGURE 3.4
The influence curve for the variance.

As a final example of computing the influence curve, consider the population median defined earlier in functional form as $T_2(F) = F^{-1}\left(\frac{1}{2}\right) = x_{.5}$. Several methods for computing the influence curve for functionals involve the inverse CDF. The most straightforward method is to differentiate both sides of the following identity. Since $F^{-1}\left(\frac{1}{2}\right) = x_{.5}$, it follows that $F\left(F^{-1}\left(\frac{1}{2}\right)\right) = F(x_{.5}) = 1/2$. Or, expressing the median of the altered CDF, $\tilde{F}$, using the same identity as

$$L = \tilde{F}\left(\tilde{F}^{-1}\left(\frac{1}{2}\right)\right) = \tilde{F}(\tilde{x}_{.5}) = (1-\varepsilon)F(\tilde{x}_{.5}) + \varepsilon\delta_{x_0}(\tilde{x}_{.5}) = 1/2.$$

Differentiating both sides of the above expression with respect to ε results in

$$\frac{dL}{d\varepsilon} = (-1)F(\tilde{x}_{.5}) + (1-\varepsilon)\frac{d\tilde{x}_{.5}}{d\varepsilon}f(\tilde{x}_{.5}) + \delta_{x_0}(\tilde{x}_{.5}) + \varepsilon\frac{d\delta_{x_0}(\tilde{x}_{.5})}{d\varepsilon} = 0,$$

where $f()$ is the PDF of x in the above expression.

Evaluating the above expression at $\varepsilon = 0$ gives us (remembering the chain rule from calculus and that $\tilde{x}_{.5} = x_{.5}$ at $\varepsilon = 0$),

$$(-1)F(x_{.5}) + \left[\frac{d\tilde{x}_{.5}}{d\varepsilon}\right]_{\varepsilon=0} f(x_{.5}) + \delta_{x_0}(x_{.5}) = 0\,.$$

Solving the above expression for the influence curve of the median, $\left[\frac{d\tilde{x}_{.5}}{d\varepsilon}\right]_{\varepsilon=0}$, we obtain

$$\left[\frac{d\tilde{x}_{.5}}{d\varepsilon}\right]_{\varepsilon=0} = \frac{1/2 - \delta_{x_0}(x_{.5})}{f(x_{.5})}\,.$$

By definition of the δ-function we have $\delta_{x_0}(x_{.5}) = \begin{cases} 0 & x_{.5} < x_0 \\ 1 & x_{.5} \geq x_0 \end{cases}$, so that our final expression for the influence curve for the median as

$$\left[\frac{d\tilde{x}_{.5}}{d\varepsilon}\right]_{\varepsilon=0} = \begin{cases} \dfrac{-1}{2f(x_{.5})} & \text{for} \quad x_0 \leq x_{.5} \\ \dfrac{1}{2f(x_{.5})} & \text{for} \quad x_0 > x_{.5} \end{cases}.$$

A plot of this influence function appears in Figure 3.5.

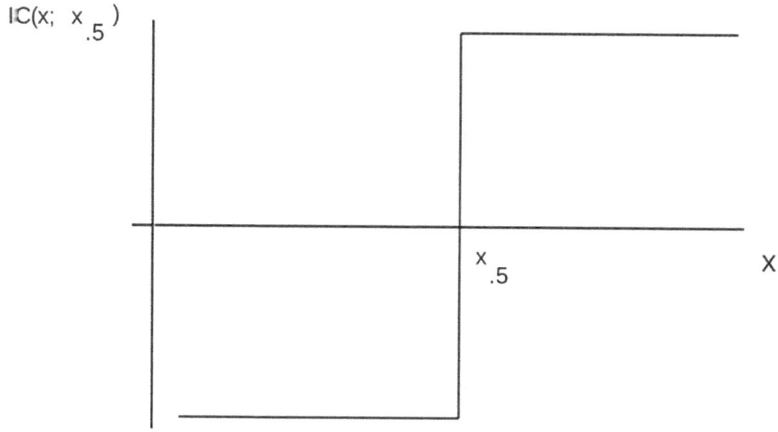

FIGURE 3.5
The influence curve for the median.

We note that the influence of the point x_0, or for simplicity, x, on the median, is either a constant positive value at $\frac{1}{2f(x_{.5})}$ if x is to the right of the median, $x_{.5}$, or a constant negative value at $\frac{-1}{2f(x_{.5})}$ if x is to the left of the median. We also note that this effect is bounded, a very different result than that obtained for the mean where the influence of point x on the mean was linear and unbounded.

The information contained in the mean and median influence curves can be used to explain several of their important features. For example, we know that the sample mean, $\bar{x}$, is the optimal estimator to use if the underlying distribution of the data is normal. In this case, sample data from a normal distribution should begin to "stack" around the true mean, μ, and to trail off symmetrically on both sides of μ. Recalling the fact that the "parameter" $(T(F))$ is simply the "estimator" $(T(F_n))$ evaluated with a very large sample $(T(F) \rightarrow T(F_n))$, the influence curve for the mean tells us that to be fully "efficient" (a concept to be defined later in this chapter), the sample mean must utilize each datum's value in computing the sample mean.

On the other hand, the sample median is not fully efficient if the data are normal, achieving an efficiency of about 64% against the sample mean, a figure to be derived later. The influence curve reveals why this is true. Each datum's value influences the sample median (at n equal to infinity) only by its position to the left or right of the mean (the mean and the median are equal in the normal distribution). This obviously provides less information about the value of the mean than the datum's exact value and accounts for the loss of efficiency of the sample median.

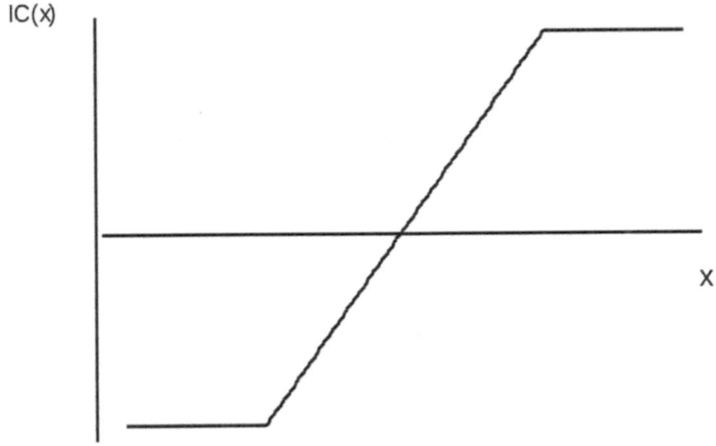

FIGURE 3.6
The influence curve for a resistant estimator.

However, there is another side of the coin. We know that the sample mean is not a resistant estimator of the mean. The sample mean can be greatly affected by outliers in the dataset. The sample median offers protection against these same outliers. The influence curve reveals why this is true. An unusually large or small value of the data will receive its full value in the computation of the sample mean and hence will have a large (in magnitude) "influence" on the value of the sample mean, as indicated by the influence curve for the mean (the sample mean at n equal to infinity). The influence of an outlier on the sample median is bounded at either a constant positive or a constant negative value, regardless of the size of the outlier. Thus, the sample median offers protection against outliers that may result from non-normal data, protection not available through the sample mean.

So the sample mean offers full efficiency against normal data but is nonresistant to outliers, while the sample median lacks the efficiency of the sample mean against normal data but is very resistant to outliers. By comparing the influence curves for the mean and median, we see the features required for an estimator to be efficient against normal data and resistant to outliers. Essentially, we would like an estimator that considers the datum's value so long as that value is "close" to the mean so that the efficiency of the sample mean may be retained. However, we need to limit outliers' influence on the estimator to achieve the desired resistance properties. In other words, we need to consider an estimator with an influence curve that has an appearance like the one in Figure 3.6.

The following sections will give estimators with influence curves similar to the one above.

Another use of the influence curve is its role in determining the distribution properties of estimators that can be written as functionals. This technique is very general and very powerful. It will enable us to determine the distributions for a broad variety of estimators, at least in the asymptotic sense, including the maximum likelihood estimators (MLEs).

Suppose x is our random variable of interest, with CDF $F(x)$, and parameter of interest, written as a functional of $T(F)$. One property possessed by $IC(x;T(F))$ is that its average value is zero, a fact that will soon be shown to be very useful. That is,

$$E(IC(x;T(F))) = \int IC(x;T(F))dF(x) = 0$$

While this result will not be proved, it is easy to demonstrate using the examples previously given.

For the first illustration, consider the functional $T_1(F) = \int x dF(x) = \mu$, where $IC(x;\mu) = x - \mu$. Then $E(IC(x;\mu)) = E(x-\mu) = 0$.

As a second illustration, consider the functional $T_4(F) = \int (x-\mu)^2 dF(x) = \sigma^2$ with $IC(x;\sigma^2) = (x-\mu)^2 - \sigma^2$. It follows that $E(IC(x;\sigma^2)) = E((x-\mu)^2 - \sigma^2) = \sigma^2 - \sigma^2 = 0$.

Finally, we have $T_2(F) = F^{-1}(1/2) = x_{.5}$ and $IC(x;x_{.5}) = \begin{cases} \dfrac{-.5}{f(x_{.5})} & x \le x_{.5} \\ \dfrac{.5}{f(x_{.5})} & x > x_{.5} \end{cases}$. It is easy to show that $E(IC(x;x_{.5})) = 0$.

The distributional properties of our estimator require a bit more effort, but the end results are extremely useful and important. Let $\hat{\theta}_n = T(F_n)$ be our estimator of the parameter $\theta = T(F)$. We will obtain the asymptotic distribution of $\hat{\theta}_n$. Recall from Section 2.4 the notion of using a first-order Taylor series to expand a vector-valued function. If the function f is an *n*-variate function of $\underline{x} = (x_1, \ldots, x_n)'$, then we can expand $f(\underline{x})$ about the point $\underline{\lambda} = (\lambda_1, \ldots, \lambda_n)'$, obtaining

$$f(\underline{x}) \approx f(\underline{\lambda}) + \sum_{i=1}^{n} (x_i - \lambda_i) \left. \frac{\partial f(\underline{x})}{\partial x_i} \right|_{\underline{x}=\underline{\lambda}}.$$

The above result is useful to motivate the notion of a "first-order Taylor series expansion" of the functional $T(G)$, where G is some CDF of x. Since G may be defined for an infinite number of values of x, say for x continuous from $-\infty$ to ∞, we can think of $T(G)$ as being an infinite-variate function. Expanding $T(G)$ about the CDF F results in

$$T(G) \approx T(F) + \int IC(x;T(F)) d(G(x) - F(x)).$$

The analogy between the above functional expansion and the Taylor series expansion of $f(\underline{x})$ is that the differential $d(G(x)-F(x))$ is analogous to $(x_i - \lambda_i)$; the influence curve, $IC(x;T(F))$ (recall that the influence curve is the derivative of the functional), is analogous to the partial derivatives of f; the integral replaces the sum because of the infinite-valued nature of the functional as opposed to the finite n-valued function $f(\underline{x})$.

Since $E(IC(x;T(F))) = \int IC(x;T(F))dF(x) = 0$, the above expression reduces to $T(G) \approx T(F) + \int IC(x;T(F))dG(x)$. This type of expansion is referred to as the Fréchet expansion of $T(G)$ (Huber, 1981). Now we replace the arbitrary CDF G with the ECDF F_n. Our expansion becomes

$$\begin{aligned} T(F_n) &\approx T(F) + \int IC(x;T(F))dF_n(x) \\ &= T(F) + \frac{\sum_{i=1}^{n} IC(x_i;T(F))}{n} \end{aligned}$$

or

$$\sqrt{n}\left(T(F_n) - T(F)\right) \approx \frac{\sum_{i=1}^{n} IC(x_i;T(F))}{\sqrt{n}}.$$

Now the individual values of the influence curve at each x_i, $IC(x_i;T(F))$ are independent and identically distributed (i.i.d.) random variables (because $x_1,..,x_n$ are i.i.d. random variables) with $E(IC(x_i;T(F))) = 0$ and

$$Var(IC(x_i;T(F))) = E(IC^2(x_i;T(F))) = E(IC^2(x;T(F))) = V^2.$$

The central limit theorem tells us that if $x_1,..,x_n$ are i.i.d. with $E(x_i) = 0$ and $Var(x_i) = E(x^2)$, then

$$\frac{\sum x_i}{\sqrt{n}} \xrightarrow{L} N(0, E(x^2)).$$

Applying this result to our situation, we obtain the asymptotic expression for the distribution of $T(F_n)$:

$$\sqrt{n}\left(T(F_n) - T(F)\right) \approx \frac{\sum IC(x_i;T(F))}{\sqrt{n}} \xrightarrow{L} N\left(0, E\left(IC^2(x;T(F))\right)\right).$$

That is, the estimator is asymptotically unbiased for the parameter, and the asymptotic variance of $\sqrt{n}T(F_n)$ is equal to the expected value of the squared influence curve.

Some examples will certainly help to clarify this result. Consider the functional $T_1(F)=\int x dF(x)=\mu$, where $IC(x;\mu)=x-\mu$ and $T_1(F_n)=\bar{x}$. Applying the above result gives the asymptotic distribution

$$\sqrt{n}(\bar{x}-\mu)\xrightarrow{L}N\left(0,E(x-\mu)^2=\sigma^2\right),$$

our basic central limit theorem result. Our theory covers this most basic result.

Consider the median as the second example, where $T_2(F)=F^{-1}(1/2)=x_{.5}$ and $IC(x;x_{.5})=\begin{cases}\dfrac{-.5}{f(x_{.5})} & x<x_{.5}\\ \dfrac{.5}{f(x_{.5})} & x>x_{.5}\end{cases}$.

Here, $T_2(F_n)=M$, the sample median. The asymptotic distribution involving M is

$$\sqrt{n}(M-x_{.5})\xrightarrow{L}N\left(0,V^2\right),$$

where $V^2=E\left(IC^2(x;x_{.5})\right)=\frac{1}{2}\left(\frac{-1}{2f(x_{.5})}\right)^2+\frac{1}{2}\left(\frac{1}{2f(x_{.5})}\right)^2=\frac{1}{4f^2(x_{.5})}$,

a useful result if one needs to perform inference using the sample median.

As a final example, consider estimating the location parameter $T(F)=\theta$ using the MLE $T(F_n)=\hat{\theta}_n$. Section 3.8 will show that all MLEs can be written as functionals; hence, this section's theory applies to all MLEs. We will show that the influence curve for the MLE is

$$IC(x;\theta)=\frac{\dfrac{\partial \ln(f(x|\theta))}{\partial\theta}}{I(\theta)},$$

where $I(\theta)=E\left(\dfrac{\partial \ln(f(x|\theta))}{\partial\theta}\right)^2$ is the Fisher information number discussed in Section 2.2. Thus, the asymptotic variance for the MLE of $T(F)=\theta$ is

$$E\left(IC^2(x;\theta)\right)=\frac{E\left(\dfrac{\partial \ln(f(x|\theta))}{\partial\theta}\right)^2}{I^2(\theta)}=I^{-1}(\theta).$$

The same result is presented in Section 2.2. And our asymptotic distribution for the MLE $\hat{\theta}_n$ is

$$\sqrt{n}\left(\hat{\theta}_n - \theta\right) \xrightarrow{L} N\left(0, I^{-1}(\theta)\right),$$

which again agrees with the results presented earlier.

The theory that we have discussed in this section applies to a wide variety of estimators, ranging from the sample mean, sample variance, and sample median to the MLEs. There are many other estimators of the location parameter, $T(F) = \theta$, that we have yet to discuss. We shall see that the power of this approach will enable us to examine the properties of most of these other estimators.

3.3 Some Finite Sample Versions of the Influence Curve

One of the difficulties with studying the influence curve is that it relates to the behavior of the parameter when subjected to a small amount of contamination (of the amount ε) at point x. What we really want to study is the behavior of the estimator when the dataset contains outliers. It was mentioned earlier in this chapter that the parameter is really an estimator evaluated at an infinite sample size. Hence, the influence curve reflects the behavior of the estimator based on a very large sample. This approach has some appeal but is not as intuitive as studying the estimator based on a finite sample size. Unfortunately, it is actually more difficult to study the robustness properties of an estimator when based on a finite sample due to the fact that one must consider the infinite number of combinations of sample sizes together with the individual sample values themselves. Nonetheless, useful, intuitive information can be extracted from the various finite sample versions of the influence curve that will be helpful in aiding our interpretation of the influence curve itself.

Many of the following finite sample versions of the influence curve are discussed in Mallows's unpublished technical report (1975) and reviewed in detail in Cook and Weisberg (1982).

The first version that we will present is the empirical influence curve, or EIC, of the estimator $T(F_n)$ at the point x, defined by replacing F by F_n in the influence curve or

$$EIC\left(x; T\left(F_n\right)\right) = \lim_{\varepsilon \to 0} \frac{T\left(\tilde{F}_n\right) - T\left(F_n\right)}{\varepsilon}$$

So, for example, if the functional is $T_1(F) = \int x dF(x) = \mu$, then the estimator is $T_1(F_n) = \int x dF_n(x) = \frac{\sum x_i}{n} = \bar{x}$. It is easy to show that if $\tilde{F}_n(x) = (1-\varepsilon)F_n(x) + \varepsilon\delta_{x_0}(x)$, then $T_1(\tilde{F}_n) = \bar{x} + \varepsilon(x_0 - \bar{x})$ and, consequently, that $EIC(x_0; T_1(F_n)) = x_0 - \bar{x}$, where x_0 is just some specific value of x. Note that this EIC is now a function of x_0 and the estimator $\bar{x}$ and converges to the influence curve for the parameter μ as n goes to infinity. This is as expected since $F_n(x) \to F(x)$ as $n \to \infty$.

As a second example, consider the variance as a functional, $T_4(F) = \int (x-\mu)^2 dF(x) = \sigma^2$. It is easy to show using the same approach as was used in deriving the influence curve for σ^2 that $EIC(x_0; T_4(F_n)) = (x_0 - \bar{x})^2 - s^2$, where $s^2 = \frac{\sum (x_i - \bar{x})^2}{n}$. Again, we note that the EIC is a function of the estimator (not the parameter) and converges to the influence curve as $n \to \infty$.

A second version of a finite sample representation of the influence curve is the sensitivity curve, according to Tukey (1970). Here, the behavior of an estimator is evaluated when an additional hypothetical observation is added. The sensitivity curve for the estimator $T(F_n) = \hat{\theta}_n$ at the point x is defined as

$$SC(x; \hat{\theta}_n) = (n+1)\left(\hat{\theta}_{n+1} - \hat{\theta}_n\right),$$

where $\hat{\theta}_n$ is the estimator of θ based on the observations $x_1, \ldots, x_n$, and $\hat{\theta}_{n+1}$ is the estimator of θ based on the original n observations plus an additional observation at the value x. That is, the sensitivity curve of the estimator is simply the standardized difference between the estimator evaluated at the original observations and the estimator evaluated at the original observations plus an additional arbitrary point.

As the first example, consider the sample mean, $\hat{\theta}_n = \bar{x}_n = \frac{\sum x_i}{n}$. Then $\hat{\theta}_{n+1} = \bar{x}_{n+1} = \frac{(\sum x_i) + x}{n+1}$ and the sensitivity curve for the sample mean is

$$\begin{aligned} SC(x; \bar{x}_n) &= (n+1)(\bar{x}_{n+1} - \bar{x}_n) = (n+1)\left(\frac{(\sum x_i) + x}{n+1} - \frac{\sum x_i}{n}\right) \\ (n+1)&\left(\frac{n}{n+1}\bar{x}_n + \frac{x}{n+1} - \bar{x}_n\right) = (n+1)\left(\frac{x}{n+1} - \frac{\bar{x}_n}{n+1}\right) \\ &= x - \bar{x}_n \end{aligned}.$$

We note that $SC(x; \bar{x}_n)$ equals $EIC(x; \bar{x}_n)$, as is often the case for other estimators as well. It is noted that the sensitivity curve for the sample mean is a function of the estimator and is linear and unbounded in x. Intuitively, the justification for the nonresistance of the sample mean to outliers is clear. The "effect" of x on the standardized difference in the sample means computed with and without x is unbounded and tends to be very large as $|x|$ increases.

Consider the sensitivity curve of the sample variance as a second example. If $T(F_n) = s_n^2 = \frac{\sum (x_i - \bar{x}_n)^2}{n}$, then it follows if x is added to the original n point dataset, that $\bar{x}_{n+1} = \left(\frac{n}{n+1}\right)\bar{x}_n + \frac{x}{n+1}$ and that

$$s_{n+1}^2 = \frac{1}{n}\left(\sum_{i=1}^{n}(x_i - \bar{x}_{n+1})^2 + (x - \bar{x}_{n+1})^2\right).$$

Substitution of the expression for $\bar{x}_{n+1}$ into the above equation, followed by some algebra, results in

$$s_{n+1}^2 = \left(\frac{n-1}{n}\right)s_n^2 + \left(\frac{1}{n+1}\right)(x - \bar{x}_n)^2.$$

Thus, it follows that the sensitivity curve for the sample variance, as a function of x, is

$$SC(x; s_n^2) = (n+1)(s_{n+1}^2 - s_n^2) = (x - \bar{x}_n)^2 - \left(\frac{n+1}{n}\right)s_n^2.$$

We note that this sensitivity curve is an unbounded, quadratic function of x and that this function converges to $IC(x;\sigma^2)$ as $n \to \infty$.

As a third example, consider the sample median. We recall that the influence curve for the median was a negative constant for x less than the median and a positive constant for x greater than the median. The sensitivity curve provides us with a useful interpretation of the influence curve. If we consider a simple example where our dataset has only four values, $x_{(1)}, x_{(2)}, x_{(3)}$, and $x_{(4)}$, then the sample median is $M_n = M_4 = \frac{x_{(2)} + x_{(3)}}{2}$. If we add a new data point, x, and vary x from $x = -\infty$ to $x = \infty$, we see that $M_5 = x_{(2)}$ for $x < x_{(2)}$; $M_5 = x$ for $x_{(2)} \le x \le x_{(3)}$; and $M_5 = x_{(3)}$ for $x > x_{(3)}$. Thus, the sensitivity curve (divided by $(n+1)$) for the sample median at $n = 4$ is defined as

$$\frac{SC(x;\dot{x}_n)}{n+1} = \begin{cases} x_{(2)} - M_4 & \text{for} \quad x < x_{(2)} \\ x - M_4 & \text{for} \quad x_{(2)} \le x \le x_{(3)} \\ x_{(3)} - M_4 & \text{for} \quad x > x_{(3)} \end{cases}.$$

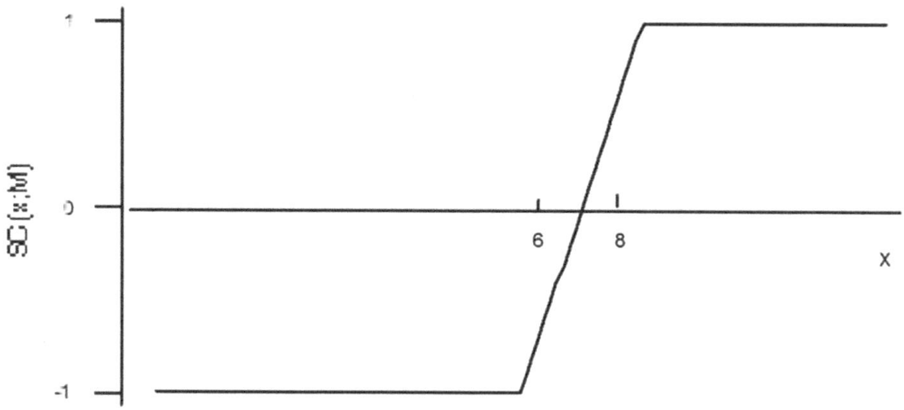

FIGURE 3.7
The sensitivity curve for the sample median based on a sample of size four.

A sketch of this sensitivity curve appears in Figure 3.7, where the four sample observations were taken to be 1, 6, 8, and 9.

The next step in the process is to allow n to get large. The difference between the two middle points decreases as the data begins to "stack up" around the true population median. Consequently, the slope of the middle portion of the above plot gets steeper. In the limit, the sample median's sensitivity curve approaches the population median's influence curve. Thus, we see that any value of x to the left of M_n causes the median to decrease by a constant amount, and any value of x to the right of M_n causes the median to increase by a constant amount. The increase or decrease amounts are bounded, no matter the x value.

The final finite sample version of the influence curve is called the finite sample influence curve. The finite sample influence curve of the estimator, $\hat{\theta}_n$, based on a sample of size n, at the point x_i is computed by comparing the estimator based on all n observations to the estimator based on $n - 1$ observations with the ith point removed. That is,

$$\mathrm{SIC}\left(x_i;\hat{\theta}_n\right) = (n-1)\left(\hat{\theta}_n - \hat{\theta}_{n-1,-i}\right),$$

where $\hat{\theta}_{n-1,-i}$ is the estimator using all the data except the ith point. Readers familiar with multiple linear regression will notice the similarity between the sample influence curve (SIC) and the influence diagnostics commonly used in regression. Both are based on this "single point deletion scheme."

To illustrate the computation of the SIC for $\hat{\theta}_n = \overline{x}_n$, we apply the definition of the SIC to obtain

$$\text{SIC}(x_i;\overline{x}_n) = (n-1)(\overline{x}_n - \overline{x}_{n-1,-i})$$
$$= (n-1)\left(\frac{\sum_{j=1}^{n} x_j}{n} - \frac{\sum_{j=1,j\neq i}^{n} x_j}{n-1}\right).$$

Now add and subtract the term $\frac{x_i}{n-1}$ within the expression in the rightmost parentheses to obtain

$$(n-1)\left(\overline{x}_n - \frac{\sum_{j=1}^{n} x_j}{n-1} + \frac{x_i}{n-1}\right) = (n-1)\left(\frac{x_i}{n-1} - \frac{\overline{x}_n}{n-1}\right) = x_i - \overline{x}_n,$$

for $i = 1,\ldots,n$.

Thus, the standardized difference between the sample means, computed with and without the data point x_i, is just $x_i - \overline{x}_n$. It is worth noting that the SIC for the sample mean is of the same form as the EIC and the sensitivity curve (SC) for the sample mean and that, as $n \to \infty$, the SIC converges to the influence curve for the population mean. Furthermore, the SIC is clearly sample dependent since it can only be computed at the n observed data points.

The SIC may be computed for other estimators, especially those that we have used to illustrate the other finite sample versions of the influence curves. It does not seem necessary to do so here. The reader surely has the idea that the three versions presented here lead to very similar results no matter what estimator is used. Mallows (1975) has shown that all three versions are asymptotically equivalent and consistent estimates of the influence curve for the appropriate population parameter. In subsequent sections of this chapter, we will use the influence curve to evaluate the robustness properties of a variety of estimators.

3.4 Robustness: General Concepts

We will now begin to develop a more structured approach toward the evaluation of the robustness properties of an estimator in the location model. As mentioned earlier, the primary tool for this evaluation is the influence curve. This approach was developed in the mid-1970s by Frank Hampel. The reader is referred to Hampel (1974) and Hampel et al. (1986) for a much more thorough discussion than will be presented here. Our goal in this section is to familiarize the user of statistics with some of the techniques helpful in evaluating the robustness properties of an estimator.

The traditional approach to mathematical statistics assumes that the data are i.i.d. observations from some idealized distribution, such as the normal. Mathematics is then used to find the optimal estimators – for example, the MLEs – of the unknown parameters. For example, the location parameter and variance are the parameters of interest in the location model. One would hope that the estimators obtained, optimal under one set of conditions, would be nearly optimal if the ideal set of conditions were not met. For example, the sample mean and the sample variance (dividing by n, not $n-1$) are the MLEs for the location model if the observations are i.i.d. from a normal distribution. The interesting question is, if the data are not normal but observations from some other symmetric, bell-shaped distribution, are these estimators still reasonable?

One hopes that some sort of continuity principal applies, where so long as the true distribution is "close" to the ideal distribution (usually the normal), the estimators are "close" to optimal. This is often considered the definition of a robust procedure that satisfies this continuity principal. Unfortunately, optimal procedures, such as MLEs, often fail to satisfy this key property.

Notice that our concern in this discussion is with the failure of the distribution assumption to hold. This, obviously, is only one of several ways in which the assumed conditions for the model may fail. Other possibilities include a nonconstant variance, nonindependent observations, nonidentical distributions, and all combinations of these violations. Some work on robust estimation has been done under each of these conditions, but the major emphasis has been on the failure of the distribution assumption, called "distributional robustness," since many experts consider it as the most important violation.

The distribution robustness properties of an estimation procedure fall into two major categories: qualitative and quantitative. In qualitative robustness, the concern is with the estimators' ability to satisfy the continuity principal mentioned above. Quantitative robustness is concerned with numerical summaries that describe certain robustness features of the estimator. These features are usually measured from the estimator's influence curve.

Qualitative robustness, as in "continuity" arguments from calculus, involves "$\varepsilon-\delta$" type discussions. One such qualitative robustness property of an estimator is that if two empirical cumulative distributions, F_n and G_n, are "close" (within ε), then the estimators based on these two ECDFs, $T(F_n)$ and $T(G_n)$, should be "close" (within δ) as well. Closeness is measured by some appropriate metric, such as the Prohorov or Levy metrics. The arguments based on this theory can be quite theoretical and are beyond the scope of this text. The interested reader is referred to the book by Huber (1981) for more details.

On the other hand, quantitative robustness utilizes features of the influence curve, such as its slope and other shape characteristics that are much easier to calculate once the influence curve is obtained. The following is a summary of the primary characteristics of interest to us.

The "gross error sensitivity" of the estimator measures the worst influence that a fixed amount of contamination can have on the value of the estimator. The gross-error sensitivity is denoted by γ^* and defined as

$$\gamma^* = \sup_x \mid (\mathrm{IC}(x;\theta)) \mid.$$

Thus, γ^* simply measures the largest value, in magnitude, that the influence curve can take, over all possible values of x. For example, for the mean where $\theta = \mathrm{T}_1(\mathrm{F}) = \mu = \int x\mathrm{dF}(x)$, recall that $\mathrm{IC}(x;\mu) = x - \mu$, an unbounded function in x. Hence, it follows that $\gamma^* = \infty$. On the other hand, for the median where $\theta = \mathrm{T}_2(\mathrm{F}) = x_{.50} = \mathrm{F}^{-1}(.5)$, we found the influence curve to be a bounded function with bounds of $1/2f(x_{.50})$ (for $x > x_{.5}$) and $-1/2f(x_{.50})$ (for $x < x_{.5}$).

It follows that $\gamma^* = \dfrac{1}{2\mathrm{f}(x_{0.5})}$. Now, if the density of x is the standard normal, then

$$\gamma^* = \sqrt{2(3.14159)}\,/\,2 = 1.253.$$

It turns out that if the $\mathrm{N}(0,1)$ distribution is used for the density, $f(x)$, then the minimum value of γ^* is 1.253 value for the median. From a practical viewpoint, γ^* will be a finite value so long as the influence curve for the estimator (at n equal to infinity) is bounded. If the influence curve is unbounded, as is the case for the mean, γ^* is infinity and the estimator lacks resistance to outliers. Smaller values of γ^* imply that the estimator is more resistant than larger values. The sample median is, therefore, very resistant to outliers, a fact that we have already observed.

A second quantitative robustness characteristic is the "local-shift-sensitivity," denoted by λ^*, which measures the effect of "wiggling around" the observations, such as those that would occur with the rounding or grouping of data. It is defined as

$$\lambda^* = \sup_{x \neq y} \frac{|\mathrm{IC}(x;\theta) - \mathrm{IC}(y;\theta)|}{|x - y|}.$$

We see that λ^* is a measure of the effect on the estimator of moving the contamination from the point x to the point y, standardized by the distance moved, $x - y$. Technically, λ^* is referred to as a first-order Lipschitz constant, but if the influence curve is continuous, then λ^* is just the maximum absolute first derivative of the influence curve. That is, $\lambda^* = \max|\mathrm{IC}'(x;\theta)|$ over all values of x, where $\mathrm{IC}'(x;\theta) = \dfrac{\mathrm{dIC}(x;\theta)}{\mathrm{dx}}$.

For example, consider the mean with $IC(x;\mu) = x - \mu$ $IC'(x;\mu) = 1$, for all x. Thus, $\lambda^* = 1$. This is the best possible value of λ^*. On the other hand, if we consider the median, it can be shown (see the problems at the end of this section) that λ^* is $+\infty$, the worst possible value. This value results from the "jump" of $1/f(x_{.5})$ that occurs at $x_{.5}$.

A third measure that is used in quantitative robustness is a measure of the percentage of contamination, free to take any values that will cause the estimator to be unreliable. This measure is called the "breakdown point" or "breakdown bound" and measures the minimum fraction of free contamination that will carry the estimator over all bounds. Thus, the breakdown point, denoted by δ^*, measures the smallest amount of free contamination that will cause the estimator to "blow up" to infinity.

For example, one data point, an extremely severe outlier, at plus infinity will cause the sample mean also to be plus infinity. Or, from the infinite sample size viewpoint, when we use the influence curve as our diagnostic tool, the mean (recall that the estimator becomes the parameter at n of infinity) will be infinite for any nonzero contamination fraction at infinity. Consequently, δ^* is zero, the worst possible value for δ^*. The sample median, on the other hand, can tolerate up to half the data at plus infinity and still remain finite. Thus, in an asymptotic sense, δ^* is 0.50, the best possible value for δ^*. A breakdown point of 0.5 clearly indicates the strong resistance property of the sample median to outliers.

The estimators' ability to identify and "throw out" outliers is measured by the "rejection point," denoted by ρ^*. The estimator will reject outliers if there is a region outside of which the influence curve is identically zero and ρ^* is defined as the boundary of the smallest region with this property. For both the mean and the median, ρ^* is plus infinity since neither of these estimators rejects outliers. Some estimators introduced in the next section will have a finite rejection point.

The fifth and final quantitative robustness property is the asymptotic variance of the estimator, $V^2 = E(IC^2(x;\theta))$, another quantity computed from the influence curve for the estimator. Values for V^2 the mean and the median were computed in Section 3.2.

The following sections will propose many different estimators for the location parameter. They will be compared through the calculation of these five quantitative properties.

3.5 Robustness and the Location Model

We will now begin to address the issue of estimating the location parameter in a robust fashion. We first alter the usual location model to reflect our belief that the data may be observations from one of many possible distributions instead of one specific distribution, such as the normal. In part, the normal distribution has long been justified as an appropriate assumed distribution because real data often exhibits many of the characteristics represented by the normal curve. In particular, data are often reasonably symmetric and "bell shaped." However, real data also often contains unusual values or outliers caused by a variety of reasons, ranging from data entry errors to equipment malfunctions. The normal distribution model is not flexible enough to account for these data anomalies. It seems ridiculous to insist on finding optimal methods for a model that is unreasonable, at best, for many situations. Our goal is then to alter the location model in such a way that it is more realistic to situations encountered in practice. Rather than restrict the distribution of the data to one specific distribution, we alter our model to the more general case, where the data are random observations from one of many possible distributions, ranging from the normal to other symmetric, "heavy-tailed" distributions that are, in some sense, "close" to the normal distribution. The altered location model is

$$y_i = \theta + \varepsilon_i,$$

where the ε_i are assumed to be i.i.d. errors from some distribution F, where F is a member of the family of symmetric, "heavy-tailed" distributions. This is a very rich family of distributions, as will be demonstrated shortly.

Heavy-tailed distributions, which include the *t*-distribution family, are distributions that place more mass in the tails of the distribution than placed by the normal distribution. Thus, the term "heavy-tailed" refers to the tail behavior with respect to the normal curve. This point was illustrated in Chapter 2. An example of an extremely heavy-tailed distribution is the Cauchy, a *t*-distribution with one degree of freedom. Figure 3.8, obtained using R, compares the Cauchy to the $N(0,1)$. The heavy-tailed nature of the Cauchy is very apparent.

The philosophy behind this choice of distributions to model outliers is clear if one observes random samples from these distributions. The R program can be easily used to demonstrate this point. Consider, for example, generating a random sample of size $n = 10$ from a standard normal, $N(0,1)$, and a *t*-distribution with one degree of freedom (a *t* with one degree of freedom is the same as a Cauchy distribution). On R, type as follows:

```
# Generate random samples and add an outlier to the
Cauchy distribution
set.seed(42) # For reproducibility
n <- 10
sample_normal <- rnorm(n, mean = 0, sd = 1)
sample_cauchy <- rcauchy(n)
sample_cauchy <- c(sample_cauchy, 15.7) # Add an outlier
# Stack the samples
```

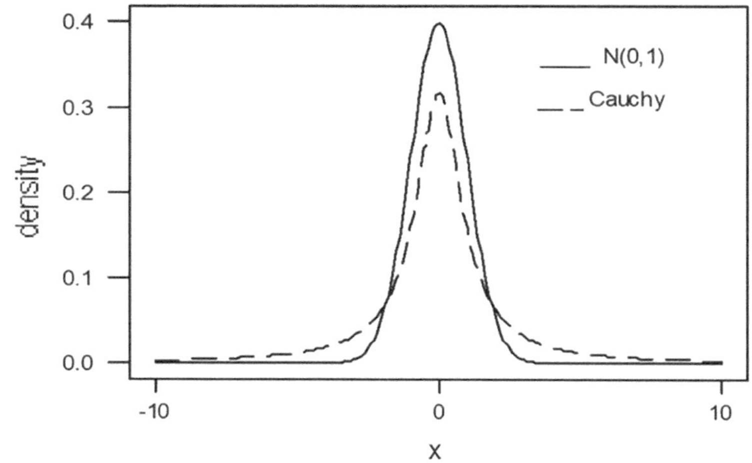

FIGURE 3.8
The Cauchy and the N(0, 1) densities.

```
data <- data.frame(Value = c(sample_normal,
  sample_cauchy),
  Distribution = c(rep("N(0,1)", n), rep("Cauchy", n +
1)))
# Calculate descriptive statistics
desc_stats <- data %>%
  group_by(Distribution) %>%
  summarise( Min = min(Value),Q1 = quantile(Value, 0.25),
    Median = median(Value),Mean = mean(Value),Q3 =
  quantile(Value, 0.75),Max = max(Value),SD = sd(Value))
print(desc_stats)
# Create horizontal dot plots
ggplot(data, aes(x = Value, y = Distribution)) + geom_
  jitter(width = 0.2, height = 0.1) +ggtitle("Horizontal
  Dot Plot by Distribution") + theme_minimal()
# Create horizontal box plots
ggplot(data, aes(x = Value, y = Distribution)) + geom_
  boxplot() +
```

The summary statistics in Figure 3.9, dot plots, and box plots reveal the nature of sampling from such a heavy-tailed distribution as the Cauchy, as the outliers are readily apparent. We see that these heavy-tailed distributions are much more likely to give rise to outliers in small samples than we can expect from the normal distribution. Thus, the altered location model provides a more realistic model for outliers of the type commonly observed in real datasets.

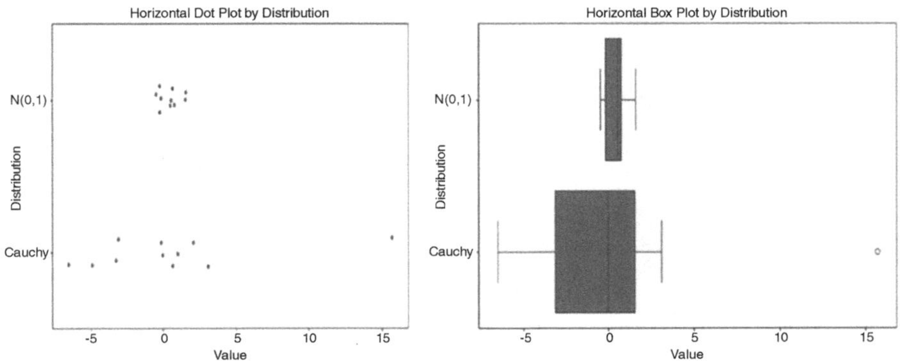

FIGURE 3.9
Dot plot and box plot for normal and Cauchy.

Some members of this family include the so-called ε-contaminated normal distributions, which are written in the form

$$CN(x|\varepsilon) = (1-\varepsilon)N(0,1) + \varepsilon h(x),$$

where $h(x)$ is a symmetric density. The above distribution has the following interpretation: with probability (1- ε), the random variable, x, comes from the standard normal, and with probability ε, x comes from the density $h(x)$.

A special case of the ε-contaminated family of distributions is the contaminated normal family, first used by Tukey (1949), where the function $h(x)$ is also a normal distribution with a variance greater than one. Thus, the contaminated normal distributions are written as

$$CN(x;\varepsilon,\sigma^2) = (1-\varepsilon)N(0,1) + \varepsilon N(0,\sigma^2),$$

where $\sigma^2 > 1$. The $N(0,\sigma^2)$ distribution is called the contaminating distribution, where ε represents the amount of contamination and σ^2 is the size of the contamination. The contaminated normals are a useful family for representing outliers in the location model and other models. Many simulation studies designed to investigate the robustness properties of various estimators have utilized the contaminated normal family. Typically, in these studies, a variety of values of ε and σ^2 are used to represent those situations, ranging from where outliers are rare and when they occur, are small, to those situations where outliers are frequent and large. For the typical "mild" outlier case, one might consider ε of 0.01 to 0.05 with σ^2 in the range of 2–5, and for the "severe" outlier case, ε from 0.10 to 0.25 and σ^2 ranging from 10 to 20. Simulation studies then often compare the performance of various estimators across a variety of mild, moderate, and severe outlier models by varying the combinations of ε and σ^2.

A plot comparing the Cauchy to a $CN(x;.25,25)$ appears in Figure 3.10. The necessary R code is as follows:

```
# Generate the x values
x <- seq(-10, 10, by = 0.1)
# Calculate the PDFs
pdf_norm_0_1 <- dnorm(x, mean = 0, sd = 1)
pdf_norm_0_5 <- dnorm(x, mean = 0, sd = 5)
pdf_cauchy <- dcauchy(x)
# Create a combination of the two normal distributions
combined_norm <- 0.75 * pdf_norm_0_1 + 0.25 *
pdf_norm_0_5
# Plot the results
data <- data.frame(x = rep(x, 3),density = c(combined_
norm, pdf_cauchy, combined_norm),distribution =
rep(c("Combined Normal", "Cauchy", "Combined Normal"),
each = length(x)))
ggplot(data, aes(x = x, y = density, color =
distribution)) +
  geom_line() + scale_color_manual(values = c("Combined
Normal" = "blue", "Cauchy" = "red")) +ggtitle("Comparison
of Combined Normal Distribution and Cauchy Distribution")
+xlab("x") +
  ylab("Density") + theme_minimal() + theme(legend.
title = element_blank())
```

This plot (Figure 3.10) reveals that the contaminated normal distribution with parameters 0.25 and 25 is nearly as heavy tailed as the Cauchy curve with the narrower center portion in the plot. We will see shortly that the Cauchy is really much heavier tailed, however.

The R program is especially convenient for generating random samples from the contaminated normal family. To generate n observations from $CN(x;\varepsilon,\sigma^2)$, we first generate n random observations from the two distributions, $N(0,1)$ and $N(0,\sigma^2)$, and then for each of the n pairs of normals, generate n Bernoulli random variables with probability $P = 1-\varepsilon$ to determine which normal random variable of the pair is to be used. For example, to generate ten random observations from a $CN(x;.25,25)$, type as follows:

```
# Set the parameters
n <- 10
epsilon <- 0.25
sigma <- 5
# Generate random observations from the two normal
distributions
set.seed(42) # For reproducibility
```

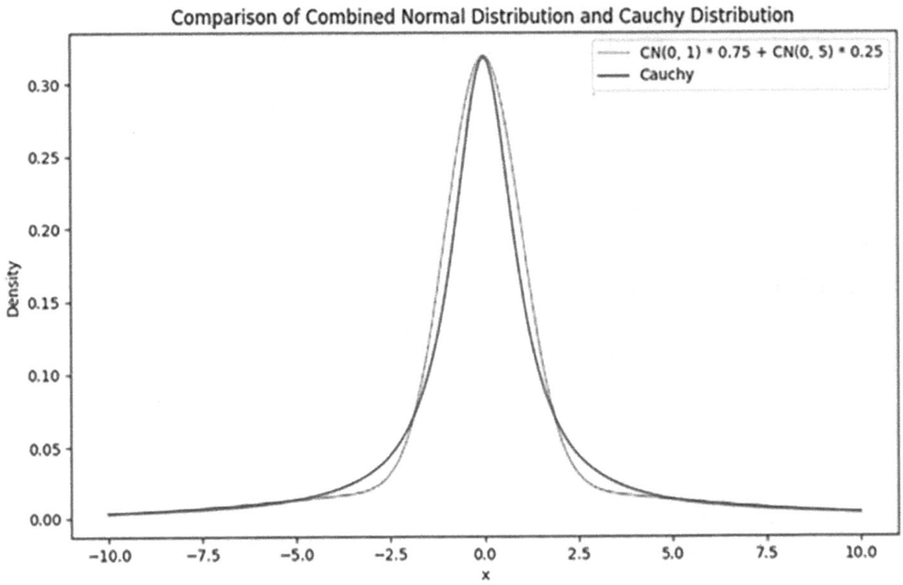

FIGURE 3.10
Plot of the Cauchy and the CN(x; 0.25, 25) densities.

```
normal_01 <- rnorm(n, mean = 0, sd = 1)
normal_0sigma <- rnorm(n, mean = 0, sd = sigma)
# Generate Bernoulli random variables
bernoulli <- rbinom(n, size = 1, prob = 1 - epsilon)
# Create the contaminated normal sample
contaminated_normal <- normal_01 * bernoulli +
normal_0sigma * (1 - bernoulli)
# Combine all into a DataFrame for better visualization
data <- data.frame(
  N_0_1 = normal_01,
  N_0_25 = normal_0sigma,
  Bernoulli = bernoulli,
  CN_0.25_25 = contaminated_normal)
# Print the result
print(data)
# Create the dot plot and box plot for the contaminated
normal sample
# Dot plot
ggplot(data, aes(x = CN_0.25_25, y = 1)) + geom_
jitter(height = 0.1) +ggtitle('Dot Plot of Contaminated
Normal Sample') +
```

```
  xlab('Value') + ylab('') +theme_minimal() +theme(axis.
text.y = element_blank(), axis.ticks.y =
element_blank())
# Box plot
ggplot(data, aes(x = CN_0.25_25, y = 1)) +geom_boxplot()
+ ggtitle('Box Plot of Contaminated Normal Sample') +
xlab('Value') + ylab('') +theme_minimal() +theme(axis.
text.y = element_blank(), axis.ticks.y =
element_blank())
```

The output is as follows:

	N(0,1)	N(0,25)	Bernoulli	CN(.25,25)
0	0.496714	-2.317088	1	0.496714
1	-0.138264	-2.328649	0	-2.328649
2	0.647689	1.209811	1	0.647689
3	1.523030	-9.566401	1	1.523030
4	-0.234153	-8.624589	1	-0.234153
5	-0.234137	-2.811438	1	-0.234137
6	1.579213	-5.064156	1	1.579213
7	0.767435	1.571237	1	0.767435
8	-0.469474	-4.540120	1	-0.469474
9	0.542560	-7.061519	0	-7.061519

It is seen from Figure 3.11 that the data contains two extreme outliers, causing the standard deviation to be greatly inflated. The outliers are nearly equally spaced around the true mean of zero, and consequently, the sample mean is not too distorted in this example.

On a more theoretical level, it may be necessary to derive the properties of the random variable x when the PDF of x is a contaminated normal. For example, suppose we need $\mathrm{E}(x)$ and $\mathrm{var}(x)$ when $x \sim \mathrm{CN}(x;\varepsilon,\sigma^2)$. To find $\mathrm{E}(x)$, recall that the differential operator, d, is a linear operator, and we note that

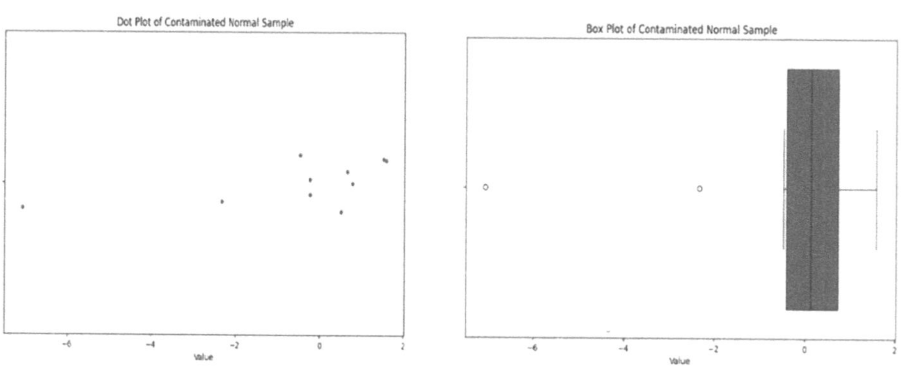

FIGURE 3.11
Dot plot and box plot for the contaminated normal.

$$\begin{aligned}E(x) &= T_1(F) = \int x dF(x) = \int x d\left[CN(x;\varepsilon,\sigma^2)\right] \\ &= \int x d\left[(1-\varepsilon)N(0,1) + \varepsilon N(0,\sigma^2)\right] \\ &= (1-\varepsilon)\int x dN(0,1) + \varepsilon \int x dN(0,\sigma^2) \\ &= 0 + 0 = 0.\end{aligned}$$

To obtain $\mathrm{var}(x)$, we note that

$$\begin{aligned}\mathrm{var}(x) &= E(x^2) - E(x) = E(x^2) \\ &= \int x^2 d\left[CN(x;\varepsilon,\sigma^2)\right] \\ &= (1-\varepsilon)\int x^2 dN(0,1) + \varepsilon \int x^2 dN(0,\sigma^2) \\ &= (1-\varepsilon) + \varepsilon\sigma^2 = 1 + \varepsilon(\sigma^2 - 1).\end{aligned}$$

Thus, it is easy to see that if $\sigma^2 > 1$ that $\mathrm{var}(x) > 1$.

Another family of heavy-tailed distributions, also introduced by Tukey, is the "normal/independent" distributions, where a standard normal (x) is divided by a random variable (y), which is independent of x – that is, $z = x / y$, where z is the normal/independent random variable. Several examples include the standard normal, where y is identically equal to one. Also included are the t-distributions, with *df* degrees of freedom, where y is the square root of a chi-square with *df* degrees of freedom divided by *df*. Thus, a Cauchy distribution, a t with one degree of freedom, is a $N(0,1)$ variable divided by the absolute value of an independent $N(0,1)$ variable, the square root of a chi-square variable with one degree of freedom.

The contaminated normals are also members of the normal/independent family with y equal to one with probability $1-\varepsilon$ and $1/\sigma$ with probability ε. The double exponential distribution is included in the family, where the density of y is $f(y) = y^{-3}\exp(-y^2/2)$. The double exponential is also commonly used in simulation studies as a model for outliers. Another member of this family is the "slash" distribution, where y is a uniform random variable on the $0-1$ interval. The slash distribution is characterized by having a center similar to the $N(0,1)$ and tails similar to the Cauchy. This distribution was first used in the Princeton Robustness Study (Andrews et al., 1972), the first major study to evaluate the robustness properties of a large variety of estimators. Some results of that study will be presented later in this chapter.

The reader will recall that the kurtosis parameter, introduced in Chapter 1, can be used as a measure of the tail behavior of the distribution of the random variable x. Recall that the kurtosis parameter is denoted by β_2 and defined as

$$\beta_2 = \frac{E(x-\mu)^4}{\left(E(x-\mu)^2\right)^2},$$

where $\mu = \mathrm{E}(x)$, the expected value of x. Since the normal curve has $\beta_2 = 3$, the "heavier-tailed" distributions introduced above should have $\beta_2 > 3$ for all distributions in the family. This is certainly true for the normal/independent family, and, thus, the contaminated normals, as well, since

$$\beta_2 = \frac{\mathrm{E}(z)^4}{\left(\mathrm{E}(z)^2\right)^2} = \frac{\mathrm{E}\left(x^4\right)\mathrm{E}\left(y^{-4}\right)}{\left(\mathrm{E}\left(x^2\right)\right)^2 \mathrm{E}\left(y^{-2}\right)^2} = \frac{3\mathrm{E}\left(y^{-4}\right)}{\mathrm{E}\left(y^{-2}\right)^2} \geq 3$$

since $\mathrm{var}\left(y^{-2}\right) = \mathrm{E}\left(y^{-4}\right) - \mathrm{E}\left(y^{-2}\right)^2 \geq 0$.

In the next section, we will begin our study of the robustness properties of estimators of the location parameter by introducing the concept of robust efficiency.

3.6 Robust Efficiency

In evaluating the properties of any estimator, we need to evaluate certain criteria that characterize its performance in a meaningful way. For example, in Chapter 2, we discussed the concepts of expected value, variance, and the mean square error of an estimator. Thus, if two estimators are both unbiased for the parameter of interest, the preferred estimator would be the one with the smaller variance.

One useful evaluation of an unbiased estimator is to compare the variance of the estimator to the variance of the "optimal" estimator for a given situation. Let $\hat{\theta}_n$ be the estimator of interest, and recall the result based on the Cramer-Rao lower bound (CRLB), from Section 2.2, which states that $\mathrm{var}\left(\hat{\theta}_n\right) \geq n^{-1} I^{-1}(\theta)$, where $I(\theta) = -E\left(\frac{\partial^2 \ln f(y|\theta)}{\partial \theta^2}\right)$ is the Fisher information number at the distribution f. We will find it useful to restate this expression so that the right-hand side involves only the Fisher information number by expressing $\mathrm{var}\left(\sqrt{n}\left(\hat{\theta}_n\right)\right) \geq I^{-1}(\theta)$. The "efficiency" of $\hat{\theta}_n$ is a measure of how close the variance of $\hat{\theta}_n$ is to the minimum possible variance, the CRLB. One measure of this is the "efficiency of $\hat{\theta}_n$ with respect to the minimum possible variance," defined as

$$\mathrm{eff}\left(\hat{\theta}_n, \mathrm{crlb}\right) = \frac{\mathrm{crlb}}{\mathrm{var}\left(\sqrt{n}\hat{\theta}_n\right)} = \frac{1}{I(\theta)\mathrm{var}\left(\sqrt{n}\hat{\theta}_n\right)}.$$

It should be clear that $0 \le \text{eff}\left(\hat{\theta}_n, \text{crlb}\right) \le 1$ and if $\text{eff}\left(\hat{\theta}_n, \text{crlb}\right) = 1$, the estimator $\hat{\theta}_n$ is said to be fully efficient for θ at the distribution f.

To illustrate the concept of efficiency, first consider determining the Fisher information number for a variety of distributions. In each of these examples, the distributions are members of the location-scale family of distributions discussed in Section 1.10. In each example, τ is the scale factor.

Consider the Gaussian (normal) density, where $f_G(y) = \frac{1}{\sqrt{2\pi}\tau}\exp\left(\frac{-(y-\theta)^2}{2\tau^2}\right)$; it can be shown that $I(\theta) = \frac{1}{\tau^2}$.

For the Cauchy density, where $f_c(y) = \frac{1}{\pi\tau}\frac{1}{\left(1+\left(\frac{y-\theta}{\tau}\right)^2\right)}$, it can be shown using some calculus and, perhaps, a table of integrals, that $I(\theta) = \frac{1}{2\tau^2}$.

And, last, for the double exponential distribution with $f_{DE}(y) = \frac{1}{2\tau}\exp\left(-\left|\frac{y-\theta}{\tau}\right|\right)$, it follows with little effort that $I(\theta) = \frac{1}{\tau^2}$.

If the random variable y were known to be distributed via some distribution f, then it would be possible to apply theorems from mathematical statistics to find the estimator that is fully efficient for this situation – that is, the estimator whose variance achieves the minimum value set by the CRLB. However, in our altered location model, we consider that the actual distribution of y is not only unknown but a member of our symmetric, heavy-tailed (with respect to the normal) family of distributions. Hence, we cannot hope to find a fully efficient estimator for all distributions in this family, as the fully efficient estimator for one distribution is not fully efficient for another distribution. Rather, our goal changes to finding an estimator that has high efficiency across all, or the "majority," of members in our family. Additionally, we hope the estimator maintains high efficiency, say, 0.90 or better, at the normal distribution. An estimator that achieves this property is said to be "efficiency robust."

To illustrate this idea, consider the three distributions mentioned above, the Gaussian, Cauchy, and double exponential, and consider the two estimators: $\bar{y}$, sample mean; and M, the sample median. Recall that $\text{var}\left(\sqrt{n} - \bar{y}\right) = \text{var}(y)$, and it can be shown that for the Gaussian distribution, $\text{var}(y) = \tau^2$; for the Cauchy distribution, $\text{var}(y)$ does not exist; and for the double exponential distribution, $\text{var}(y) = 2\tau^2$.

Thus, the efficiency values for $\bar{y}$ at these three distributions are as follows:

$$\text{eff}_G\left(\bar{y}, \frac{1}{\tau^2}\right) = \frac{1}{\frac{1}{\tau^2} * \tau^2} = 1$$

$$\text{eff}_C\left(\bar{y}, \frac{1}{2\tau^2}\right) = \frac{1}{\frac{1}{2\tau^2}*\text{undefined}} = \text{undefined}$$

$$\text{eff}_{DE}\left(\bar{y}, \frac{1}{\tau^2}\right) = \frac{1}{\frac{1}{\tau^2}*2\tau^2} = 1/2.$$

We note that the sample mean is fully efficient compared to normal data, a fact that is well known. The sample mean should behave badly when data are from a Cauchy distribution, a fact demonstrated from the small example in the previous section. The sample mean is only 50% fully efficient compared to the double exponential distribution.

On the other hand, the asymptotic variance of the sample median was determined in Section 3.2 to be $\text{var}(M) = \frac{1}{4\text{nf}^2(\theta)}$, or $\text{var}\left(\sqrt{n}M\right) = \frac{1}{4\text{f}^2(\theta)}$, where the location parameter θ and the 50th percentile coincide due to the symmetry of the distributions in our symmetric, heavy-tailed family. The efficiency of M with respect to the CRLB for each of these three distributions is as follows:

$$\text{eff}_G\left(M, \frac{1}{\tau^2}\right) = \frac{1}{\frac{1}{\tau^2}*\frac{1}{4\left(\frac{1}{2\pi\tau^2}\right)}} = \frac{2}{\pi} = .64$$

$$\text{eff}_C\left(M, \frac{1}{2\tau^2}\right) = \frac{1}{\frac{1}{2\tau^2}*\frac{1}{4\left(\frac{1}{\pi^2\tau^2}\right)}} = \frac{8}{\pi^2} = .81.$$

$$\text{eff}_{DE}\left(M, \frac{1}{\tau^2}\right) = \frac{1}{\frac{1}{\tau^2}*\frac{1}{4\left(\frac{1}{4\tau^2}\right)}} = 1$$

We observe that the sample median is fully efficient against the double exponential distribution, has relatively high efficiency against the Cauchy distribution, and is 64% as efficient as the sample mean if data are from the normal distribution. We conclude that the sample median is quite efficient against extreme heavy-tailed distributions and moderately efficient against the nominal distribution in the family, the normal distribution.

It appears that the sample median is "close" to being efficiency robust, while the sample mean is not efficiency robust, especially if our family includes the Cauchy distribution. These observations reinforce those made earlier in this chapter: namely, that the sample mean is not resistant to outliers, while the sample median is. An interesting question that remains is whether any estimators can protect against the outliers that would surely result from samples from heavy-tail distributions while maintaining a high efficiency against the normal distribution. Several possibilities will be discussed in the next section.

3.7 L-Estimators in the Location Model

Among the many robust estimators proposed for the location parameter are linear combinations of order statistics, called l-estimators, and those similar to MLEs, called M-estimators. L-estimators are likely the oldest alternative estimators to the sample mean, which is a member of the l-estimator family. M-estimators are relatively new, at least compared to the l-estimators, to the robust scene, introduced by Huber in 1964. However, not all modern robust procedures fall into one of these two categories. Later in this chapter, we will present the so-called high breakdown estimators introduced in the late 1980s. In this section, we will present the l-estimators.

L-estimators are characterized as those estimators that can be written as linear combinations of the order statistics, $y_{(1)},\dots,y_{(n)}$ so that $\hat{\theta}_n$ is an l-estimator if it can be written as $\hat{\theta}_n = \sum c_i y_{(i)}$, for constants $c_1,\dots,c_n$. We require that $0 \le c_i \le 1$ and $\sum c_i = 1$. Thus, the sample mean, $\bar{y}$, is an l-estimator with each $c_i = 1/n$. The sample median is also an l-estimator since it is either a single-order statistic if nn is odd or the average of two order statistics if n is even. Another example, due to Tukey, is the trimean defined as $.5(M + .5(Q_L + Q_U))$, the average of the median with the average of the quartiles. Each of the quantities in this expression is either an order statistic or the average of two order statistics. Each of these estimators is an unbiased estimator of θ in our altered location model.

Another commonly used l-estimator, one possessing desirable robustness properties, is the $\alpha(100\%)$-trimmed mean. While several definitions of this estimator may be found in the literature, we will consider this version: delete the $[\alpha n]$ smallest and largest ordered observations and take the arithmetic mean of the remaining $n-2[\alpha n]$ observations, where [] denotes the largest integer $\le \alpha$. We can write the $\alpha(100\%)$-trimmed mean as

$$T_5(F_n) = \bar{y}_T = \frac{1}{n-2[\alpha n]}\left(y_{(1+[\alpha n])} + \dots + y_{(n-[\alpha n])}\right).$$

Evidence (Stigler, 1977) suggests that the trimmed mean has been used as early as 1820. It is known to be less dependent on outliers. The robustness properties of this estimator can be determined using the influence curve approach introduced earlier in this chapter. That is, we first must express the estimator in its asymptotic form as a functional, find its influence curve and calculate the quantitative robustness properties from the influence curve.

The trimmed mean, expressed as a parameter, is simply the $\alpha(100)\%$ trimmed population mean,

$$T_5(F) = \frac{1}{1-2\alpha}\int_{\alpha}^{1-\alpha} F^{-1}(t)dt.$$

Perhaps a more insightful expression of $T_5(F)$ is to substitute y for $F^{-1}(t)$; then $F(y) = t$ and $f(y)dy = dt$. Then we can write $T_5(F)$ as

$$T_5(F) = \frac{1}{1-2\alpha}\int_{F^{-1}(\alpha)}^{F^{-1}(1-\alpha)} ydF(y) = \int_{y_\alpha}^{y_{1-\alpha}} yf_\alpha(y)dy,$$

where, for example, y_α is the α percent point of the density $f(y)$.

It is seen that $T_5(F)$ is simply the expected value of the random variable y with respect to the "$\alpha(100)\%$ trimmed density," the density of y properly standardized to account for the fact that $2\alpha 100\%$ of the density has been trimmed off. That is, $f_\alpha(y) = \dfrac{f(y)}{1-2\alpha}$ for $y_\alpha \le y \le y_{1-\alpha}$. An example comparing $f(y)$ to $f_\alpha(y)$ appears in the plot in Figure 3.12, which displays $f(y)$ and $f_\alpha(y)$ for the case where the density is a $N(0,1)$ and $\alpha = .05$.

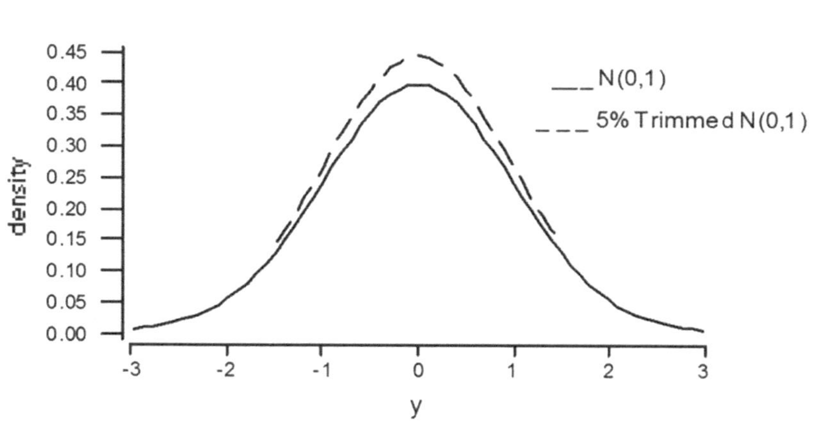

FIGURE 3.12
The 5% trimmed N(0, 1) and the N(0, 1).

In Figure 3.12, the topmost density is $f_\alpha(y)$. It is left to the reader to verify that $f_\alpha(y)$ is indeed a density function.

The trimmed estimator is obtained by the empirical substitution principle, by substituting $F_n(x)$ into the above expression for $F(x)$. The influence curve (see Staudte and Sheather (1990) for details) for $T_5(F)$ is

$$IC(x;T_5(F)) = \begin{cases} F^{-1}(\alpha)/(1-2\alpha) & x \leq F^{-1}(\alpha) \\ x/(1-2\alpha) & F^{-1}(\alpha) \leq x \leq F^{-1}(1-\alpha) \\ F^{-1}(1-\alpha)/(1-2\alpha) & F^{-1}(1-\alpha) \leq x \end{cases}.$$

A sketch of this influence curve is in Figure 3.13.

We note that the "influence" of an extreme outlier on the value of a trimmed mean is not zero, as one might at first believe. To convince yourself of this fact, compute the SC for the trimmed mean using the same arguments as those used for the sample median.

Clearly, the trimmed mean is a robust estimator of θ, provided that the percentage of outliers in either tail of the data does not exceed α. It is interesting to note that the $\alpha(100\%)$-trimmed mean converges to the sample median as the percentage of trimmed observations approaches 50%.

A second estimator, similar in nature to the trimmed mean, is the $\alpha(100\%)$ Winsorized mean, defined by altering the original observations, by replacing the $[n\alpha] = h$ smallest and largest observations with the next smallest and next largest observations, respectively. That is, replace the first h-ordered observations by $y_{(h+1)}$ and the last h ordered observations by $y_{(n-h)}$. Thus,

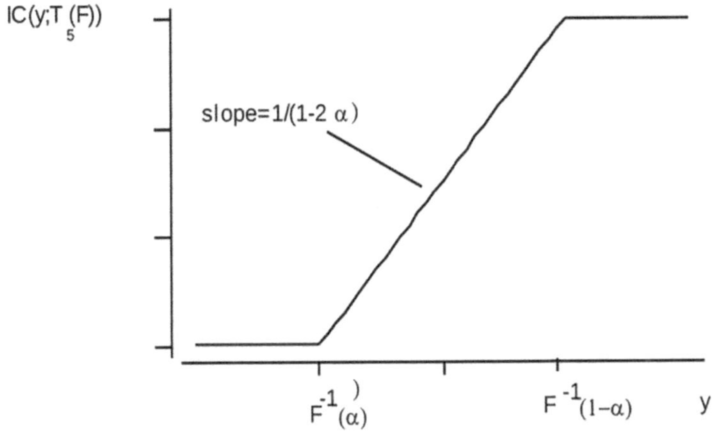

FIGURE 3.13
Influence curve for the alpha percent trimmed mean.

these two values are counted h+1 times. The $\alpha(100\%)$ Winsorized mean is computed as the mean of this altered sample – that is,

$$T_6(F_n) = \bar{y}_w = \frac{1}{n}\left((h+1)y_{(h+1)} + y_{(h+2)} + \ldots + y_{(n-h-1)} + (h+1)y_{(n-h)}\right).$$

It is clear that the Winsorized mean will be a resistant estimator so long as all outliers are contained in the h smallest and h largest observations. To examine the robustness properties of the Winsorized mean, we first must express it as a functional, asymptotic Winsorized mean and then derive its influence curve. The functional form of the Winsorized mean (see Saudte and Sheather, 1990) is

$$T_6(F) = \int_{\alpha}^{1-\alpha} F^{-1}(t)dt + \alpha\left(F^{-1}(\alpha) + F^{-1}(1-\alpha)\right)$$

and has the following influence curve:

$$IC(y;T_6(F)) = y \text{ for } |y| \le F^{-1}(1-\alpha) \text{ and}$$
$$\left(F^{-1}(1-\alpha) + \alpha / f\left(F^{-1}(1-\alpha)\right)\right)\operatorname{sgn}(y) \text{ for } |y| > F^{-1}(1-\alpha).$$

Graphically, the influence curve, as seen in Figure 3.14, features similar robustness characteristics as the trimmed mean: namely, the bounded nature of the curve reflects the resistance of the Winsorized mean to outliers. The slope of the middle portion of the curve is 1, compared to that for the

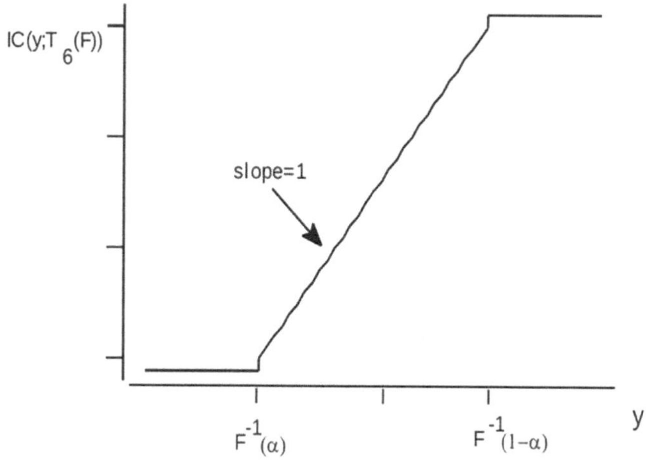

FIGURE 3.14
Influence curve for the alpha percent Winsorized mean.

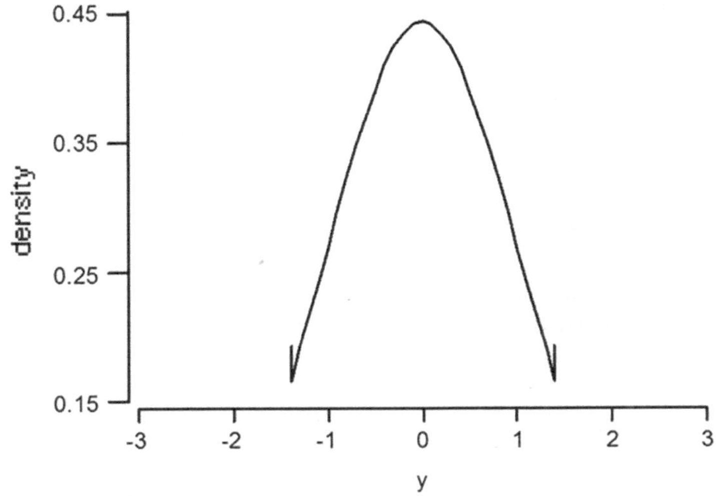

FIGURE 3.15
The 5% Winsorized N(0, 1).

trimmed mean of $\frac{1}{1-2\alpha}$. The discontinuities in the influence curve occur at $F^{-1}(\alpha)$ and result from the fact that α% of the area of $F(y)$ has been stacked at these points.

Analogous to the trimmed functional, one can think of the Winsorized parameter as the $\alpha(100)\%$ Winsorized (population) mean of y with respect to the $\alpha(100)\%$ Winsorized density, $f_{\alpha w}(y)$. This density is plotted for the case where $f(y)$ is $N(0,1)$ and $\alpha = .05$ in Figure 3.15.

The robustness properties of the trimmed and the Winsorized means will be compared later in this chapter.

3.8 M-Estimators in the Location Model

Another form of robust estimators are the M-estimators, introduced by Huber (1964). The name for these estimators derives from their similarity to the MLEs (the "m" actually means "maximum likelihood type" estimators). Indeed, as we will show later in this section, the MLEs are actually M-estimators.

In the subsequent derivations, for the sake of simplicity, we will consider the location model with the scale parameter, τ, equal to one. Later, we will relax this assumption and consider the estimation of the scale parameter as well.

In the location model, the M-estimator is defined as the value of θ, called $\hat{\theta}$, that minimizes with respect to θ^* the following defining expression

$$\sum_{i=1}^{n} \rho(y_i - \theta),$$

where the function $\rho(r)$ is some appropriately chosen function, called the "loss function." This loss function is usually continuous and convex. The properties of the M-estimator will depend on the choice of the ρ function. Many choices are possible, and several will be presented and discussed later in this section.

An equivalent expression for the M-estimator can be found from the above expression by taking the derivative with respect to θ, setting this derivative equal to zero, and solving for the unknown value of θ. That is, we want to find $\hat{\theta}$ that satisfies the "defining equation"

$$\sum_{i=1}^{n} \psi(y_i - \theta) = 0,$$

where the function ψ, called the " ψ-function," is the derivative of ρ with respect to the argument of ρ. That is, $\psi(r) = \dfrac{\partial \rho(r)}{\partial r}$, where $r_i = y_i - \theta$. Even though the derivative of the defining expression is taken with respect to θ, the reader should keep in mind that the ψ function results by taking the derivative of ρ with respect to the argument of ρ. Thus, in our expression above, $r_i = y_i - \theta$, and applying the "chain rule" from calculus results in

$$\frac{\partial \rho(r_i)}{\partial \theta} = \frac{\partial \rho(r_i)}{\partial r_i} \frac{\partial r_i}{\partial \theta} = \psi(r_i)(-1),$$

and the "–1" vanishes when the derivative is set equal to zero.

M-estimators are a very rich class of estimators that includes the MLEs. To see this, let $y_1, \ldots, y_n$ represent a random sample from a continuous distribution with location parameter θ, – that is, $y_i \sim f(y;\theta) = f(y - \theta)$. Then the log-likelihood function for this sample is

$$\ln L(\theta \mid \underline{y}) = \sum_{i=1}^{n} \ln f(y_i - \theta) = -\sum_{i=1}^{n} \rho(y_i - \theta),$$

where $\rho(y_i - \theta) = -\ln f(y_i - \theta)$. The MLE is that value of θ maximizing the log-likelihood function and is found in the usual way by solving

$$\frac{\partial \ln L(\theta \mid \underline{y})}{\partial \theta} = 0$$

or, equivalently,

$$\sum \frac{f'(y_i - \theta)}{f(y_i - \theta)} = 0,$$

where $f'(y_i - \theta) = \frac{\partial f(y_i - \theta)}{\partial \theta}$. Of course, this is the method discussed in Section 2.2.

The relationship between the MLEs and the M-estimators is established by letting $\rho(y_i - \theta) = -\ln f(y_i - \theta)$ and noting that maximizing the log-likelihood is the same as minimizing the negative of this function. Since

$$\psi(r_i) = \frac{\partial \rho(r_i)}{\partial r_i} = \frac{-\partial \ln f(y_i - \theta)}{\partial r_i} = \frac{-\partial \ln f(r_i)}{\partial \theta}\frac{\partial \theta}{\partial r_i} = \frac{f'(r_i)}{f(r_i)},$$

it follows that an MLE is an M-estimator with $\psi(r_i) = \frac{f'(r_i)}{f(r_i)}$, where $r_i = y_i - \theta$.

Several examples will prove enlightening here. Consider first the case where the density is normal with unknown mean, θ, and known variance of one. The log density function is

$$\ln f(y_i;\theta) = \text{constant} - \frac{(y_i - \theta)^2}{2},$$

resulting in a ρ function of

$$\rho(y_i - \theta) = -\ln f(y_i;\theta) = -\text{constant} + \frac{(y_i - \theta)^2}{2}.$$

Since the constant in the above expression has no effect on the location of the minimum of the function, we simply drop it from the expression. Thus, the ρ function is

$$\rho(y_i - \theta) = \frac{(y_i - \theta)^2}{2},$$

commonly referred to as the "quadratic loss function." A sketch of this function appears in Figure 3.16.

The MLE is found by taking the derivative of the log-likelihood function with respect to the unknown parameter θ and setting the result equal to zero

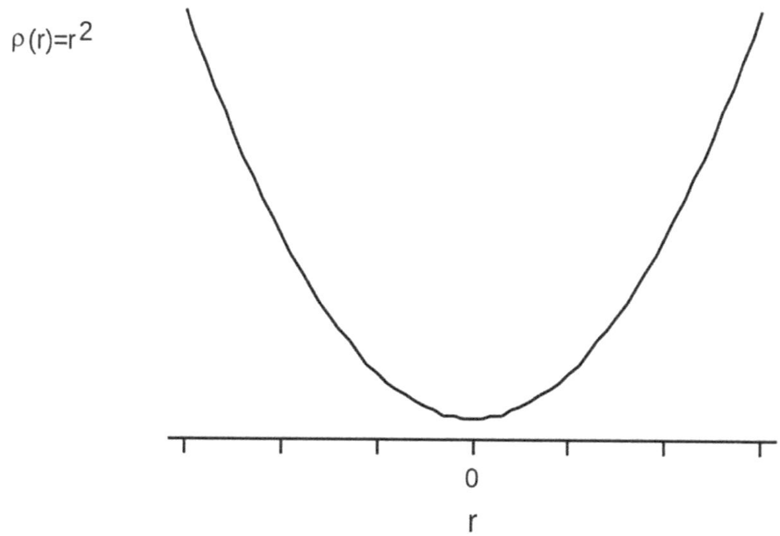

FIGURE 3.16
The quadratic loss function.

and solving or, equivalently, taking the derivative of $\rho(r)$ with respect to r and setting $\sum\psi(r_i)=0$ results in

$$\psi(r)=\frac{\partial\rho(r)}{\partial r}=\frac{\partial\frac{r^2}{2}}{\partial r}=r$$

and

$$\sum\psi(\mathrm{r}_i)=\sum\psi(\mathrm{y}_i-\theta)=\sum(\mathrm{y}_i-\theta)=0,$$

which implies

$$\hat{\theta}=\frac{\sum \mathrm{y}_i}{\mathrm{n}}=\bar{\mathrm{y}},$$

the sample mean. Note that $\psi(r)=r$, the identity ψ function, sketched in Figure 3.17, is similar in appearance to the influence curve for the mean (asymptotic).

Thus, we see that if $\mathrm{y}\sim\mathrm{N}(\theta,1)$, the MLE for θ is the sample mean, a result we discussed in Chapter 2. Additionally, if the M-estimator of θ is desired and the ρ function $\rho(r)=\frac{r^2}{2}$ is used or, equivalently, the ψ function $\psi(r)=r$,

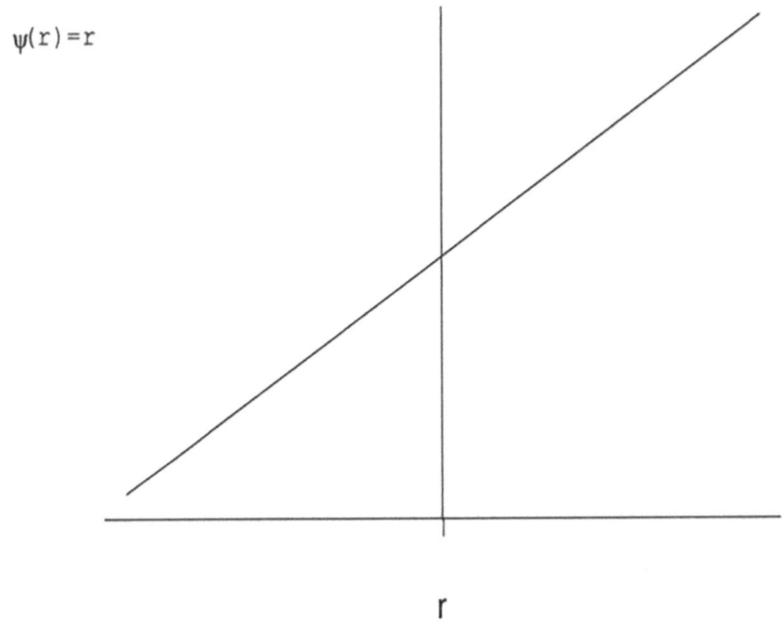

FIGURE 3.17
The identity Psi function.

the "identity ψ function" is used, the resulting estimator is $\bar{y}$. The sample mean can be thought of in several different ways: the method of moments estimator, the ls estimator, the MLE (for normal data), and the M-estimator if the identity ψ is used.

As a second example, consider the double exponential distribution where

$$f(y;\theta) = f(y-\theta) = \frac{1}{2}\exp(-|y-\theta|)$$

with unknown location parameter θ and known scale parameter of $\tau = 1$. The log-likelihood function is

$$\ln f(y-\theta) = \text{cons tan}t - |y-\theta|,$$

resulting in the ρ function of (again, the constant is ignored) $\rho(y-\theta) = |y-\theta|$,

$$\rho(y-\theta) = |y-\theta| \text{ or } \rho(r) = |r|,$$

the "absolute value loss function." The absolute value loss function is plotted in Figure 3.18.

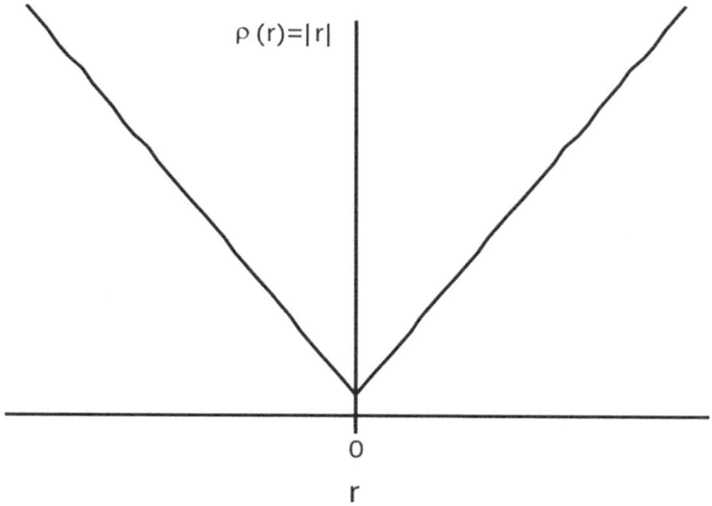

FIGURE 3.18
The absolute value loss function.

As in the previous example, the MLE is found by taking the derivative of the log-likelihood function with respect to θ and setting equal to zero or, equivalently, taking $\psi(r) = \frac{\partial \rho(r)}{\partial r} = \frac{\partial |r|}{\partial r} = \text{sgn}(r) = \begin{cases} -1 & r < 0 \\ 1 & r > 0 \end{cases}$, setting $\sum \psi(r_i) = 0$, and solving. We obtain

$$\sum_{i=1}^{n} \text{sgn}(r_i) = 0,$$

where $r_i = y_i - \theta$. Any value of $\hat{\theta}$ that results in the number of negative residuals equaling the number of positive residuals is a solution to this problem. If n is odd, only one solution exists. If n is even, any value between the two middle-order statistics will suffice. One such value is the sample median, M; by convention, M is considered the solution to this problem.

Thus, the sample median is the MLE if the sample is from the double exponential distribution; it is also the M-estimator if the y function is chosen as $\psi(r) = \text{sgn}(r)$, the sign ψ function, regardless of the distribution of the data. A sketch of this y function appears in Figure 3.19.

It is noted that this function is similar in appearance to the influence curve for the median (asymptotic), a fact we will justify shortly.

Recall that the robustness properties of an estimator are derived from the influence curve for the asymptotic version of the estimator. Since the defining equation for the M-estimator can be written as

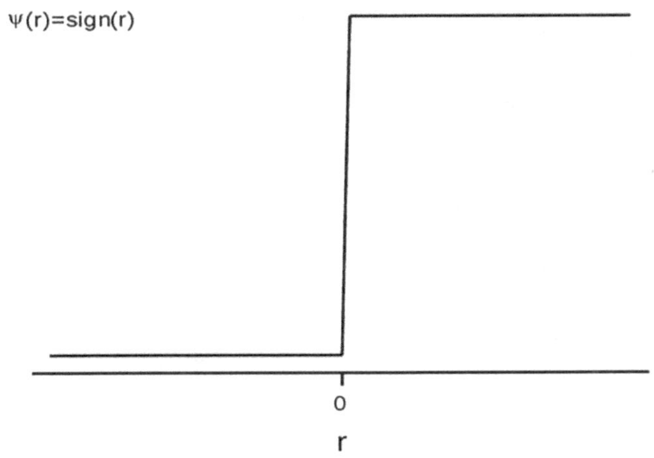

FIGURE 3.19
The sign Psi function.

$$\sum_{i=1}^{n} \psi(y_i - \theta) = \int \psi(y-\theta)dF_n(y) = 0,$$

it follows by substitution of $F(y)$ for $F_n(y)$ that the defining equation for the asymptotic M-estimator, the m parameter, denoted by $\theta = T_m(F)$, is

$$\int \psi(y - T_m(F))dF(y) = 0.$$

We note that in general we cannot write the m parameter explicitly as a functional, as is the case for the mean where $T_1(F) = \int ydF(y)$. This implies that the properties of the M-estimator will have to be computed indirectly, as is the case for the median.

For example, to find the influence curve for the M-estimators, we first recall that the altered distribution is of the form

$$\tilde{F}(y) = (1-\varepsilon)F(y) + \varepsilon\delta_{y_0}(y),$$

and by definition, the m parameter for this altered distribution, $T_m(\tilde{F})$, satisfies

$$\int \psi\left(y - T_m(\tilde{F})\right)d\,\tilde{F}(y) = 0.$$

Using the fact that the differential is a linear operator, we obtain

$$(1-\varepsilon)\int \psi\left(y - T_m(\tilde{F})\right)dF(y) + \varepsilon\int \psi\left(y - T_m(\tilde{F})\right)d\delta_{y_0}(y) = 0.$$

Taking the partial derivative of both sides of the above expression with respect to ε yields

$$(-1)\int \psi\left(y-T_m(\tilde{F})\right)dF(y)+(1-\varepsilon)\int \psi'\left(y-T_m(\tilde{F})\right)\left[-\frac{dT_m(\tilde{F})}{d\varepsilon}\right]dF(y)$$
$$+\int \psi\left(y-T_m(\tilde{F})\right)d\delta_{y_0}(y)+\varepsilon\frac{d\int \psi\left(y-T_m(\tilde{F})\right)d\delta_{y_0}(y)}{d\varepsilon}=0 \quad ,$$

where $\psi'(r)=\dfrac{\partial\psi(r)}{\partial r}$. Setting $\varepsilon=0$ in the left-hand side of the above expression results in

$$(-1)\int \psi\left(y-T_m(F)\right)dF(y)$$
$$+\int \psi'\left(y-T_m(F)\left[-\frac{dT_m(\tilde{F})}{d\varepsilon}\right]_{\varepsilon=0}\right) dF(y)^2\psi\left(y_0-T_m(F)\right)+0=0.$$

Noting that the first term on the left-hand side is zero by definition of the *m* parameter and solving for the influence curve, we obtain

$$\left[\frac{dT_m(\tilde{F})}{d\varepsilon}\right]_{\varepsilon=0}=\frac{\psi\left(y_0-T_m(F)\right)}{\int \psi'\left(y-T_m(F)\right)dF(y)}=\frac{\psi\left(y_0-T_m(F)\right)}{E\left(\psi'\left(y-T_m(F)\right)\right)}.$$

Or we write

$$IC\left(y;T_m(F)\right)=\frac{\psi(r)}{E\left(\psi'(r)\right)},$$

where $r=y-T_m(F)$.

Thus, we see that the influence curve for the M-estimator is proportional to the ψ function used in defining the M-estimator. Consequently, in studying the robustness properties of the M-estimator, we want to look closely at the properties of its influence curve or, more directly, the properties of its ψ function.

The asymptotic distribution of the M-estimator is easily determined from our functional theory developed earlier. If $T_m(F)$ is the m parameter and $T_m(F_n)$ the M-estimator, then it follows that

$$\sqrt{n}\left(T_m(F_n)-T_m(F)\right)\rightarrow N\left(0,V^2\right),$$

$$\text{where } V^2 = E\left(IC^2\left(y;T_m(F)\right)\right) = \frac{E\left(\psi^2(r)\right)}{\left[E\left(\psi'(r)\right)\right]^2}.$$

One obvious application of the above results is to consider the MLEs and their relationship to the M-estimators again. Recall that for $y \sim f(y-\theta)$, the MLE for $\theta, \hat{\theta}$, can be thought of as an M-estimator with ψ function equal to

$$\psi(y-\theta) = \frac{\partial \ln f(y-\theta)}{\partial \theta}.$$

Now the influence function for θ requires $E\left(\psi'(r)\right)$, where $r = y-\theta$. We have

$$E\left(\psi'(y-\theta)\right) = E\left(\psi'(r)\right) = E\left(\frac{\partial \psi(r)}{\partial r}\right) = E\left(\frac{\partial\left[\dfrac{\partial \ln f(y-\theta)}{\partial \theta}\right]}{\partial r}\right)$$

$$= E\left(\frac{\partial\left[\dfrac{\partial \ln f(y-\theta)}{\partial \theta}\right]}{\partial r}\frac{\partial \theta}{\partial \theta}\right) = E\left(\frac{\partial^2 \ln f(y-\theta)}{\partial \theta^2}\right)\frac{\partial \theta}{\partial r} = -E\left(\frac{\partial^2 \ln f(y-\theta)}{\partial \theta^2}\right) = I(\theta),$$

where $I(\theta)$ is the Fisher information number.

This implies that the influence function for the MLE $\hat{\theta}$ is

$$IC\left(y;\hat{\theta}\right) = \frac{\psi(r)}{E\left(\psi'(r)\right)} = \frac{\dfrac{\partial \ln f(y-\theta)}{\partial \theta}}{I(\theta)},$$

and, as shown earlier,

$$\sqrt{n}\left(T(F_n) - T(F)\right) = \sqrt{n}\left(\hat{\theta}_n - \theta\right) \to N\left(0, E\left(IC^2\right)\right),$$

where IC denotes the influence function. In this case,

$$E\left(IC^2(y;\theta)\right) = \frac{E\left(\dfrac{\partial \ln f(y-\theta)}{\partial \theta}\right)^2}{I(\theta)^2} = \frac{I(\theta)}{I(\theta)^2} = I^{-1}(\theta)$$

is the asymptotic variance of the MLE, the well-known result stated previously in Chapter 2. Thus, the theory of functionals has enabled us to find the asymptotic distribution of the M-estimator in general and the MLEs as well as a specific application.

We have now established a number of very important facts. First, the M-estimators and their influence functions are defined in terms of a ψ function. Second, the robust properties of an estimator are measured in terms of the estimator's influence function. Thus, we will develop appropriate M-estimators by choosing ψ functions so that the desirable robustness properties of the M-estimator are achieved, including continuous estimators with small asymptotic variances in the neighborhood about the idealized distribution; small gross-error sensitivity and local-shift sensitivity; and a large breakdown point with a finite rejection point.

In general, the ψ function should be continuous, odd, and bounded. This ensures that the M-estimator will be continuous and have a breakdown point of 50%, the highest achievable value (see Huber (1981) for a proof of this result). After Huber (1964) introduced the basic notion of the M-estimator for the location model, many other researchers proposed their own versions of the "ideal ψ function." In fact, a number of these functions were studied in detail in the Princeton Robustness Study (see Andrews et al., 1972). We will review only a few of the most commonly used ψ functions. We also assume that the residuals have a scale of one. Later in Chapter 4, we will show how the residuals can be rescaled to achieve this assumption.

The first ψ function introduced, and still a very popular one, is the Huber ψ function (Huber 1964), denoted by ψ_H, and defined as

$$\Psi_H(r) = \begin{cases} -C_H & r < -C_H \\ r & |r| \le c_H \\ C_H & r > C_H \end{cases},$$

where c_H is a constant appropriate for ψ_H. A plot of ψ_H with $c_H = 1.5$ appears in Figure 3.20.

It is clear from the plot that the M-estimator based on ψ_H combines the properties of the sample mean and the sample median and is the MLE for a density with a Gaussian center and double exponential tails, called the "least favorable distribution" by Huber (1964). It should also be noted that Ψ_H, the influence function for this M-estimator with the exception of the constant $E(\psi'(r))$ is similar in appearance to the influence function for the trimmed mean. We shall see later in this section that this expresses the asymptotic similarity between these two estimators. However, they may give very different results on any finite sample problem.

Usually, the constant c_H, called the "tuning parameter," is chosen to be a value between 1 and 2, depending on how much efficiency (asymptotic) one desires at the nominal Gaussian distribution. For example, it can be shown

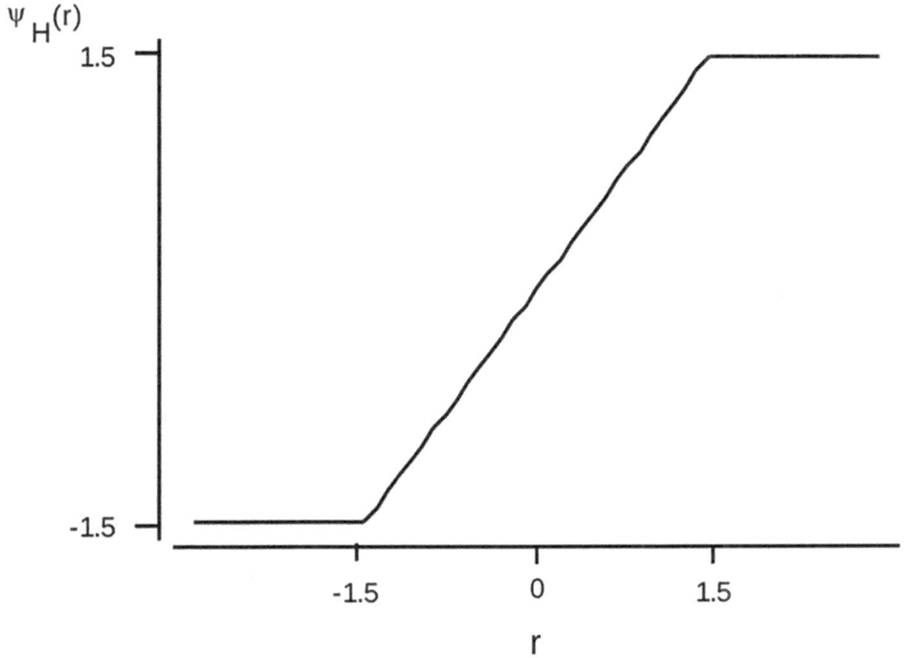

FIGURE 3.20
The Huber Psi function with constant 1.5.

(see the exercises at the end of this chapter) that if c_H is set at 1.345, that this M-estimator, commonly referred to as a Huber estimator, has 95% relative efficiency (asymptotic) compared to the sample mean, at the Gaussian distribution. One can see that as c_H increases, the efficiency of the Huber estimator will approach one because Ψ_H will approach the identity ψ function of the sample mean, and consequently, the Huber estimator will approach the sample mean.

However, the gain in efficiency will be offset by the loss in resistance to outliers that accompanies the sample mean. On the other hand, a small value of c_H has the opposite effect: an estimator that is extremely resistant to outliers, approaching the sample median, with the corresponding loss of efficiency associated with the sample median. Hence, the goal in choosing the tuning parameter for the Huber estimator, or the tuning parameter for any of the other Ψ function, is to pick the parameter to achieve a high degree of efficiency, say 95%, while maintaining a high resistance to outliers.

We note that ψ_H does not achieve a finite rejection point, so the estimator does not reject extreme data values. Functions that redescend to zero will achieve a finite rejection point. These ψ functions are commonly referred to as "redescending ψ functions." Several redescending ψ functions are given in the following.

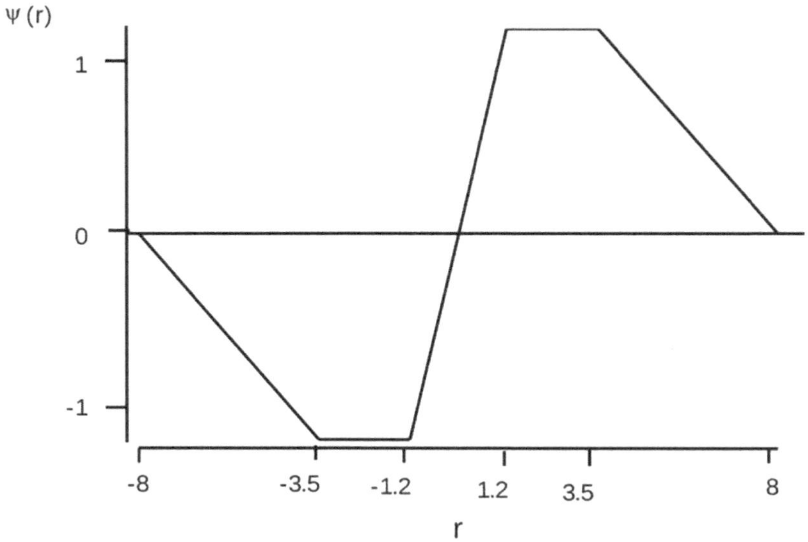

FIGURE 3.21
The Hampel Psi function with $a = 1.2$, $b = 3.5$, and $c = 8.0$.

Hampel (1968) introduced the ψ function:

$$\psi(r) = \text{sgn}(r)\begin{cases} |r| & 0 \le |r| < a \\ \dfrac{a-|r|}{c-b}a & a \le |r| < b \\ & 0r| < c \\ 0 & |r| \ge c \end{cases},$$

where the constants a, b, and c are the tuning parameters. Figure 3.21 gives a sketch of the Hampel ψ function for $a = 1.2$, $b = 3.5$, and $c = 8.0$.

It is readily observed that this ψ function possesses properties similar to that of the Huber ψ function, except it redescends back to zero, thereby resulting in "zero" influence for values of r that are less than $-c$ or greater than c.

Another redescending ψ function, introduced by Andrews (1972), is defined as

$$\psi(r) = \begin{cases} c\sin\left(\dfrac{r}{c}\right) & |r| \le c\pi \\ 0 & |r| > c\pi \end{cases},$$

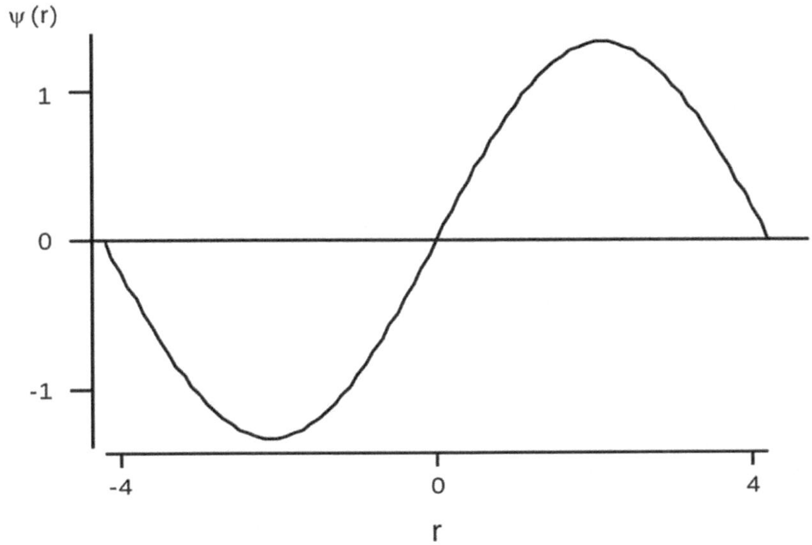

FIGURE 3.22
The sine Psi function with $c = 1.34$.

where c is the tuning parameter. A plot of this ψ function with $c = 1.34$ is presented in Figure 3.22.

The sine ψ function appears as a smooth version of the Hampel function. The M-estimator based on the sine ψ function with $c = 1.34$ results in an estimator that is 95% as efficient as the sample mean when the data is from the Gaussian distribution.

A third redescending ψ function, introduced by Beaton and Tukey (1974) is the "bisquare ψ function," denoted by $\psi_b(r)$, and is defined as

$$\psi_b(r) = \begin{cases} r\left(1 - \left(\dfrac{r}{c_b}\right)^2\right)^2 & |r| \le c_b \\ 0 & |r| > c_b \end{cases},$$

where the tuning constant, $c_b = 4.685$, which results in a 95% efficient estimator compared to the sample mean against Gaussian data. A sketch of ψ_b is in Figure 3.23.

The same relationship exists between the tuning parameter of an M-estimator based on a redescending ψ function and the estimator's efficiency and resistance properties. That is, as this constant increases, so does the efficiency, but the resistance to outliers decreases. Additionally, since observations can actually be removed by use of this form of ψ function, it is logical

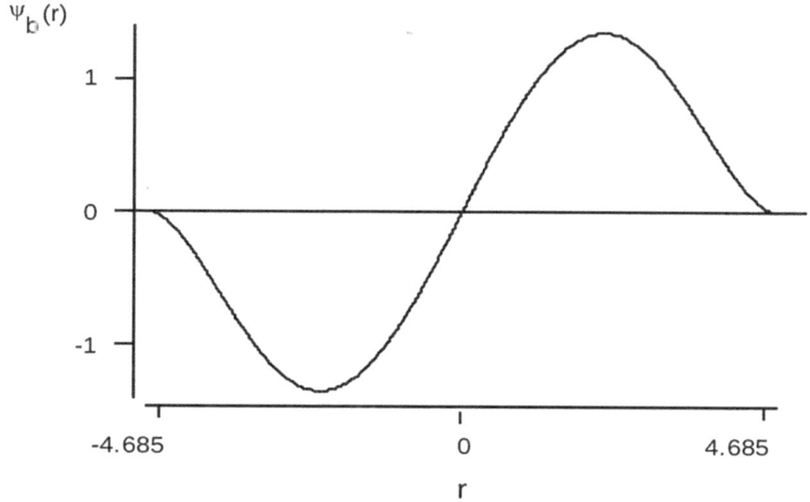

FIGURE 3.23
The bisquare Psi function with $c = 4.685$.

that the tuning parameter will be larger than that needed for the Huber ψ function. This is indeed the case, as a comparison of the tuning parameters across these four ψ functions will reveal.

Many other forms of ψ functions have been introduced, and the interested reader is referred to Holland and Welsch (1977) for a presentation of these four functions and several others. It has been observed that very little difference results either in the theoretical or the empirical properties of M-estimators based on ψ_H or other similar non-redescending functions or based on ψ_b or other similar redescending functions.

On the other hand, there may be a considerable difference between M-estimators based on non-redescending and redescending ψ functions, especially if modest to severe outliers are present in the dataset. The redescending functions can allow a point to have zero influence on the estimator, a property not achieved by the non-redescenders.

In Table 3.1, several estimators of θ are compared by listing their asymptotic quantitative robustness properties. The estimators are the sample mean, $\bar{y}$; the sample median, M; the 5% trimmed mean, $\bar{y}_t$; the 5% Winsorized mean, $\bar{y}_w$; the Huber M-estimator using $c_H = 1.5$, $\hat{\theta}_{M-H}$; the Hampel M-estimator using $a = 2.4$, $b = 4.5$, and $c = 9.5$, $\hat{\theta}_{M-\text{Hampel}}$. Recall that the quantitative robustness properties are derived from the influence curve and are σ^2, the asymptotic variance of $\sqrt{n}$ times the estimator; γ^*, the gross-error sensitivity; δ^*, the breakdown point; λ^*, the local-shift sensitivity; and ρ^*, the rejection

TABLE 3.1

Several Estimators of θ

Estimator	σ^2	γ^*	δ^*	λ^*	ρ^*
$\bar{y}$	**1.000**	∞	0.00	**1.00**	∞
M	1.571	**1.25**	**0.50**	∞	∞
$\bar{y}_w$	1.014	2.13	0.05	∞	∞
y_t	1.026	1.83	0.05	1.11	∞
$\hat{\theta}_{\text{M-H}}$	1.037	1.73	**0.50**	1.15	∞
$\hat{\theta}_{\text{M-Hampel}}$	1.026	1.86	**0.50**	1.10	**6.41**

point. In verifying the numbers in Table 3.1, the reader needs to remember that they come from the estimator's influence function and, for the M-estimator, the influence curve is equal to the ψ function divided by the constant $E(\psi'(r))$. Additionally, the estimators are compared at the $N(0,1)$ distribution, a point of reference for our altered location model.

In Table 3.1, the boldface values represent the optimal value for each characteristic. For example, against the $N(0,1)$ distribution, the sample mean has the lowest possible variance and is fully efficient, while the sample median is the most resistant to outliers, having the smallest gross-error sensitivity. It is noted that the breakdown point for the trimmed and Winsorized means is equal to the trimming fraction, α, since these estimators will be resistant to outliers so long as there are fewer outliers than $h = \lfloor \alpha n \rfloor$.

The asymptotic nature of these values is clearly apparent in the value of ∞ for the local-shift sensitivity for M and y_w. This is due to the "jump" present in the influence function for both estimators and the corresponding slope of ∞ at this jump. For finite sample problems, the jump does not exist (the reader should recall the SC for the sample median), and λ^* is some finite value. Thus, $\lambda^* = \infty$ is the result of a limit as $n \to \infty$ and is an artifact of the asymptotic nature of these calculations and not truly reflective of the way these estimators work for practical problems.

It is seen from the above table that the Huber and Hampel M-estimators are approximately equally efficient (although values of the tuning parameters may be found to match these efficiencies exactly) and are nearly identical in their resistance properties, at least at the $N(0,1)$ distribution. The trimmed and Winsorized means are also similar to each other, as are the M-estimators. This situation will change when the nominal distribution changes from the standard Gaussian to other heavy-tailed distributions in our family, a result we shall see in the next chapter. The reader interested in deriving the values presented in Table 3.1 should turn to the problems at the end of this chapter.

Exercises

1. Show that the altered CDF, $\tilde{F}(x)$, satisfies the properties of a CDF.
2. Consider the quantile of x of order p, written as a functional as $T_3(F) = F^{-1}(p) = q_x(p)$.
   a) Find the influence curve of $q_x(p)$.
   b) Let $T_3(F_n) = F_n^{-1}(p)$ be the estimator of $q_x(p)$. What is the asymptotic distribution of $F_n^{-1}(p)$? Be sure to specify all parameters of this distribution.
   c) Suppose you desire to test $H_0: q_x(p) = q_0$ versus $H_1: q_x(p) \neq q_0$. Set up the test statistic that you would use. Please specify all the terms involved in this test as completely as you can.
3. Consider the two contaminated normal distributions, $CN(x;.25,25)$ and $CN(x;.1,9)$.
   a) Use R to obtain plots of the PDFs for these contaminated normal distributions.
   b) Use whatever tools you can think of to compare these contaminated normal distributions to the standard normal distribution. Don't get too carried away. Just a couple of tools are needed. What do you conclude?
4. Consider the contaminated normal family of distributions as presented in Section 3.5.
   a) Derive $E(x)$ and $Var(x)$ if x follows the contaminated normal distribution, $CN(x;\varepsilon,\sigma^2)$.
   b) Show that $\beta_2 > 3$ if $\sigma > 1$ and $0 < \varepsilon < 1$. (Don't use the proof in Section 3.5, but derive β_2 for the contaminated normal distributions, and show why it must be greater than 3).
   c) Compute the variance (asymptotic) of $\sqrt{n}\bar{x}$ and $\sqrt{n}M$ for the following contaminated normal distributions.
      i) $CN(x;.01,9)$
      ii) $CN(x;.1,9)$
      iii) $CN(x;.1,25)$

   What can you conclude based on your results?
5. Show that the influence curve for the median (of a distribution symmetric about 0) is equal to $\psi(x)/E\psi'(x)$, where $\psi(x) = \begin{cases} -1 & x < 0 \\ 1 & x > 0 \end{cases}$.
   Hint: To compute $E\psi'(x)$, first find $E\psi_\varepsilon'(x)$, where

$$\psi_\varepsilon(x) = \begin{cases} -1 & x < -\varepsilon \\ x/\varepsilon & -\varepsilon \le x \le \varepsilon. \\ 1 & x > \varepsilon \end{cases}$$

Then $E\psi'(x) = \lim_{\varepsilon \to 0} E\psi_\varepsilon'(x)$.

6. Consider the Cauchy distribution defined as

$$f_c(x) = \frac{1}{\pi\left(1 + (x-\theta)^2\right)},$$

for $-\infty < x < \infty$.

   a) Derive the influence curve for the maximum likelihood estimate of θ. Actually, all you need is the ψ function of the MLE. We'll call this the Cauchy ψ function. I'll give you the fact that $E(\psi'(r)) = .5$. Do not derive this. See Section 3.8 of your book.
   b) Use R to plot the ψ function obtained in part a). Overlay on this plot the bisquare ψ function with a constant of 4.5. By comparing these two curves, comment on the robustness properties of an M-estimator based on the choice of the Cauchy ψ function.

7. Derive the values given in the table in Section 3.8 of the book for σ^2, γ^*, and λ^* for the median and $\hat{\theta}_{M-H}$, the Huber M-estimator with a constant of 1.5. Compare these numbers for the median and $\hat{\theta}_{M-H}$, and interpret their meaning. Hint: You'll need to use integration by parts to find σ^2 for $\hat{\theta}_{M-H}$.

8. a) Show that the M-estimator based on the Huber ψ function with $c_H = 1.345$ results in 95% (asymptotic) efficiency with respect to the sample mean if the data is from a normal distribution.
   b) Repeat part a) above with the bisquare ψ function with a constant 4.685.

   (This is a very hard problem and may be skipped).

9. Recall that the random variable x is said to be a member of the "location-scale" family if its CDF, $F(x)$, can be written as $F\left(\frac{x-\theta}{\tau}\right)$, where θ and τ represent the location and scale parameters, respectively. Suppose that $x \sim U(a,b)$, where $U(a,b) = \begin{cases} \frac{1}{b-a} & a \le x \le b \\ 0 & \text{elsewhere} \end{cases}$.

   a) Is x a location-scale random variable? Justify your result.

b) Consider the sample median, M, and the sample mean, $\overline{x}$, as estimators of $E(x)$. What is the efficiency of the sample median with respect to the sample mean for data from a $U(a,b)$? Assume that the sample size is large. Hint: $E(x)=\frac{b+a}{2}$ and $Var(x)=\frac{(b-a)^2}{12}$. Which estimator is best?

10. a) Derive the general expression for the influence curve for the *p*th percentile – that is, $F^{-1}(p)=x_p$, where $F(x)$ is the CDF of the random variable *x*.

   b) What is the general expression for the asymptotic variance of the sample estimate of x_p – that is, let $\hat{x}_p = F_n^{-1}(p)$ be the sample estimate of x_p, what is the asymptotic distribution of $\hat{x}_p$?

   c) Suppose that we estimate x_p for $p=1/3$ by $x_{(i)}$, where $i=\left[\frac{n}{3}+1\right]$. Use the information in part b) to set up the test $H_0: x_{.3333}=q_0$ versus $H_1: x_{.3333}\neq q_0$. Assume *n* is large enough. Give the test statistic and distribution, and indicate how you would compute the *p*-value for this test.

11. Consider computing an M-estimator based on the following ρ function, for some $c>0$ and $r_i = y_i - \theta$. Note that for simplicity, the scale parameter is one.

   a) What is the ψ function corresponding to the ρ function listed above?

   b) Suppose the distribution of *y* is $N(0,1)$. Give the numerical values of all five quantitative robustness properties, including the asymptotic variance. Pick as the value of *c* the value that results in a 95% efficient M-estimator using this ψ compared to the sample mean. If you cannot solve for the value of *c*, use $c=2$ in all subsequent calculations.

   c) A mathematical statistics book claims that the ψ function used above results in the trimmed mean. Is the book correct or incorrect in its statement? Very briefly discuss.

Multiple-Choice Questions

1) In a simple linear regression model, which of the following represents the slope of the regression line?

   a) β_0

   b) β_1

   c) ε

   d) y_i

   Answer: b)

2) What is the primary method used to estimate the coefficients in a simple linear regression model?
   a) Maximum likelihood estimation
   b) Method of moments
   c) Least squares
   d) Bayesian estimation

   Answer: c)

3) Which assumptions are NOT made in a simple linear regression model?
   a) The error term has a mean of zero.
   b) The error term has a constant variance.
   c) The error terms are independent.
   d) The independent variable is normally distributed.

   Answer: d)

4) If the true relationship between x and y is quadratic, which of the following statements about using a simple linear regression model is true?
   a) The simple linear regression model will fit the data perfectly.
   b) The residual plot will show a random scatter.
   c) The residual plot will likely show a pattern.
   d) The coefficient of determination, R^2, will be one.

   Answer: c)

5) What does the coefficient of determination, R^2, represent in a simple linear regression model?
   a) The proportion of the variance in the dependent variable that is predictable from the independent variable.
   b) The strength of the correlation between the dependent and independent variables.
   c) The average of the residuals.
   d) The significance level of the regression coefficients.

   Answer: a)

6) In the context of simple linear regression, what does homoscedasticity refer to?
   a) The residuals are normally distributed.
   b) The variance of the residuals is constant for all values of x.
   c) The mean of the residuals is zero.
   d) The residuals are independent.

   Answer: b)

7) What is the expected value of the error term ε in a simple linear regression model?

a) $E(\varepsilon)=\beta_0$

b) $E(\varepsilon)=\beta_1$

c) $E(\varepsilon)=0$

d) $E(\varepsilon)=y_i$

Answer: c)

8) Which plot is used to visually assess the appropriateness of the simple linear regression model by plotting residuals versus fitted values?

a) Scatterplot

b) Q-Q plot

c) Residual plot

d) Histogram

Answer: c)

9) If the p-value for the slope β_1 in a simple linear regression model is 0.03, what can we conclude at the 0.05 significance level?

a) We fail to reject the null hypothesis.

b) There is no linear relationship between x and y.

c) The slope β_1 is not significantly different from zero.

d) We reject the null hypothesis that the slope β_1 is zero.

Answer: d)

10) Which transformation method is specifically discussed in the chapter for dealing with nonlinearity in the data for simple linear regression?

a) Logarithmic transformation

b) Tukey's method

c) Polynomial transformation

d) Inverse transformation

Answer: b)

4

Robust Estimation in the Location Model: Computer-Intensive Methods

Summary

In Chapter 3, some robust estimators were given to estimate the location parameter. Methods for evaluating the robust properties of these and other estimators were also presented. Chapter 4 is concerned, in part, with the numerical methods necessary to obtain the M-estimator. Several methods are presented and compared. Other robust estimation techniques are also studied in this chapter, emphasizing the so-called high breakdown bound estimators. The theoretical properties of all the robust methods outlined in this text are asymptotic (large sample) in nature. Small-sample properties of these methods are difficult or impossible to evaluate without the use of the computer. The chapter concludes with a presentation and study of two computer-intensive techniques for evaluating the small-sample properties of our robust procedures, the Monte Carlo and bootstrap methods.

Learning Objectives

By the end of this chapter, readers should be able to:

- Explain M-estimators and how they are used in robust estimation for location parameters, including how these estimators adjust for data scale and heavy-tailed distributions.
- Become proficient in using numerical techniques such as Newton-Raphson and iteratively reweighted least squares (IRLS) to compute M-estimators for location models, understanding the conditions for convergence and potential pitfalls.
- Describe and compare different robust estimation techniques, such as high breakdown bound estimators like least median of squares (LMS) and least trimmed squares (LTS), and understand their advantages and limitations, especially in the presence of outliers.
- Use Monte Carlo methods to evaluate the small-sample properties of robust estimators, particularly in scenarios where theoretical small-sample results are difficult to derive analytically.
- Use bootstrap resampling techniques to estimate the sampling variance and confidence intervals of robust estimators, particularly in small-sample scenarios, and understand the differences between bootstrap and Monte Carlo methods.

DOI: 10.1201/9781003565321-4

4.1 Computation of M-Estimators in the Location Model

In order to apply our theory of M-estimators to practical problems, we must first alter our definition of M-estimators to account for the scale of the data, assumed to be one in our previous presentation in Chapter 3. In our family of heavy-tailed distributions, let s denote the scale parameter. We used the symbol τ as the scale parameter in Chapter 1. In applications, s will be unknown and must be estimated. If the data distribution were Gaussian, then the scale parameter would be the standard deviation – that is, s would be σ, and we would estimate σ by the sample standard deviation, s. There are two problems with using the sample standard deviation here in our altered model. First, the data are not assumed to be normally distributed, and the standard deviation is not the appropriate scale estimator for distributions other than the normal. Second, since the data are assumed to follow a heavy-tailed distribution, outliers must be a concern, and it is well known that the sample standard deviation is nonresistant to outliers, which are inflated by their presence in the dataset. Consequently, a robust scale estimator is required for our problem.

One such estimator is the median absolute deviation, or "mad," defined as

$$\hat{s} = \text{mad} = \frac{\text{med}\left|y_i - \hat{\theta}\right|}{0.6745}.$$

The use of the mad is motivated by its obvious resistance properties, resulting from using the median operator on the absolute value of the residuals. Typically, $\hat{\theta}$ is taken as the sample median so that the mad becomes the median absolute deviation about the median. The constant 0.6745 results from the desire to obtain a consistent estimator of σ if the data are normal. To see this, note that if $y = \theta + \varepsilon$, where $\varepsilon \sim N(0,\sigma^2)$, it follows that $\text{med}(\varepsilon) = 0$ and that $\text{med}(|\varepsilon|) = 0.6745\sigma$, the 75th percentile of the $N(0,\sigma^2)$ distribution. In other words, $\sigma = \dfrac{\text{med}(|\varepsilon|)}{0.6745}$. Consequently, σ may be estimated by $\text{mad} = \hat{\sigma} = \dfrac{\text{med}_i(|r_i|)}{0.6745}$, where $r_i = y_i - M$. Since M is a consistent estimator for θ, it follows that the mad is a consistent estimator for σ. Additionally, empirical studies have shown that the mad is very resistant and a reasonably efficient estimator of scale when the data are observations from our heavy-tailed distribution family members.

Suppose then that $\hat{s}$ is some "good" estimator of the scale of the data, such as the *mad*. We alter the defining equation for the M-estimator to account for the scale by solving the equation:

$$\sum_{i=1}^{n}\psi\left(\frac{y_i-\theta}{\hat{s}}\right)=0,$$

where the function ψ is of the type presented in the previous section, such as the Huber or bisquare functions. Of course, if $\psi(r)=r$, the identity ψ function, the resulting M-estimator is just the sample mean, as discussed in the previous section. For all other ψ functions, we must solve a nonlinear equation in one unknown, and numerical techniques must be employed. Several such methods may be applied, but only two methods will be presented here: the Newton-Raphson technique and the IRLS method. While both methods are often used in practice, there are strong justifications for the use of the latter method.

The Newton-Raphson technique is a long-established numerical method used to find a nonlinear equation's solution ("root"). To present the method in a more general context, suppose that the function g is some function of θ and θ_1 is that value of θ where $g(\theta_1)=0$. That is, θ_1 is the root of $g(\theta)$.

Of course, if g is a linear function of θ, say $g(\theta)=a+b\theta$, then the root is easily found by setting $a+b\theta$ equal to zero and solving for θ_1 as $\theta_1=-\frac{a}{b}$. For nonlinear functions, one obvious approach is to approximate $g(\theta)$ by a linear function, using the first-order univariate Taylor series, expanding $g(\theta)$ about some initial guess of θ_1, denoted by $\hat{\theta}_0$. We obtain

$$g(\theta_1)\approx g(\hat{\theta}_0)+(\theta_1-\hat{\theta}_0)g'(\hat{\theta}_0),$$

where $g'(\hat{\theta}_0)=\left.\frac{dg(\theta)}{d\theta}\right|_{\theta=\hat{\theta}_0}$. Setting the above equation equal to zero and solving for θ_1 gives us

$$\hat{\theta}_1=\hat{\theta}_0-\frac{g(\hat{\theta}_0)}{g'(\hat{\theta}_0)},$$

where $\hat{\theta}_1$ denotes the estimate of the root, θ_1.

An example of the Newton-Raphson method applied to the function $g(x)$ (note that g is now a function of x for this illustration) is demonstrated graphically in Figure 4.1. Here, the root of the function $g(x)$ is at the location of x where $g(x)=0$, say, the point a. Suppose the initial estimate of this point is a_0, a value slightly to the right of a. Obviously, $g(a_0)$ is not zero, and a_0 is not a root of $g(x)$. The updated estimate, a_1, is obtained by approximating $g(x)$ at the point a_0 with a straight line that is tangent to $g(a_0)$; this is the "tangent line" and has a slope $g'(a_0)$. The formula for this line is the first formula given above. The value of a_1 is found by locating where this straight line crosses the x-axis, given by the second formula above. If the function $g(x)$ were truly linear, the value a_1 would indeed be the root of $g(x)$. However, for a nonlinear function, such as the one in the Figure 4.1, the updated estimate is also not a root of $g(x)$, and the process is repeated a second time. The tangent line is placed at $g(a_1)$ and a_2 is found, the value where the tangent line crosses the x-axis. Again, a_2 is not a root of $g(x)$. We note, though, that the sequence of estimates, a_0, a_1, and a_2, are getting closer to the root, a. If the process were to be repeated one more time, it would be clear that the next estimate would be very close to the true root. Several additional iterations might be required to achieve the desired level of convergence, such as $|at - a(t-1)| < 0.001$, where t represents the iteration number.

We note that this method works very fast, especially if the initial guess is "close" to the root of the function. The method possesses a "quadratic rate of convergence," the fastest rate possible. It should also be noted that the method is a "locally convergent method" in that the initial value, or the "starting value," must be "close" to the final root for convergence to be guaranteed. This fact follows from the notion that every function, $g(x)$, with root a, is "approximately linear" within a small neighborhood of a. If the starting value is within this neighborhood, convergence to a is guaranteed using the Newton-Raphson method.

The method may not converge if the initial value is outside this neighborhood. For example, for the function $g(x)$ depicted in Figure 4.1, suppose the algorithm begins with the initial value, a_0, far to the left of the root, a. It is possible that the method will not converge. This is precisely what happens in Figure 4.2, where the starting value a_0 results in updated estimates, a_1, a_2 and so on that fail to converge to the true root a.

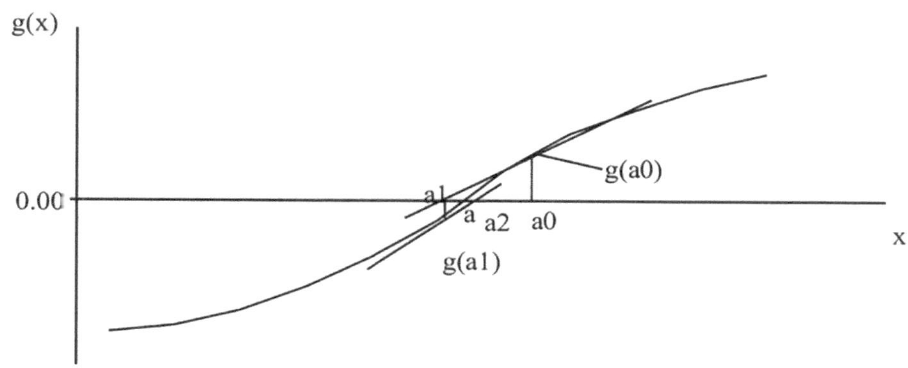

FIGURE 4.1
Illustrating the Newton-Raphson method converging to the root of the defining equation.

It can be seen in the above figure that if the starting value is any value within the interval 1 to u, the Newton-Raphson will converge to the root. The interested reader may refer to any good book on numerical methods for a thorough explanation of the properties of this method. In our robust estimation problem, the function of interest, $g(\theta)$, is

$$g(\theta)=\sum_{i=1}^{n}\psi\left(\frac{y_i-\theta}{\hat{s}}\right).$$

Let $\hat{\theta}_0$ denote our initial guess of the root, θ_1. Then the Newton-Raphson procedure gives the updated estimate as

$$\hat{\theta}_1=\hat{\theta}_0+\hat{s}\frac{\sum_{i=1}^{n}\psi\left(\frac{y_i-\theta}{\hat{s}}\right)}{\sum_{i=1}^{n}\psi'\left(\frac{y_i-\hat{\theta}_0}{\hat{s}}\right)},$$

where $\psi'(r)=\dfrac{d\psi(r)}{dr}$. We note that the denominator of the right-hand expression in the above equation results from our need to calculate

$$\frac{dg(\theta)}{d\theta}=\frac{d\sum_{i=1}^{n}\psi\left(\frac{y_i-\theta}{\hat{s}}\right)}{d\theta}=\frac{d\sum_{i=1}^{n}\psi\left(\frac{y_i-\theta}{\hat{s}}\right)}{dr}\frac{d\left(\frac{y_i-\theta}{\hat{s}}\right)}{d\theta}=-\frac{1}{\hat{s}}\sum_{i=1}^{n}\psi'\left(\frac{y_i-\theta}{\hat{s}}\right),$$

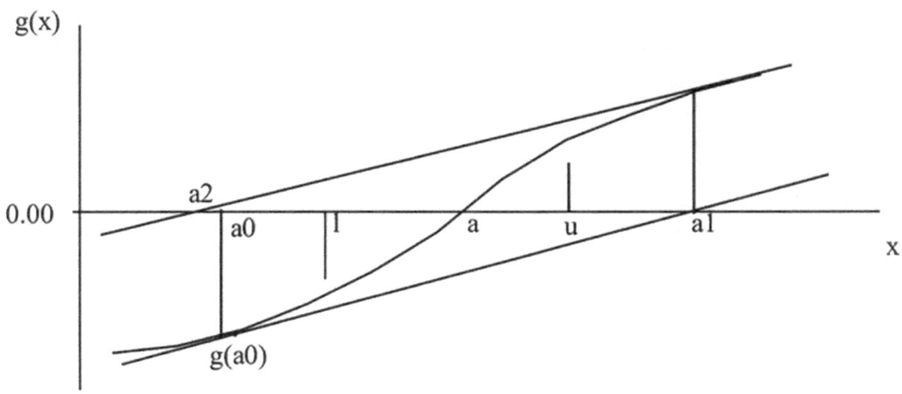

FIGURE 4.2
An example where the Newton-Raphson method fails to converge.

where $r = \frac{y_i - \theta}{\hat{s}}$.

The complete Newton-Raphson algorithm is then:

(1) Choose $\hat{\theta}_0$, say $\hat{\theta}_0 = M$, and compute ŝ, say $\hat{s} = \text{mad} = \frac{\text{med}\left|y_i - \hat{\theta}_0\right|}{0.6745}$;

(2) Compute the rescaled residuals, $r_i' = \frac{y_i - \hat{\theta}_0}{\hat{s}}$;

(3) Compute the updated estimate,

$$\hat{\theta}_1 = \hat{\theta}_0 + \hat{S}\frac{\sum \Psi\left(\frac{y_{i-\hat{\theta}_0}}{\hat{S}}\right)}{\sum_{i=1}^{n} \Psi'\left(\frac{y_{i-\hat{\theta}_0}}{\hat{S}}\right)}.$$

If we quit here, the estimate $\hat{\theta}_1$ is called a "1-step M-estimator."

(4) If $\hat{\theta}_1 \approx \hat{\theta}_0$, the algorithm stops, and $\hat{\theta}_1$ is declared the M-estimate, $\hat{\theta}$.

Otherwise, replace $\hat{\theta}_0$ by $\hat{\theta}_1$, and return to step (2) above.

Typically, about 5–10 iterations are required for the algorithm to converge. We note that the scale estimate, ŝ, is not iterated in this version of the algorithm. Updating ŝ could be accomplished by placing the computation of ŝ in step (2). Another point worth mentioning concerns the stopping rule used to terminate the algorithm. One rule that works well is to use an "absolute relative change-stopping rule," terminating the process when $\frac{\left|\hat{\theta}_1 - \hat{\theta}_0\right|}{\hat{\theta}_0} < 0.001$.

A second numerical procedure commonly used to obtain the solution to the sample M-estimator defining equation is the IRLS method. To see how the method works, note that the sample-defining equation for the M-estimator can be rewritten as

$$\sum \psi\left(\frac{y_i - \theta}{\hat{s}}\right) = 0 \Leftrightarrow \sum \psi\left(\frac{y_i - \theta}{\hat{s}}\right)\frac{\frac{y_i - \theta}{\hat{s}}}{\frac{y_i - \theta}{\hat{s}}} = \sum w_i \frac{y_i - \theta}{\hat{s}} = 0,$$

$$\Leftrightarrow \sum w_i (y_i - \theta) = 0$$

where $w_i = \psi\left(\frac{y_i - \theta}{\hat{s}}\right) / \frac{y_i - \theta}{\hat{s}}$ We see that the above equation is just the weighted least squares "normal equation," which is easily solved to obtain

$$\hat{\theta} = \frac{\sum w_i y_i}{\sum w_i}.$$

The closed-form representation for the M-estimator is very appealing. However, the w_i are unknowns, being functions of the unknown values of θ and $\hat{s}$. A reasonable way to proceed then is to make an initial guess of θ, say $\hat{\theta}_0$, and to use the above equation to "update" the estimate. The procedure may be iterated until a suitable convergence criterion is reached.

The complete IRLS algorithm is then as follows:

1. Obtain an initial guess of θ, say $\hat{\theta}_0 = M$, and compute an estimate of s, say
$$\hat{s} = \text{mad}.$$
2. Compute the rescaled residuals, $r_i' = \frac{y_i - \hat{\theta}_0}{\hat{s}}$ and $w_i = \psi(r_i') / r_i'$, $i = -1, \ldots, n$ for some appropriately chosen psi function.
3. Compute $\hat{\theta}_1 = \frac{\sum w_i y_i}{\sum w_i}$.
4. If $\hat{\theta}_1 \approx \hat{\theta}_0$, the algorithm stops and $\hat{\theta}_1$ is declared as the M-estimate, $\hat{\theta}$. Otherwise, replace $\hat{\theta}_0$ by $\hat{\theta}_1$ and return to step 2.

The above method may be altered like the Newton-Raphson method in that the scale estimate may be updated at each iteration, computing $\hat{s} = \text{med}\left|y_i - \hat{\theta}_0\right| / 0.6745$ in step 2. Usually, a relative change-stopping rule is used to check for convergence. Convergence to the unique root is guaranteed with a monotone psi function such as the Huber function. However, similar to the Newton-Raphson method, the IRLS procedure may converge to one of several non-desirable roots if a redescending psi function is used.

The appealing nature of the IRLS procedure results from the representation of the M-estimator as a weighted least squares estimator, where the weights, $w_i = \psi(r_i') / r_i'$ must satisfy the following conditions:

1. $0 \leq w_i \leq 1$. This will be achieved, provided that $\left|\psi(r_i')\right| \leq \left|r_i'\right|$, a property met by all reasonable psi functions.

2. The w_i are determined from the data, where "good" data should receive a weight of approximately one, and "bad" data, – that is, outliers – should receive a weight of approximately zero.
3. The w_i are functions of the residuals determined from the M-estimate. That is, $r' = y_i - \hat{\theta} / \hat{s}$, where $\hat{\theta}$ is the M-estimate. At the convergence of the IRLS algorithm, the w_i will meet this condition.

It is intuitively appealing to think of the M-estimator as a weighted least squares estimator based on some appropriately chosen weight function. For example, the sample mean can be thought of as a weighted least square estimator with a weight function of $w_i = 1$ for all values of r'. Thus, an outlier, which usually results in a large value of r', receives full weight in the calculation of the sample mean. Hence, the sample mean is nonresistant to outliers. Increasingly larger outliers will result in increasingly larger values of r', resulting in greater distortion to the sample mean. This accounts for the lack of resistance of the sample mean to outliers. An estimator that is resistant to outliers requires a weight function that gives a decreasingly smaller weight to increasingly larger outliers.

The weight functions associated with the Huber and bisquare psi functions clearly illustrate this point. The Huber weight function is

$$w_i = \psi(r_i')/r_i' = \begin{cases} 1 & |r_i'| \le c_H \\ c_H / |r_i'| & |r_i'| > c_H. \end{cases}$$

A plot of the Huber weight function is given in Figure 4.3. The resistance of the M-estimator, based on the Huber weight function, is apparent from the shape of the function.

The bisquare weight function, commonly referred to as the biweight function, has the form:

$$w_i = \psi(r_i')/r_i' = \begin{cases} \left(1 - \left(r'/c_b\right)^2\right)^2 & |r_i'| \le c_b \\ 0 & |r_i'| > c_b, \end{cases}$$

with the plot of this function in Figure 4.4.

A comparison of the latter two figures reveals the biweight function is much harsher toward outliers than the Huber weight function since it gives much lower weight to large residuals (in magnitude) and, in fact, gives zero weight to residuals exceeding c_b, the bisquare constant. These weight functions are compared in Figure 4.5.

Another feature of the IRLS algorithm is that the weights provide a useful diagnostic for the determination of outliers in the dataset. Compared

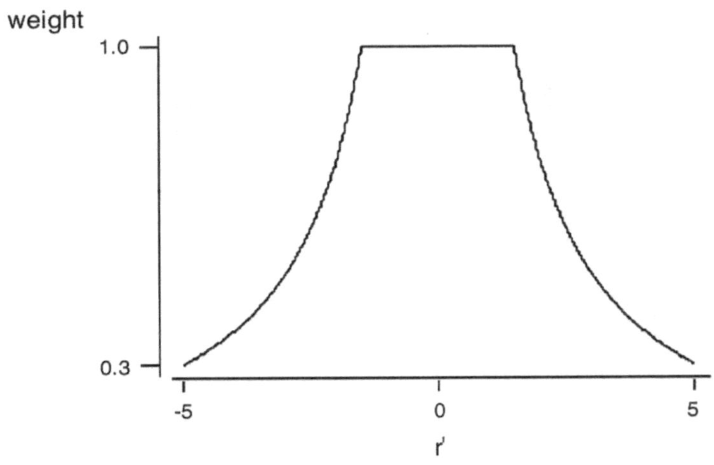

FIGURE 4.3
Huber weight function using constant 1.5.

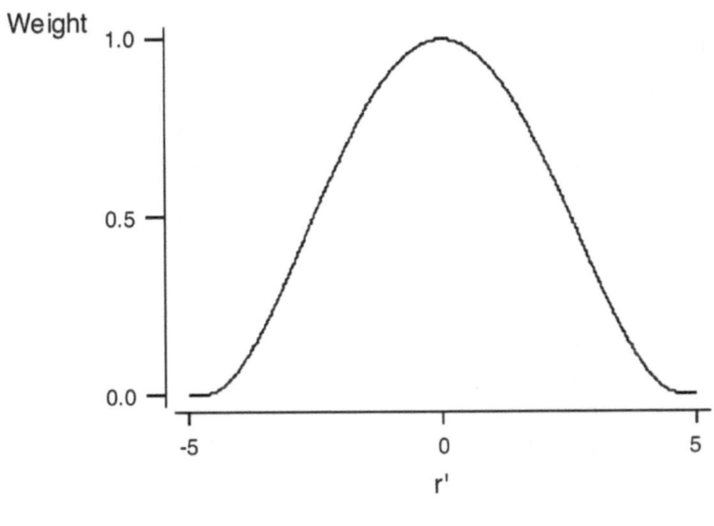

FIGURE 4.4
Bisquare weight function using constant 4.685.

to the weights given to the other observations, those observations that receive small weights have been detected as outliers. A simple "weight index plot," where the weights (w_i) are plotted versus the index (i) of the observations, is often useful as a visual representation of "unusual" or influential data points.

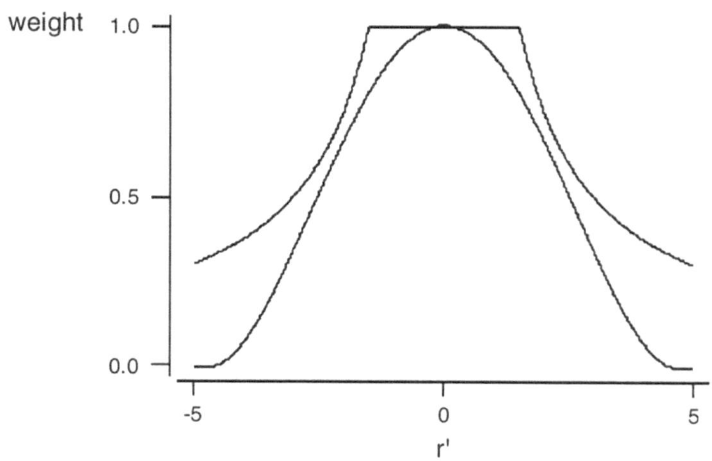

FIGURE 4.5
Comparison of the Huber and bisquare weight functions with constants 1.5 and 4.685.

4.2 Robust Inference for M-Estimators in the Location Model

We have obtained an estimator of θ in the location model that is resistant to outliers. The M-estimators also possess other desirable statistical properties. For example, the asymptotic distribution of θ_m is known to be

$$\sqrt{n}\left(\hat{\theta}_m - \theta\right) \xrightarrow{L} N\left(0, V^2\right),$$

where $V^2 = s^2 \dfrac{E\left(\psi^2(r)\right)}{\left[E\left(\psi'(r)\right)\right]^2}$ and s^2 is the squared scale parameter of the distribution of y. We note that s^2 equals σ^2 if the distribution of y is normal. Of course, this asymptotic distribution is an asymptotic result, and one must wonder what the properties of the M-estimator are for finite samples. One primary need for knowing the finite sample behavior of M-estimators is to be able to perform inference on θ using the M-estimator results. For example, how should a 95% confidence interval for θ be constructed using the M-estimator, or how can tests of hypotheses on θ be made using the M-estimator? Perhaps the more important question is, what are the statistical properties of these procedures?

It is important to realize that there is a need for robust inferential (i.e., confidence intervals and hypothesis tests) methods, especially if one feels that real data is likely to come from distributions contained in the heavy-tailed distribution family. It turns out that all tests based on LS estimates, including the classical *t*-tests, such as the two-sample *t*-test, and *F*-tests, such as the *F*-tests used in regression and analysis of variance are, at best, only robust of validity and not robust of efficiency. Here, robust of validity refers to the accuracy of probability statements used by the procedure. For example, an $\alpha = 0.05$ level test is truly a 0.05 level test if the probability of the test making a type I error (rejecting the null hypothesis when the null hypothesis is true) is indeed 0.05 and not some other value like 0.25. For confidence intervals, a procedure is robust of validity if a 95% confidence interval accurately reflects the true proportion of confidence intervals computed by this method that contains the true value of the unknown parameter.

Empirical studies have shown that tests based on the nonresistant LS estimates tend to be conservative in their probability statements if the samples are drawn from distributions in the heavy-tailed family. Thus, for example, if a *t*-test is made using data containing outliers, the nominal or advertised significance level tends to be an upper bound on the true significance level of the test. This conservatism is acceptable to statisticians, and the procedure is said to be robust of validity.

On the other hand, robust of efficiency refers to the power of the testing procedure or the narrowness of the confidence interval method. A testing procedure that is robust of efficiency maintains high power values across a variety of distributional situations, including heavy-tailed distributions, those distributions likely to give rise to outliers in actual datasets. A confidence interval procedure that is robust of efficiency maintains narrow confidence intervals, on average, across this same variety of situations.

The classical inferential procedures, based on LS estimates, taught in all statistics courses and used by the majority of practitioners, can be shown to be robust of validity in that the actual significance level of the method, say the *t*-test, is bounded above by the nominal (or, stated level), but the power of the test is much lower than it is in the "ideal" case where the error distribution is assumed to be $N(0,\sigma^2)$. Consequently, it is desirable to have an estimation procedure that results in estimates that are resistant to outliers and provides for powerful tests and narrower confidence intervals.

We will consider several possible ways to make inferential statements using the M-estimators. One obvious choice would be to use moment estimates of the asymptotic variance of the M-estimator, V^2. That is, use $\hat{v}$ in our computation of the $\text{var}(\hat{\theta}_m)$, where

$$\hat{V}^2 = \left(\frac{n^2}{n-1}\right)\frac{\hat{s}^2 \sum \psi^2(r_i')}{\left(\sum \psi'(r_i')\right)^2},$$

where $\hat{s}$ is the mad, $r_i' = \dfrac{y_i - \hat{\theta}_m}{\hat{s}}$, and $\psi'(r)$ denotes the derivative of the psi function. Note that if the identity psi function is used, then $\psi(u) = u, \psi'(u) = \dfrac{\partial \psi(u)}{\partial u} = 1$, and $\hat{V}^2$ reduces to

$$\hat{V}^2 = \frac{\sum(y_i - \bar{y})^2}{n-1},$$

the usual estimate of σ^2, appropriate for use as the numerator of the variance of $\bar{y}$, the M-estimator resulting from using the identity psi function.

For nonidentity psi functions, this estimate of variance for $\hat{\theta}_m$ is $\frac{\hat{V}^2}{n}$, and a proper large sample test would be of the form to test the hypotheses $H_0 : \theta = \theta_0$ versus an appropriate upper-tailed, lower-tailed, or two-tailed alternative hypothesis used as the test statistic

$$z = \frac{\hat{\theta}_m - \theta_0}{\hat{V} / \sqrt{n}} \sim N(0,1),$$

where the large sample distribution of z is the standard normal. However, the question remains as to what is the sampling distribution of z for small to modest-sized samples. Research has shown that this "z-test" is inferior to the following procedure.

Another approach to this problem has been proposed by Birch and Myers (1983). They simply replace n in the above expression by the sum of the weights resulting from using the IRLS procedure – that is, they replace n by $\sum w_i$, the "effective sample size" for m estimation using the IRLS procedure. Their work demonstrated that the test statistic, the "*t*-test based on robust weights,"

$$t_m = \frac{\hat{\theta}_m - \theta_0}{\hat{V} / \sqrt{\sum w_i}} = \frac{\hat{\theta}_m - \theta_0}{se\left(\hat{\theta}_m\right)}$$

follows an approximate *t*-distribution with $n-1$ degrees of freedom. Furthermore, empirical studies conducted by Birch and Myers showed that t_m satisfies both robustness of validity and efficiency and can be much more powerful than the classical *t*-test for data from heavy-tailed distributions.

A $(1-\alpha)100\%$ confidence interval for θ can be constructed in an obvious way using the above formula for the *t*-test based on robust weights as

$$\hat{\theta}_m \pm t_{\left(1-\frac{\alpha}{2},\, n-1\right)} se\left(\hat{\theta}_m\right),$$

where $t_{1-\frac{\alpha}{2},\, n-1}$ is the $\left(1-\frac{\alpha}{2}\right)100\%$ point of the *t*-distribution.

This confidence interval procedure maintains validity (accurate confidence levels) and efficiency (narrow average length of confidence intervals) across a broad range of heavy-tailed distributions.

4.3 High Breakdown Bound Estimators: LMS and LTS

In this section, we present another form of estimator for the location parameter θ in our altered location model. Like M-estimators, they provide resistance to outliers and may be extended to the multiple linear regression model.

Recall that the LS estimator in the location model is the value $\hat{\theta}_{\text{LS}}$ such that the sum of squared residuals is minimized. That is,

$$\sum\left(y_i - \hat{\theta}\right)^2$$

is minimized for $\hat{\theta} = \hat{\theta}_{\text{LS}} = \bar{y}$. The M-estimators have attempted to make a robust LS estimator by replacing the squaring operation in the above sum with some other suitable function. In the terminology of Section 3.9, the function $\rho(r) = r^2$, where r is the residual, is replaced by some other form of the convex loss function, such as the one corresponding to the Huber psi function.

Rousseeuw (1987) suggested that robustness could be achieved if the median operation replaced the summation operation in the above expression. That is, choose the value of $\hat{\theta}$ so that the objective function

$$\text{med}\left(r_1^2, \ldots, r_n^2\right),$$

is minimized overall $\hat{\theta}$, where $r_i = y_i - \hat{\theta}$. Rousseeuw called this estimator the "least median of squares estimator" or LMS.

The LMS estimator is desirable because of its resistance to outliers. This procedure achieves the highest breakdown possible of 50%, a value shared by the M-estimators. On the other hand, the LMS estimator is a very inefficient estimator in the location model, at least compared to the M-estimators discussed in the previous sections. In fact, it can be see Rousseeuw and Leroy (1987), who state that the asymptotic efficiency of the LMS estimator relative to the sample mean for normal data is only about 1.4%. We recall that the sample median achieves about 64% efficiency, and the Huber M-estimator using the constant 1.345 achieves about 95% efficiency when compared to the sample mean for normal data.

The main use of the LMS estimator is not for estimating θ in the location model but for estimating the regression coefficients in the multiple regression model (MLR model). In MLR, the LMS estimator maintains its 50% breakdown point (asymptotic value) and is thus considered the best choice for the starting value for the M-estimator. We will come back to this issue in Chapter 6.

Another related estimator, also introduced by Rousseeuw, is called the least trimmed squares (LTS) estimator and is defined as the value of $\hat{\theta}$ for which the objective function

$$\sum_{i=1}^{h}\left(r^2\right)_{(i)}$$

is minimized, where $\left(r^2\right)_{(1)},\ldots,\left(r^2\right)_{(n)}$ are the n ordered squared residuals and h is the number of ordered squared residuals considered after trimming with $h \le n$. The best resistance properties are achieved when h is close to $\frac{n}{2}$, in which case, the breakdown bound is approximately 50% (and equals 50% asymptotically). When h is half the data (or half the data plus one), the asymptotic efficiency of the LTS estimator relative to the sample mean for normal data is about 7.1%, slightly better than the LMS estimator.

Computation of the LMS and the LTS estimators can be obtained exactly for the location model and only approximately for the MLR model. The user will note that the objective functions for the LMS or LTS estimators are non-differentiable, so defining equations for either estimator cannot be found. Consequently, we cannot use Newton-Raphson or IRLS-type methods to find the estimates. Fortunately, a straightforward algorithm enables the exact solution to obtain either objective function. The algorithm for LMS in the location model is as follows:

Step 1: Obtain the order statistics $y_{(1)},\ldots,y_{(n)}$.

Step 2: Compute $h=\left[\frac{n}{2}\right]+1$, where [] represents the "greatest integer" $\le\frac{n}{2}$. Thus, h represents at least half the sample size. Determine the $(n-h+1)$ intervals,

$$\left\{y_{(1)},y_{(h)}\right\},\left\{y_{(2)},y_{(h+1)}\right\},\cdots,\left\{y_{(n-h+1)},y_{(n)}\right\}$$

and their $(n-h+1)$ interval lengths,

$$\left(y_{(h)}-y_{(1)}\right),\left(y_{(h+1)}-y_{(2)}\right),\cdots,\left(y_{(n)}-y_{(n-h+1)}\right).$$

Note that each of the above intervals contains exactly h observations.

Step 3: Compute the LMS estimator of θ, $\hat{\theta}_{LMS}$, as the midpoint of the interval with the shortest length. If there are several intervals that are tied at being the shortest, then $\hat{\theta}_{LMS}$ is the average of their midpoints.

Computation of the LTS estimate proceeds exactly as in the LMS except at step 2 of the LMS algorithm; rather than computing the interval length for each of the $(n-h+1)$ intervals, we obtain the interval means,

$$\overline{y}^{(1)}, \overline{y}^{(2)}, \ldots, \mathrm{y}^{-(n-n+1)},$$

and the interval sum of squares,

$$SQ^{(1)} = \sum_{i=1}^{h}\left(y_{(i)} - \overline{y}^{(1)}\right)^2, SQ^{(2)} = \sum_{i=2}^{h+1}\left(y_{(i)} - \overline{y}^{(2)}\right)^2, \ldots, SQ^{(n-h+1)}$$
$$= \sum_{i=n-h+1}^{n}\left(y_{(i)} - \overline{y}^{(n-h+1)}\right)^2.$$

Then we replace step 3 of the LMS algorithm with the following step.

Step 3: The LTS estimate, $\hat{\theta}_{\mathrm{LTS}}$, is the interval mean $\overline{\mathrm{y}}^{(j)}$ with the smallest corresponding interval sum of squares $\mathrm{SQ}^{(j)}$.

The user will note the extremely unstable nature of the LMS estimate. Small changes in the value of x will result in rather large changes in the LMS estimate. We conclude that the LMS estimator is very sensitive to local shifting or rounding of the data. That is, very slight adjustments to the data may give rise to quite different values for the estimator, which is a very undesirable situation. We note the unstable nature of the LTS estimator, behaving in a similar manner, though not as unstable as the LMS estimator.

4.4 The Monte Carlo Method

The concept of repeated sampling is essential to understanding the role played by the Monte Carlo method in statistical theory. Many estimators' properties can be considered "repeated sampling properties." The repeated sampling concept is used to study the sampling properties of estimators or inferential procedures. One identifies a population of interest, draws a sample from this population with replacement (sampling without replacement is recommended only when the population size is much larger than the sample size), and evaluates the estimator or performs the inferential procedure based on this random sample. This process may then be repeated many times: drawing samples and performing the operation (computing the estimator or the inferential procedure) for each sample. As the number of repetitions of this sampling procedure approaches infinity, a proper summary of the operation performed on each sample approaches some theoretical value.

Probability statements concerning a random variable, whether the random variable is an estimate or a test statistic, or operations such as the expectation and variance can be interpreted in terms of operations on the repeated samples. For example, the expected value of an estimator of θ, $E(\hat{\theta})$, can be thought of simply as the average of an infinite number of values of $\hat{\theta}$, each value obtained from one repetition of the repeated sampling process. As another example, the power of a testing procedure, the two-sample *t*-test, for example, is the probability that the computed test statistic will exceed some predetermined cutoff point. This probability can be interpreted as the proportion of times and the relative frequency in a very large number of samples in which the observed test statistic exceeds the cutoff point. As the number of samples approaches infinity, this relative frequency approaches the power of the test. Thus, the expectation operator and probability statements are "repeated sampling concepts."

Often, this repeated sampling approach is used as a concept only to help students visualize the sampling distribution of a sample statistic or help interpret the notion of a probability statement. For many of the common statistics, their properties can be determined analytically. As an example, the expected value of the sample mean is analytically determined to be μ, the expected value of the random variable. There is no need to draw repeated samples from some populations in order to learn this result. On the other hand, not all estimators or inferential procedures can be described analytically. For example, the M-estimator or the maximum likelihood estimator (MLE) have properties that are asymptotic in nature only. These formulas are only approximate for a finite sample size. For example, for the M-estimator we know from the theory of functionals that

$$z = \frac{\hat{\theta}_m - \theta}{V / \sqrt{n}} \rightarrow N(0,1)$$

as n tends to infinity. The distribution of z for $n = 10$ (even for normal data) is unknown. And, of course, we must work in the finite sample size world in the practice of applied statistics.

While the properties of many procedures cannot be evaluated analytically for small samples, they can be estimated very accurately through the Monte Carlo method. Repeated samples from prespecified distributions may be obtained using the computer and random number generators. The estimator may be computed for each sample, and its behavior may be summarized across the repeated samples. Of course, due to time and space constraints, it is not possible to obtain the "infinite" number of samples required to represent the exact sampling distribution of an estimator, but a large enough number of samples can be obtained to provide an accurate representation of the sampling distribution and, consequently, an accurate representation of the properties of the estimator.

An outline of how this would be accomplished for the M-estimator, say using Huber's psi function with constant 1.5, is as follows. For some prespecified cumulative distribution, say $F(x)$, generate a random sample of size $n, y_1, \ldots, y_n$. Compute $\hat{\theta}$ based on this sample. This process is repeated a large number of times, say r times. Here, r is referred to as the "number of Monte Carlo replications." We will obtain a series of estimates:

Sample	1	2	. . .	r
estimate	$\hat{\theta}_1$	$\hat{\theta}_2$	. . .	$\hat{\theta}_r$

Now, $\hat{\theta}_1, \hat{\theta}_2, \ldots, \hat{\theta}_r$ represent a random sample of size r from the sampling distribution of $\hat{\theta}$, with cumulative distribution $F(\hat{\theta})$. Thus, the ECDF (recall that the empirical cumulative distribution of this collection of estimates would place a mass of $\frac{1}{r}$ at each value of $\hat{\theta}_i$) of the r estimates, denoted by $F_r(\hat{\theta})$, represents an estimate of the CDF, $F(\hat{\theta})$.

We can summarize the behavior of $\hat{\theta}$ over the r Monte Carlo replications by computing

$$\bar{\hat{\theta}} = \frac{\sum_{i=1}^{r} \hat{\theta}_i}{r} \text{ and } s^2(\hat{\theta}) = \frac{\sum_{i=1}^{r} \left(\hat{\theta}_i - \bar{\hat{\theta}}\right)^2}{r-1},$$

where $\bar{\hat{\theta}}$ and $s^2(\hat{\theta})$ are the Monte Carlo mean and variance of $\hat{\theta}$, respectively.

We know from the Glavenko-Cantelli theorem, introduced in Chapter 1, that the ECDF approaches the CDF (cumulative distribution function) as the sample size approaches infinity. Thus, it follows that

$$F_r(\hat{\theta}) \to F(\hat{\theta})$$

as the number of Monte Carlo replications, r, approaches infinity. Of course, this is as expected since as r approaches infinity, the Monte Carlo method is simply the repeated sampling concept for generating the sampling distribution of $\hat{\theta}$.

Thus, as r approaches infinity, it follows that

$$\bar{\hat{\theta}} \to E(\hat{\theta}) \text{ and } s^2(\hat{\theta}) \to \text{Var}(\hat{\theta}) \cdot$$

To see this, recall that the sample mean of r numbers can be expressed in terms of the ECDF as a functional, $\bar{\hat{\theta}} = \sum \frac{\hat{\theta}_i}{r} = \int \hat{\theta} dF_r(\hat{\theta})$, which converges in probability to $\int \hat{\theta} dF(\hat{\theta}) = E(\hat{\theta})$, as r goes in infinity. A similar argument holds for the Monte Carlo variance.

If we perform a Monte Carlo experiment with $r = 1000$, for example, we know that the Monte Carlo mean and variance of $\hat{\theta}$ are "close" to the expected value and variance of $\hat{\theta}$, respectively. Consequently, even though we may not have a formula for computing the expected value and variance of $\hat{\theta}$, we can find these values for every distribution function $F(y)$ through Monte Carlo simulation.

In addition, all other properties of $\hat{\theta}$ that come from the sampling distribution of $\hat{\theta}, F(\hat{\theta})$, can be estimated using this technique. For example, we can estimate any percent point of $F(\hat{\theta})$, say $F_{1-\frac{\alpha}{2}}(\hat{\theta})$, the $\left(1-\frac{\alpha}{2}\right)100\%$ point of $F(\hat{\theta})$, by using the $\left(1-\frac{\alpha}{2}\right)100\%$ point of the Monte Carlo distribution of $\hat{\theta}, F_{1-\frac{\alpha}{2}}(\hat{\theta})$. In this way, we can study the properties of confidence intervals and proposed test procedures.

Clearly, one of the positive features of the Monte Carlo method is the ease with which it can be applied. A statistical software package such as R can generate random data from a wide variety of distributions. With minor programming skills and a little imagination, conducting a Monte Carlo experiment is easy. On the negative side, the Monte Carlo results are rather limited in that the results apply only to the situations studied through the design of the Monte Carlo. If results are obtained for sizes 10 and 20 samples, it is unclear how these results apply to other sample sizes. If the study used normal random numbers and certain results were obtained, it is not clear how these results extend to other distributions.

The Monte Carlo method will be demonstrated using R. Suppose the researcher wishes to compare a variety of estimators of θ in the location model. Some estimators of interest include the sample mean, the median, the *m*-estimators, the Huber with constant 1.345 (the 95% efficient value against the sample mean for normal data), and the LMS and LTS estimators. It is interesting to study these estimators across a wide variety of distributions of errors, ranging from the normal, a distribution unlikely to generate extreme outliers, to heavier-tailed distributions.

The contaminated normal family, introduced in Section 3.5, provides a nice way of generating data with a gradation of mild heavy-tailed to extreme heavy-tailed behavior. For example, a contaminated normal $CN(y;\varepsilon,\sigma^2)$, with $\varepsilon = 0.05$ and $\sigma^2 = 9$ is only slightly heavier tailed than the normal, while choosing $\varepsilon = 0.1$ with $\sigma^2 = 25$ results in moderate heavy-tail behavior compared to the normal and choosing $\varepsilon = 0.25$ with $\sigma^2 = 100$ results in a very heavy-tailed distribution with a high percentage of extreme outliers. Another extremely heavy-tailed distribution worth considering is the Cauchy.

Another consideration in the design of the Monte Carlo experiment is the sample size. Perhaps a good choice would be to evaluate samples of sizes 10, 20, and 40, which seem to extend across a range of small or moderate sample sizes.

4.4.1 A Mini R Experiment

To keep the results manageable, our Monte Carlo will consider only two distributions, the standard normal and a contaminated normal with $\varepsilon = 0.1$ and $\sigma^2 = 9$, a moderately heavy-tailed distribution. We will consider only the n = 20 case.

The Monte Carlo will require a special program to compute the above five estimators for each Monte Carlo repetition of each distribution. See the coming R code for details.

The R code to obtain the $N(0,1)$ results with $n = 20$ is presented for $r = 1000$ repetitions.

```
# Ensure reproducibility
set.seed(123)
# Function to generate standard and contaminated normal samples
generate_samples <- function(n, epsilon = 0, sigma2 = 1)
{ mixture <- rbinom(n, 1, epsilon) rnorm(n, mean = 0,
sd = sqrt((1 - mixture) * 1 + mixture * sigma2))}
# Define estimators
estimators <- list( mean = mean, median = median, huber =
function(x) rlm(x ~ 1, psi = psi.huber, k = 1.345)$coeffi-
cients[1], lts = function(x) ltsReg(x ~ 1)$coefficients[1])
# Monte Carlo simulation function
monte_carlo <- function(n, r, epsilon, sigma2, estimators)
{ results <- setNames(vector("list", length(estimators)),
names(estimators))
   for (i in 1:r) { samples <- generate_samples(n, epsilon, sigma2)
     results[[i]] <- lapply(estimators, function(f) f(samples)) }
   return(results)}
# Parameters for the simulation
params <- list(n = 20, r = 1000, epsilon = 0.1, sigma2 = 9)
# Execute the Monte Carlo simulation
mc_results <- do.call(monte_carlo, c(params, list(estimators)))
# Checking and preparing data for plotting
if (all(sapply(mc_results, length) == params$r)) { results_df
<- as.data.frame(mc_results)
```

```
   results_melted <- melt(results_df, variable.name = "Estima-
tor", value.name = "Estimate")
# Plotting using ggplot2 for horizontal box plots without grid
lines
   ggplot(results_melted, aes(y = Estimator, x = Estimate)) +
geom_boxplot() + ylab("Estimator") +
lab(expression(hat(theta))) +
   ggtitle("Boxplots of the Five Estimators Based on 1000
Repetitions from N(0,1)") + theme_minimal() + theme(
      panel.grid.major = element_blank(),
      panel.grid.minor = element_blank(),
      panel.background = element_blank(),
      axis.line = element_line(colour = "black") )}
```

Upon completion of the macro, 1,000 values for each sample mean, the sample median, the Huber *m*estimate with the constant of 1.345, the LMS, and the LTS are given, respectively. The correlation test for normality strongly supports the normality assumption for all estimators, with the smallest correlation coefficient between the observed sample estimates and the normal scores of 0.997 resulting in the sample mean.

The numerical summaries for this situation are as follows (edited for better display):

```
# Define a function to calculate descriptive statistics
describe_stats <- function(data) { n <- length(data)
  mean_val <- mean(data)
  median_val <- median(data)
  sd_val <- sd(data)
  sem_val <- sd_val / sqrt(n) # Standard Error of the
  Mean
  c(N = n, MEAN = mean_val, MEDIAN = median_val, STDEV =
sd_val, SEMEAN = sem_val)}
# Apply the function to each estimator's results and
collect them into a data frame
descriptive_stats <- sapply(mc_results, describe_stats)
# Convert the results into a more readable format
descriptive_stats_df <- t(descriptive_stats)
rownames(descriptive_stats_df) <- names(mc_results)
# Print the results
print(format(descriptive_stats_df, scientific=FALSE,
digits = 5))
          N     MEAN        MEDIAN      STDEV       SEMEAN
  mean    1000 -0.0137277  -0.0089653  0.2969753  0.0093912
  median  1000 -0.0201564  -0.0295232  0.2867721  0.0090685
  huber   1000 -0.0170253  -0.0127615  0.2534464  0.0080147
  lts     1000 -0.0175601  -0.0137247  0.2597738  0.0082148
```

It is again seen that all estimates are reasonably "close" to the parameter of interest, $\theta = 0$, as all estimators are within two standard error units of zero.

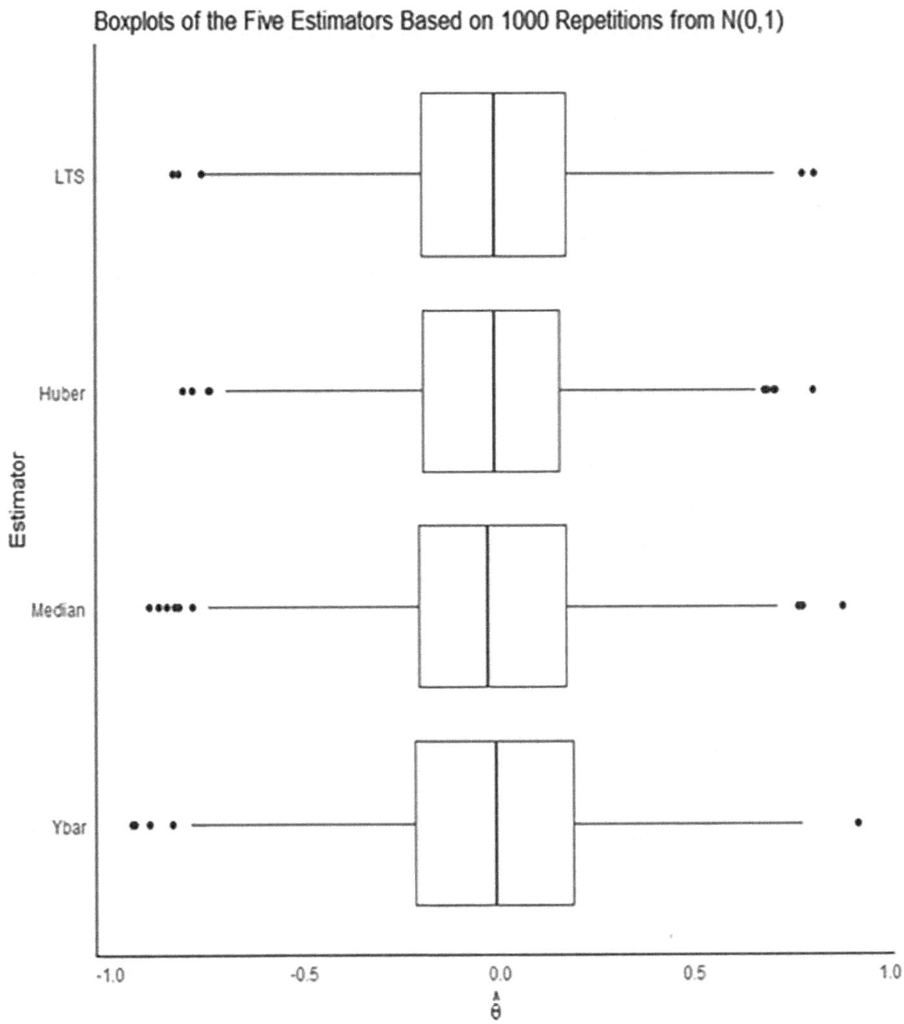

FIGURE 4.6
Box plots of the five estimators.

The sample mean's standard deviation further validates our Monte Carlo experiment. The box plots for the estimated sampling distributions are in Figure 4.6.

We note that the estimated sampling distributions for the sample mean, sample median, and the Huber $m$estimate are very "normal" in appearance, while the latter two estimators have a less than distinctive normal shape. Of course, theory tells us that the sampling distribution of the sample mean is normal regardless of sample size when the observations are from a normal population, as they are here. For the other four estimators, the asymptotic sampling distributions are normal.

The box plots above reveal very little difference between the mean and the Huber sampling distributions, while the median has an obviously larger interquartile range (IQR). Both the LMS and the LTS estimates appear to be symmetric about zero, as are the other three estimators, but they both have much larger IQRs.

Consequently, the normality assumption cannot be refuted by this sampling distribution. The Monte Carlo results support our theory so far. In fact, the correlation test supports the normality assumption for all five estimators, with the smallest correlation coefficient value of 0.9987 for the LTS estimator, still far above the 0.05 cutoff value of 0.9841.

It is also noted that, as expected, the sample mean has the smallest Monte Carlo standard deviation. The Monte Carlo standard deviation is an estimate of the standard deviation of the estimator, $\mathrm{SD}\left(\hat{\theta}\right) = \sqrt{\mathrm{Var}\left(\hat{\theta}\right)}$. For the sample mean, the theoretical value for the standard deviation is $\sigma / \sqrt{n} = 1/\sqrt{20} = 0.2236$, extremely close to the Monte Carlo result of 0.22974, a relative error of only 2.7%. This difference is due to the Monte Carlo sampling variability. Unfortunately, the theoretical values for the other four estimators cannot be compared to the Monte Carlo results since variance expressions for these estimators are asymptotic results only. In our experiment, n is only 20. But, since our Monte Carlo standard deviation for the sample mean was so close to the theoretical result, we will assume that the Monte Carlo results for the other estimators are close to their theoretical but unknown results. The efficiencies of the four alternative estimators with respect to the sample mean may then be computed and compared to the theoretical results. The efficiencies are computed as

$$\mathrm{eff}\left(\hat{\theta}, \bar{y}\right) = \frac{\mathrm{est\ Var}\left(\hat{y}\right)}{\mathrm{est\ Var}\left(\hat{\theta}\right)} = \frac{\mathrm{stdev}(\bar{y})^2}{\mathrm{stdev}(\hat{\theta})^2}.$$

The four efficiencies are 0.70, 0.94, 0.26, and 0.23 for the median, Huber with constant 1.345, LMS, and LTS estimators, respectively. The theoretical (asymptotic) efficiencies are 0.64, 0.95, 0.014, and 0.071, respectively. It looks as though the asymptotic results are accurate for the Huber M-estimator at $n = 20$, and much less so for the other LMS and LTS estimators. The results for the contaminated normal situation, $\mathrm{CN}(y; 0.1, 9)$, are interesting in that they are expected to shed some light on the lack of resistance to outliers of the sample mean and the desirable robustness properties of the Huber

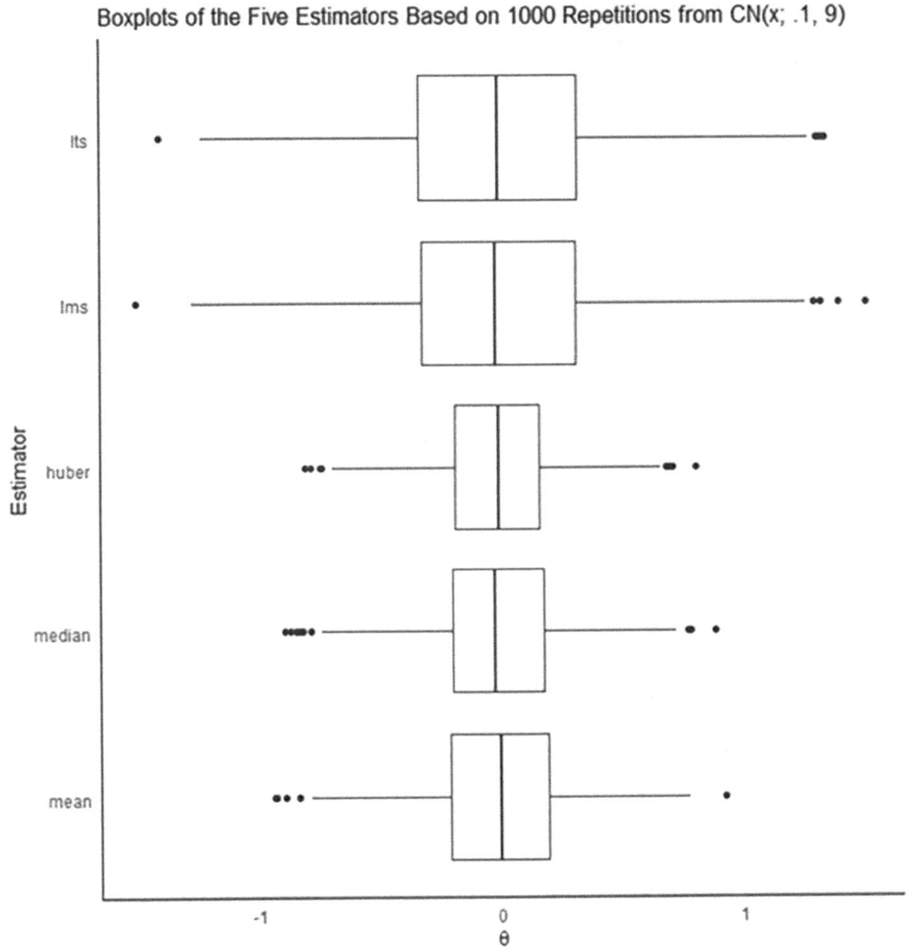

FIGURE 4.7
Box plots of the five estimators based on the contaminated normal.

M-estimator. The graphical results, beginning with the box plot, are in Figure 4.7.

The box plots reveal that the Huber M-estimator has the smallest IQR, with the largest IQR belonging to the LMS and LTS estimators (see the second box plot figure). The sample mean, sample median, and the Huber M-estimator all had several outliers among the 1,000 sample estimates. The correlation test for normality strongly supports the normality assumption for all estimators, with the smallest correlation coefficient between the observed sample estimates and the normal scores of 0.997, resulting in the sample mean. The numerical summaries for this situation are as follows (edited for better display):

```
> print(summary_stats)
              N       MEAN    MEDIAN   STDEV  SEMEAN
ybar       1000 -0.00274   0.00376 0.30357 0.00960
median     1000 -0.01087 -0.01012 0.29970 0.00948
Huber      1000 -0.00740 -0.00726 0.26505 0.00838
LMS        1000 -0.01000 -0.00060 0.44620 0.01410
LTS        1000 -0.01140 -0.01210 0.47880 0.01510
```

It is again seen that all estimates are reasonably "close" to the parameter of interest, $\theta = 0$, as all estimators are within two standard error units of zero. The sample mean's standard deviation further validates our Monte Carlo experiment. The theoretical value for the standard deviation for the sample mean based on a random sample of size 20 for a $\mathrm{CN}(y;0.1,9)$ is given as

$$\mathrm{SD}(y) = \frac{\sigma_y}{\sqrt{n}} = \frac{\sqrt{(1-\varepsilon)(1)+\varepsilon(\sigma^2)}}{\sqrt{n}} = \frac{\sqrt{.9+.1(9)}}{\sqrt{20}} = 0.3,$$

a value very close to the Monte Carlo result of 0.3036, a 1.2% relative error. Further verification is provided by the standard deviation for the sample median, where the asymptotic value is given by $\mathrm{SD}(M) = \dfrac{1}{2\cdot\sqrt{n}f(0)} = 3003,$ where $f(0)$ denotes the density of y evaluated at $y = 0$ (from the variance expression for M in Section 3.6). The Monte Carlo result of 0.2997 is within 2% relative error. The efficiency robustness of the Huber M-estimator is now apparent, as its Monte Carlo standard deviation is the smallest among the five estimators. In fact, the efficiency values of the four estimators, compared to the sample mean computed by the efficiency formula given earlier, are 1.025, 1.312, 0.463, and 0.4020 for the sample median, Huber M-estimator, LMS, and LTS estimators, respectively.

Consequently, we have demonstrated that the Huber M-estimator with tuning constant 1.345 is approximately 95% as efficient as the sample mean when the sample size is 20, and the data are drawn from a normal distribution. Moreover, it achieves approximately 131% efficiency relative to the sample mean under a contaminated normal distribution with a contamination proportion of 0.1 and a contamination variance of 9, representing a moderately heavy-tailed distribution. The sample median is about 70% as efficient as the sample mean for normal data, n = 20, and 103% as efficient for the contaminated normal case. The LMS and LTS estimators are only about 25% as efficient for the normal data, n = 20, and about 42% as efficient for the contaminated normal distribution. Clearly, the Huber M-estimator is the superior estimator in the situation examined here.

Of course, our study could easily be extended along the lines suggested earlier, covering more distributions, sample sizes, and estimators. The Monte Carlo method is a useful and easily implemented methodology. A final word of caution is in order, however. The Monte Carlo results are extremely sensitive to errors in programming. The user of this approach must exert great care to validate the Monte Carlo results, hopefully through preliminary experiments to discover errors before major time and effort is spent collating incorrect results. Monte Carlo validation can be accomplished by comparing the results to known theoretical results, such as those illustrated above, or by comparing the results of other published Monte Carlo results.

4.5 The Bootstrap Method

The Monte Carlo technique is a great aid in the study of the properties of a statistical procedure. The Monte Carlo method works by resampling from some specific CDF, F(y). Of course, to apply this technique, one needs to identify the CDF of interest. The resampling philosophy can also help us to extract additional information from a single sample of data. Resampling from the ECDF, rather than the CDF, is referred to as the bootstrap method. One useful application of the bootstrap method is enabling the user to approximate the sample variance of an estimator, say the LMS estimator, based on a small sample size. Since no small-sample variance formulas exist for many estimators, including the LMS estimator, an approximate variance for a sample estimator is extremely valuable information.

To set up the problem, suppose we have the estimate $\hat{\theta}$, and we need to obtain $\mathrm{se}(\hat{\theta})$. Unfortunately, we have no formula for this standard error, as would be the case for the M-estimators, LMS, LTS, or even the median. One approach to the problem is mimicking the Monte Carlo method using resampling. However, instead of resampling from the CDF, $F(y)$, which we do not know, we resample from the ECDF, $\mathrm{F_n}(\mathrm{y})$, which we do know. The resampling is performed "with replacement" so that each resampled observation is sampled from an independent and identical distribution.

An overview of the bootstrap method is as follows. Given a sample of the response variable y, denoted as $\mathrm{y_1,\ldots,y_n}$, with ECDF $\mathrm{F_n}(\mathrm{y})$, select n new observations, $\mathrm{y_1^*,\ldots,y_n^*}$, called the "bootstrap sample," from $\mathrm{y_1,\ldots,y_n}$ using sampling with replacement. Compute the estimate of θ, say $\hat{\theta}^*$, based on this bootstrap sample. Repeat this process B times, where B represents the number of bootstrap samples obtained from the initial sample $\mathrm{y_1,\ldots,y_n}$. Upon completion of this resampling process, we will have obtained B estimates of θ, all from one sample of data. Our estimates are summarized in Table 4.1.

TABLE 4.1
Bootstrapping Estimators

Bootstrap Sample	1	2	3	......	B
Estimate	$\hat{\theta}_1^*$	$\hat{\theta}_2^*$	$\hat{\theta}_3^*$	......	$\hat{\theta}_B^*$

We can then compute the bootstrap mean and bootstrap variance of our estimator as

$$\bar{\hat{\theta}}^* = \frac{\sum_{i=1}^{B} \hat{\theta}_i^*}{B} \text{ and } s^2\left(\hat{\theta}^*\right) = \frac{\sum_{i=1}^{B}\left(\hat{\theta}_i^* - \bar{\hat{\theta}}^*\right)^2}{B-1}$$

as our summary statistics of the bootstrap method. Then we can use $\bar{\hat{\theta}}^*$ as an estimate of $E\left(\hat{\theta}\right)$. That is, $\hat{E}\left(\hat{\theta}\right) = \bar{\hat{\theta}}^*$. Additionally, $s^2\left(\hat{\theta}^*\right)$ is an estimate of $\text{Var}\left(\hat{\theta}\right)$. That is, $\widehat{Var}\left(\hat{\theta}\right) = s^2\left(\hat{\theta}^*\right)$. We call the $\sqrt{\widehat{\text{Var}}\left(\hat{\theta}\right)}$ the bootstrap standard error of $\hat{\theta}$, and it is an estimate of $\sqrt{\text{Var}\left(\hat{\theta}\right)}$.

Some comments concerning the bootstrap are in order. First, it is obvious that if we sample with a replacement from the original sample, we will have repeated observations in the bootstrap sample. This can create problems with the bootstrap procedure if the sample size, n, is small.

Second, we can think of the B bootstrap estimates, $\hat{\theta}_1^*, \ldots, \hat{\theta}_B^*$ as giving rise to the "bootstrap ECDF" of $\hat{\theta}$, denoted by $F_B\left(\hat{\theta}^*\right)$. This ECDF may be thought of as an estimate of the ECDF of $\hat{\theta}$, an estimate of an estimate, so to speak. Thus, $F_B\left(\hat{\theta}^*\right)$ is an estimate of $F_r\left(\hat{\theta}\right)$, the ECDF of $\hat{\theta}$, obtained from some repeated sampling scheme, such as the Monte Carlo method. Since ECDFs converge to CDFs as the number of observations (in this case, an observation is a $\hat{\theta}^*$) increase, it follows that $F_B\left(\hat{\theta}^*\right)$ behaves like $F_r\left(\hat{\theta}\right)$, as B increases. Generally, B of 100–250 is considered large enough for $F_B\left(\hat{\theta}^*\right)$ to be a proper estimate of $F_r\left(\hat{\theta}\right)$. These ideas can be expressed in terms of functional notation. We have for the bootstrap mean:

$$\bar{\hat{\theta}}^* = \frac{\sum \hat{\theta}_i^*}{B} = \int \hat{\theta}_i^* dF_B\left(\hat{\theta}^*\right) \overset{B\to\infty}{} \int \hat{\theta}\, dF_r\left(\hat{\theta}^*\right) = \hat{E}\left(\hat{\theta}\right),$$

a justification that the bootstrap mean of $\hat{\theta}$ is an estimator of $E(\hat{\theta})$. In a similar fashion, we can show that the bootstrap variance of $\hat{\theta}$ is an estimator of $\text{Var}(\hat{\theta})$.

In fact, the bootstrap ECDF, $F_B(\hat{\theta}^*)$, can be used to estimate other properties of $F(\hat{\theta})$, the (often) unknown sampling distribution of $\hat{\theta}$. For example, we can estimate the shape of $F(\hat{\theta})$ by using the shape of $F_B(\hat{\theta}^*)$. Furthermore, we can use appropriate percent points of $F_B(\hat{\theta}^*)$ to construct confidence intervals for θ. There are a variety of techniques developed to construct confidence intervals using the bootstrap sample estimates. One method, the most straightforward one, called the bootstrap percentile method, defines the $(1-\alpha)100\%$ confidence interval estimate for θ as $\left(F_{B,\frac{\alpha}{2}}(\hat{\theta}^*), F_{B,1-\frac{\alpha}{2}}(\hat{\theta}^*)\right)$, where $F_{B,\frac{\alpha}{2}}(\hat{\theta}^*)$ and $F_{B,1-\frac{\alpha}{2}}(\hat{\theta}^*)$ are the $\alpha/2$ and the $1-\alpha/2$ percent points of $F_B(\hat{\theta}^*)$. Usually, B of 1,000 or more is considered appropriate for this method to be effective. A second method, the bias-corrected procedure, is also used to obtain confidence intervals based on the bootstrap. The interested reader is referred to Efron and Trishirani (1993) for more information on this technique.

In summary, the bootstrap and Monte Carlo methods involve resampling. In the Monte Carlo method, we resample from the true CDF of $y, F(y)$, to obtain the r sample estimates, $\theta_1,\ldots,\theta_r$. Together, these estimates constitute the ECDF of $\hat{\theta}, F_r(\hat{\theta})$, an estimate of the CDF of $\hat{\theta}$, $F(\hat{\theta})$. Summary statistics expressed in terms of functionals from $F_r(\hat{\theta})$ are estimates of the corresponding functionals from $F(\hat{\theta})$.

In the bootstrap method, we resample from the ECDF of $y, F_n(y)$, based on a random sample of y, to obtain the B bootstrap estimates $\theta_1^*,\ldots,\theta_B^*$. Together, these estimates constitute the bootstrap ECDF of $\hat{\theta}, F_B(\hat{\theta}^*)$, an estimate of the ECDF of $\hat{\theta}$, which in turn is an estimate of the CDF of $\hat{\theta}$. Thus, $F_B(\hat{\theta}^*)$ is also an estimate of this CDF. Analogously, summary statistics expressed in terms of functionals from $F_B(\hat{\theta}^*)$ are estimates of the corresponding functionals from $F(\hat{\theta})$.

The R program is especially well suited for demonstrating the bootstrap method as a technique for estimating θ based on some $\hat{\theta}$, estimating the standard error of $\hat{\theta}$, and obtaining the percentile confidence interval estimate of θ.

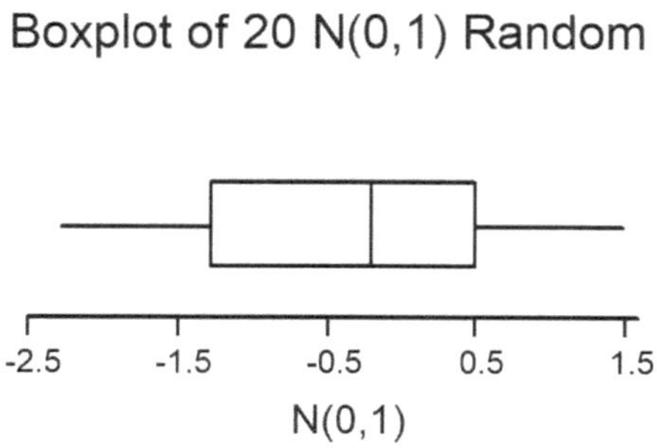

FIGURE 4.8
Box plot of N(0, 1).

4.5.1 A Mini R Experiment

To demonstrate the bootstrap resampling method, suppose we want to obtain estimates of the standard error for our five estimators of the location parameter. We will begin our experiment by generating a sample of size 20 from a $N(0,1)$ distribution; see Figure 4.8. The estimators we will compute are the sample mean, median, Huber M-estimate with constant 1.345, LMS, and LTS. We will resample $B = 500$ times. We will be interested in comparing the bootstrap results with those obtained through the analytic, though asymptotic, formulas and with those results obtained through our Monte Carlo study of this same example in the last section. Obtain a sample from the $N(0,1)$ distribution to begin the experiment. On R, type as follows:

```
# Generate a sample of size 20 from a standard normal distri-
bution sample_data <- rnorm(20, mean = 0, sd = 1)
```

The five estimators can be computed for this data using the macros introduced in the previous section. On R, type as follows:

```
    set.seed(123) # Ensure reproducibility
    c1 <- rnorm(20, mean = -0.392, sd = 1.087)
    # Calculate summary statistics
  summary_stats <- function(data) {
    n <- length(data)
    mean_val <- mean(data)
    median_val <- median(data)
    sd_val <- sd(data)
    sem_val <- sd_val / sqrt(n) # Standard Error of the
    Mean
    c(N = n, Mean = mean_val, Median = median_val, StDev =
    sd_val, SEMean = sem_val)}
    > print(c1_stats_df)
            N       Mean     Median    StDev    SEMean
    N(0,1)  20 -0.2380549 -0.2615761 1.057287 0.2364166
```

Note that there are no outliers in this dataset. The sample mean is −0.2380549, slightly off from the true value of 0. The sample variance is 1.057, which is very close to the true standard deviation of 1.

Next, we will get the Huber M-estimator, using a constant of 1.345.

```
# Compute the Huber M-estimator with a constant of 1.345
huber_model <- rlm(c1 ~ 1, psi = psi.huber, k = 1.345)
huber_estimate <- coef(huber_model)
huber_se <- summary(huber_model)$sigma / sqrt(length(c1))
# Compute 95% Confidence Interval for Huber M-Estimator
t_value <- qt(0.975, df = length(c1) - 1)
ci_lower <- huber_estimate–t_value * huber_se
ci_upper <- huber_estimate + t_value * huber_se
```

The Huber M-estimate is

```
Huber M estimate: 0.1475469
```

The standard error for the Huber is

```
Standard error: 0.2011423
```

The 95% confidence interval for theta is: (huber−t*se, huber+t*se):

```
95% Confidence Interval for Theta: -0.2734488 to
0.5685426
```

To get the LMs and LTs estimates:

```
# Compute LMS and LTS estimators
lms_lts_model <- lqs(c1 ~ 1, method = "lms")
lts_model <- lqs(c1 ~ 1, method = "lts")
```

The LMs and LTs are, respectively:

```
LMS estimate: -0.0798521
LTS estimate: 0.1438687
```

We recall that the small-sample formulas for the standard error for the median, M-estimator, LMS, and LTS estimators are unknown. The above *CI* and *se* results are only a good guess at what they should be. The results above in bold will be studied more closely using the bootstrap method. To perform the bootstrap resampling on this data, use the following R code:

```
set.seed(123)
c1 <- rnorm(20, mean = 0, sd = 1)
# Perform the bootstrap
B <- 500
boot_results <- bootstrap_estimators(c1, B)
```

TABLE 4.2

Comparison of the Bootstrap Means and Standard Errors to the Five Estimates and Standard Errors Based on One Sample of Size 20 from N(0, 1)

Estimator	Estimate	Std. Error	Bootstrap Mean	Bootstrap Std. Error
ybar	−0.392	0.243	−0.394	0.224
Median	−0.206	0.292	−0.286	0.305
Huber	−0.390	0.260	−0.353	0.255
LMS	0.048	1.278	−0.191	0.561
LTS	0.071	1.280	−0.154	0.570

The bootstrap means and standard deviations for the five estimators are given.

```
Estimator      N     Mean     Median   StDev
ybar          500  -0.3940  -0.4006  0.2244
median        500  -0.2856  -0.2059  0.3048
huber         500  -0.3533  -0.3525  0.2546
LMS           500  -0.1914  -0.2089  0.5608
LTS           500  -0.1537  -0.1079  0.5699
```

We are particularly interested in the values under the "mean" and "stdev" results. The numbers under the mean column are the bootstrap means of the five estimators based on the original 20 observations. The numbers under the stdev column are the bootstrap estimates of the standard error of each of the five estimators. These column entries may be compared to the bold values presented above. For ease of comparison, all numbers appear in Table 4.2.

The "Std. Error" column in Table 4.2 was computed using the available standard error formulas for the estimators. For example, the standard error for ybar is s/sqrt(*n*); the formulas for the median, LMS, and LTS estimators are based on asymptotic results since finite sample size formulas do not exist for these estimators, and the standard error for the Huber M-estimator is based on the formula presented earlier in Section 4.2.

Perhaps the most useful of the numbers in the above table are the bootstrap standard errors for each estimator. These are estimates of the true standard errors. For example, we know that the actual standard deviation for the sample mean is $\sigma / \sqrt{n} = 1 / \sqrt{20} = 0.224$. For our random sample, we estimate 0.224 with the standard error of $s / \sqrt{n} = 1.087 / \sqrt{20} = 0.243$. The bootstrap standard error is 0.224, which is really an estimate of 0.243, the standard error of the sample mean. There is a relative absolute error of the bootstrap standard error of approximately 7.8% with respect to the observed standard error. While this error is slightly larger than we might like, the bootstrap standard error of the sample mean is close enough to the observed

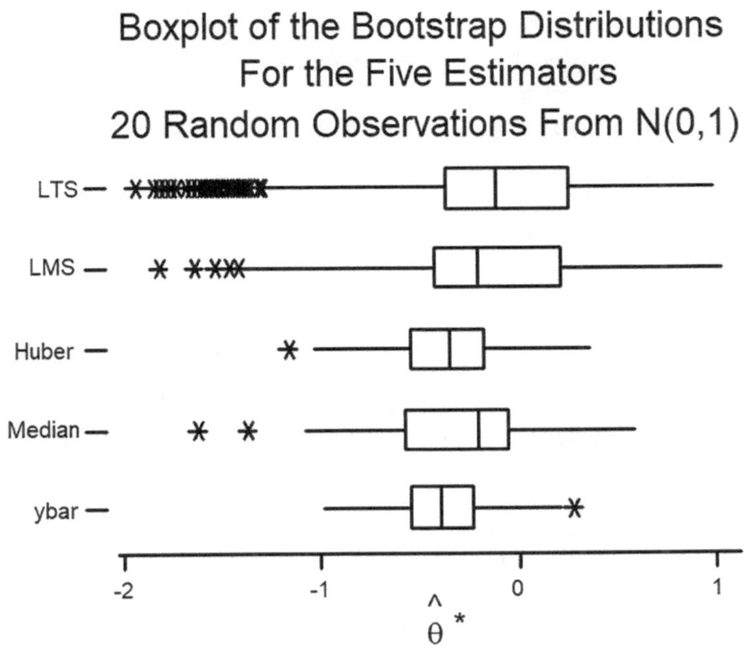

FIGURE 4.9
Box plot of the bootstrap distributions for the five estimators.

standard error to suggest that the other bootstrap standard errors are close to the observed standard errors of the other four estimators. Since no exact finite sample formulas exist for these other estimators, the bootstrap results may provide more accurate estimates of standard errors than those provided by the asymptotic formulas, which are approximate formulas for $n = 20$, the situation in this example. For example, the asymptotic formulas for the standard errors for the LMS and LTS estimators result in much larger values than those provided by the bootstrap standard errors. The standard error formula for the Huber M-estimator, $se(\hat{\theta}) = \hat{V} / \sqrt{\sum w_i}$, gives a result very close to the bootstrap standard error, with a relative absolute error of 1.9%, while the relative error for the standard error for the median is about 9%.

The five estimators' bootstrap distributions can be displayed using the box plot in Figure 4.9.

We see that the sample mean and the Huber have distinctly normal-shaped bootstrap distributions, while the distributions for the other estimators show strong heavy-tailed tendencies, a fact revealed clearly by the box plots. The bootstrap distributions are reasonably symmetric in all cases, suggesting that the bootstrap percentile method for constructing confidence intervals is a reasonable procedure.

TABLE 4.3

Comparison of 95% Confidence Intervals Obtained by Analytic Methods and the Bootstrap Percentile Method

Estimator	Analytic Lower Limit	Analytic Upper Limit	Bootstrap Lower Limit	Bootstrap Upper Limit
ybar	−0.901	0.117	−0.851	0.031
Median	−0.779	0.367	−0.897	0.166
Huber	−0.934	0.153	−0.873	0.096
LMS	−2.457	2.553	−1.439	0.705
LTS	−2.435	2.577	−1.667	0.707

Application of the bootstrap percentile method to obtain 95%, for example, confidence intervals for θ would be formed from the 2.5th and 97.5th sample percentiles of the bootstrap distributions. For $B = 500$, this amounts to using the average of the 12th and 13th ordered values and the average of the 488th and 489th ordered values of θ for each estimator. Table 4.3 lists the 95% confidence limits for each of the five estimators obtained by the bootstrap percentile method and obtained via formulas.

A fairly strong agreement exists with the confidence interval for the sample mean computed using the standard formula: $\left(\bar{y} \pm t_{\left(1-\frac{\alpha}{2},n-1\right)} \dfrac{s}{\sqrt{n}}\right)$ and the 95% bootstrap interval computed by the percentile method. At least, they are close enough to demonstrate that the bootstrap approach provides a sensible alternative to using the formulas when the formulas exist. There is strong agreement between these two approaches for the Huber estimate as well, again suggesting that the formula given for the confidence interval in Section 4.2 works well for finite samples. On the other hand, there is less agreement with the confidence interval based on the median for the two methods and even less agreement for the LMS and LTS estimators. This evidence further supports the claim that the standard error formulas for the median, LMS, and LTS estimators are unreasonable for small sample sizes.

However, exact confidence intervals may be obtained for the median for small sample sizes. The method described in most books on nonparametric statistics will not be reviewed here. It is also of interest at this point to compare the standard error results obtained by the analytic formulas and those obtained by the bootstrap method for each of the five estimators for this example to the standard errors obtained by the Monte Carlo technique applied to this same example. The Monte Carlo results for the $N = 20$, $N(0,1)$ situation were presented in Section 4.4. Since the number of Monte Carlo replications was 1,000, we could consider the Monte Carlo generated standard errors to be very accurate estimates of the true but unavailable (except for the sample mean) standard errors. Table 4.4 presents all three sets of standard errors.

TABLE 4.4

Comparison of Standard Errors for the Five Estimators Obtained by Analytic Methods, the Bootstrap Method, and the Monte Carlo Method

Estimator	Analytic Std. Error	Bootstrap Std. Error	Monte Carlo Std. Error
ybar	0.243	0.224	0.230
Median	0.292	0.305	0.275
Huber	0.260	0.255	0.237
LMS	1.278	0.561	0.449
LTS	1.280	0.570	0.475

TABLE 4.5

Summary of the Five Estimates and Their Standard Errors for the Rat Data

Estimator	ybar	Median	Huber	LMS	LTS
Estimate	7.60	5.00	5.28	4.50	4.50
Std. error	2.11	2.65*	1.03	3.46	3.46

A comparison of the three columns above reveals that the bootstrap standard errors are closer to the Monte Carlo standard errors than to the analytic standard errors for all the estimators except for the median. For the LMS and LTS estimators, the bootstrap standard errors are far superior to those obtained by the asymptotic formulas.

The bootstrap method provides a measure of variability and a mechanism for obtaining confidence intervals based on any estimator and is especially helpful in those cases where the analytic methods do not exist. The procedure can be easily applied to any real dataset. For example, let us reconsider the "rat data" used in previous sections. Table 4.5 summarizes the five different estimates we are considering and their standard errors. These standard errors were all obtained using the appropriate asymptotic formulas, except for the sample mean, which uses the exact formula. All tabled values were presented in previous sections of this chapter.

The * in the above table indicates that the median, standard error were computed using the asymptotic formula $var(M) = \frac{1}{4f^2(y_{0.5})n}$, where $f(y)$ is assumed to be the normal curve, and σ is estimated by the sample standard deviation of 6.69 for the rat data.

A summary of the bootstrap mean and standard errors are presented in Table 4.6. The tabled data was obtained by using the rat data with $B = 200$. The results are as follows.

We note that the bootstrap mean and standard error match very closely the results in the previous table for ybar obtained through the formulas. The

TABLE 4.6

Summary of the Bootstrap Estimates and Standard Errors for the Five Estimators for the Rat Data

Estimator	ybar	Median	Huber	LMS	LTS
BS estimate	7.49	5.26	5.54	4.72	4.83
BS std. error	1.92	1.67	1.45	1.72	2.19

TABLE 4.7

Comparison of 95% Confidence Intervals Obtained by Analytic Methods and the Bootstrap Percentile Method

Estimator	ybar	Median	Huber	LMS	LTS
CI	(2.8, 12.4)	(−.2, 10.2)	(3.0, 7.6)	(−2.3, 11.3)	(−2.3, 11.3)
BS CI	(4.1, 12.2)	(3.5, 12.7)	(3.7, 12.5)	(3.0, 13.0)	(2.7, 17.7)

bootstrap standard error results are again much smaller than those given by the asymptotic formulas for the median, LMS, and LTS estimators, while the results for the Huber M-estimator are larger.

A comparison of the 95% confidence intervals for these five estimators based on the (asymptotic for the median, LMS, and LTS) distribution of the estimators and the bootstrap percentile method is given in Table 4.7:

We note that the differences are rather severe, except for ybar, where the agreement is quite close. The agreement between the two methods is poor for the upper confidence limit for the Huber M-estimate. However, looking at the 200 ordered values for this estimator reveals that the 194th element is 9.7, much closer to the 7.6 obtained from the formula. Perhaps better results would have been obtained had we used $B = 1000$ instead of $B = 200$.

The bootstrap method's popularity is growing rapidly, primarily due to its application in statistical models more complicated than the location model. For example, the bootstrap method is used to find confidence intervals for estimates from multiple regression models and from nonparametric regression procedures such as local polynomial regression models, among many other situations. We will discuss the bootstrap method applied to multiple linear regression models in Chapter 7. The reader is referred to Efron and Tibshriani (1993) and Shao and Tu (1995) for more information.

Exercises

1. Consider the following hypothetical dataset:

   0, 10, 12, 13, 14, 14, 15, 17, 23, 50

   a) Use R or SPSS to display this data. Discuss your observations.

   b) Compute the following estimates of θ:

i. Compute the mean, median, 10% trimmed mean, and the 10% Winsorized mean.

ii. Compute the M-estimate using Huber's psi function with a constant 1.5, mad, and one iteration of IRLS with the mean as the starting value.

iii. Repeat ii) above using one iteration of the Newton-Raphson technique instead of IRLS.

c) Compute 95% confidence intervals for θ using as estimates the mean, median, and M-estimates in part ii) in part b). Compare these intervals, and discuss your results.

2. The compliance with proposed labeling regulations was studied by determining the amount of shrimp in shrimp cocktail (King, F. J. and Ryan, J. J. 1976. "Collaborative Study of the Determination of the Amount of Shrimp in Shrimp Cocktail," *J. Off. Anal. Chem.* 59, 644–649.). The response is the amount of shrimp as a percentage of the declared total weight of the contents of the commercial container. The measurements are made independently by different laboratories. Differences in responses are due to a number of factors, including differing amounts of shrimp in the container and difficulties in distinguishing small pieces of shrimp from horseradish. The results represent 18 different laboratories using the same method on the same size container of shrimp cocktail:

32.2, 33.0, 30.8, 33.8, 32.2, 33.3, 31.7, 35.7, 32.4

31.2, 26.6, 30.7, 32.5, 30.7, 31.2, 30.3, 32.3, 31.7

A reliable measure of this measurement method's standard error is desired.

a) Use your software package to display this data.

b) Obtain the mean, median, and the m estimate based on the Huber psi with a constant of 1.5 using the IRLS method. Do not iterate the mad, and use an absolute relative change-stopping rule of 0.001 (0.1%).

c) Obtain the standard error of your estimates obtained in part b) above.

d) Obtain 95% confidence intervals for θ based on your estimates obtained in part a) above. Compare these intervals, and discuss your results.

3. Consider the phosphorus data from the exercises in Chapter 1 and the yield variable. Let's do some m estimation of the location parameter for this variable.

a) Obtain a dot plot of the data. Comment on characteristics such as center, spread, and unusual features.

b) Obtain a plot of the defining equation based on the Huber psi function with a constant 1.5. What are the roots of the defining equation? Discuss.

c) Obtain a plot of the defining equation based on the bisquare psi function with a constant 4.5. What are the roots of the defining equation? Discuss.

d) Find the M-estimate of the location parameter based on the Newton-Raphson method and the Huber psi function with a constant 1.5. Do not iterate the mad, and use an absolute relative change-stopping rule of 0.001 (0.1%). Discuss.

e) Find the M-estimate of the location parameter based on the Newton-Raphson method and the bisquare psi function with a constant 4.5. Do not iterate the mad, and use an absolute relative change-stopping rule of 0.001 (0.1%). Discuss.

f) Find the M-estimate of the location parameter based on the IRLS method and the Huber weight function with a constant 1.5. Do not iterate the mad, and use an absolute relative change-stopping rule of 0.001 (0.1%). Discuss.

g) Find the M-estimate of the location parameter based on the IRLS method and the biweight function with a constant 4.5. Do not iterate the mad, and use an absolute relative change-stopping rule of 0.001 (0.1%). Discuss.

h) Compare the results above as they relate to the ability of the Newton-Raphson and IRLS methods to find the proper roots using the Huber and bisquare psi functions. Do they find the proper roots? Are there multiple roots? Does the IRLS take more iterations to converge than the Newton-Raphson method? Discuss.

i) Compare your results in parts f) and g). What data points are outliers? Use weight index plots to display and determine which points are outliers. Discuss.

j) Compare your results of parts f) and g) to the sample mean and sample median. Discuss the differences in light of outliers that may have been found in part i) above.

k) Obtain 95% confidence intervals for the location parameter using the usual t-distribution and robust methods (confidence interval based on robust weights). Compare them. Which appears to be best? Why?

l) Obtain the LMS and LTS estimators. Discuss.

m) Plot the objective functions for the LMS and LTS estimators. Discuss.

n) Plot the sensitivity curves for the LMS and LTS estimators. Discuss.

o) Perform a bootstrap analysis for the mean, median, the Huber M-estimator with constant 1.5 (four iterations using IRLS), LMS, and LTS estimates. Use 500 bootstrap samples. Obtain the bootstrap means and standard errors. Compare your bootstrap results to the actual estimates and their analytic standard errors in a table as seen in Section 4.5 of your book. Obtain the 95% bootstrap percentile confidence intervals for each of these estimators. In a table as seen in Section 4.5, compare your bootstrap results to the confidence intervals obtained by the analytic methods. Discuss.

4. Perform a Monte Carlo experiment to compare the five estimators: mean, median, Huber M-estimator with constant 1.5 (four iterations with IRLS), LMS, and LTS. Generate samples of size 20 from a CN(y; 0.25, 100) distribution using 1,000 replicates. The main results that we are seeking are the efficiencies of the four estimators with respect to the mean. Comment on your results.

5. Write a new macro using your software package that will accomplish the following.

   For a given set of data, you will obtain the LMS and LTS estimates of the center, the bootstrap standard errors (based on 200 samples), and the 95% bootstrap confidence intervals obtained via the bootstrap percentile method. Demonstrate your code by applying it to your response variable from your dataset. You may construct the program by using pieces of programs that already exist from your book. Your solutions should include a listing of your detailed code.

6. Consider again computing an M-estimator based on the following ρ function,

$$\min_{\theta}\sum_{i=1}^{n}\rho(y_i-\theta),$$

where

$$\rho(r_i)=\begin{cases}\dfrac{r_i^2}{2} & \text{if} -c\le r_i\le c\\[2ex] \dfrac{c^2}{2} & \text{if} \left|r_i\right|>c\end{cases},$$

where for some $c>0$ and $r_i=y_i-\theta$. Note that for simplicity, the scale parameter is one.

a) What is the ψ function corresponding to the ρ function listed above?

b) Consider now the general form of the *m*estimate where scale must be estimated as well as θ. Then $\hat{\theta}$ is the solution to

$$\sum_{i=1}^{n} \psi\left(\frac{y_i - \theta^*}{\hat{s}}\right) = 0,$$

where the ψ is defined in a) above.

Multiple-Choice Questions

1. What does the scale parameter "s" in the M-estimators of the location model signify?
   a) Mean of the data
   b) Median of the data
   c) Standard deviation of the data
   d) Variance of the data

   Answer: c)
2. Which robust scale estimator is described as resistant to outliers?
   a) Mean absolute deviation
   b) Standard deviation
   c) Median absolute deviation
   d) Variance

   Answer: c)
3. What is the primary purpose of using the constant 0.6745 in the calculation of mad?
   a) To standardize the distribution
   b) To normalize the residuals
   c) To ensure consistency when estimating standard deviation under normality
   d) To minimize the mean squared error

   Answer: c)
4. Which of the following is a characteristic of the M-estimators?
   a) They minimize the sum of squared residuals.
   b) They maximize the likelihood function.
   c) They are sensitive to outliers.
   d) They use a psi function to limit the influence of outliers.

   Answer: d)
5. What method is used to solve the defining equation for M-estimators?
   a) Maximum likelihood estimation
   b) Newton-Raphson technique
   c) Least squares fitting

d) Analytical solution

Answer: b)

6. Which of the following methods is NOT directly used to estimate M-estimators?
   a) Iterated reweighted least squares
   b) Least trimmed squares
   c) Bootstrap
   d) Least median of squares

   Answer: c)

7. In the context of robust estimators, what is the primary function of the Huber psi function?
   a) It provides a maximum likelihood estimate.
   b) It reduces the impact of outliers by adjusting weights.
   c) It calculates the median of absolute deviations.
   d) It linearizes the residuals.

   Answer: b)

8. What does the bisquare weight function achieve in robust estimation?
   a) It increases the efficiency of estimators.
   b) It gives zero weight to large residuals.
   c) It maximizes the likelihood of the data.
   d) It linearizes nonlinear models.

   Answer: b)

9. Which estimator is known for its high breakdown point but low efficiency in robust statistics?
   a) Huber M-estimator
   b) Least squares estimator
   c) Least median of squares estimator
   d) Median absolute deviation estimator

   Answer: c)

10. What role does the constant "c" play in the Huber psi function?
    a) It determines the point at which weights for residuals are adjusted.
    b) It controls the convergence rate of the estimator.
    c) It specifies the minimum number of iterations for convergence.
    d) It sets the threshold for outlier detection.

    Answer: a)

5

Comparing Two or More Groups of Data: Exploratory, Classical Inference, and Other Forms of Analysis

Summary

We are now ready to apply the exploratory data analysis (EDA) and robust methods presented in the previous chapters to one group or two or more groups. From a statistical viewpoint, the extension is from the location model, where one group or treatment is studied, to the one-way analysis of variance (ANOVA) model, where k groups or treatments are studied. We will find that graphical displays of raw data and residuals also play an important role in the one-way ANOVA models. As in the location model, alternative estimation procedures will be needed since outliers are a common, natural occurrence in this and other statistical models. Transformations of raw data to achieve desired model assumptions may be necessary, and a variety of transformation techniques will be presented in this chapter.

Learning Objectives

By the end of this chapter, readers should be able to:

- Comprehend the purpose, structure, and assumptions of the one-way ANOVA model.
- Conduct and interpret one-way ANOVA tests to compare treatment means.
- Identify when classical assumptions are violated, and apply robust or nonparametric methods like the Kruskal-Wallis test.
- Use different transformations, including the Box-Cox method, to meet ANOVA assumptions.
- Perform and interpret multiple comparison procedures after significant ANOVA results.
- Use box plots and residual plots to assess data distributions, outliers, and the fit of the ANOVA model.
- Apply ANOVA techniques to real-world datasets, understanding their practical implications.
- Understand the limitations of classical ANOVA and the importance of robust techniques in handling non-normal data or outliers.

DOI: 10.1201/9781003565321-5

5.1 One-Way Analysis of Variance: The Model and Classical Analysis

The one-way ANOVA model is a statistical model designed to compare the mean response across k "treatments" or "populations." The following is a brief outline and description of the model and the classical analysis used for this model. The material in this section is similar to that presented in any introductory text in statistics; see, for example, Iman (1994) and Moore (1995).

The classical one-way ANOVA model expresses the response variable, y, from some treatment, say the ith treatment, as a random observation about the treatment mean. If there are k treatments and the random observations all have similarly shaped distributions, we can write the one-way ANOVA model as:

$$y_{ij} = \mu_i + \varepsilon_{ij},$$

where i indicates the treatment, $i = 1,\ldots k$ and j represents an observation receiving the ith treatment, $j = 1,\ldots,n_i$, where n_i is the number of observations receiving the ith treatment. The total sample size is denoted by $N = \sum_{i=1}^{k} n_i$. It is often assumed that the random terms, the ε_{ij}, are normally distributed random variables with variance σ^2, constant over the k treatments. The model above is often referred to as the "means model" since it models observations as random responses about their respective means.

This means the model may also be expressed as the "one-way ANOVA effects model" in the form:

$$y_{ij} = \mu + \alpha_i + \varepsilon_{ij},$$

where α_i is defined as $\alpha_i = \mu_i - \mu$, μ is the "grand population mean" defined as $\mu = \dfrac{\sum_{i=1}^{k} n_i \mu_i}{N}$. The α_is are called the "effects" and indicate whether the ith treatment mean is greater than the grand population mean, in which case $\alpha_i > 0$; less than μ, with $\alpha_i < 0$; or equal to μ, with $\alpha_i = 0$. The effects model is especially helpful in analyzing variance models such as factorial, nested, and split-plot models. In the one-way ANOVA case, the means model is adequate for our purposes and will be used exclusively throughout the remainder of this chapter.

The least squares estimate (which is the maximum likelihood estimate under the normality and constant variance assumptions of our model) of the ith treatment mean, μ_i, is the ith sample mean, denoted by $\bar{y}_i$, the average of those responses receiving the ith treatment. These estimates enable the total variability of the response to be partitioned into the well-known sum of squares for the error and the model terms. In one-way ANOVA, these formulas are

$$SS_{total} = \sum_{i,j}\left(y_{ij} - \bar{y}\right)^2, SS_{error} = \sum_{i,j}\left(y_{ij} - \overline{y_i}\right)^2, \text{ and } SS_{model} = \sum_{i=i}^{k} n_i\left(\overline{y_i} - \bar{y}\right)^2.$$

It is also well known that

$$SS_{total} = SS_{error} + SS_{model}.$$

The SS_{total} measures the total variability of the data, the SS_{model} measures the variability between the sample means, and the SS_{error} measures the (pooled) variability of the data within each of the k treatments. Thus, total variability is partitioned into two pieces, SS_{model} and SS_{error}. The phrase "analysis of variance" refers to the analysis of these sources of variability. The degrees of freedom for these three sums of squares are $N-1$, $N-k$, and $k-1$ for the total, error, and model components, respectively.

The hypothesis of equality of treatments means, $H_0: \mu_1 = \mu_2 = \ldots = \mu_k$, with H_1: not H_0, is tested by the usual F-test as

$$F = \frac{MS_{model}}{MS_{error}} = \frac{SS_{model} / df}{SS_{error} / df_{error}},$$

a test statistic with an F-distribution, under the null hypothesis and the assumptions of normality and constant variance, with $k-1$ and $N-k$ degrees of freedom for the numerator and denominator, respectively. This F-test is often referred to as the "overall" or "model" F-test.

Multiple comparison procedures, such as Fisher's protected least significant difference (LSD) method or Tukey's honestly significant difference (HSD) method, may be used to characterize the differences between the k treatments. These procedures allow for the equality of all pairs of treatments to be tested while controlling the overall error rate (sometimes called the experiment-wise or simultaneous error rate) to be some small value acceptable to the researcher, such as 0.05.

In Fisher's protected LSD method, for example, the overall error rate is controlled by the model F-test described above. If this test statistic is insignificant, at a given α level, all k means are declared equal. If this test is significant at a given α level, this method then rejects $H_0: \mu_i = \mu_j$, for all pairs of treatments i and j, if

$$\left|\bar{y}_i - \bar{y}_j\right| > LSD_{ij} = t_{1-\alpha/2,\, d_{error}} \sqrt{MS_{error}\left(\frac{1}{n_i} + \frac{1}{n_j}\right)},$$

where α is the overall error rate. A complete discussion of this and other multiple comparison methods may be found in many texts, including Schabenberger and Pierce (2002); Neter, Kutner, Natcheshim, and Wasserman (1996); and Ott and Longnecker (2001).

Of course, the validity of these methods is dependent on the assumptions. The assumptions are verified using the graphical aids introduced for the location model in Chapter 1. Helpful plots are side-by-side box plots or dot plots of the raw data or of the residuals, where the residual for the jth subject receiving treatment i is defined as $r_{ij} = y_{ij} - \bar{y}_i$. If the graphical procedures support normality and constant variance assumptions, the residuals from each of the k treatments could also be pooled and considered as a single sample from one location model. The symmetry or Q-Q plots could be applied to the pooled residuals to further check the normality assumption. However, if the plots reveal strong non-normal tendencies and/or strong heteroscedasticity, we ought to consider alternative methods for estimating and testing parameters. There are two basic approaches to alter the classical analysis of one-way ANOVA data. One method relies on altering the data so that the altered data conforms to the classical model. The classical model is then used to analyze the altered or transformed data. The second approach leaves the data unaltered or in its natural state but changes the classical model to account for the unusual features in the data. In such cases, the estimation technique employed will differ from the classical method of least squares. Several possible approaches involve nonparametric methods.

One popular nonparametric method is called the Kruskal-Wallis test. Here, the data from the individual k treatments are combined into one group, ordered from smallest to largest, and each ordered data value is assigned a rank from 1 to N. This is an example of the "rank" transformation of the data. These ranks are then used to construct the Kruskal-Wallis test statistic, which is used to test the equality of means. The only assumption required for this test is that the data are random observations from continuous distributions with the same shape. For more details on this test, see Neter, Kutner, Natcheshim, and Wasserman (1996).

Another nonparametric approach, equivalent to the Kruskal-Wallis method, is to simply replace the original observations, y_{ij}, with their ranks, as described above, and perform the classical one-way ANOVA with the ranks instead of the original data. Perhaps this method is more in the spirit of the first approach mentioned above, where the data is transformed to better conform to the classical model. The rank transformation often results in reducing the influence of outliers and making sample variances (of the ranked data) more constant across the k treatments.

Another example of the second approach (that uses the raw data but an altered model) is to alter the error structure for the classical model from the normal distribution to a heavy-tailed distribution, a model that allows for outliers and nonconstant sample variances (resulting from outliers in the data). The estimation technique could be M-estimation rather than least squares. That is, the sample means for each treatment could be replaced by the M-estimate for each treatment. A method for performing robust one-way ANOVA based on M-estimation will be given in detail later in the text as part of the discussion of the general linear model (GLM).

Other alterations of the classic one-way ANOVA model may need to be considered. For example, suppose the response variable is the number of successes in n independent trials (a binomial random variable). In that case, the one-way ANOVA model may be altered to accommodate a binomial distribution error structure. Appropriate adjustments for the binomial and other distributions that are members of the exponential family of distributions are discussed under the topic of the GLM. The final chapter of the text is devoted to this topic.

An example of these basic tools applied to the one-way model is now illustrated.

5.1.1 The Mill Data

The following data in Table 5.1 (taken from Koopmans, 1987) represents the percentage of wastage per week in cloth at five mills producing cloth for Levi Strauss in Albuquerque, New Mexico. Wastage is due to a variety of sources, such as defects in cloth or leftover material (cut from patterns) that cannot be reused. Negative values of wastage result when workers can make the product requiring less material than specified by the computerized layouts of the pattern. Consistently small values of wastage would exhibit proper control by a mill over the production process. On the other hand, large variability of the wastage values would indicate strong inconsistencies in the production process, perhaps resulting in an inferior, less desirable product. Therefore, it is desired for a mill to have both a small average value of wastage and small variability. The following analysis will attempt to determine which of the five mills (labeled 1, 2, 3, 4, and 5) give the best balance of performance in both categories over the given time.

A summary of the data follows:

```
> # Calculate descriptive statistics for each mill
> desc_stats <- mill_data_long %>% + group_by(Mill) %>% +
  summarise(+ N = n(),+ MEAN = mean(Wastage, na.rm = TRUE),
+       MEDIAN = median(Wastage, na.rm = TRUE),
+       STDEV = sd(Wastage, na.rm = TRUE),
+       SEMEAN = sd(Wastage, na.rm = TRUE) / sqrt(N),
+       MIN = min(Wastage, na.rm = TRUE),
+       MAX = max(Wastage, na.rm = TRUE),
+       Q1 = quantile(Wastage, 0.25, na.rm = TRUE),
+       Q3 = quantile(Wastage, 0.75, na.rm = TRUE)
+   )
> # Print the descriptive statistics
> print(desc_stats)
```

TABLE 5.1
The Mill Data

Obs.	Mill1	Mill2	Mill3	Mill4	Mill5
1	0.12	1.64	1.21	1.15	2.40
2	1.01	−0.60	0.97	1.02	−0.37
3	0.15	−1.16	0.74	0.38	0.82
4	−0.20	−0.13	−0.21	0.83	0.92
5	−0.30	0.40	1.01	0.66	−0.93
6	−0.07	1.70	0.47	1.02	0.80
7	0.32	0.38	0.46	0.88	1.58
8	0.27	0.43	0.39	0.27	2.23
9	−0.32	1.04	0.36	0.51	0.31
10	−0.17	0.42	0.96	1.12	1.68
11	0.24	0.85	0.98	0.59	1.13
12	0.03	0.63	0.65	1.30	1.23
13	0.35	0.90	0.57	0.68	1.69
14	−0.08	0.71	0.51	1.45	
15	1.94	0.43	0.34	0.52	
16	0.28	1.97	−0.08	0.73	
17	1.30	0.30	−0.39	0.71	
18	4.27	0.76	0.09	0.34	
19	0.14	7.02	0.15	0.07	
20	0.30	0.85			
21	0.24	0.60			
22	0.13	0.29			

```
Mill     N  MEAN  MEDIAN STDEV  SEMEAN    MIN   MAX     Q1      Q3
mill1   22 0.452 0.195   1.00   0.214   -0.32 4.27 -0.045 0.315
mill2   22 0.883 0.615   1.54   0.327   -1.16 7.02  0.385 0.888
mill3   22 0.483 0.47    0.440 0.0939 -0.39 1.21   0.245 0.85
mill4   22 0.749 0.71    0.366 0.0780  0.07 1.45   0.515 1.02
mill5   22 1.04  1.13    0.956 0.204  -0.93 2.4    0.8   1.68
```

We note that there is a considerable difference in the standard deviations between the five groups. For example, the largest standard (std.) deviation is 1.535/0.366 = 4.19 times larger than the smallest std. deviation. This strongly suggests that the assumption of common variance is violated. The outliers, of course, are part of the cause of this problem. The classical one-way ANOVA is as follows:

```
> # Run ANOVA
> anova_model <- aov(Wastage ~ Mill, data = mill_data_long)
> summary(anova_model)
```

```
Df Sum Sq Mean Sq F value Pr(>F)
Mill 4 4.51 1.1273 1.16 0.334
Residuals 90 87.49 0.9721
15 observations deleted due to missingness
> post_hoc <- glht(anova_model, linfct = mcp(Mill = "Tukey"))
> summary(post_hoc)
              Simultaneous Tests for General Linear Hypotheses
      Multiple Comparisons of Means: Tukey Contrasts
      Fit: aov(formula = Wastage ~ Mill, data =
      mill_data_long)
      Linear Hypotheses:
                              Estimate Std. Error t value Pr(>|t|)
      mill2 - mill1 == 0 0.43091 0.29728 1.450 0.596
      mill3 - mill1 == 0 0.03089 0.30879 0.100 1.000
      mill4 - mill1 == 0 0.29667 0.30879 0.961 0.871
      mill5 - mill1 == 0 0.58542 0.34491 1.697 0.439
      mill3 - mill2 == 0 -0.40002 0.30879 -1.295 0.693
      mill4 - mill2 == 0 -0.13423 0.30879 -0.435 0.992
      mill5 - mill2 == 0 0.15451 0.34491 0.448 0.991
      mill4 - mill3 == 0 0.26579 0.31989 0.831 0.920
      mill5 - mill3 == 0 0.55453 0.35488 1.563 0.524
      mill5 - mill4 == 0 0.28874 0.35488 0.814 0.925
      (Adjusted p values reported -- single-step method)
```

One can see that the model F-test failed to be rejected, so there is no need to use the pairwise comparisons in our analysis. However, we did it only for illustration, and we found no significant differences between the five mills in average run-up, as in Figure 5.1. Figure 5.2, a scatterplot of the residuals by mill, clearly shows a similar spread to the data and the presence of several outliers.

To examine the residuals more closely, look at a normal probability plot (Q-Q plot), as in Figure 5.3.

```
# Calculate residuals from the ANOVA model
residuals_data <- residuals(anova_model)
# Generate a data frame for plotting
qq_data <- data.frame(sample_quantiles = residuals_data)
# Creating the Q-Q plot manually using ggplot2
qq_plot <- ggplot(qq_data, aes(sample = sample_quan-
tiles)) +
  stat_qq() + # This adds the Q-Q plot points
  stat_qq_line(color = "black") + # This adds the Q-Q
line
  ggtitle("Normal Probability Plot of the Residuals") +
  xlab("Theoretical Quantiles") + ylab("Residuals") +
  theme_minimal() # Applying a minimal theme for
aesthetics
# Print the Q-Q plot
print(qq_plot)
```

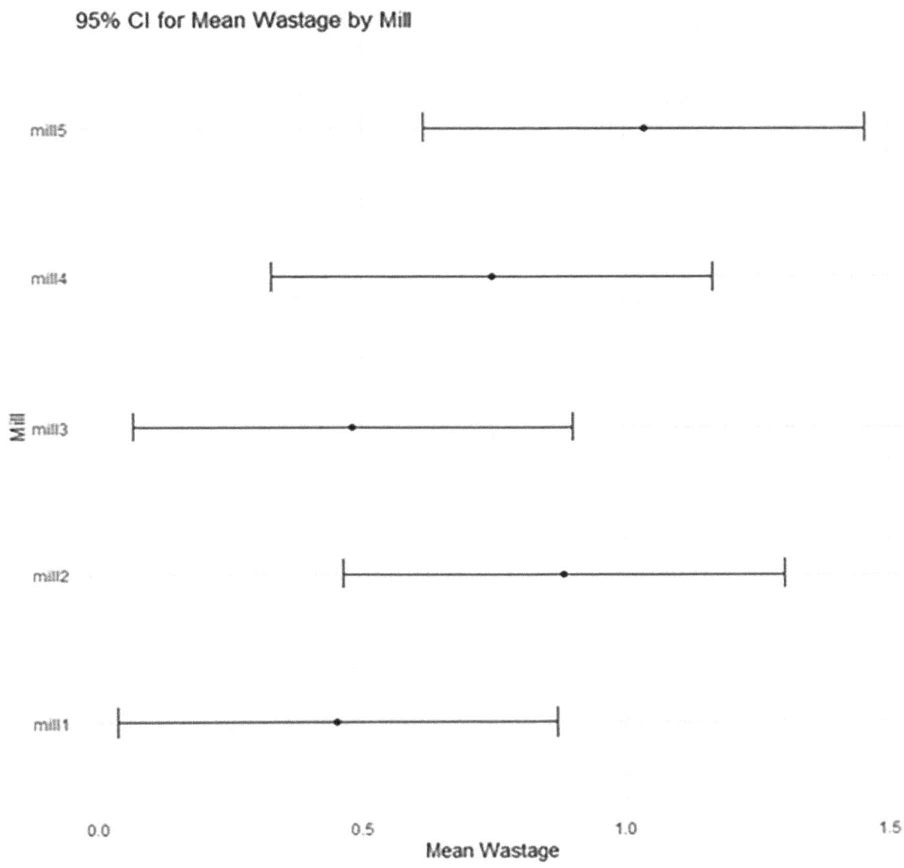

FIGURE 5.1
95% CI for the mean wastage by mill.

Several severe outliers are clearly indicated among the residuals. Otherwise, the residuals appear to be extremely "normal."

5.1.2 Discussion: A Summary of Our Analysis So Far

A number of graphical and analytical statistical methods were used to analyze the mill data. Displays such as box plots revealed some very unusual values of wastage (outliers) for several mills. For example, mills 1 and 2 had nine of the total ten outliers present in the entire dataset. Classical one-way ANOVA was unable to detect significant differences in mean wastage values between the mill groups because of the presence of outliers. The sample means for mills 1 and 2 are increased due to the several large extreme outliers that were observed in these groups. Also, the sample standard deviations for these two mills are

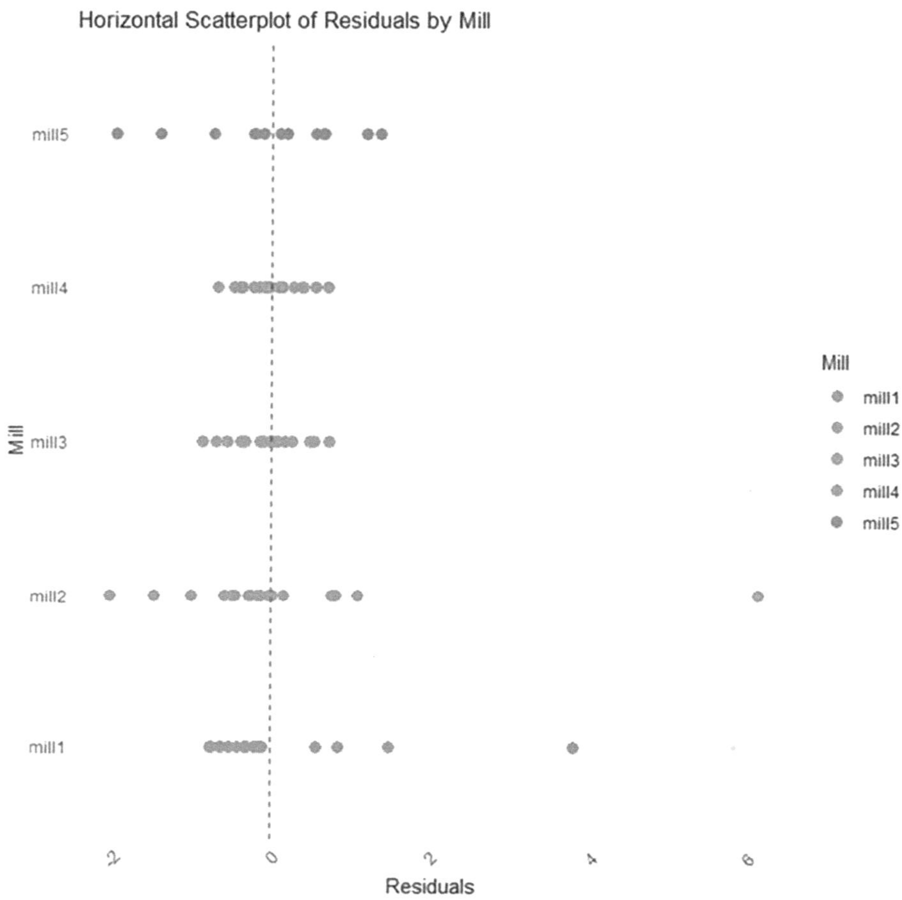

FIGURE 5.2
A scatterplot of the residuals by mill.

inflated. Mill 5 seems to be the mill with the poorest performance. The box plot of the mill data suggests differences between mills that are not supported by the inferential analysis. Of course, the validity of the classical inferential technique is questioned due to the lack of evidence justifying the normality and constant variance assumptions. The next section will return to this data to see if alternative statistical methods will provide additional information.

5.2 Transformations in One-Way ANOVA

Transformations of data to improve the data's applicability to the classical model have been a standard technique of applied statisticians for many years. For example, count data is often transformed by the square-root

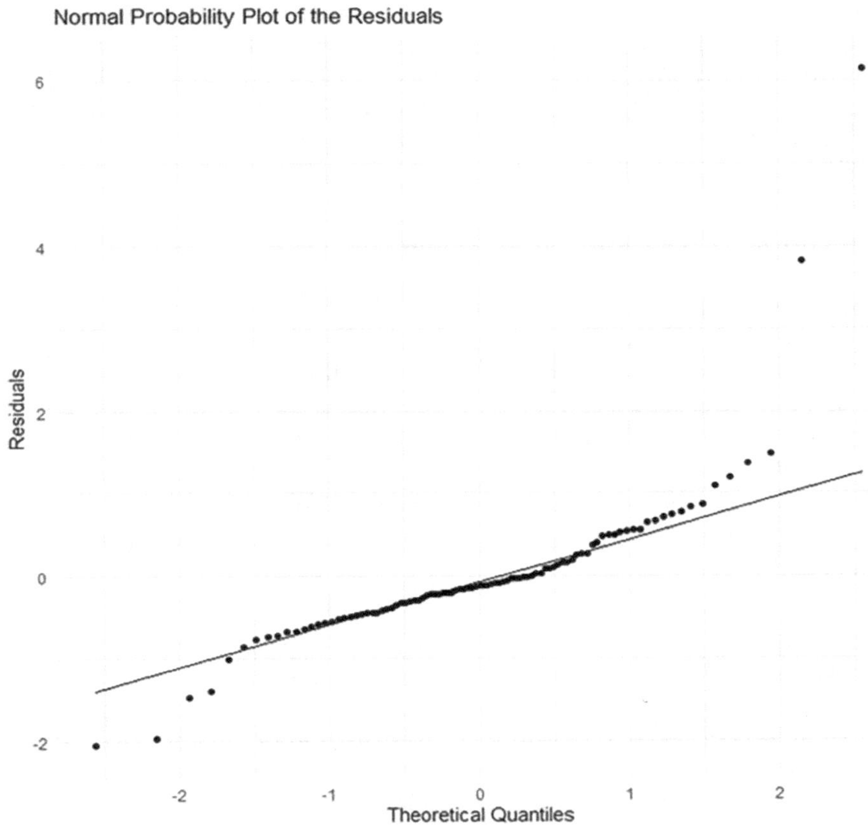

FIGURE 5.3
The Q-Q plot for the residuals.

transformation as a first step in the data analysis. In this section, we will propose a simple method for determining the proper transformation to use on a given dataset. Recall from the discussion of transformations in Section 1.8 that transformations are applied to data to achieve a number of desired goals, including symmetry of data within each group or treatment (to achieve normality) and constant variance across the k groups. The approach we will discuss is based on John Tukey (1977). A related method by Box and Cox (1964) is presented in the next section.

Tukey's method (1977) considers the simple form of the power transformations, introduced in Section 1.8, as

$$y_{ij}^{*} = \begin{cases} y_{ij}^{\lambda} & \lambda > 0 \\ \ln\left(y_{ij}\right) & \lambda = 0 \\ -y_{ij}^{\lambda} & \lambda < 0, \end{cases}$$

where the random variable y_{ij} is assumed to be positive. Problems arise in using the power transformations when the data are both positive and negative (See Tukey (1977) for suggested remedies to this problem). Tukey proposed a method to find the value of λ that achieved constant variance across the k groups. The method to accomplish this is termed the "spread versus level" plot and is described as follows.

5.2.1 Spread versus Level Plot Algorithm

1) Start at λ equal 1.
2) Compute a measure of center and spread for each treatment. Tukey recommends resistant measures such as the median, M_i, and the interquartile range, IQR_i, for $i = 1,...,k$. The sample mean and the sample standard deviation for each group could also be used if no outliers are present in the data.
3) Plot $\log(IQR_i)$ (on the y-axis) versus $\log(M_i)$, for $i = 1,...,k$.
4) Fit a line to this scatter of k points, called the slope of the line b.
5) Compute y^*_{ij} as described above using $\lambda = 1 - b$.

Some comments are in order before we demonstrate the technique. First, in applications, we try to use "nice" values of λ, so if λ=0.42, we probably will use λ=0.5, the square-root transformation, or λ= 0.33, the cube-root transformation. Second, the success of the transformation in achieving a constant sample variance depends a great deal on the ability of the line in the spread versus level plot to characterize the scatter of points. If the fit is very good, then the transformed data should exhibit very constant variances across the k treatments. And, if the fit is very poor, then no value of λ will accomplish our goal. Finally, in fitting the line, we must be aware that any outliers present on the spread versus level plot can greatly distort the slope estimate and, consequently, the final value of λ. It is recommended that a "robust method" be used to obtain the slope.

5.2.2 The Plankton Data

We will illustrate the spread versus level plot technique for determining the value of λ using the "plankton" data from Chapter 4 of Tukey's EDA book. The data in Table 5.2, originally obtained from Winsor and Clarke (1940), analyzing the number of plankton caught using different kinds of nets, is presented in the following.

Exploring this data and quantifying the differences between the three kinds of nets is interesting. Our exploratory analysis is easily accomplished on R. The following annotated R session presents some of our

TABLE 5.2

The Plankton Data

Haul	Kind 1	Kind 4	Kind 3
1	387	6,060	18,900
2	428	7,600	22,300
3	470	7,900	27,800
4	497	8,260	28,100
5	537	8,600	28,800
6	540	8,900	29,000
7	620	9,250	31,400
8	760	9,830	32,800
9	845	10,200	32,800
10	895	11,000	34,600
11	1,020	15,500	39,500
12	1,050		43,300

exploratory tools applied to the plankton data as well as an application of the spread versus level technique to determine whether a transformation is needed and, if so, which one. Obtain box plots of the raw data as in Figure 5.4.

The box plots clearly reveal that the medians, as do the IQRs, follow a strong increasing pattern. An analysis of the frequency of plankton by kind of net would be facilitated if the data differed in only one of these characteristics.

Obtain the letter values to be used in the spread versus level plot.

```
> # Calculate letter values for each kind
> lv_kind1 <- calculate_letter_values(kind1)
> lv_kind4 <- calculate_letter_values(kind4)
> lv_kind3 <- calculate_letter_values(kind3)
> # Print results
> print("Letter values for Kind 1:")
[1] "Letter values for Kind 1:"
> print(lv_kind1)
  DEPTH  LOWER  UPPER      MID SPREAD
1      M 580.00  580.0 580.000   0.00
2      H 490.25  857.5 673.875 367.25
3      E 432.20 1007.5 719.850 575.30
4      1 387.00 1050.0 718.500 663.00
> print("Letter values for Kind 4:")
[1] "Letter values for Kind 4:"
> print(lv_kind4)
```

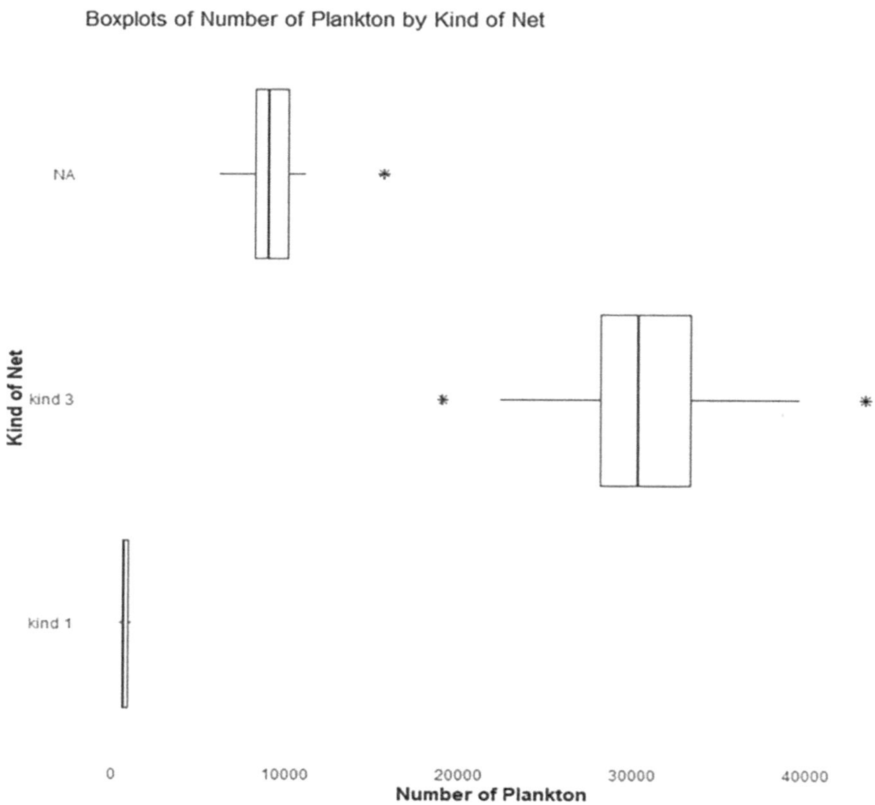

FIGURE 5.4
The box plots of the plankton data.

```
  DEPTH LOWER UPPER     MID SPREAD
1     M  8900  8900  8900.0      0
2     H  8080 10015  9047.5   1935
3     E  7600 11000  9300.0   3400
4     1  6060 15500 10780.0   9440
> print("Letter values for Kind 3:")
[1] "Letter values for Kind 3:"
> print(lv_kind3)
  DEPTH LOWER UPPER     MID SPREAD
1     M 30200 30200 30200.0      0
2     H 28025 33250 30637.5   5225
3     E 22850 39010 30930.0  16160
4     1 18900 43300 31100.0  24400
```

The numbers in bold above are the medians and IQRs for the three groups. Perform the spread versus level plot method steps as in Figure 5.5.

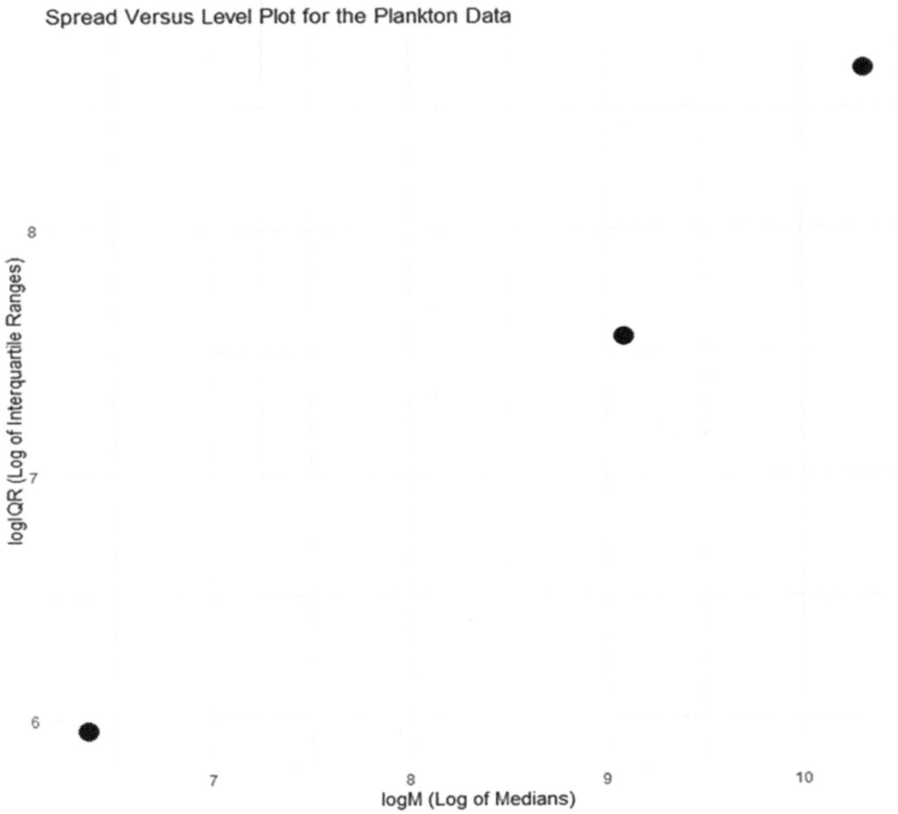

FIGURE 5.5
The spread versus level plot.

Fit a line to this scatter of points using the following R code:

```
> # Fit the linear regression model
> model <- lm(logIQR ~ logM, data = data)
> # Display the summary of the model
> summary_model <- summary(model)
> print(summary_model)
Call:
lm(formula = logIQR ~ logM, data = data)
Residuals:
        1       2         3
0.05002 -0.16183 0.11181
Coefficients:
            Estimate Std. Error t value Pr(>|t|)
(Intercept) 1.66027 0.62040 2.676 0.2277
logM 0.66742 0.07092 9.411 0.0674 .
---
```

```
Signif. codes: 0 '***' 0.001 '**' 0.01 '*' 0.05 '.' 0.1'' 1
Residual standard error: 0.203 on 1 degrees of freedom
Multiple R-squared: 0.9888, Adjusted R-squared: 0.9777
F-statistic: 88.58 on 1 and 1 DF, p-value: 0.06739
```

The regression equation is:

$$\text{logIQR} = 1.660 + 0.667\ \text{logM}$$

This suggests a value of $\lambda = 1 - 0.667 = 0.333$. In the spirit of using common values of λ from the ladder of transformations, the value of $\lambda = 0.25$, the quarter-root transformation, is suggested, with $\lambda = 0.5$, the square-root transformation also suggested. We could also try $\lambda = 0.33$, a less common power value than the other two. The reader will be asked to consider this value as an exercise. Since the R^2 is 98.9 for the LS fit, we expect this transformation will work very well.

```
# Transform data from wide to long format
plankton_long <- tidyr::pivot_longer(plankton_data, cols =
  c(kind1, kind4, kind3), names_to = "kind", values_to =
  "counts")
plankton_long <- plankton_long %>% dplyr::filter(!is.
  na(counts))
# Calculate the quarter root of counts
plankton_long$fourthroot <- plankton_long$counts^(1/4)
# Create the box plot of the quarter root of counts
boxplot <- ggplot(plankton_long, aes(x = kind, y = fourth-
  root)) + geom_boxplot(outlier.color = "black") +
  coord_flip() + # This makes the box plot horizontal
    labs(title = "Boxplots of the Quarter Root of the Number
  Plankton by Kind of Net",x = "Kind of Net", y = "Quarter
  Root of Counts")+ theme_minimal()
print(boxplot)
```

Figure 5.6 strongly suggests constant spread across the three kinds of nets. It is seen that the square-root transformation is not as successful at achieving constant spread across the three treatments.

Further analysis of the plankton data could then be conducted on the quarter-root of plankton counts, with the square-root transformation being a possible alternative. An example of a statistical analysis of this data would be a one-way ANOVA. The quarter root–transformed data is used here since it achieves near-constant variance across the three treatments. Fortuitously, the quarter-root transformation has also achieved near-symmetric data with each group. The improvement in constant spread across the three groups is seen in the data in Table 5.3 where the standard deviations for the three treatments are given for both the raw and the transformed ($\lambda = 0.25$) data:

We see that the ratio between the largest and smallest spread across the three groups changes from 28.58 to 1.68 for the raw and transformed data, respectively. We conclude that the spread versus level plot method has been

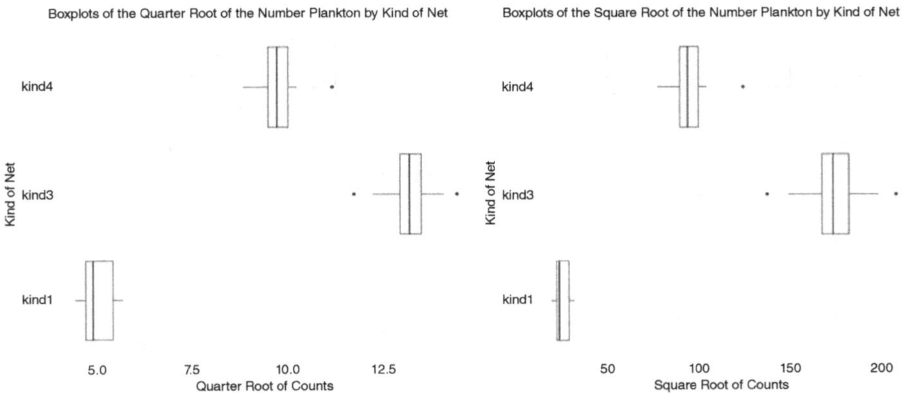

FIGURE 5.6
The box plots for the transformed data.

TABLE 5.3
Comparison of Standard Deviations for Raw Data and Transformed Data

Treatment	Raw Data	Transformed Data (λ = 0.25)
Kind 1	234	0.438
Kind 2	2,438	0.595
Kind 3	6,689	0.737

successful in determining the proper power transformation to achieve our goal of constant spread across the groups.

5.2.3 The Mill Data Revisited

As a second example of the application of the spread versus level technique, consider the "mill data" from the previous section again. Recall that the wastage of cloth is being examined across five different mills. Box plots of the raw data suggested three groupings among the mills and revealed fairly strong heterogeneity of sample standard deviations, with the largest standard deviation being about four times larger than the smallest standard deviation. The presence of a number of moderate and extreme outliers is most likely the cause of much of this difference in variability. The F-test for one-way ANOVA failed to detect a significant difference between mean responses across the five mills.

Using the R code, we obtain the IQRs of 0.390, 0.520, 0.605, 0.505, and 0.880 and M's of 0.195, 0.615, 0.470, 0.710, and 1.13 for the mills 1 through 5, respectively. The IQRs, a measure of spread that is insensitive to these outliers, still reveal strong heterogeneity in the spreads across the five mills, with the largest IQR being 0.880/0.390 = 2.25 times larger than the smallest IQR. While differences in spreads appear to exist across the five mills, we also note that the relationship between the IQRs and M's is not increasing. For example, Mill 3 has the second-largest median but the fourth-largest IQR. One might speculate that a power transformation will be unable to achieve constant variance across the mills. The annotated R analysis follows:

```
> # Perform the linear regression (spread versus level plot)
> regression_model <- lm(logIQR ~ logM, data = mill_stats)
> # Summary of the regression model
> summary(regression_model)
Call:
lm(formula = logIQR ~ logM, data = mill_stats)
Coefficients:
            Estimate Std. Error t value Pr(>|t|)
(Intercept)  -0.3377     0.1250  -2.701   0.0737 .
logM          0.4312    0.1471    2.931   0.0609 .
---
Signif. codes: 0 '***' 0.001 '**' 0.01 '*' 0.05 '.' 0.1'' 1
Residual standard error: 0.1914 on 3 degrees of freedom
Multiple R-squared: 0.7412, Adjusted R-squared: 0.6549
F-statistic: 8.592 on 1 and 3 DF, p-value: 0.06094
> # Create the spread versus level plot
> ggplot(mill_stats, aes(x = logM, y = logIQR)) +
+   geom_point() +
+   geom_smooth(method = "lm", se = FALSE, color = "black") +
+   labs(+title = "Spread Versus Level Plot",+subtitle = "for
  the Mill Data",+ x = "log(Median)",+ y = "log(IQR)"
+   ) +
+   theme_minimal()
'geom_smooth()' using formula 'y ~ x'
> # Print the regression model details
> print(regression_model)
Call:
lm(formula = logIQR ~ logM, data = mill_stats)
Coefficients:
(Intercept)         logM
    -0.3377       0.4312
```

The regression equation is

$$\mathbf{logIQR = -0.338 + 0.431\ logM}$$

The R^2 of 74.12% suggests a modest fit of the line to the data as in Figure 5.7. We would expect the lambda of 1 – 0.431 = 0.569 to work only moderately well in achieving constant variance across the five mills. We will use the square-root transformation. Since several observations are negative, we'll

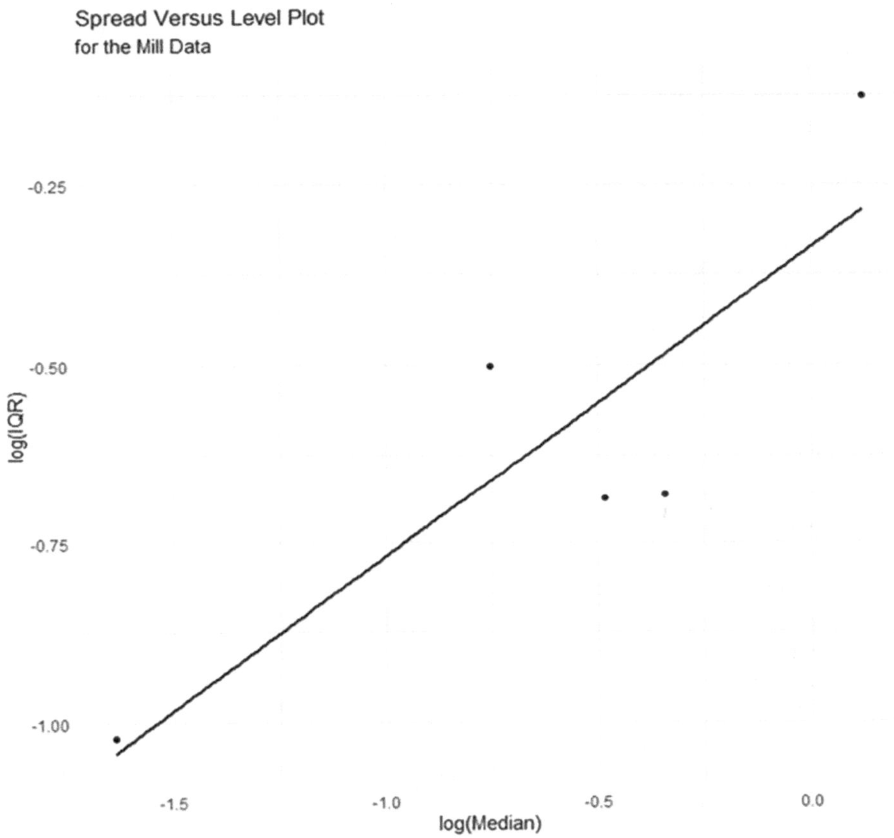

FIGURE 5.7
The estimated linear regression line (spread vs. level plot).

take the square root of the positive part of each observation and attach the sign after the square root has been taken.

We see that the IQRs are nearly constant for four of the five mills. However, outliers are a serious problem. Note that these IQRs are no more constant than the raw data. Additionally, the outliers seem to be more severe here than in the raw data. Try one-way ANOVA on the square-root data to see if the classical F-test can detect differences in the five groups.

```
> # Transform the wastage data
> long_data <- long_data %>%
+   mutate( + sqrwast = sign(Wastage) * sqrt(abs(Wastage))
+   )
> # Perform the one-way ANOVA
> anova_result <- aov(sqrwast ~ Mill, data = long_data)
```

```
> summary(anova_result)
            Df Sum Sq Mean Sq F value Pr(>F)
Mill         4   2.88  0.7198   1.938 0.111
Residuals   90  33.43  0.3714
15 observations were deleted due to missingness
```

We see that the F-test is insignificant, as in the raw data case.

```
> # Display the summary statistics for each group
INDIVIDUAL 95 PCT CI'S FOR MEAN
LEVEL     N         MEAN          STDEV
mill1    22        0.3845         0.6538
mill2    22        0.7189         0.7509
mill3    22        0.5625         0.5015
mill4    22         0.835         0.2336
mill5    22        0.8316         0.7692
```

In summary, the mill data demonstrates that the power family of transformations is not always successful in achieving constant variance across groups. The spread versus level technique works best when there is a monotonic relationship between the sample standard deviations and the sample means, a relationship that does not hold for the mill data.

On the other hand, the outliers in the mill data are a far more serious problem than the heterogeneity of variance in terms of damaging our ability to detect differences in mean wastage values between the five mills. Even if a transformation could correct the nonconstant variance problem, it is doubtful that the classical inference procedure could find differences between the mean response for the mills if outliers were still present.

One useful inference procedure that is robust to outliers that utilize "transformed" data is the "altered" Kruskal-Wallis method, mentioned previously in Section 5.1. Here, the data from the k treatments are combined into one group, and each observation is assigned a rank, a value from one to N. The raw data in each group are replaced by their respective ranks. The classical one-way ANOVA method is then applied to the ranked data, resulting in a sum of squares for model, error, and total. The F-test is then computed as the ratio of the mean square for the model divided by the mean square error. Empirical studies suggest that the "rank" transformation often removes outliers, stabilizes variance across groups, and achieves symmetry within each group. See Conover and Iman (1981) for a discussion of the properties of this technique.

The mill data will be used to demonstrate the one-way ANOVA based on ranks. An annotated session on R follows.

```
> # Rank the data within the entire dataset
> long_data <- long_data %>%
+   mutate(Rank = rank(Wastage))
> # Create a horizontal side-by-side box plot of the ranked
  data by mill
```

```
> ggplot(long_data, aes(x = Rank, y = Mill)) +
+   geom_boxplot() +
+   labs( +      title = "Boxplots of the Ranks of Wastage by
  Mill",
+     x = "Ranks", +      y = "Mills" +    ) + + theme_minimal()
```

Note that the box plots in Figure 5.8 reveal three distinct groupings of mills: {mill 1}, {mills 2, 3, and 4}, and {mill 5}. Additionally, the IRQs of the ranked data are nearly constant, and there has been a decrease in the number and severity of the outliers. Perform a one-way ANOVA on the ranked data, along with Fisher's LSD multiple comparison method.

```
> # Perform the one-way ANOVA on the ranked data
> anova_result_ranked <- aov(Rank ~ Mill, data = long_data)
```

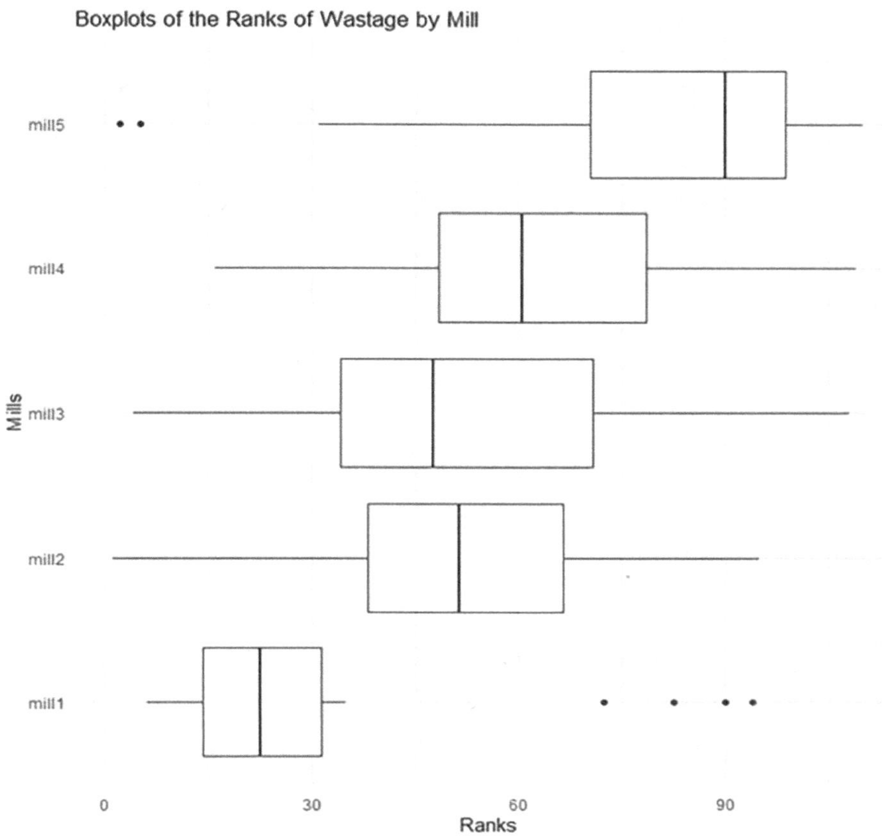

FIGURE 5.8
The box plots of the ranks of wastage by mill.

```
> summary(anova_result_ranked)
             Df Sum Sq Mean Sq F value    Pr(>F)
Mill          4  27531    6883   8.668 4.38e-06 ***
Residuals   105  83370     794
---
Signif. codes: 0 '***' 0.001 '**' 0.01 '*' 0.05 '.' 0.1'' 1
```

Note that the classical F-test applied to the ranked data has detected a significant difference between the five mills in terms of mean wastage. Fisher's LSD procedure helps to quantify these differences.

```
> # Apply Fisher's LSD test if the ANOVA F-test is significant
> if (summary(anova_result_ranked)[[1]][["Pr(>F)"]][1] < 0.05) {
+   lsd_test <- LSD.test(anova_result_ranked, "Mill", p.adj = "none")
+   print(lsd_test)+ } else {+ cat("The ANOVA F-test is not significant;
Fisher's LSD test is not applied.\n")+ }
$statistics
   MSerror  Df Mean       CV  t.value      LSD
  793.9994 105 55.5 50.77116 1.982815 16.84597
$parameters
        test p.ajusted name.t ntr alpha
  Fisher-LSD      none   Mill   5  0.05
$means
          Rank      std  r       se      LCL      UCL Min Max    Q25   Q50
mill1 31.22727 27.20195 22 6.007569 19.31537 43.13917   6  94 14.250 22.50
mill2 51.29545 27.17126 22 6.007569 39.38356 63.20735   1  95 38.125 51.50
mill3 52.06818 30.51007 22 6.007569 40.15628 63.98008   4 108 34.125 47.75
mill4 63.59091 24.99364 22 6.007569 51.67901 75.50281  16 109 48.500 60.50
mill5 79.31818 30.59783 22 6.007569 67.40628 91.23008   2 110 70.500 90.00
         Q75
mill1 31.375
mill2 66.375
mill3 70.750
mill4 78.500
mill5 98.750
$comparison
NULL
$groups
          Rank groups
mill5 79.31818      a
mill4 63.59091     ab
mill3 52.06818      b
mill2 51.29545      b
mill1 31.22727      c
```

The Fisher's protected LSD based on ranks reveals the following groupings: {mill 1, mill 3}<{mill 3, mill 2, mill 4}<{mill 2, mill 4, mill 5}. The groupings above indicate that there is considerable overlap between the groups (mills), with no strong distinct groupings. However, it does appear that Mill 1 and Mill 5 are clearly distinct, with Mill 5 having a significantly greater wastage.

In summary, the rank transformation has resulted in data within each mill group that has nearly constant standard deviations across the mills, with the largest standard deviation being only 31.75/19.51 = 1.63 times larger than the smallest standard deviation. The box plots also reveal that the data within each group is nearly symmetric, with fewer and less severe outliers than with the raw data or the square root of the data. When applied to the ranks, the classical inferential technique reveals differences in the mills that could not be detected before the transformation was applied, clearly indicating the usefulness of the rank transformation.

Of course, a statistical analysis of data does little good unless the results of the analysis are acted upon to form conclusions that may be expressed to the client. In our mill data example, the client may be the manager in charge of production over the five mills. A summary of our results that would be of interest to the manager is as follows:

Mill 1: The run-up process is in good control a majority of the time. However, a disturbingly large proportion of outliers (18%) suggest occasional breakdowns in the process. Check for the source of these outliers – perhaps a new machine operator or problems with equipment may account for these problems.

Mill 2: The run-up process is inconsistent, too often producing large values of wastage. In fact, 23% of the observations are outliers (several outliers are too small values). This process needs to be checked carefully for the source of the outliers (equipment, operators, etc.).

Mill 3: The wastage process is in control with a small mean and small standard deviation and no outliers.

Mill 4: The run-up process consistently produces large wastage values (the mean is fairly large, and the standard deviation is small). At least there are no outliers (no occasional breakdowns in the process). Whatever the mill does wrong to lead to large wastage values, it does it consistently from run to run.

Mill 5: The wastage process is inconsistent here, producing the largest average amount of wastage. This mill must consider examining its entire process to see where improvements can be made.

5.2.4 Theoretical Justification

The theoretical justification for the success of the spread versus level method is provided by the theory of "variance stabilizing transformations." This theory says that if the variance of y_{ij} is some function, h, of the ith treatment mean, say, $\mathrm{var}(y_{ij}) = h(\mu_i)$, then we can find a suitable function, g, so that the transformed variable, $y^*_{ij} = g(y_{ij})$, has a variance that is constant, or "stable," across the i groups.

In Chapter 2, we showed that the approximate variance of a function of a single random variable can be obtained by using a Taylor series expansion of that function about the mean of that variable, a technique referred to as the "delta method." We showed that

$$\mathrm{var}\left(\mathrm{g}\left(\mathrm{y}_{\mathrm{ij}}\right)\right) \approx \mathrm{var}\left(\mathrm{y}_{\mathrm{ij}}\right)\left(\mathrm{g}'(\mu_{\mathrm{i}})\right)^2 = \mathrm{h}(\mu_{\mathrm{i}})\left(\mathrm{g}'(\mu_{\mathrm{i}})\right)^2.$$

Thus, choosing g so that $\mathrm{var}\left(\mathrm{g}\left(\mathrm{y}_{\mathrm{ij}}\right)\right)$ is a constant implies that we want g so that $\mathrm{h}(\mu_{\mathrm{i}})\left(\mathrm{g}'(\mu_{\mathrm{i}})\right)^2$ is constant. Or, equivalently, choose g so that $\mathrm{g}'(\mu_{\mathrm{i}}) = \dfrac{\sqrt{\mathrm{constant}}}{\sqrt{\mathrm{h}(\mu_{\mathrm{i}})}}$. Solving for g by integration over y, we choose g so that $\mathrm{y}^*_{\mathrm{ij}} = \mathrm{g}\left(\mathrm{y}_{\mathrm{ij}}\right)$ is proportional to $\int \dfrac{\mathrm{dy}_{\mathrm{ij}}}{\sqrt{\mathrm{h}\left(\mathrm{y}_{\mathrm{ij}}\right)}}$, the key relationship in determining the choice of g.

Two common data transformations are the log and the square-root functions. They arise from the variance stabilizing transformation theory in the following way. Consider the case where the variance of y is the square of the mean of y. That is, $\mathrm{var}\left(y_{ij}\right) = h(\mu_i) = \mu_i^2$. This is the case for the exponential distribution with probability density function (PDF) $f\left(y_{ij}\right) = \lambda_i e^{-\lambda_i y_{ij}}$, for $\mathrm{y}_{\mathrm{ij}} > 0$. Here, $\mathrm{E}\left(\mathrm{y}_{\mathrm{ij}}\right) = \mu_{\mathrm{i}} = \dfrac{1}{\lambda_{\mathrm{i}}}$ and $\mathrm{Var}\left(\mathrm{y}_{\mathrm{ij}}\right) = \mu_{\mathrm{i}}^2 = \dfrac{1}{\lambda_{\mathrm{i}}^2}$. The transformation, g, is chosen to be proportional to $\int \dfrac{dy_{ij}}{\sqrt{h\left(y_{ij}\right)}} = \int \dfrac{dy_{ij}}{\sqrt{y_{ij}^2}} = c.\log\left(y_{ij}\right)$, where c is some appropriate constant. Hence, for simplicity, choose $\mathrm{g}\left(\mathrm{y}_{\mathrm{ij}}\right) = \log\left(\mathrm{y}_{\mathrm{ij}}\right)$, a power transformation with $\lambda = 0$.

As another example, consider the Poisson distribution with parameter λ_i. Then it is well known that $\mathrm{E}\left(\mathrm{y}_{\mathrm{ij}}\right) = \mu_{\mathrm{i}} = \lambda_{\mathrm{i}} = \mathrm{Var}\left(\mathrm{y}_{\mathrm{ij}}\right)$. Hence, $\mathrm{var}\left(\mathrm{y}_{\mathrm{ij}}\right) = \mathrm{h}(\mu_{\mathrm{i}}) = \mu_{\mathrm{i}}$. Thus, g is chosen to be proportional to $\int \dfrac{\mathrm{dy}_{\mathrm{ij}}}{\sqrt{\mathrm{h}\left(\mathrm{y}_{\mathrm{ij}}\right)}} = \int \dfrac{dy_{\mathrm{ij}}}{\sqrt{\mathrm{y}_{\mathrm{ij}}}} = \mathrm{c}\cdot\sqrt{\mathrm{y}_{\mathrm{ij}}}$. Or, for simplicity, choose $\mathrm{g}\left(\mathrm{y}_{\mathrm{ij}}\right) = \sqrt{\mathrm{y}_{\mathrm{ij}}}$, a power transformation with $\lambda = .5$. The Poisson is often used to model count data, and the square-root transformation is frequently suggested as the appropriate data transformation for count data.

These are two of many situations where transformations are suggested by variance stabilizing theory. Two other examples that arise from this theory are the $\sin^{-1}\left(\sqrt{\hat{\mathrm{p}}}\right)$ transformation used on sample proportions, $\hat{\mathrm{p}}$ (often called the "arcsin" transformation) and the $\dfrac{1}{2}\log\left(\dfrac{1+\mathrm{r}}{1-\mathrm{r}}\right)$ transformation on the sample correlation (often called the "Fisher Z transformation"). The reader may examine these two examples in problems at the end of this chapter.

The spread versus level plot technique is related to the variance stabilizing theory in the following way. In general, if $\text{var}(y_{ij}) = h(\mu_i) = \beta_0 \mu_i^{2\beta_1}$ – that is, the variance of y_{ij} is some power of the mean of y_{ij} (the "power variance model"), where β_0 and β_1 are unknown parameters, then the variance stabilizing theory states that the proper choice of g is

$$g(y_{ij}) = \int \frac{dy_{ij}}{\sqrt{\beta_0 y_{ij}^{2\beta_1}}} = c y_{ij}^{1-\beta_1},$$

where c is some constant. Thus, we choose $\lambda = 1 - \beta_1$. The spread versus level plot is one way of obtaining an estimate of λ. We see that if $\text{var}(y_{ij}) = h(\mu_i) = \beta_0 \mu_i^{2\beta_1}$, then

$$\text{stdev}(y_{ij}) = \sqrt{\text{var}(y_{ij})} = c\mu_i^{\beta_1},$$

which implies

$$\ln(\text{stdev}(y_{ij})) = \ln(c) + \beta_1 \ln(\mu_i),$$

where c is some constant. Thus, for the power variance model, the log of the standard deviation of y is a linear function of the log of the mean of y. In the spread versus level plot technique, the IQR_i and M_i are estimating the standard deviation and mean of y_{ij}, and it follows that an estimate of λ is $\lambda = 1 - \hat{\beta}_1$, where $\hat{\beta}_1$ is the slope estimate from the spread versus level plot.

In terms of data analysis, the transformation of data accomplishes several goals. One, mentioned earlier, is to simplify the relationship between the population means and variances. In the one-way ANOVA setting, if variances are constant across treatments, then differences in treatments can be quantified completely by differences in the treatment means. Another advantage of using transformed data instead of raw data with its nonconstant variances is that if statistical inference is to be performed, it is more likely that the transformed data meets the assumptions required by the inferential techniques. For example, in the classical analysis in a one-way ANOVA situation, the assumptions of normality and common variance are required for the validity of the inferential procedure. The spread versus level plot technique is designed to enable the transformed data to satisfy these conditions. However, as the mill data and the theoretical justification demonstrate, the technique is useful only under certain limited situations. The rank transformation is also explored as a method to achieve the required assumptions for the classical inferential method. The Box-Cox transformation is another transformation technique to be discussed in the next section.

5.3 The Box-Cox Transformation

A second method for transforming data is based on the so-called Box-Cox transformation, introduced by Box and Cox (1964). This transformation is also in the form of the power transformation, although a slightly different version is used than that introduced in the previous section. The motivation of the theory for determining the proper transformation and illustration of the method will be first applied to the location model. The technique will then be extended to the one-way classification situation.

In the location model, let y be our random variable with cumulative distribution function (CDF) $F(y)$ and PDF $f(y)$. Box and Cox proposed the family of transformations of the form

$$y^* = \begin{cases} \dfrac{y^\lambda - 1}{\lambda} & \lambda \neq 0 \\ \ln(y) & \lambda = 0. \end{cases}$$

We note the similarity between the transformations suggested by Box and Cox and those used in the last section suggested by Tukey.

Box and Cox considered the value of λ as that power required so that the resulting transformed variable, y^*, follows a normal distribution with a mean of μ and a variance of σ^2. That is, what value of λ results in $y^* \sim N(\mu, \sigma^2)$? Thus, the value of λ would result in a "normalizing" transformation.

Under these conditions, our location model becomes

$$y_i^* = \mu + \varepsilon_i,$$

where $\varepsilon_i \sim N(0, \sigma^2)$ and $i = 1, \ldots, n$. Box and Cox proposed to find the proper value of λ by the method of maximum likelihood estimation. We need the joint density of the y^* sample, which is

$$\begin{aligned} f_{y^*}(y_1^*, \cdots, y_n^*) &= \prod_{i=1}^{n} \frac{1}{\sqrt{2\pi}\sigma} \exp\left(-\frac{(y_i^* - \mu)^2}{2\sigma^2} \right) \\ &= (2\pi)^{-\frac{n}{2}} \sigma^{-n} \exp\left(-\frac{\sum_{i=1}^{n} (y_i^* - \mu)^2}{2\sigma^2} \right). \end{aligned}$$

Using the change of variable theorem from calculus, we can express the density of the y sample as

$$f_y(y_1, \ldots, y_n) = f_{y^*}(y_1^*, \ldots, y_2^*) J(\lambda; y^*) = (2\pi)^{-\frac{n}{2}} \sigma^{-n} \exp\left(-\frac{\sum_{i=1}^{n} (y_i^* - \mu)^2}{2\sigma^2} \right) \prod_{i=1}^{n} (y_i^{\lambda - 1}),$$

where $J(\lambda;y^*)$ is the Jacobian of the transformation and is equal to

$$J(\lambda;y^*) = \prod_{i=1}^{n} \left| \frac{dy_i^*}{dy_i} \right| = \prod_{i=1}^{n} (y_i)^{\lambda-1}$$

since we are assuming that the y variable is a positive random variable.

The log-likelihood function of the unknown parameters (μ, σ^2, λ) is then

$$\ln L(\mu, \sigma^2, \lambda) = -\frac{n}{2}\ln(2\pi) - \frac{n}{2}\ln(\sigma^2) - \frac{1}{2\sigma^2}\Sigma(y_i^* - \mu)^2 + (\lambda - 1)\sum_{i=1}^{n} \ln(y_i).$$

At a fixed value of λ, we can take derivatives of $\ln L(\mu, \sigma^2, \lambda)$ with respect to the unknown parameters μ and σ^2, set each equation equal to zero, and solve for the maximum likelihood estimators (MLEs). We obtain

$$\frac{\partial \ln L(\mu, \sigma^2, \lambda)}{\partial \mu} = \frac{\Sigma(y_i^* - \mu)}{\sigma^2} = 0$$

$$\frac{\partial \ln L(\mu, \sigma^2, \lambda)}{\partial \sigma^2} = -\frac{n}{2}\frac{1}{\sigma^2} + \frac{1}{2\sigma^4}\Sigma(y_i^* - \mu)^2 = 0.$$

For fixed λ, the MLEs are

$$\hat{\mu}_{(\lambda)} = \bar{y}^* \text{ and } \hat{\sigma}^2_{(\lambda)} = \frac{\sum_{i=1}^{n}(y_i^* - \bar{y}^*)^2}{n}.$$

Thus, for a fixed λ, the maximized log-likelihood is

$$\begin{aligned}\ln L(\lambda) &= \text{constant} - \frac{n}{2}\ln\left(\hat{\sigma}^2_{(\lambda)}\right) + \ln\left(J(\lambda;y^*)\right) \\ &= \text{constant} - \frac{n}{2}\ln\left(\hat{\sigma}^2_{(\lambda)}\right) + (\lambda - 1)\sum_{i=1}^{n}\ln(y_i).\end{aligned}$$

We now must determine the value of λ, denoted by $\hat{\lambda}$, which maximizes $\ln L(\lambda)$, denoted by $\ln L(\hat{\lambda})$. This can be done numerically by plotting $\ln L(\lambda)$ against λ for a trial series of values of λ. Since the "constant" term does not depend on the value of λ, it may be ignored in the calculation of $\ln L(\lambda)$.

The success of the Box-Cox procedure at achieving normality and constant variance with the power transformation depends a great deal on the behavior of the original data. The need for a transformation to achieve normality of the data is reflected in the behavior of the plot of $\ln L(\lambda)$ versus λ. A "peaked" shaped curve indicates that likelihood is maximized at $\hat{\lambda}$ and falls away quickly in a small neighborhood around $\hat{\lambda}$, suggesting that the value of $\hat{\lambda}$ will likely work well in achieving our goals. A "flat" curve, on the other hand, indicates that many different values of λ would work equally well, and it is doubtful that any value will work very well.

The need for the transformation, a value of λ other than one, can be more formally measured by the use of a confidence interval estimate for λ. The method is based on maximum likelihood theory and the use of the likelihood ratio test. Without delving into the details of these methods, an α level test of the hypotheses $H_0 : \lambda = \lambda_0$ versus $H_1 : \lambda \neq \lambda_0$, for arbitrary values of λ_0, is accomplished using the statistic $2\left(\ln L\left(\hat{\lambda}\right) - \ln L(\lambda_0)\right)$, where $\ln L(\lambda_0)$ is the log-likelihood evaluated at the arbitrary value of $\lambda = \lambda_0$, and $\ln L\left(\hat{\lambda}\right)$ is the maximum value of the log-likelihood. Likelihood theory states that this test statistic follows an asymptotic chi-square distribution with 3 – 2 = 1 degree of freedom since one less parameter needs to be estimated to obtain $\ln L(\lambda_0)$ than $\ln L\left(\hat{\lambda}\right)$.

Therefore, a 5% level test would be to reject $H_0 : \lambda = \lambda_0$ if this test statistic exceeds the $(1-\alpha)100\%$ point of a chi-square distribution with one degree of freedom, $X^2_{1-\alpha,1}$. A little algebra reveals that this decision rule is equivalent to failing to reject H_0 if $\ln L(\lambda_0) \geq \ln L\left(\hat{\lambda}\right) - .5X^2_{1-\alpha,1}$. Thus, a 5% level test is this:

fail to reject H_0, if $\ln L(\lambda_0) \geq \ln L(\hat{\lambda}) - .5(3.84) = \ln L(\hat{\lambda}) - 1.92$. Graphically, this test is equivalent to drawing a horizontal line at $\ln L\left(\hat{\lambda}\right) - 1.92$ on the plot of $\ln L(\lambda)$ versus λ, and at the intersection of this line with the curve $\ln L(\lambda)$, drop vertical lines down to the λ axis. Then the $(1-\alpha)100\%$ confidence interval for λ is represented by all values of λ captured within the intersection of these two vertical lines with the λ axis. If λ_0 is contained within this interval, the hypothesis $H_0 : \lambda = \lambda_0$ is not rejected. These concepts will now be demonstrated in the next two examples.

5.3.1 The Volcano Data Application

The Box-Cox transformation method for the location model will now be demonstrated using the R program with the volcano data, introduced in Chapter 1. Recall that this dataset consisted of the heights of 219 of the world's highest volcanoes. A thorough exploratory data analysis of the data is presented in Chapter 1. The method of mid-summaries was used to determine that the original data could be transformed from being strongly skewed to the right to symmetric data by the square-root transformation. It will be

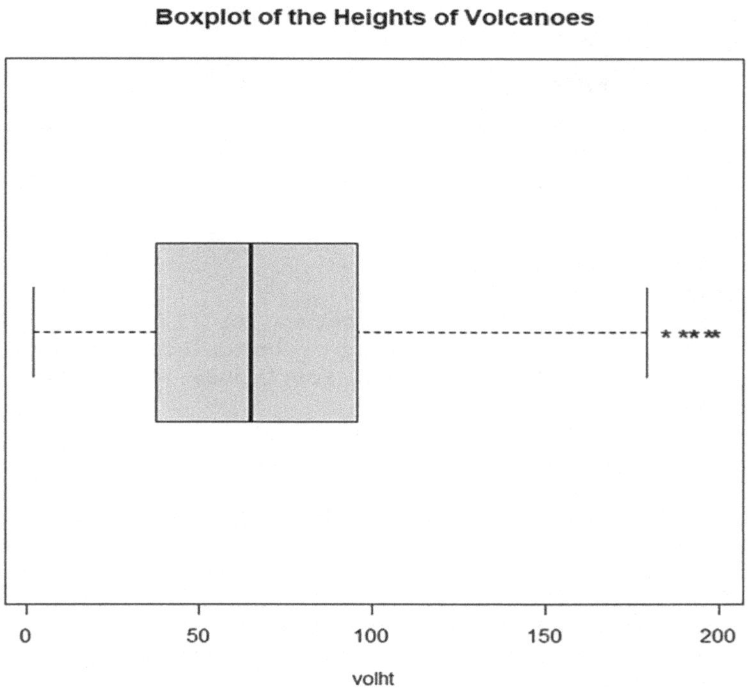

FIGURE 5.9
The box plot of the heights of volcano data.

interesting to determine the Box-Cox transformation for this dataset and to compare the results with the method of mid-summaries.

The R code has been written to perform the Box-Cox procedure for the location model. The annotated R session analyzing the volcano data is as follows.

Let's look again at the volcano data, where the box plot is given in Figure 5.9.

```
# Create the box plot
boxplot(volcano_heights, main = "Boxplot of the Heights of
  Volcanoes", ylab = "volht",
      xlab = "", outline = TRUE, boxwex = 0.75, staplewex =
      0.75,
      outpch = "*", outcol = "black", ylim = c(0, 200), las =
1, cex.axis = 0.8, col = "gray", border = "black", whisklty =
2, whiskcol = "black", staplecol = "black")
```

The original data is clearly skewed to the right, as shown in Figure 5.9. To get the optimal values for lambda using the Box-Cox method, let's simulate lambda and the 95% CI for it as follows:

```
> # Perform the bootstrap
> set.seed(123) # for reproducibility
> bootstrap_results <- boot(data = data$Value, statistic =
  boxcox_lambda_logL, R = 1000)
> # Calculate the 95% Confidence Interval from the bootstrap
  results
> bootstrap_ci_lambda <- boot.ci(bootstrap_results, index = 1,
  type = "bca") # Lambda CI
> bootstrap_ci_logL <- boot.ci(bootstrap_results, index = 2,
  type = "bca")  # LogL CI
> # Display results
> cat("Bootstrap 95% Confidence Interval for lambda:", bootstrap_
  ci_lambda$bca[4], "to", bootstrap_ci_lambda$bca[5], "\n")
Bootstrap 95% Confidence Interval for Lambda: 0.3030303 to
  0.5454545
> cat("Bootstrap 95% Confidence Interval for log likelihood:",
  bootstrap_ci_logL$bca[4], "to", bootstrap_ci_logL$bca[5],
  "\n")
Bootstrap 95% Confidence Interval for log-likelihood:
  -538.1284 to -489.892
> cat("Optimal lambda based on bootstrap:", mean(bootstrap_
  results$t[,1]), "\n")
Optimal Lambda based on bootstrap: 0.458101
> cat("Optimal log-likelihood based on bootstrap:",
  mean(bootstrap_results$t[,2]), "\n")
Optimal log-likelihood based on bootstrap: -512.2165
```

We see that the maximum value of lambda occurs approximately at 0.5, suggesting the square-root transformation.

Figure 5.10 reveals that the maximum is achieved at lambda = 0.5. Thus, the Box-Cox method suggests the square-root transformation. Dropping vertical lines at the intersection of the horizontal line gives the 95% CI for lambda of approximately (0.3, 0.55).

We note that the Box-Cox method determines that the square-root transformation is the appropriate "normalizing" transformation for the volcano data. The reader may refer to Section 1 of Chapter 1 to confirm that the square-root transformation does result in data that is very "normal" in appearance. In this instance, both methods, the method of mid-summaries and the Box-Cox method, give the same value of $\hat{\lambda}$. This is often the case. The 95% confidence interval for lambda is approximate (0.3, 0.55), suggesting that a value of lambda other than one is preferred. Transformed data should appear more "normal" than raw data. We have seen in Chapter 1 that this is indeed the case. In addition, we know from Chapter 1, where we first analyzed the volcano data, that the log transformation skews the data too much to the left. The 95% confidence interval does not support a value of lambda equal to zero or the log transformation.

The Box-Cox transformation method can easily be extended from the location model to other general linear model family models. This

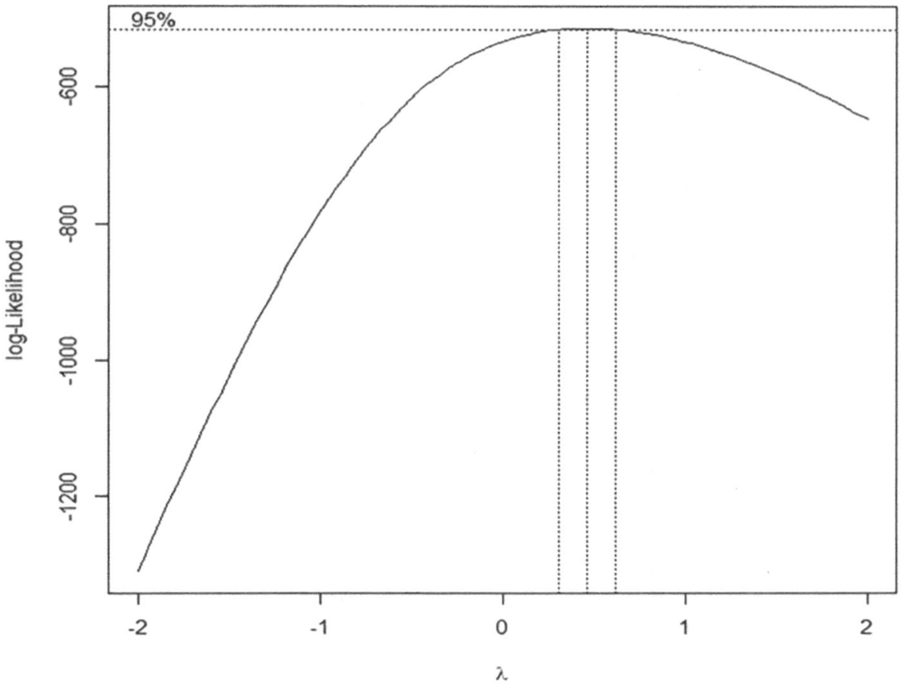

FIGURE 5.10
Box-Cox procedure applied to the volcano data; horizontal line is used to get a 95% CI for lambda.

extension will be made to the one-way ANOVA model in the remainder of this section.

The goal of the Box-Cox transformation in one-way ANOVA is to find that value of λ so that the resulting transformed data, y^*_{ij}, follows a normal distribution with constant variance across the k treatments. The transformed responses, y^*_{ij}, result from applying the Box-Cox transformation to the raw data. Thus, the means model for the transformed data is

$$y^*_{ij} = \mu_i + \varepsilon_{ij}.$$

Setting up and evaluating the likelihood equations proceeds exactly as in the location model, so we will not repeat that process here. Following the location model derivation, for a fixed value of λ, the MLEs of the k unknown μ_is and σ^2 are easily obtained as

$$\hat{\mu}_{i(\lambda)} = \bar{y}^*_i \text{ and } \hat{\sigma}^2_{(\lambda)} = \frac{\sum_{i=1}^{k} \sum_{j=1}^{n_i} (y^*_{ij} - \bar{y}^*_i)^2}{N}.$$

The usual estimates of these parameters in the one-way ANOVA model assume normality and common variance, except they are based on the transformed data, not the raw data. Substitution of these estimates into the log-likelihood expression results in

$$\ln L(\lambda) = \text{constant} - \frac{N}{2}\ln\left(\hat{\sigma}^2_{(\lambda)}\right) + (\lambda - 1)\sum_{i,j}\ln\left(y_{ij}\right).$$

An expression similar to the one obtained in the location model, with appropriate adjustments needed to reflect the differences in the two models. The MLE of $\lambda, \hat{\lambda}$, is that value that maximizes $\ln L(\lambda)$. It is convenient to find $\hat{\lambda}$ by searching through a trial series of values of λ. The technique will now be demonstrated on R using the plankton data.

5.3.2 The Plankton Data Using R

Recall the plankton data from the previous section, where the response is the number of plankton caught using three kinds of nets. Exploratory data methods revealed that this data contained a very high degree of heterogeneity of variance, with the largest of the three standard deviations being 28.58 times larger than the smallest standard deviation. The spread versus level plot technique was used on this data to determine that a $\lambda = 0.25$ should provide data with a more stable variance. Further analysis revealed that the case with the largest standard deviation was only 1.68 times larger than the smallest standard deviation for the quarter root–transformed data. It was also demonstrated that the square-root transformation was not as successful at achieving constant variance.

We will further analyze the plankton data using the Box-Cox technique using R. Let's visualize the data firstly as follows:

```
# Set up the plotting area to have 3 panels (1 row, 3 columns)
  for Histogram for kind1, Kind3, and kind4.
par(mfrow = c(1, 3))
hist(plankton_data$kind1, main = "Histogram of Kind1", xlab =
  "Counts", col = "blue", border = "black", ylim = c(0, 5))
hist(plankton_data$kind4, main = "Histogram of Kind4", xlab =
  "Counts", col = "white", border = "black", ylim = c(0, 5))
hist(plankton_data$kind3, main = "Histogram of Kind3", xlab =
  "Counts", col = "red", border = "black", ylim = c(0, 5))
# Reset par to default
par(mfrow = c(1, 1))
```

One can see from Figure 5.11 that the histograms, side-by-side box plot, and observations for Kind 1, Kind 3, and Kind 4 are not normal. Let's set up the trial values of lambda as –2, –1, –0.5, –0.25, –0.15, –0.05, 0.1, 0.15, 0.25, 0.5, 0.75, 1, and 2 as follows:

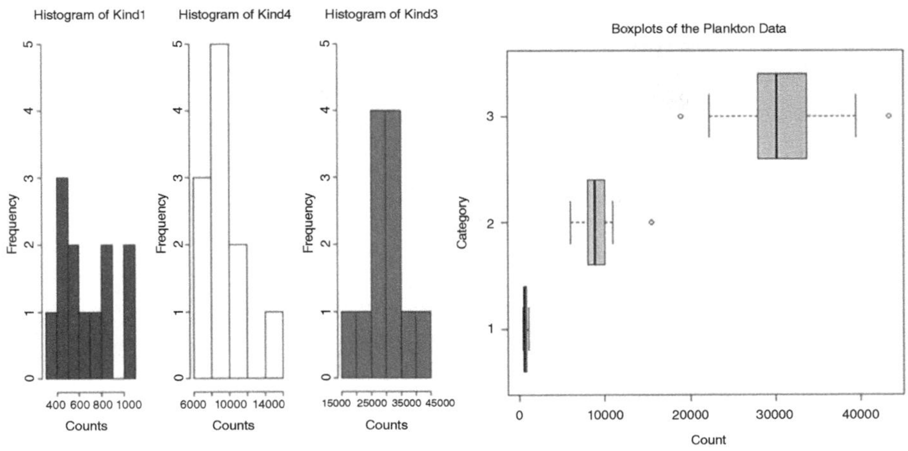

FIGURE 5.11
The histograms and the side-by-side box plot for the plankton data.

```
> # Example data setup
> data <- data.frame(raw_data = rnorm(100, mean = 50, sd =
  10)) # simulate normal data
> # Define lambda values
> lambda_values <- c(-2, -1, -0.5, -0.25, -0.15, -0.05, 0,
  0.1, 0.15, 0.25, 0.5, 0.75, 1, 2)
> # Perform Box-Cox transformation and store log-likelihoods
> boxcox_results <- boxcox(raw_data ~ 1, lambda = lambda_val-
  ues, data = data, plot = FALSE)
> # Extract the log-likelihoods for each Lambda
> log_likelihoods <- boxcox_results$y
> # Find the Lambda that maximizes the log likelihood
> max_index <- which.max(log_likelihoods)
> max_lambda <- lambda_values[max_index]
> max_log_likelihood <- log_likelihoods[max_index]
> # Output the results similar to R
> results_df <- data.frame(Lambda = lambda_values, LogL =
  log_likelihoods)
> print(results_df)
 ROW Lambda      LogL
   1 -2.000 -398.817
   2 -1.000 -320.102
   3 -0.500 -282.398
   4 -0.250 -265.660
   5 -0.150 -260.284
   6 -0.050 -256.281
   7  0.000 -254.933
   8  0.100 -253.703 #this is the maximum value
   9  0.150 -253.819
```

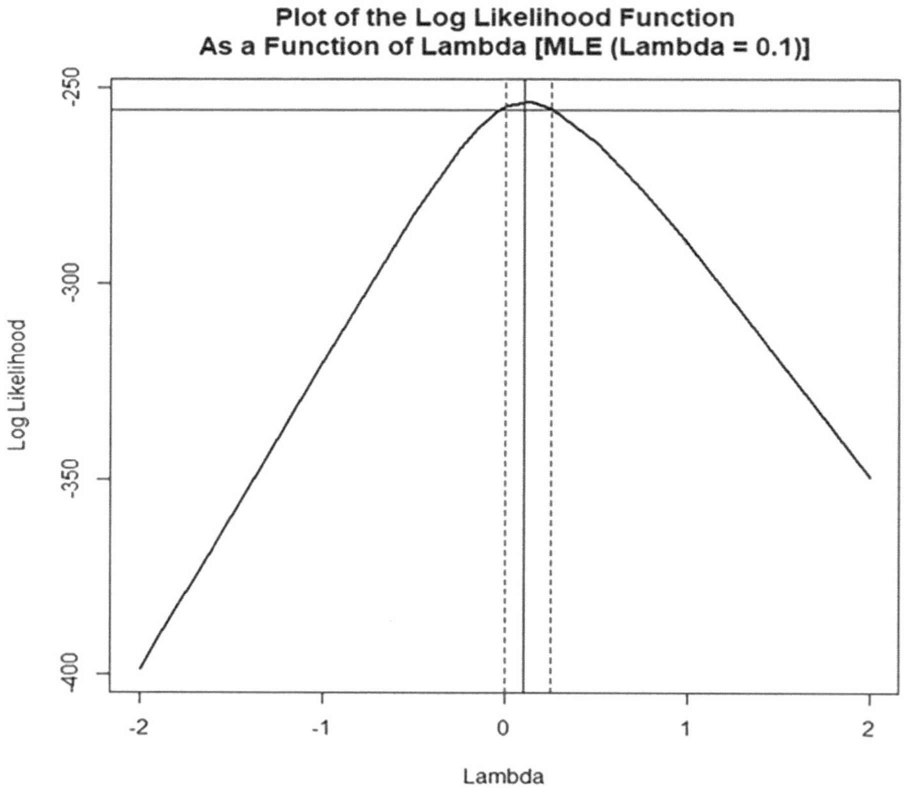

FIGURE 5.12

Box-Cox procedure applied to the plankton data; horizontal line is used to get a 95% CI for lambda.

```
  10  0.250 -255.347
  11  0.500 -264.241
  12  0.750 -276.492
  13  1.000 -290.045
  14  2.000 -349.249
```

We note from Figure 5.12 that the maximum log-likelihood is obtained at lambda = 0.1 and that the 95% CI is approximately from (0, 0.25). Let's evaluate the three values of Lambda: 0, 0.1, and 0.25.

```
# Load necessary libraries
library(readxl)
library(dplyr)
# Load the data from the Excel file
plankton_data <- read_excel("Plankton_Data.xlsx")
```

```
# Prepare the plotting area for 4 plots in 2x2 layout
par(mfrow = c(2, 2)) # This will create a 2x2 plotting area
# Combine the original data into a single dataframe for
plotting
combined_data_original <- data.frame( count = c(plankton_
data$kind1, plankton_data$kind4, plankton_
data$kind3),category = factor(rep(1:3, each =
nrow(plankton_data))) # Create factor variable for each
category)
# Plotting the original data box plots horizontally
boxplot(count ~ category, data = combined_data_original,
main = "Boxplots of the Original Plankton Data", ylab = "Cat-
egory", xlab = "Count", names = c("1", "2", "3"),col =
c("gray", "gray", "gray"),outline = TRUE, notch = FALSE,
horizontal = TRUE, las = 1)
# Apply the log transformation
plankton_data$log_kind1 <- log10(plankton_data$kind1)
plankton_data$log_kind4 <- log10(plankton_data$kind4)
plankton_data$log_kind3 <- log10(plankton_data$kind3)
# Combine the log-transformed data into a single dataframe for
plotting
combined_data_log <- data.frame(log_count = c(plankton_
data$log_kind1, plankton_data$log_kind4, plankton_data$log_
kind3), category = factor(rep(1:3, each =
nrow(plankton_data))))
# Plotting the log-transformed box plots horizontally
boxplot(log_count ~ category, data = combined_data_log, main =
"Boxplots of the Log of the Plankton Data", ylab =
"Category",xlab = "Log Count",names = c("1", "2", "3"),col =
c("gray", "gray", "gray"), outline = TRUE, notch = FALSE,
horizontal = TRUE, las = 1)
# Apply the tenth root transformation
plankton_data$tenth_root_kind1 <- plankton_data$kind1^(1/10)
plankton_data$tenth_root_kind4 <- plankton_data$kind4^(1/10)
plankton_data$tenth_root_kind3 <- plankton_data$kind3^(1/10)
# Combine the tenth root-transformed data into a single
dataframe for plotting
combined_data_tenth_root <- data.frame(tenth_root_count =
c(plankton_data$tenth_root_kind1, plankton_data$tenth_root_
kind4, plankton_data$tenth_root_kind3),category =
factor(rep(1:3, each = nrow(plankton_data))))
# Plotting the tenth root-transformed box plots horizontally
boxplot(tenth_root_count ~ category, data = combined_data_
tenth_root, main = "Boxplots of the Tenth Root of the Data",
ylab = "Category", xlab = "Tenth Root of Count", names =
c("1", "2", "3"),col = c("gray", "gray", "gray"),outline =
TRUE, notch = FALSE, horizontal = TRUE, las = 1)
# Apply the quarter-root transformation
```

```
plankton_data$quarter_root_kind1 <- plankton_data$kind1^(1/4)
plankton_data$quarter_root_kind4 <- plankton_data$kind4^(1/4)
plankton_data$quarter_root_kind3 <- plankton_data$kind3^(1/4)
# Combine the quarter root-transformed data into a single
dataframe for plotting
combined_data_quarter_root <- data.frame( quarter_root_count =
c(plankton_data$quarter_root_kind1, plankton_data$quarter_
root_kind4, plankton_data$quarter_root_kind3),category =
factor(rep(1:3, each = nrow(plankton_data))))
```

The box plots in Figure 5.13 indicate that the lambda of 0.25 better stabilizes the IQRs for the plankton data. Also, we find that the largest IQR is 1.42 times larger than the smallest for lambda of 0.25 but 2.14 times larger for lambda of 0.1. However, when standard deviations are compared, the largest standard deviation is 1.1 times larger than the smallest for lambda = 0.1 but

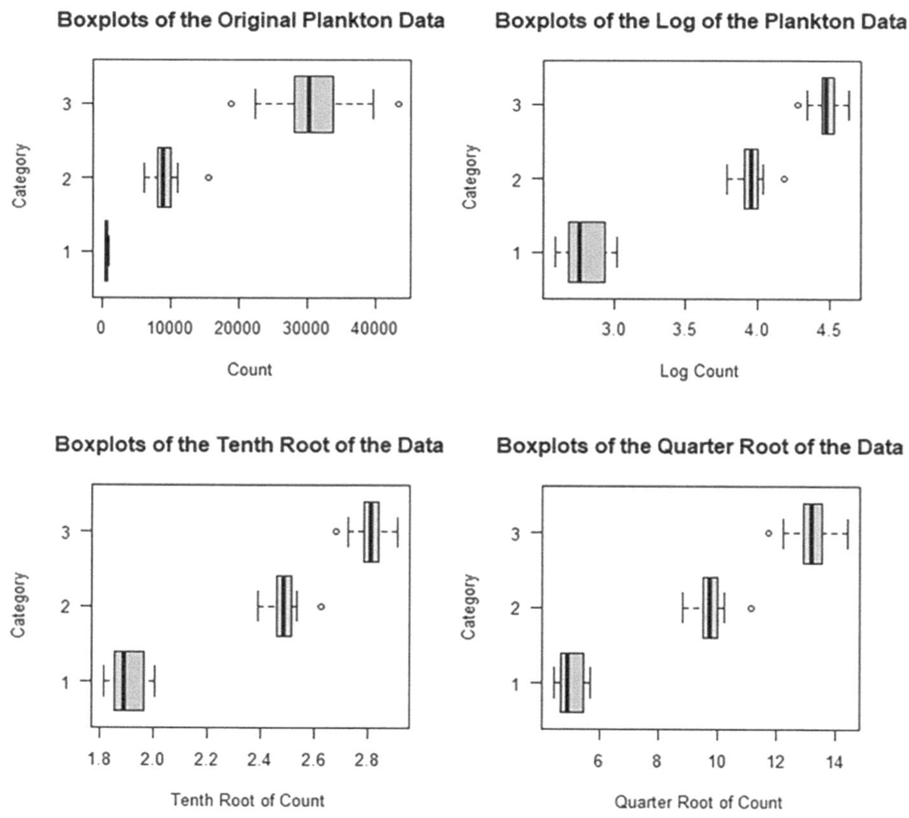

FIGURE 5.13
Box plots for different transformations of the plankton data.

1.68 times larger for the lambda = 0.25 case. The difference in choice of proper lambda is due to the outliers in groups 2 and 3, causing inflation in the standard deviations for the second and third groups.

We note that the Box-Cox method results in a λ of 0.1, while the spread versus level plot method selected the value of 0.25. The difference is due to the fact that the Box-Cox method stabilizes spread based on standard deviations, while the spread versus level plot method uses the IQRs. Since there are several outliers in the plankton data (in both the raw and transformed scales), there are expected differences in the final selected values of λ between these two methods.

The Box-Cox technique may also be used to transform data in the simple and multiple linear regression models, models to be presented in Chapters 6 and 7.

Exercises

1. Consider the plankton data from Section 3.2. Apply the Tukey spread versus level plot method, described in Section 3.2, to this data, except use s_i and $\bar{y}_i$ as the measures of spread and center, respectively, instead of IQR_i and M_i. Compare the value of λ that you obtain with the value obtained by using the Box-Cox method.
2. Consider the plankton data example presented in this chapter.
   a) Transform the raw data using the power family with $\lambda = .33$. Obtain side-by-side box plots of the transformed data.
   b) Compare the results obtained in part a) with the results in the text for $\lambda = .25$ and $\lambda = .5$. Which transformation is the most successful at achieving constant variance across the three levels of the treatment variable?
3. Consider the random variable y, where $\mathrm{E}(\mathrm{y}) = \theta$ and $\mathrm{Var}(\mathrm{y}) = \theta(1-\theta)$.
   a) Find the variance stabilizing transformation, g, so that $g(y)$ has a constant variance.
   b) Suppose that the random variable y follows a binomial distribution with parameters n and p. That is, $\mathrm{y} \sim \mathrm{Bi}(\mathrm{n},\mathrm{p})$. Then y is the number of "successes" in n independent trails, where the probability of success on any one trail is p. Define the sample proportion, $\hat{\mathrm{p}}$, as $\hat{\mathrm{p}} = \frac{\mathrm{y}}{\mathrm{n}}$. Find the transformation, g, so that $\mathrm{g}(\hat{\mathrm{p}})$ has constant variance (independent of p).
4. The sample correlation coefficient, r, between n pairs of observations (x_i, y_i) is defined as

$$r = \frac{\sum_{i=1}^{n}\left(x_i - \bar{x}\right)\left(y_i - \bar{y}\right)}{\sqrt{\sum_{i=1}^{n}\left(x_i - \bar{x}\right)^2 \sum_{i=1}^{n}\left(y_i - \bar{y}\right)^2}}.$$

It has been shown that $E(r) \approx \rho$ and $\mathrm{Var}(r) \approx \frac{(1-\rho^2)^2}{n}$, where ρ is the population correlation coefficient; find the transformation, g, so that $g(r)$ has a constant variance (independent of ρ).

5. Consider the phosphorus data from Chapter 1. Let yield be the response variable, and fertilize (the rate at which phosphorus fertilizer, 0, 30, 60, and 120 kilograms per hectare has been applied to the same plots of land under maize cultivation over 30 years) be the treatment variable. Note that fertilization can be thought of as a categorical variable with four levels. Answer the following questions using your software package.
   a) Display the response variable by each value of the treatment variable. Use dot plots and box plots as illustrative in the mill data example in the text. Comment on your observations.
   b) Obtain the classical one-way ANOVA analysis of this data. Use Fisher's protected LSD to compare mean response across treatment levels. Discuss your results.
   c) Obtain the residuals from the classical one-way ANOVA analysis. Display these residuals using a plot of the residuals versus treatment level, the probability plot of residuals, and the box plot of the residuals. Comment on your observations with emphasis on the assumptions of normality and a constant variance.
   d) Use the spread versus level algorithm to determine whether nonconstant variance is present in this data. If so, what transformation is suggested to remove this heterogeneity of variance?
   e) Repeat parts a), b), and c) above using the transformed data. Compare your results to those obtained for the raw data.
   f) Consider the rank transformation of the raw data. Perform the one-way ANOVA analysis of this data using the rank method as discussed in the text. Compare your results to those obtained in part e) above.
   g) Apply the Box-Cox method to this data. What transformation is suggested? Compare this transformation to the one obtained using the spread versus level plot technique (in part d) above). If this transformation is different from that obtained in part d), compare the one-way ANOVA results using the Box-Cox transformed data to the results obtained in f) above.
6. **Toxicity data**: The toxicity data (from Case Studies in Biometry: Chapter 2: Assessing Toxicity of Pollutants in Aquatic Systems by A. John Bailer and James T. Oris) concerns the modeling of the ability of *Ceriodaphnia dubia* to reproduce in the presence of varying levels of a pollutant. The variables are label = comment, animal = a unique identifier

associated with each *C. dubia* tested, conc = concentration of pollutant (micro grams/L), brood1 = number of young produced in the first brood, brood2 = number of young produced in the second brood, brood3 = number of young produced in the third brood, total = sum of young produced in the three broods (= brood1 + brood2 + brood3). The data is on the book website under the name toxicity_data.xlsx. Consider the response variable total and the treatment variable conc, a categorical variable with five levels. Answer the following questions on R. Do parts a) through g) of question 5 above.

7. **Bolts data:** A manufacturer of automotive accessories provides hardware – for example, nuts, bolts, washers, and screws – to fasten the accessory to the car or truck. Hardware is counted and packaged automatically. Specifically, bolts are dumped into a large metal dish. A plate that forms the bottom of the dish rotates counterclockwise. This rotation forces bolts to the outside of the dish and up along a narrow ledge. Due to the vibration of the dish caused by the spinning bottom plate, some bolts fall off the ledge and back into the dish. The ledge spirals up to a point where the bolts are allowed to drop into a pan on a conveyor belt. As a bolt drops, it passes by an electronic eye that counts it. When the electronic counter reaches the preset number of bolts, the rotation is stopped, and the conveyor belt is moved forward.

   There are several adjustments on the machine that affect its operation. These include a speed setting that controls the speed of rotation (SPEED1) of the plate at the bottom of the dish, total number of bolts (TOTAL) to be counted, a second-speed setting (SPEED2) that is used to change the speed of rotation (usually slowing it down) for the last few bolts, the number of bolts to be counted at this second speed (NUMBER2), and the sensitivity of the electronic eye (SENS). The sensitivity setting is to ensure that the correct number of bolts are counted. Too few bolts packaged cause customer complaints. Too many bolts packaged increase costs. The correct number of bolts was counted for each run conducted in this experiment. From an engineering standpoint, the sensitivity should not affect the time to count bolts if the correct number of bolts is counted. The measured response is the time (TIME), which, in seconds, it takes to count the desired number of bolts. In order to put times on an equal footing, the response to be analyzed is the time to count 20 bolts (T20BOLT). The data is found on the website under bolts_data.xlsx. Let t20bolt be the response variable and speed1, a categorical variable with three levels, be the treatment variable.

   Answer the following questions on R. Do parts a) through g) of question 5 above.

Multiple-Choice Questions

1. What does the one-way ANOVA model compare?
   a) The mean response across multiple populations
   b) The variance of a single treatment
   c) The distribution of errors in a model
   d) The correlation between two variables

   Answer: a)
2. Which of the following is a nonparametric alternative to the one-way ANOVA?
   a) T-test
   b) Kruskal-Wallis test
   c) Pearson correlation
   d) Linear regression

   Answer: b)
3. Which of the following methods can be used to handle outliers in one-way ANOVA?
   a) Leave them as they are.
   b) Use transformations.
   c) Exclude the outliers.
   d) All of the above.

   Answer: d)
4. What is the purpose of transforming the data in the context of one-way ANOVA?
   a) To increase the sample size
   b) To make the data fit the assumptions of the ANOVA model
   c) To change the hypothesis being tested
   d) To increase the number of treatments

   Answer: b)
5. What is Fisher's LSD method used for in the context of ANOVA?
   a) To test for normality
   b) To test for variance equality
   c) To conduct multiple comparisons of means
   d) To determine sample size

   Answer: c)

6. What does the presence of outliers in ANOVA suggest?
   a) A well-fitting model
   b) A need for transformations
   c) Perfect data symmetry
   d) Equal variances across treatments

   Answer: b)
7. What does a "spread versus level" plot help determine in the context of transformations?
   a) The relationship between two variables
   b) The need for a data transformation
   c) The best type of regression model
   d) The best way to calculate means

   Answer: b)
8. What is the main assumption behind the use of classical one-way ANOVA?
   a) The response variable is categorical.
   b) The response variable is normally distributed with equal variance across treatments.
   c) The response variable is ordinal.
   d) The sample size is always equal.

   Answer: b)
9. Which of the following is true about the Kruskal-Wallis test?
   a) It assumes normality.
   b) It is robust to outliers.
   c) It requires equal sample sizes.
   d) It is a parametric test.

   Answer: b)
10. What is the key outcome of the Box-Cox transformation in one-way ANOVA?
    a) It eliminates outliers.
    b) It stabilizes variance and normalizes the data.
    c) It increases the number of treatments.
    d) It decreases the mean.

    Answer: b)

6

Simple Linear Regression: Classical, Robust, and Nonparametric Methods

Summary

This chapter will present one of the most frequently used and misused statistical models, the simple linear regression (SLR) model. In the first section, the basic SLR model will be presented. We will discuss the estimation and testing of parameters and review basic model evaluation methods. The estimation technique that will be emphasized is the method of least squares. The second section will present the need to consider transformations in the SLR model and will illustrate two useful methods: one due to Tukey and the other applications of the Box-Cox method for determining the proper transformation.

The effects of outliers on the SLR model will be presented in Section 6.3, along with several versions of robust alternatives to the usual methods of estimation and testing parameters in the SLR model. The chapter will conclude by presenting several nonparametric or model-free methods, such as kernel regression. Each section will include a detailed illustration of the SPSS steps and/or R codes.

Learning Objectives

By the end of this chapter, readers should be able to:

- Understand the basics of the SLR model, including its assumptions, estimation techniques, and interpretation.
- Learn about the effects of outliers and how robust estimation techniques can address these effects in SLR.
- Explore transformation methods like Tukey's and Box-Cox methods to linearize relationships and meet SLR assumptions.
- Apply nonparametric methods for regression, including kernel regression, when standard assumptions are violated.
- Gain proficiency in implementing SLR analyses using SPSS and R for both classical and robust methodologies.
- Develop skills to evaluate model quality through visual and numerical assessments, including scatterplots, residual analysis, and goodness-of-fit metrics.
- Understand the theoretical foundation and practical application of inference methods, including hypothesis testing and confidence intervals for regression coefficients.

DOI: 10.1201/9781003565321-6

- Analyze real-world data using SLR and make informed decisions based on the results.

6.1 The Simple Linear Regression Model

This section will review the basics of the simple linear regression (SLR) model. Throughout the section, the reader is assumed to be already familiar with the concepts of SLR. The material is reviewed in this section to acquaint students and researchers with the notation and terminology of the SLR that will be used in the remaining sections.

One common problem in research is determining whether a functional relationship exists between two variables and, if so, identifying and quantifying this relationship. For example, we may desire to study the relationship between the amount of money our company spends on advertising our product and the amount of product sales. Hopefully, the more we tell people about the product, the more they want to buy it. Regression models can be used to quantify and evaluate the relationship between these variables.

We label the two variables as x and y, where the "x" variable is often called the "independent" variable or "regressor" variable. The "y" variable is often called the "dependent" variable or "response" variable. These two variables are distinguished by the fact that the x variable is available, and can be controlled by the researcher, whereas the y variable reacts or changes due to the changes in the x variable.

In our example above, the x variable is the amount of money spent on advertising, and the y variable is the amount of sales of our product. It is desired to express y as some function of x, say $y = f(x)$, where f is some arbitrary function.

A common statistical model appropriate for the above situation is

$$y = f(x) + \varepsilon,$$

where ε represents the random error term, usually assumed to have a mean of zero and a variance of σ^2.

Thus, according to our model, the expected value of y, at the fixed value of x, denoted by $E(y|x)$, equals $f(x)$, and the variance of y, at the fixed value of x, denoted by $\text{Var}(y|x)$, is σ^2. It is noted that while the mean of y is a function of x and varies according to the function $f(x)$, the variance is constant across all values of x. This is the homogeneity of variance assumption, similar to that made in the one-way classification model of Chapter 5.

It is certainly possible that several other regressor variables, in addition to the variable x, are acting together to drive the response variable (y). In this case, our model might become

$$y = f(x_1, x_2, \ldots, x_k) + \varepsilon,$$

where $x_1, x_2, \ldots, x_k$ represent k regressors variables. Chapter 7 will consider this situation in more detail, while this chapter will address the one variable case.

While it is possible to deal with this statistical model using the single variable x and the arbitrary function $f(x)$, we will first address the situation where the function $f(x)$ is defined as the linear function, $f(x) = \beta_0 + \beta_1 x$. Here, $f(x)$ is recognized as the formula for the straight line in x, with the coefficients β_0 and β_1 representing the intercept and the slope of this line, respectively. Our statistical model now becomes

$$y = f(x) + \varepsilon = \beta_0 + \beta_1 x + \varepsilon,$$

commonly referred to as the "simple linear regression," or SLR, model. The "simple" refers to the use of one regressor in the model (as opposed to using k regressors as in "multiple" regression), and "linear" refers to the "linearity in the regression coefficients." By this, we mean that each coefficient, β_0 or β_1, enters the SLR model as "$+\beta_j$ times something."

In the SLR model, we note that β_0 enters the equation as β_0 times one, and β_1 enters the equation multiplied by x. As a counterexample of a linear model, consider the function

$$f(x) = \frac{\beta_0}{1 + exp(-\beta_1 x)}.$$

Here, the coefficient β_1 multiplies x but also appears in the exponent of exp and the denominator. This form of $f(x)$ is a nonlinear function, and special techniques will be required to deal with this form of regression model, some of which will be discussed in this chapter.

It follows from our SLR model that $E(y|x) = \beta_0 + \beta_1 x$ and $var(y|x) = \sigma^2$. A SLR model analysis is performed after $n(x, y)$ pairs of observations are taken on n independent experimental units.

Our data is represented as the n pairs (x_i, y_i), for $i = 1, \ldots, n$. The first step in the analysis is graphical, where these n pairs of points are plotted using the usual x-y coordinate system. The result is a scatter of n points, often called a "scatterplot" or "scatter diagram." Visual inspection of this scatterplot helps to affirm the use of the SLR model and the validity of the resulting statistical analysis. According to the SLR model, the data should behave like random observations about the true line represented by $f(x) = \beta_0 + \beta_1 x$. Thus, we would hope that our scatterplot also exhibits this tendency. If so, we continue with the analysis, which usually includes obtaining estimates of the model coefficients, performing inferences on one or both of the coefficients, and evaluating quantitative measures to express the quality of the model.

If the scatterplot indicates that the data does not conform to the straight-line model, then several options are available to the researcher. One option, if appropriate, would involve transforming either the y and/or the x variable so that the transformed data more closely conforms to the simple linear regression model. In this case, the analysis proceeds as before but on the transformed data. A second option is to use another model, such as a nonlinear or multiple linear regression model or consider an error distribution other than the normal, such as a heavy-tailed, binomial, or Poisson distribution. These options will be discussed in more detail later in this text.

6.1.1 Regression Coefficients: Estimation and Inference

If the SLR model has been judged as appropriate, a contention supported by the scatterplot, the analysis estimates the coefficients. The method of ordinary least squares (OLS) is usually used to accomplish this task. This method chooses the values of the coefficients so that the quantity $\sum_{i=1}^{n}(y_i - \hat{y}_i)^2$ is minimized. Here, $\hat{y}_i$ denotes the model's fitted value or "estimated value" to the response at x_i. That is, $\hat{y}_i = \hat{\beta}_0 + \hat{\beta}_1 x_i$, where $\hat{\beta}_0$ and $\hat{\beta}_1$ denote the OLS estimates of the coefficients. The quantity $(y_i - \hat{y}_i)^2$, denoted by r_i, is called the residual for the ith observation. Obviously, r_i, indicates the degree to which the model is able to fit y_i, with small values indicating a "good" fit and large values indicating a "poor" fit.

The minimized quantity, $\sum_{i=1}^{n}(y_i - \hat{y}_i)^2 = \sum_{i=1}^{n}(r_i)^2$, represents the model's total inability to fit the data across all n observations.

The OLS estimates of the coefficients are found by taking the two partial derivatives of $\sum_{i=1}^{n} r_i^2$ with respect to β_0 and β_1, setting these equations equal to zero, and solving for $\hat{\beta}_0$ and $\hat{\beta}_1$. These estimates may be expressed as

$$\hat{\beta}_0 = \bar{y} - \hat{\beta}_1 \bar{x} \text{ and } \hat{\beta}_1 = \frac{\sum_{i=1}^{n}(x_i - \bar{x})(y_i - \bar{y})}{\sum_{i=1}^{n}(x_i - \bar{x})^2} = \frac{S_{xy}}{S_{xx}}.$$

One can show that both estimators are unbiased estimators of their respective parameters (under the assumption that the SLR model is the correct or "true" model) and that the variances of these estimators are $Var\left(\hat{\beta}_0\right)=\sigma^2(\frac{1}{n}+\frac{\overline{x}^2}{S_{xx}})$ and $Var\left(\hat{\beta}_1\right)=\frac{\sigma^2}{S_{xx}}$.

If, in addition to the above assumptions on the error term, ε, it is assumed that the errors are independent observations from a normal distribution – that is, $\varepsilon \sim N\left(0,\sigma^2\right)$ – then the least squares estimators are also the maximum likelihood estimators (MLEs), and the sampling distribution of each estimator is $\hat{\beta}_0 \sim N\left(\beta_0,\mathrm{Var}\left(\hat{\beta}_0\right)\right)$ and $\hat{\beta}_1 \sim N\left(\beta_1,\mathrm{Var}\left(\hat{\beta}_1\right)\right)$.

Hypothesis tests on β_0 and β_1 and confidence interval estimates of β_0 and β_1 can then easily be constructed based on basic distribution theory concepts. For example, if $\hat{\sigma}$ denotes the appropriate estimate of σ, then one can show that

$$\frac{\hat{\beta}_o-\beta_0}{\hat{\sigma}\sqrt{\frac{1}{n}+\frac{\bar{x}^2}{S_{xx}}}} \sim \mathrm{t}_{n-2} \text{ and } \frac{\hat{\beta}_1-\beta_1}{\hat{\sigma}\sqrt{\frac{1}{S_{xx}}}} \sim \mathrm{t}_{n-2},$$

where t_{n-2} denotes a *t*-distribution with $n-2$ degrees of freedom.

Tests of hypotheses involving β_0 and β_1 can then be accomplished in an obvious way. For example, to test $H_0: \beta_1 = 0$ versus the alternative $H_1: \beta_1 \neq 0$, we can use the test statistic

$$\mathrm{t}_{\beta_1} = \frac{\hat{\beta}_1}{\mathrm{se}(\hat{\beta}_1)},$$

where $\mathrm{se}\left(\hat{\beta}_1\right)=\frac{\hat{\sigma}}{\sqrt{S_{xx}}}$. The p-value would be obtained from a *t*-distribution with $n-2$ degrees of freedom. Additionally, $(1-\alpha)100\%$ confidence intervals for these parameters can be obtained using the formulas:

$$\text{for } \beta_0 : \hat{\beta}_0 \pm (\mathrm{t}_{1-\frac{\alpha}{2},n-2})\,\mathrm{se}\,(\hat{\beta}_0),$$

where $\mathrm{se}\left(\hat{\beta}_0\right)=\hat{\sigma}\sqrt{\frac{1}{n}+\frac{\overline{x}^2}{S_{xx}}}$; and

$$\text{for } \beta_1 : \hat{\beta}_1 \pm \left(\mathrm{t}_{1-\frac{\alpha}{2},n-2}\right)\mathrm{se}\,(\hat{\beta}_1),$$

where $\mathrm{se}(\hat{\beta}_1)=\hat{\sigma}\sqrt{\frac{1}{S_{xx}}}$.

6.1.2 Estimation of Mean Response

Once the parameters are estimated, the properties of the SLR model's fitted values can also be expressed. For example, the fit at the *i*th observation is $\hat{y}_i = \hat{\beta}_0 + \hat{\beta}_1 x_i$.

Under the assumptions of the model (independent normal observations with constant variance), we can show that

$$\hat{y}_i \sim N\left(E(y_i|x_i) = \beta_0 + \beta_1 x_i,\ \mathrm{var}(\hat{y}_i)\right), \text{ where } \mathrm{var}(\hat{y}_i) = \sigma^2\left(\frac{1}{n} + \frac{\left(x_i - \bar{x}\right)^2}{S_{xx}}\right).$$

Using this information, it is easy to construct a $(1-\alpha)100\%$ confidence interval for $E(y|x_0) = \mu_{y|x_0}$, at any arbitrary value of $x = x_0$ using the formula

$$\hat{y}_0 \pm \left(t_{1-\frac{\alpha}{2}, n-2}\right) se(\hat{y}_0),$$

where $\hat{y}_0 = \hat{\beta}_0 + \hat{\beta}_1 x_0$ and $se(\hat{y}_0) = \hat{\sigma} \cdot \sqrt{\frac{1}{n} + \frac{\left(x_0 - \bar{x}\right)^2}{S_{xx}}}.$

It is easy to see that the size of the standard error of the estimate is a nonlinear function of the value x_0, increasing as $(x_0 - \bar{x})$ increases.

A $(1-\alpha)100\%$ prediction interval for an individual response at x_0 is found by using the formula

$$\hat{y}_0 \pm t_{1-\frac{\alpha}{2}, n-2}\, \hat{\sigma} \sqrt{1 + \frac{1}{n} + \frac{(x_0 - \bar{x})^2}{S_{xx}}}.$$

For details concerning the derivation of these formulas, the reader is referred to Myers (1990).

6.1.3 The Analysis of Variance Approach to SLR

The analysis of variance approach to the SLR model partitions the total variability of the response, $SS_{total} = \sum_{i=1}^{n}(y_i - \bar{y})^2$, into two pieces, one due to the SLR model, $SS_{model} = \sum_{i=1}^{n}(\hat{y}_i - \bar{y})^2$, the "sum of squares due to the model," and another due to the inability of the SLR model to fit the data, $SS_{error} = \sum_{i=1}^{n}(y_i - \hat{y}_i)^2$, the "sum of squares due to error." This partitioning

TABLE 6.1

ANOVA Table for the SLR Model

Source	df	SS	MS	F
Model	1	SS_{model}	MS_{model}	F_{model}
Error	$n-2$	SS_{error}	MS_{error}	
Total	$n-1$	SS_{total}		

leaves us with the familiar identity, the partitioning of the total sum of squares, as $SS_{total} = SS_{model} + SS_{error}$. The degrees of freedom (df) also partition into: $df_{total} = df_{model} + df_{error}$ or $(n-1) = 1 + (n-2)$, for total, model, and error, respectively. The sum of squares and degrees of freedom can conveniently be summarized in the ANOVA table given in Table 6.1.

The estimate of $\sigma^2, \hat{\sigma}^2$, used in many of the formulas presented above, is obtained by $\hat{\sigma}2 = MS_{error}$.

The F-statistic in the above table, F_{model}, is formed by the ratio $F_{model} = \frac{MS_{model}}{MS_{error}}$ and is appropriate for testing the hypotheses: $H_0 : \beta_1 = 0$ versus $H_1 : \beta_1 \neq 0$. This test statistic, which follows an F-distribution with one and $n-2$ degrees of freedom under a true H_0 is related to the *t*-test for β_1 for the above two-sided alternative in that $t_{\beta_1}^2 = F$, with both tests having identical p-values. For the two-sided alternative, these two test statistics are equivalent.

Another commonly used numerical measure computed from the ANOVA table is the R^2, the coefficient of determination, computed by the formula: $R^2 = \frac{SS_{model}}{SS_{total}}$.

Due to the partitioning of the sum of squares, it is clear that the R^2 must satisfy $0 \leq R^2 \leq 1$ or express it in terms of a percent, $0\% \leq R^2 \leq 100\%$. Large values of R^2 are indicative of a strong linear relationship between the x and y variables.

6.1.4 Assessment of Model Quality

Assessment of the SLR model is important for at least two reasons. As mentioned at the beginning of this section, it is important to determine whether the SLR model is appropriate for a given set of data. If not, either the model or the data must be changed appropriately. If the model appears to have a straight-line tendency, it is important to measure the quality of the fit. The better the model, the more faith one has in applying the model's results.

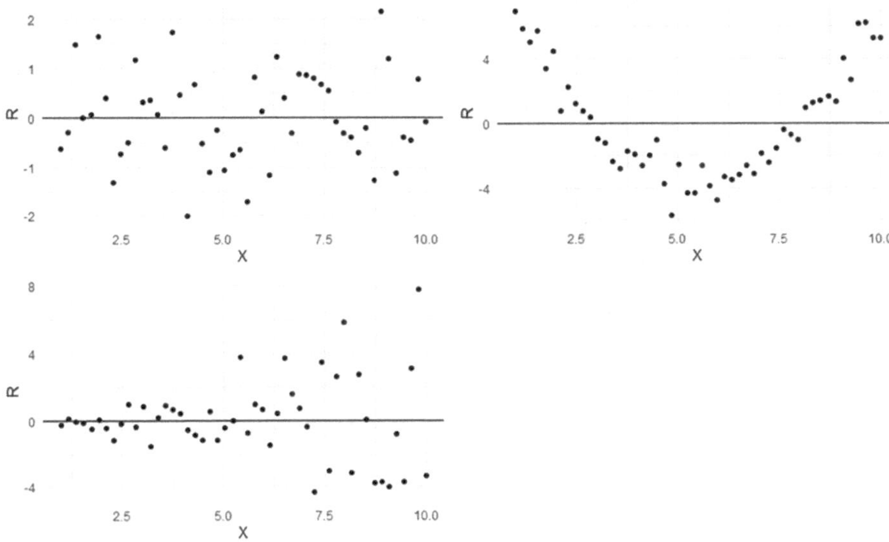

FIGURE 6.1
Residual plots of residuals versus x for a situation where the SLR model is appropriate (top-left) and where the SLR model is inappropriate due to the wrong curve (top-right) and due to nonconstant variance (bottom).

There are two primary ways to assess the quality of the SLR model relative to a given set of data: visual and numerical. Visual assessment includes the initial scatterplot of y versus x, where the straight-line tendency suggests that the SLR model may be used.

Other useful scatterplots, often referred to as residual plots, are the scatterplot of the residuals versus the x values – that is, r_i versus x_i; and the scatterplot of the residuals versus the fits – that is, r_i versus $\hat{y}_i$. The "ideal" plot in both cases should represent a random, patternless scatter of points around the residual value of zero. Departures from this ideal suggest changing the model or the data.

For example, Figure 6.1 shows three scatterplots of r_i versus x_i. The first scatterplot reveals an example of the ideal plot, while the other two plots indicate departures in the data from the SLR model. The second plot indicates a curved relationship between y and x rather than a straight-line relationship. The third plot reveals nonconstant variance across the x values, violating the constant variance assumption.

Numerical assessment of the SLR model usually begins with the R^2, looking for large values. The reader needs to be reminded, however, that the size of the R^2 alone cannot indicate the appropriateness of the SLR model. It is

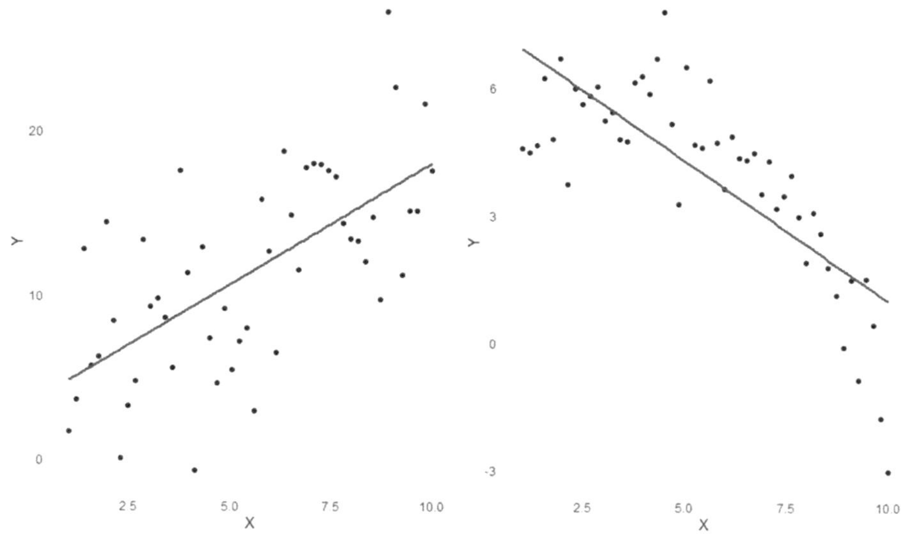

FIGURE 6.2
Scatterplot of y versus x. The left plot indicates a situation where the R^2 is small, but the SLR model appears to be appropriate. The right plot indicates a situation where the R^2 is large, but the SLR model is inappropriate.

possible for the R^2 to be small, say about 20% – 30%, and the model to be appropriate. It is also possible for the R^2 to be large and the SLR model to be inappropriate. In Figure 6.2, the left scatterplot of y versus x represents the first case, and the second scatterplot represents the second.

Another numerical measure is the test statistic F_{model} or, equivalently, t_{β_1}. Large test statistic values indicate a highly significant slope term in the SLR model. Again, however, the appropriateness of the SLR model is not addressed by the value of either test statistic. These tests are more indicative of the linear trend in the data as opposed to the value of the SLR model. In the right scatterplot in Figure 6.2, we see an example of a situation where the F_{model} will be large and highly significant, but the data would be better fit using another model.

For each of these commonly used measurements, it is apparent that the scatterplot must accompany the analysis in order for them to be appropriately interpreted. This fact alone makes it imperative that graphical analysis, the exploratory data analysis, be presented along with the quantitative analysis. R or SPSS can be used to perform this task, which is especially well suited to perform, as will be demonstrated in the following example.

TABLE 6.2

The Survival Time Data

Fish No.	Survival Time (y)	Concentration (x)	Group
1	328.1	100,000	1
2	373.25	100,000	1
3	274.16	100,000	1
4	417.83	63,096	2
5	552.08	63,096	2
6	488.65	63,096	2
7	676.08	39,811	3
8	812.83	39,811	3
9	961.61	39,811	3
10	1,496.24	25,119	4
11	1,137.63	25,119	4
12	1,244.51	25,119	4
13	2,147.83	15,849	5
14	1,663.41	15,849	5
15	1,963.36	15,849	5
16	2,798.98	10,000	6
17	3,334.26	10,000	6
18	3,556.31	10,000	6

6.1.5 Real Application: The Survival Time Data

The susceptibility of catfish to a certain chemical pollutant was determined by immersing individual fish in two liters of an emulsion containing the pollutant and measuring the survival time in minutes. The data in Table 6.2 give the survival time (y) in minutes and the concentration (x) of the pollutant in parts per million in 18 fish. Since there are three observations per dose, we set a group variable to distinguish the six distinct doses.

The 18 survival times are placed in the SPSS. The annotated SPSS session analyzing the survival time data is as follows:

User should go for Analyze—Descriptive—Descriptive

The output is given in Table 6.3.

The scatterplots of survival time versus dose and log dose of pollutant are in Figure 6.3.

We note that the data follows a strong nonlinear trend with signs of non-constant variance (heterogeneity) across the six doses. Thus, the SLR model is inappropriate for this data. One approach to dealing with this data is to transform either one or both of the variables. The log (base 10) transformation is applied to both for reasons to be made clear in the next section. We now note that the log survival time versus log concentration is **strongly linear,**

TABLE 6.3
Descriptive Statistics

Statistics			
		Survival Time	**Dose**
N	Valid	18	18
Mean		1,345.9511	42,312.50
Std. Error of Mean		245.14757	7,544.697
Median		1,049.6200	32,465.00
Mode		274.16[a]	10,000[a]
Std. Deviation		1,040.07306	32,009.438
Variance		1,081,751.962	1,024,604,110.853
Minimum		274.16	10,000
Maximum		3,556.31	100,000
Percentiles	25	470.9450	15,849.00
	50	1,049.6200	32,465.00
	75	2,009.4775	63,096.00

a. Multiple modes exist. The smallest value is shown.

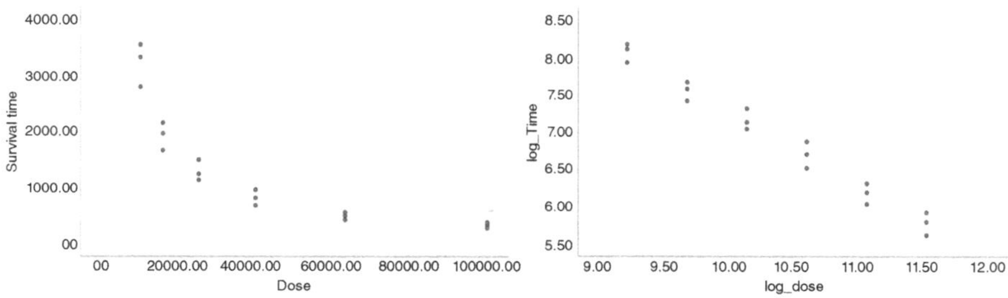

FIGURE 6.3
Left is the scatterplot of survival time versus dose of pollutant, and the right is the scatterplot of log survival time versus log dose.

with constant variance across the log concentrations, as in the right panel in Figure 6.3. The SLR model may be applied to the transformed data. Further analysis using SPSS is shown in the following diagram, Figure 6.4.

The estimated regression equation is:

$$\textbf{Log\_time} = \mathbf{7.491{-}0.998\ log\_dose}$$

The fit in Figure 6.5 looks great; now look at the residuals. See the residual plot in Figure 6.6 for the survival time data.

Do the residuals look like a random, patternless scatter of points with constant variance across the x values? The answer is yes!

See Figure 6.7.

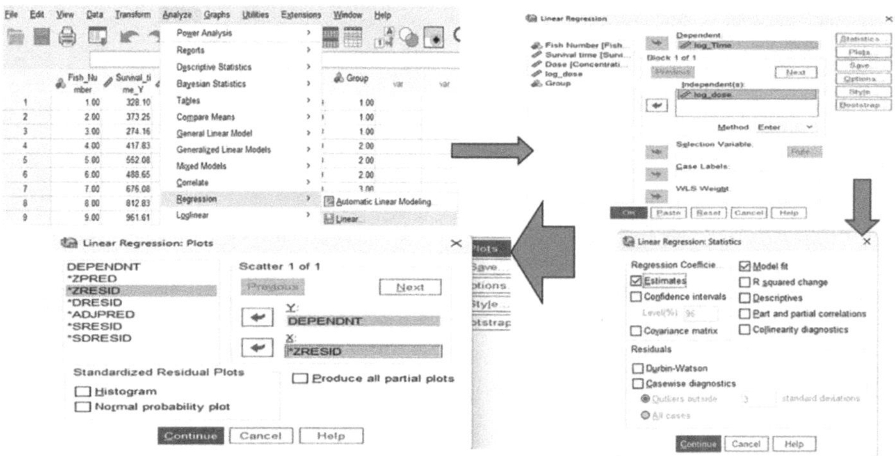

FIGURE 6.4
SPSS steps for the SLR.

TABLE 6.4
Estimated Coefficients and ANOVA

		Coefficients[11]			
	Unstandardized Coefficients		**Standardized Coefficients**		
Model	**G**	**Std. Error**	**Beta**	***t***	**Sig.**
1 (Constant)	7.491	0.174		42.975	<0.001
log_dose	−0.998	0.039	−0.988	−25.840	<0.001
a. Dependent Variable: log_Time					
$s = 0.05597$ $R^2 = 97.7\%$ $R^2(adj) = 97.5\%$					
		ANOVA[a]			
Model	**Sum of Squares**	**df**	**Mean Square**	**F**	**Sig.**
1 Regression	2.092	1	2.092	667.723	<0.001[b]
Residual	0.050	16	0.003		
Total	2.142	17			
a. Dependent Variable: log_Time					
b.Predictors: (Constant), log_dose					

All indicators suggest that the SLR model accurately describes the log survival time versus log concentration relationship. The next section will give several methods for determining the proper transformation of the variables.

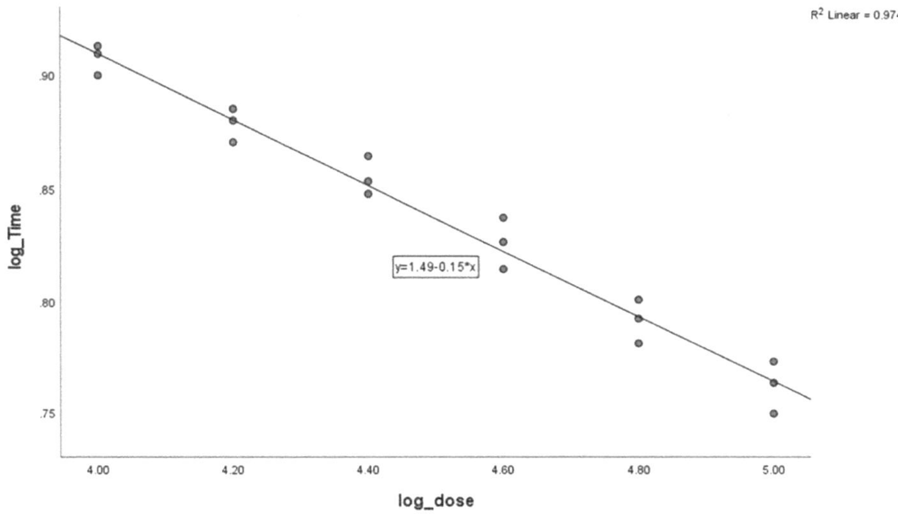

FIGURE 6.5
Plot of log survival time versus log dose and the least squares line.

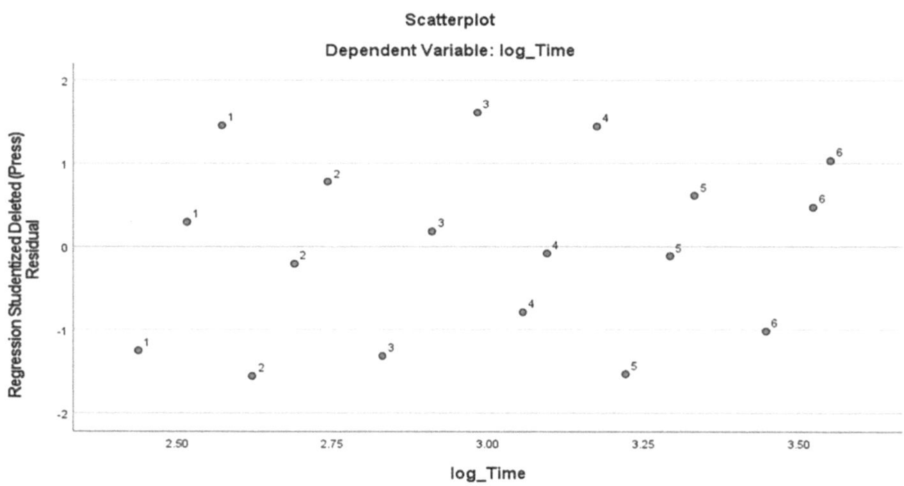

FIGURE 6.6
Residual plot for the survival time data.

6.2 Transformations to the SLR Model

It may be the case that a scatterplot of y versus x will reveal a strong non-linear tendency, in which case the SLR model cannot be used. We saw this feature in the scatterplot of the survival data, in subsection 6.1.5. Suppose the

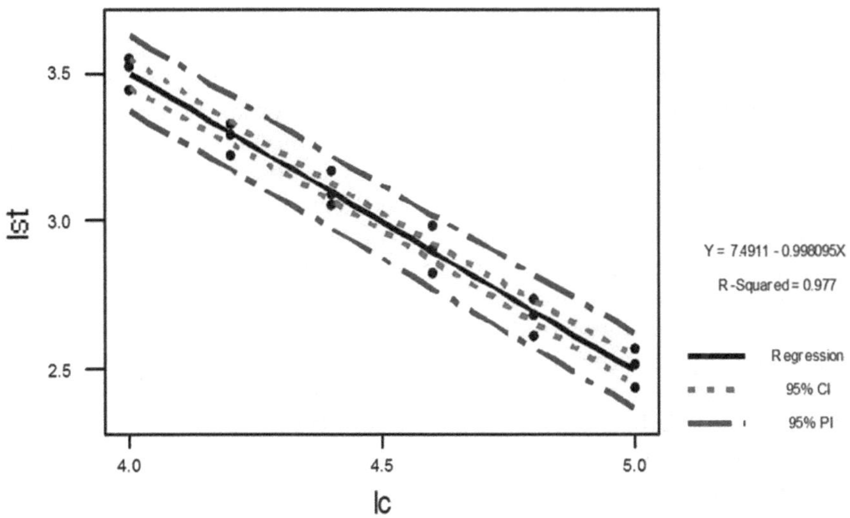

FIGURE 6.7
Plot of data and the least squares line.

nonlinearity is of some appropriate form. In that case, it may be possible to transform either the *y*, *x*, or both variables so that the SLR model is justifiable for the transformed variables.

Two very good methods exist for determining the proper transformation of the *y* and/or the *x* variable. According to Tukey (1977), one method is primarily based on the visual interpretation of the scatterplot. The second method, another application of the Box-Cox procedure introduced in Chapter 5, relies more on theory (maximum likelihood theory). Both methods work very well under certain circumstances and often give similar results. This section will present, discuss, and demonstrate both methods using SPSS.

6.2.1 Tukey's Method

As in other methods introduced by Tukey (such as the method of mid-summaries and the spread versus level plot), the form of transformations on y and x are the simple power family of transformations. That is, we will seek to find the values of λ_y and λ_x so that a straight-line relationship exists between the transformed variables y^* and x^*, where

$$y^* = \begin{cases} y^{\lambda_y} & if \quad \lambda_y > 0 \\ \log(y) & if \quad \lambda_y = 0 \\ -y^{\lambda_y} & if \quad \lambda_y < 0 \end{cases} \qquad and \qquad x^* = \begin{cases} x^{\lambda_x} & if \quad \lambda_x > 0 \\ \log(x) & if \quad \lambda_x = 0, \\ -x^{\lambda_x} & if \quad \lambda_x < 0 \end{cases}$$

where y and x are both positive or equivalently, −y and −x are both negative. The power values of λ_y and λ_x are selected from the "ladder of transformations" (see Chapter 5), which includes the values $-2, -1, -.5, 0, .5, 1$, and 2. The proper power values are selected by moving up or down the ladder in order to give the desired straight-line configuration between y^* and x^*. The "art" of the technique, of course, is in the selection of the proper power values. The curves that can be linearized by this method are illustrated at the top of Figure 6.8, where four types of curves are displayed: types I, II, III, and IV. These curves are all either monotonic, increasing, or decreasing and are either concave up or concave down with no points of inflection. Thus, the power family of transformations can be used to linearize a rather limited class of curves relating to y and x.

The mechanism for determining the proper power values can best be remembered by examining the "segmented circle" depicted in the bottom display of Figure 6.8. Here, the four types of curves are seen to represent the plots in the four quadrants of the plane. Curve type I is seen as the curve in the first quadrant and so on. If these four curves represent the plot of y^* versus x^*, then it is clear that the values of λ_y and λ_x need to be adjusted. For example, for curve type I, the value of λ_y and/or the value of λ_x need to be increased in order to linearize the curve. That is, we must move up the ladder of transformation to a larger value of λ_y and/or λ_x. The symbols on the "segmented circle" of $\lambda_y \uparrow$ and $\lambda_x \uparrow$ indicate this upward movement. For example, if the plot of the raw data, y versus x where $\lambda_y = 1$ and $\lambda_x = 1$, resembles curve type I, then a plot of y^2 versus x or a plot of y versus x^2 may be linear.

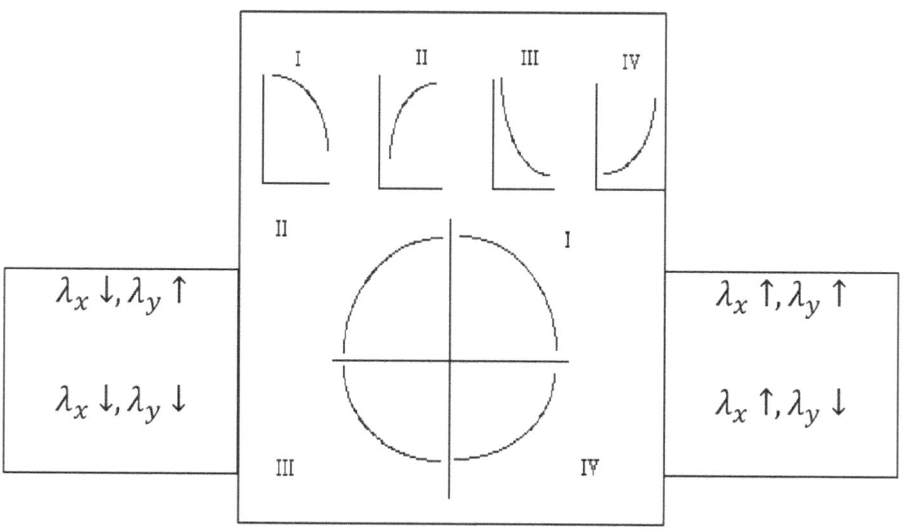

FIGURE 6.8
The curves at the top represent the curves the power family can linearize. The "segmented circle" at the bottom gives the direction of movement of the power values along the ladder of transformation.

The degree of movement up or down the ladder is usually accomplished by educated trial and error. For example, in attempting to straighten a type I curve, if the power values have not been increased enough, then a plot of the y^* versus x^* will still be nonlinear, resembling a curve type I.

On the other hand, if the power values are increased by too much, the plot of y^* versus x^* will be nonlinear but now resembling one of the other three types of curves. It is seen from the segmented circle that each curve type has an indication of movement along the ladder of transformation, either up (indicated by the "↑" symbol) or down (indicated by the "↓" symbol). Thus, when the plot of y^* versus x^* indicates a particular curve type, this is an indication that the values of λ_y and/or λ_x need to be altered by movement in the appropriate direction indicated by the symbols given in Figure 6.8 for that curve type.

It is very easy to remember the appropriate movement along the ladder to move toward a straight-line relationship. Consider, for example, the curve in the first quadrant, curve type I. This quadrant is defined as all values in the *x*-*y* plane where $x > 0$ and $y > 0$. Note from Figure 6.8 that the movement is up the ladder for both power values, as indicated by $\lambda_y \uparrow$ and $\lambda_x \uparrow$. So, the device to remember is to replace the "> 0" by the "↑" symbol and the "< 0" by the "↓" symbol. As a final example, quadrant II is defined as all values for which $y > 0$ and $x < 0$. Consequently, the proper movement along the ladder of transformations is to increase $\lambda_y \left(\lambda_y \uparrow\right)$ and to decrease $\lambda_x \left(\lambda_x \downarrow\right)$.

For each of the four types of curves, there are usually three different sets of power values resulting in a straight line: transform x only $\left(\lambda_x \neq 1, \lambda_y = 1\right)$, transform y only $\left(\lambda_x = 1, \lambda_y \neq 1\right)$, and transform both *x* and $y\left(\lambda_x \neq 1,\ \lambda_y \neq 1\right)$. Scatterplots for each set of power values should be examined closely for other statistical considerations. For example, one set of values may result in more constant variance across the *x* variable than the other sets. Also, which equation to use often depends on nonstatistical considerations, such as which model makes the most sense from a subject matter viewpoint.

6.2.2 Tukey's Method for SLR Using the Survival Time Data

Tukey's method will be demonstrated using SPSS, and the survival time data will be introduced. Recall that y is the survival time of catfish when immersed in water containing x amount of the pollutant (dose). It was stated in Section 5.1 that $\log(y)$ and $\log(x)$ resulted in a straight-line model. The following is an annotated SPSS session used to determine these transformations.

Throughout the SPSS session, λ_x will be represented by *px* and λ_y by *py*.

Step 1: Plot the raw data. Determine the type of curve.

Step 2: The graph menu was used to obtain the plot in Figure 6.9.

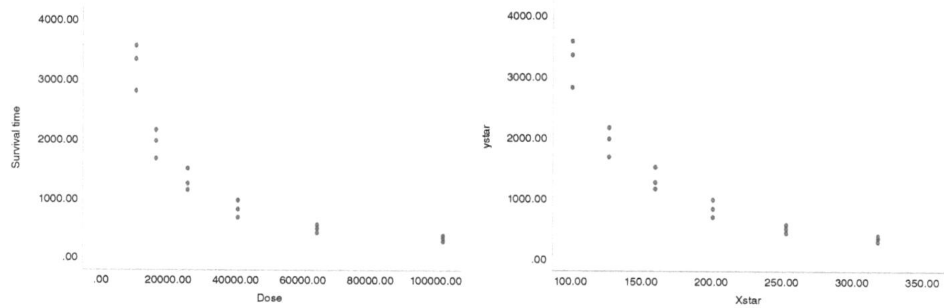

FIGURE 6.9
Left is the scatterplot of survival time versus dose of pollutant, and the right is the scatterplot of y^* versus x^* where $px = 0.5$ and $py = 1$.

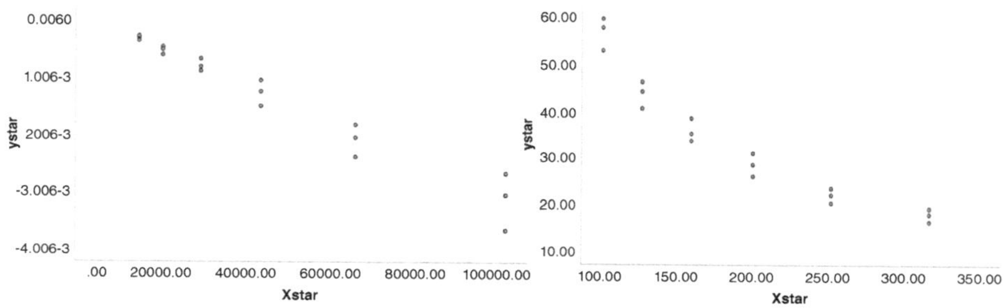

FIGURE 6.10
Left is the scatterplot of y^* versus x^* where $px = 1$ and $py = -1$, and the right is the scatterplot of y^* versus x^* where $px = 0.5$ and $py = 0.5$.

It is clear that this is a type III curve. We must move down the ladder of transformation.

Step 3: Try $px = 0.5$ and $py = 1$. The graph menu was used to obtain Figure 6.9.

The plot still indicates a type III curve, implying that we have not moved down the ladder far enough. So, let's try $px = -1$, $py = 1$.

We note from Figure 6.10 that the plot of y versus $-(1/x)$ is a straight line (the "–" sign is used to maintain the negative slope in the scatterplot). The results are not totally satisfactory since we still have strong heterogeneity of variance across the dose levels. Hence, it is better to try transforming y only, using $px = 1$ and $py = -1$. The results here are similar to those obtained for $px = -1$ and $py = 1$; only now is the heterogeneity of variance increasing across dose levels rather than decreasing. Try transforming both x and y. Try $px = 0.5$, $py = 0.5$.

The plot is still a type III curve. We need to move further down the ladder. How about px = 0.5 and py = 0?

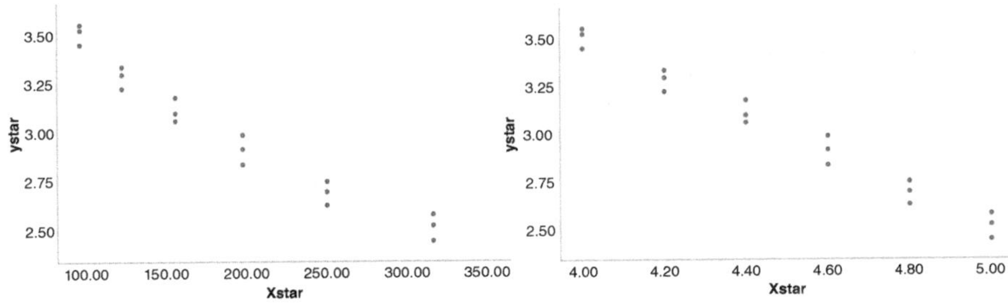

FIGURE 6.11
Left is the scatterplot of y^* versus x^* where $px = 0.5$ and $py = 0$, and the right is the scatterplot of y^* versus x^* where $px = 0.0$ and $py = 0.0$.

The type III nature of the curve is still apparent. Try $px = 0$ and $py = 0$. We note from Figure 6.11 both a straight line and constant variance across the dose levels, as in the right panel in Figure 6.11. Stop here, and use log(y) versus log(x).

6.2.3 The Box-Cox Method

A second method that can be applied to our SLR model is extending the versatile Box-Cox procedure, a technique presented in Chapter 5 (in Section 5.3). We will follow the same notation in this section as used earlier in describing the Box-Cox method for the location and one-way ANOVA models. Recall that the goal of the Box-Cox method is to find the proper power value so that the response variable achieves the desired assumptions of the model. For our SLR model, this means that we want to find the value of λ, λ_y so that the transformed response variable, y^*, has a linear relationship with the x variable and meets the assumptions of normality with constant variance. That is, find λ_y so that

$$y^* = \beta_0 + \beta_1 x + \varepsilon,$$

where $\varepsilon \sim N\left(0, \sigma^2\right)$ and $y^* = \begin{cases} \dfrac{y^{\lambda_y} - 1}{\lambda_y} & \lambda_y \neq 0 \\ \log(y) & \lambda_y = 0, \end{cases}$

where y is positive; we note that the Box-Cox method deals only with transforming the response variable, not the regressor variable. The method for finding the proper power value for y is the maximum likelihood method.

The procedure for the location model is very similar to that described in Section 5.3 and will not be repeated here. It is easy to show, following the derivation in Section 4.3, that for a fixed value of λ_y, the MLEs of the intercept and slope parameters for the SLR model are

$$\hat{\beta}_0 = \bar{y}^* - \hat{\beta}_1 \bar{x} \text{ and } \hat{\beta}_1 = \frac{S_{xy}}{S_{xx}},$$

where $S_{xy} = \sum(x_i - \bar{x})(y_i^* - \bar{y}^*)$. The MLE for σ^2 is $\hat{\sigma}^2_{\lambda_y} = \frac{\sum(y_i^* - \hat{y}_i^*)^2}{n}$, where $\hat{y}_i^* = \hat{\beta}_0 + \hat{\beta}_1 x_i$.

Substitution of these estimates into the log-likelihood expression results in

$$\ln L(\lambda_y) = \text{constant} - \frac{n}{2}\log(\hat{\sigma}^2_{\lambda_y}) + (\lambda_y - 1)\sum_{i=1}^{n}\log(y_i),$$

an expression similar to those obtained for $\ln L(\lambda_y)$ the location and one-way ANOVA models in Section 5.3. The MLE of λ_y, $\hat{\lambda}_y$ is the value that maximizes $\ln L(\lambda_y)$. This is conveniently accomplished by searching through a trial series of values of λ_y. The technique will be demonstrated in SPSS using survival data.

6.2.4 SPSS Session: Box-Cox Method for the SLR Model Using the Survival Data

The SPSS can be used to apply the Box-Cox procedure to the SLR model. The results will be

The max ln likelihood and corresponding lambda are

ROW maxlnl lambda
1 – 94.319 – 0.7

The graph menu was used to obtain the plot of the log-likelihood as a function of lambda for the survival data in Figure 6.12.

In Figure 6.12, the 95% CI for lambda is approximately (–1,–0.4).

Is –0.5 a good transformation of y? Let's see. The graph menu was used to obtain Figure 6.13 for plotting the y^* versus dose, where lambda is –0.5 for the survival data.

The resulting plot is certainly a better fit than that for the original data, but it is neither straight nor has constant variance across the dose groups. Let's combine the Box-Cox method with the Tukey method by recognizing this curve as a type III curve.

Could it be improved by transforming the dose? Try log(dose).

The result is a type I curve. The power value on x must be increased. Try 0.5.

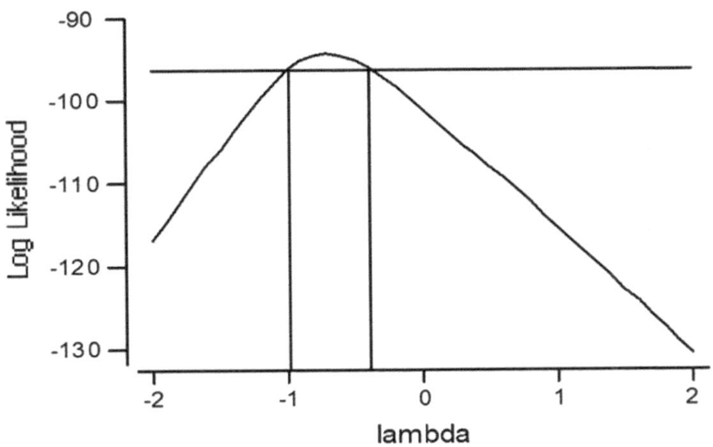

FIGURE 6.12
The log-likelihood as a function of lambda for the survival data.

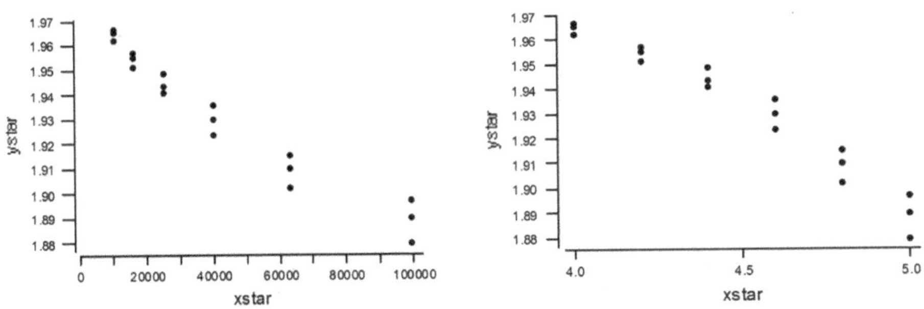

FIGURE 6.13
Left is the plot of y^* versus dose where lambda is −0.5 and the right is the plot of y^* versus x^* for $py = -0.5$ and $px = 0$.

One can see that the line is straight, as in Figure 6.14, but nonconstant variance is still a problem. Looks as if the log(y) versus $\log(x)$ is still the best combination. We note that the Box-Cox procedure has not given the same result as that obtained by the Tukey method. This is due, in part, to the fact that the Box-Cox method only transforms the y variable and not the x variable. Tukey's procedure then provides for greater flexibility in the choice of transformations. On the other hand, the Box-Cox procedure provides confidence intervals for the power value, thus enabling a range of power values to be considered.

The preceding example serves to illustrate an important concept, one that is emphasized throughout this text. A sophisticated method applied to the data as part of the statistical analysis does not guarantee that the method

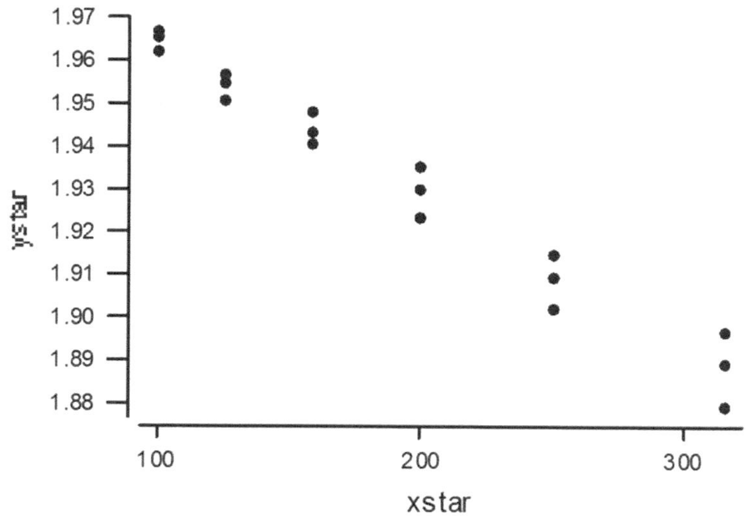

FIGURE 6.14
Plot of y^* versus x^* for $py = -0.5$ and $px = 0.5$.

improves or is even appropriate for the analysis of that data. Failure of data to meet certain model assumptions may invalidate the resulting analysis. This is one justification for the user to be aware of methods that are insensitive to restrictive model assumptions and utilize techniques that will provide valid conclusions under less restrictive conditions.

Other transformation methods are available to the researcher. For example, the Box-Cox procedure may be adjusted to transform both the y and the x variables. One way to accomplish this is suggested in problem 8 of this chapter. Another technique, due to BoxTidwell (see Myers, 1990), also allows both y and x to be transformed. The interested reader is referred to Ryan (1997) for more information on using transformations in regression.

6.3 Robust Estimation in the SLR Model

The reader will recall our discussion in Chapter 3 concerning the effects of outliers on estimating the location parameter, θ, in the location model. For example, the sample mean, the least squares estimate of θ, was shifted in the direction of the outliers. This concept was referred to as the lack of resistance of the sample mean. In addition, the standard error of the sample mean was inflated, resulting in a loss of power to the t-test, the most commonly used inferential procedure to test hypotheses about θ. Outliers can also have an adverse effect on the estimation and testing of parameters in the SLR model. For this reason, the ability of the researcher to identify outliers and to deal appropriately with them is fundamental to the analysis of the SLR model.

Modeling outliers can be accomplished in a manner similar to the method used in the location model. That is, the distribution of the error term is no longer assumed to be the normal distribution but rather a member from the family of symmetric, heavy-tailed distributions. Thus, our altered SLR model is

$$y_i = \beta_0 + \beta_1 x_i + \varepsilon_i,$$

for $i = 1, \ldots, n$, where the ε_i are i.i.d. with distribution f, a member of the symmetric, heavy-tailed family of distributions. We desire an estimation technique that will provide us with robust statistical estimates across all distributions in this family. That is, our estimator should be resistant to outliers and possess desirable statistical properties such as unbiasedness with small variance. The M-estimators satisfy this goal.

The M-estimators of β_0 and β_1, denoted by $\hat{\beta}_{m_0}$ and $\hat{\beta}_{m_1}$, are the solutions to the minimization of

$$\sum_{i=1}^{n} p\left(\frac{y_i - \hat{\beta}_{m0} - \hat{\beta}_{m1} x_i}{\hat{s}}\right),$$

where the function $\rho()$ is some appropriately chosen smooth convex loss function of the type described in Chapter 3 and $\hat{s}$ is some estimate of the scale of the residuals.

Equivalently, the M-estimators of intercept and slope may be found by taking the derivatives of the above expression with respect to both β_0 and β_1, setting each of these derivatives equal to zero, and solving.

That is, the solutions to the following "robust normal equations" are the M-estimators $\hat{\beta}_{m_0}$ and $\hat{\beta}_{m_1}$:

$$\sum_{i=1}^{n} \psi\left(\frac{y_i - \hat{\beta}_{m0} - \hat{\beta}_{m1} x_i}{\hat{s}}\right) = 0$$

$$\sum_{i=1}^{n} \psi\left(\frac{y_i - \hat{\beta}_{m0} - \hat{\beta}_{m1} x_i}{\hat{s}}\right) x_i = 0 \cdot$$

We can see that we need to solve two nonlinear equations in two unknowns. This is similar to the situation we encountered in Chapter 4 when solving for the M-estimator in the location model, where we needed to solve one nonlinear equation in one unknown. We used the Newton-Raphson and the iteratively reweighted least squares (IRLS) techniques. However, only the IRLS procedure will be discussed here due to its intuitive appeal. To apply the IRLS method, rewrite the above two equations as follows:

$$\sum w_i r_i^* = 0$$
$$\sum w_i x_i r_i^* = 0,$$

where $r_i = y_i - \hat{\beta}_{m0} - \hat{\beta}_{m1} x_i$ and $w_i = \dfrac{\psi\left(\frac{r_i}{\hat{s}}\right)}{\frac{r_i}{\hat{s}}}$. The $r_i^* = \frac{r_i}{\hat{s}}$, used in computing the weights, w_i, are called the "rescaled residuals." Because these two equations are now linear in the two unknowns, the unknowns can be obtained in closed form as

$$\hat{\beta}_{m1} = \frac{\sum (x_i - \bar{x}_w) w_i (y_i - \bar{y}_w)}{\sum w_i (x_i - \bar{x}_w)^2} = \frac{S_{xwy}}{S_{xwx}},$$
$$\hat{\beta}_{m0} = \bar{y}_w - \hat{\beta}_{m1} \bar{x}_w$$

the well-known weighted least squares estimates in the SLR model, where $\bar{x}_w = \dfrac{\sum w_i x_i}{\sum w_i}$ and $\bar{y}_w = \dfrac{\sum w_i y_i}{\sum w_i}$. Unfortunately, the w_i are also functions of the unknown estimators $\hat{\beta}_{m_0}$ and $\hat{\beta}_{m_1}$.

Consequently, the above equations contain unknowns on both the right- and left-hand sides. This dilemma is solved by iterating on the unknown estimators through the following IRLS algorithm:

1. Obtain initial estimates of the intercept and slope parameters denoted by $\hat{\beta}_{00}$ and $\hat{\beta}_{10}$, respectively. For example, the LS estimates may be used as the initial values of the parameters, although more resistant estimates are preferred, such as the least median of squares (LMS) or the least absolute value (LAV) estimates.
2. Compute the residuals, $r_i = y_i - \hat{\beta}_{00} - \hat{\beta}_{10} x_i$, for $i = 1, \ldots, n$, and obtain an estimate of the scale parameter, $\hat{s}$. Often $\hat{s}$ is computed as the rescaled mad. That is, $\hat{s} = \dfrac{\text{median}(|r_i - \text{median}(r_i)|)}{.6745}$. Obtain the rescaled residuals, $r_i^* = \frac{r_i}{\hat{s}}$, for $i = 1, \ldots, n$.

3. For a predetermined weight function, such as the Huber weight function with a tuning constant of 1.5, compute the n weights as
$$w_i = \frac{\psi(r_i^*)}{r_i^*}, i = 1,\ldots,n.$$
4. Compute the updated estimates as follows:
$$\hat{\beta}_{11} = \frac{S_{xwy}}{S_{xwx}} \text{ and } \hat{\beta}_{01} = \bar{y}_w - \hat{\beta}_{11}\bar{x}_w.$$
5. If $\hat{\beta}_{11} \approx \hat{\beta}_{10}$ and $\hat{\beta}_{01} \approx \hat{\beta}_{00}$, the iteration cycle terminates by declaring $\hat{\beta}_{m1} = \hat{\beta}_{11}$ and $\hat{\beta}_{m0} = \hat{\beta}_{01}$. Otherwise, replace $\hat{\beta}_{10}$ by $\hat{\beta}_{11}$ and $\hat{\beta}_{00}$ by $\hat{\beta}_{01}$, and return to step 2.

At the convergence of the IRLS algorithm, the M-estimates of β_0 and β_1 are obtained. Convergence problems may occur, such as converging to "local minima" instead of the global minima, as discussed in Chapter 4. However, these problems can be reduced by using highly resistant starting values, such as the LMS estimates, and beginning the iteration cycle using a non-redescending psi function, such as the Huber psi. Upon convergence with the Huber psi, the Tukey bisquare psi may be used for several more iterations to "fine-tune" the estimates.

As discussed in Chapter 4, the n weights provide a useful diagnostic for identifying the outliers. Consider the weights as a sample of size n and use the EDA techniques discussed in Chapters 1 and 2 to analyze them. For example, box plots and dot plots are useful displays of the weights, as is the "weight index" plot, the plot of w_i versus i, for $i = 1,\ldots,n$. These displays intend to reveal which observations are receiving unusually low weights. These points have been declared as outliers by the M-estimation technique. Of course, once these points have been identified, the researcher must determine the cause of such anomalous observations and adjust any correctable errors.

To demonstrate the M-estimation technique in SLR, R has been used to perform the IRLS method. The program is executed to obtain the Huber M-estimates of the intercept and slope. If we perform only one iteration, the M-estimator is called a "one-step" M-estimator. We usually continue to iterate until the maximum absolute relative change between the two coefficients is small, say less than 0.01. The code is executed to obtain the bisquare estimates.

6.3.1 The Wild Point Data Using R

The following example intends to show how the IRLS algorithm works to determine which points are outliers and which are "inliers"; those observations are consistent with the pattern expressed by the majority of the data.

TABLE 6.5

The Wild Point Data

	Obs. 1	Obs. 2	Obs. 3	Obs. 4	Obs. 5	Obs. 6
y	1	2	3	4	12	6
x	1	2	3	4	5	6

The dataset of interest is the "wild point" dataset of the following six pairs of points as in Table 6.5.

It is seen that the above data is perfectly linearly related with a true intercept of $\beta_0 = 0$ and $\beta_1 = 1$, except point 5. There, the y observation is 12 instead of 5, an error of some magnitude. Of course, observation 5 is the "wild point." This observation ought to have some negative effect on the OLS estimates of intercept and slope but should be detected as an outlier by the M-estimation technique, properly down-weighted so that the resulting M-estimates are close to the true values of 0 and 1 for the intercept and slope, respectively.

The following R code is used to plot the scatterplot for the data and look for outliers.

```
# Load necessary library
library(ggplot2)
# Wild Point Data
x <- c(1, 2, 3, 4, 5, 6)
y <- c(1, 2, 3, 4, 12, 6)
data <- data.frame(x = x, y = y)
# Plotting the scatterplot
ggplot(data, aes(x = x, y = y)) + geom_point(color =
  'blue') + labs(title = 'Scatterplot for Wild Point
  Data', x = 'X', y = 'Y') + theme_minimal() +
  ggsave('Wild_Point_Data_Scatterplot.png')
```

From Figure 6.15, it is obvious that observation 5 is a huge outlier. **How does it affect the LS analysis?**

Get the LS analysis. The regression equation is

$$y = -0.93 + 1.60x.$$

```
Predictor      Coef     Stdev    t-ratio       p
Constant     -0.933     2.735      -0.34   0.750
x            1.6000    0.7024       2.28   0.085

s = 2.938 R-sq = 56.5% R-sq(adj) = 45.6%
```

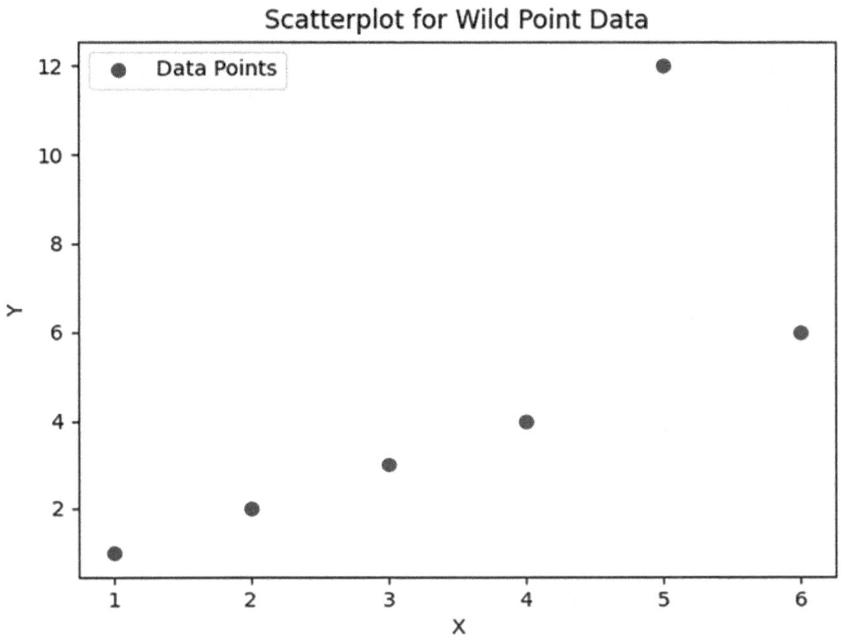

FIGURE 6.15
Plot of the wild point data.

```
Analysis of Variance

SOURCE          DF        SS         MS         F         p
Regression       1    44.800     44.800      5.19     0.085
Error            4    34.533      8.633
Total            5    79.333
```

It is seen that the OLS estimate of intercept is –0.933, and the slope is 1.600, which are far from the true values of 0 and 1. Note also that the test on the slope is not significant at the 0. 05 level, a very misleading result since we know that the true slope is 1. To obtain a robust analysis, run the following R code:

```
# Ordinary Least Squares (OLS) regression
ols_model <- lm(y ~ x, data = data)
ols_fitted <- ols_model$fitted.values
# Robust regression using Huber's T norm
huber_model <- rlm(y ~ x, data = data, psi = psi.huber)
huber_fitted <- huber_model$fitted.values
# Plotting the data points and regression lines
```

```
ggplot(data, aes(x = x, y = y)) + geom_point(color =
'blue', label = 'Data Points') + geom_line(aes(y = ols_
fitted), color = 'red', label = 'OLS Fit') + geom_
line(aes(y = huber_fitted), color = 'green', label =
'Huber Fit') + labs(title = 'Wild Point Data With the LS
and M Estimate Lines', x = 'X', y = 'Y') + theme_mini-
mal() + theme(legend.position = "top") + scale_color_
manual(values = c("blue", "red", "green"))
# Save the plot
ggsave('Wild_Point_Data_LS_M_Estimate_Lines.png')
```

Figure 6.16 gives the LS and the M-estimate fits.

The effect of the outlier on the OLS line is clear. Further analysis of the weights is often useful in clarifying anomalous observations. Some EDA displays, such as the box plot of the weights and the weight index plots, are especially helpful. The low weight is an extreme outlier compared to the other weights. Clearly, observation 5 is an outlier. Figure 6.17 also shows the weight index plot.

The above displays show that one outstandingly small weight value belongs to observation 5, as in Figure 6.17. After one iteration of the bisquare, we converged to zero and one as the estimates for the intercept and slope. These are the ideal values. Note that the weight for the outlier is zero, and those for the other five points are not one, yet the desired estimates are achieved, resulting in a perfect fit to the good five points and a clear distinction of the outlier through the "small weight, large residual" combination.

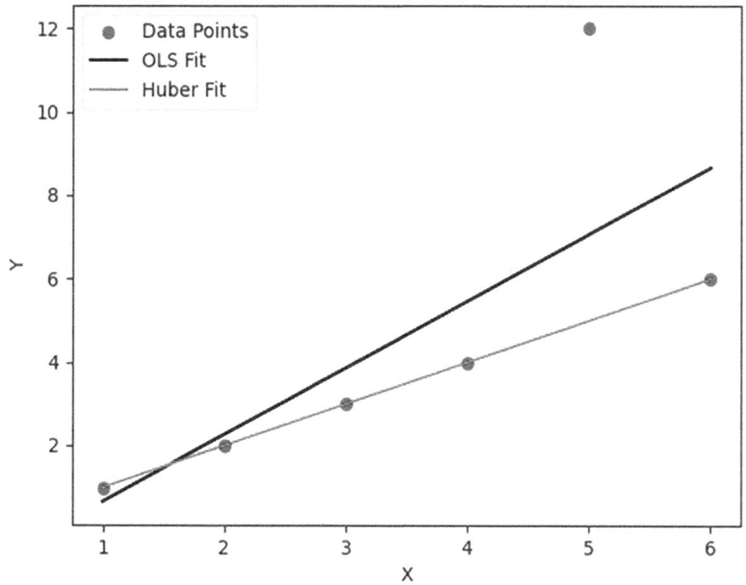

FIGURE 6.16
Wild point data with the LS and Huber_M estimate lines.

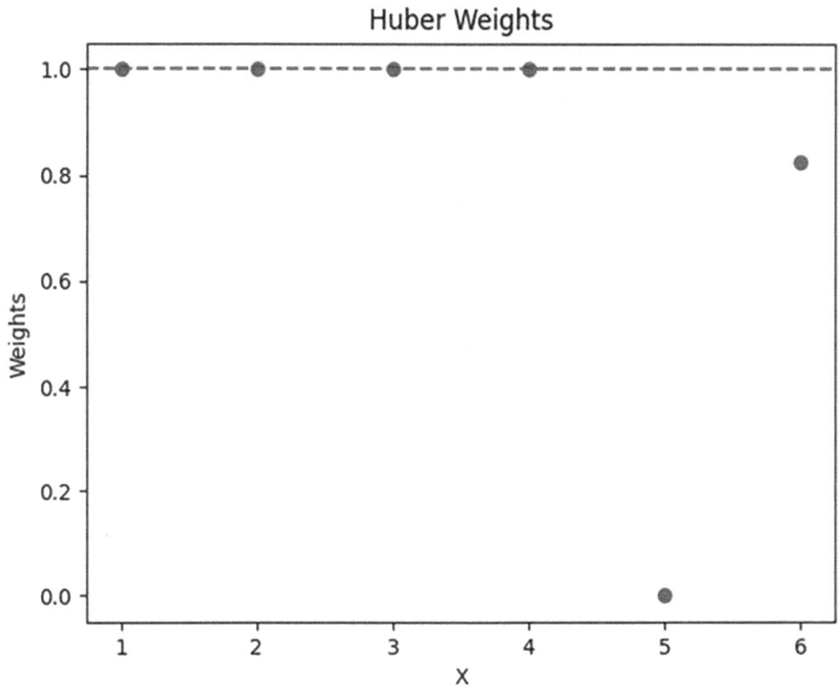

FIGURE 6.17
Weight index plot for the wild point data.

6.3.2 The Potato Leafhopper Application

To illustrate the usefulness of robust regression and several graphical diagnostic aids, a more realistic example is presented. The example is taken in part from Birch and Fleisher (1982). As part of a developing alfalfa integrated pest management program in Virginia, the relationship between relative and absolute density estimates of adult potato leafhopper (Empoasca fabae) was deemed valuable.

Regression analyses were needed to relate the mean field counted sweep-net catch from a given field (y) to the mean absolute density estimate from that field (x variable). The data appears in Table 6.6. Sweep-net counts had been obtained with pendular sweeps of a net. Absolute density estimates were obtained from suction sampling using a special vacuum cleaner. Further details concerning the sampling procedure may be found in Birch and Fleischer (1982). Each replicate resulted in a single x, y data point and consisted of a unique field or a field sampled on different dates.

We will use R program to examine this data. The annotated R session follows:

```
# Plot the original data ggplot(df, aes(x = absolute,
y = sweep)) + geom_point() + labs(title = 'Scatterplot
```

TABLE 6.6

Observations for Mean Absolute Density Estimates (x) and Mean Field Counted Sweep-Net Catch (y) of Adult Potato Leafhoppers in Alfalfa

Replicate method	Harvest quality	Stand density	absolute density	sweep
1	Strip	Good	39.96	1.55
2	Strip	Good	35.65	3.88
3	Strip	Good	56.43	2.74
4	Clean	Good	27.13	1.35
5	Clean	Good	33.66	1.89
6	Clean	Good	35.98	2.76
7	Clean	Poor	71.93	4.36
8	Strip	Good	54.45	3.45
9	Clean	Good	14.02	0.88
10	Clean	Good	12.85	0.88
11	Clean	Good	11.68	1.02
12	Clean	Good	17.52	1.19
13	Clean	Good	14.00	0.96
14	Clean	Good	14.00	0.93
15	Clean	Good	29.00	1.46
16	Clean	Good	29.00	1.08
17	Clean	Good	20.15	0.78
18	Clean	Good	14.89	0.46
19	Clean	Good	10.51	0.58
20	Clean	Good	10.51	0.40
21	Clean	Good	18.88	0.85
22	Clean	Good	18.88	0.99
23	Clean	Good	2.92	0.20
24	Clean	Good	11.34	0.65
25	Clean	Good	6.42	0.36
26	Clean	Good	43.41	2.40
27	Clean	Good	35.48	1.66
28	Clean	Good	8.91	0.35
29	Clean	Good	4.21	0.38
30	Clean	Poor	158.40	2.66

```
of Original Data', x = 'Absolute Density', y = 'Sweep
Density') + theme_minimal()
```

The scatterplot (Figure 6.18) reveals a strong linear trend except for one point, observation 30, where the x value was 158. The least squares analysis gives the following estimated regression equation:

Sweep = 0.715 + 0.0251 Absolute

```
Predictor       Coef       Stdev     t-ratio        p
Constant      0.7152      0.2136        3.35    0.002
```

```
absolute     0.025103     0.005216            4.81      0.000

s = 0.8333      R-sq = 45.3%      R-sq(adj) = 43.3%
Analysis of Variance

SOURCE          DF          SS        MS          F          p
Regression       1      16.086    16.086      23.17      0.000
Error           28      19.442     0.694
Total           29      35.527
```

```
Unusual Observations
Obs.absolute     sweep      Fit    Stdev.Fit     Residual    St.Resid
  2        36    3.880    1.610        0.156        2.270       2.77R
  7        72    4.360    2.521        0.272        1.839       2.33R
 30       158    2.660    4.692        0.693       -2.032     -4.39RX

R denotes an obs. with a large st. resid.
X denotes an obs. whose X value gives it large influence.
```

The OLS coefficients are 0.715 for intercept and 0.025 for slope. The analysis identified three observations as possible outliers: 2, 7, and 30. Let's examine the residuals from the OLS fits. Obtain a scatterplot of the residuals versus the x variable using this code:

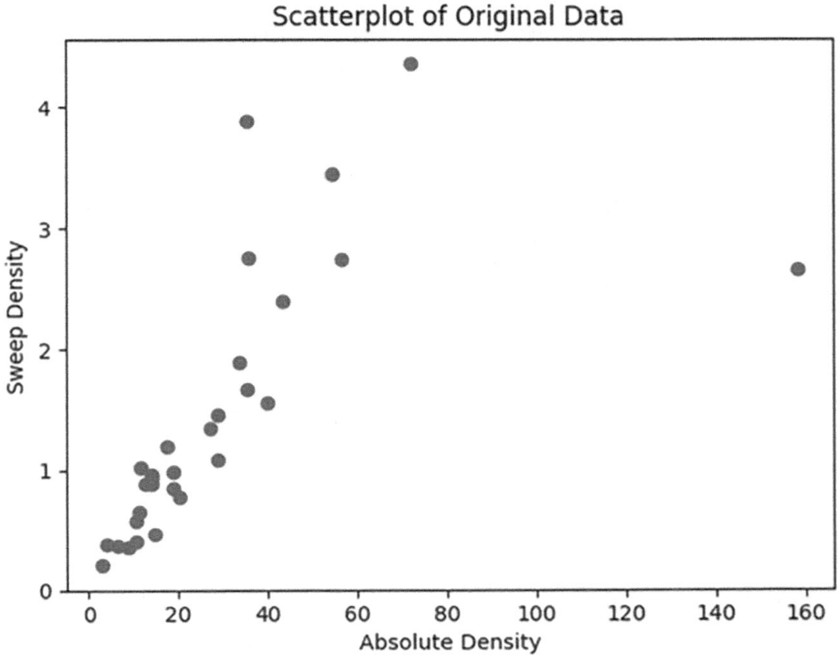

FIGURE 6.18
Scatterplot for the potato leafhopper data.

```
# Calculate residuals
residuals <- residuals(model)
# Combine the data and residuals into a data frame df_
residuals <- data.frame(absolute = df$absolute, residu-
als = residuals)
# Plot the residuals ggplot(df_residuals, aes(x = abso-
lute, y = residuals)) + geom_point() + geom_
hline(yintercept = 0, color = 'red', linetype =
'dashed') + labs(title = 'Residual Plot', x = 'Absolute
Density', y = 'Residuals') + theme_minimal()
# Normal Probability Plot of Residuals qqnorm(residuals,
main = "Normal Probability Plot") qqline(residuals,
col = "red")
```

Look for a random, patternless scatter of points indicative of a good fit. The graph menu was used to obtain Figure 6.19.

We see a strong upward pattern forced by points 2, 3, 6, 7, 8, and 26, with point 30 far removed on the lower right. Another useful display of the residuals is the normal probability plot, discussed in Chapter 2.

Look for a straight-line pattern. Again, the points indicated by replicate numbers 2, 3, 6, 7, 8, 26 and 30 poorly fit by the SLR model using the LS.

To obtain the M-estimates using Huber weight function with a constant of 1.5, use R as follows:

```
# Fit the robust regression model using Huber's M-Esti-
mator robust_model <- rlm(sweep ~ .—intercept, data =
df, psi = psi.huber) # Print the model summary
summary(robust_model)
```

The Huber m intercept and slope estimates are 0.2132 and 0.0432, respectively, quite different from the LS estimates. To compare LS and the M fits, overlay the two fitted lines in the scatterplot.

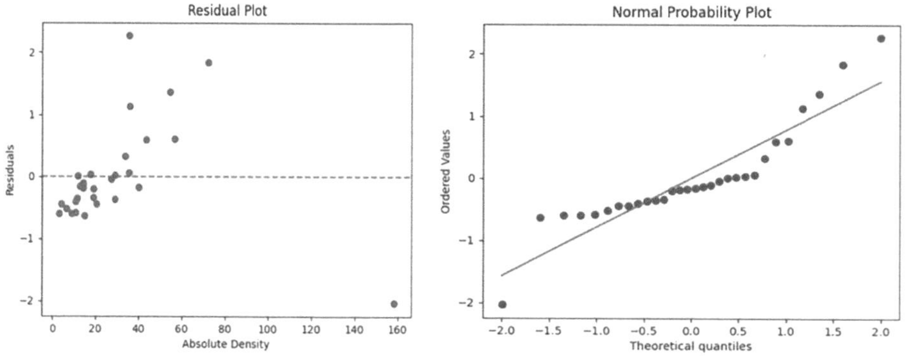

FIGURE 6.19
Residual and normality plots for the potato leafhopper data.

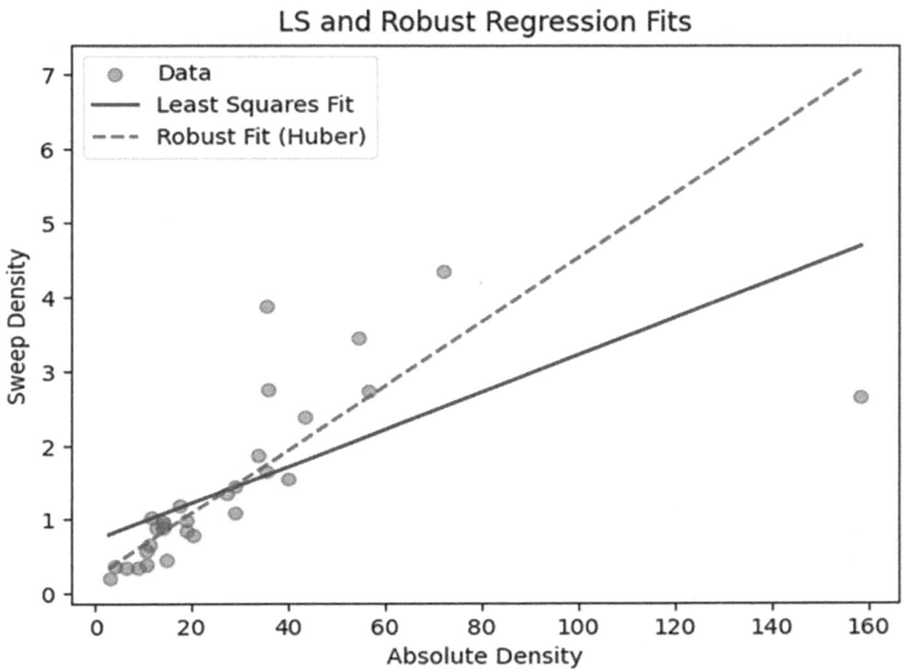

FIGURE 6.20
LS and robust regression fits for the potato leafhopper data.

It is seen from Figure 6.20 that the "robust" line fits the majority of the points better than the LS line, which is clearly pulled down by the replicate 30. Let's compare the residuals from the two fits.

Side-by-side box plots of the residuals in Figure 6.21 reveal that the residuals from the robust line are tighter than those for the LS line, except at one extreme point (observation 30). That is, the robust fit is better at the majority of the points and has done a better job of isolating the extreme outliers. Examine the weights more closely to determine the "outliers" by getting the "weight index" plot.

```
# Extract weights
weights_clean <- robust_model_clean$w
# Create a data frame for the weights
df_weights <- data.frame( Observation = 1:length(weights_
clean), Weight = weights_clean )
# Weight Index Plot
ggplot(df_weights, aes(x = Observation, y = Weight)) +
geom_point() + labs(title = 'Weight Index Plot for Clean
Data', x = 'Observation', y = 'Weight') +
theme_minimal()
```

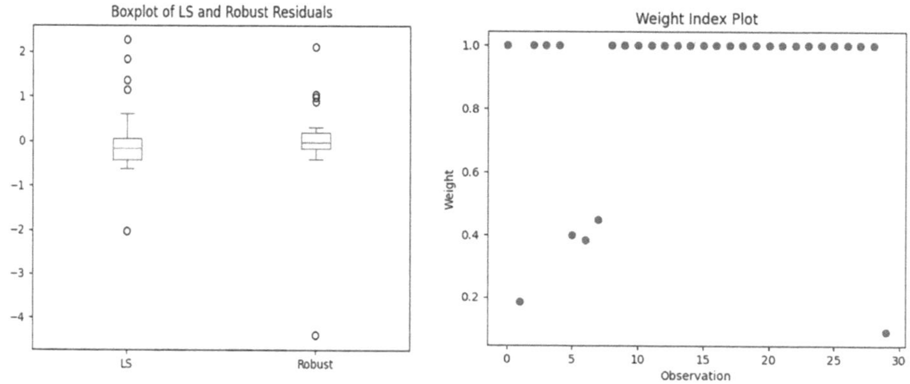

FIGURE 6.21
Box plots and index plot.

Also, from the index plot in Figure 6.21, one can see that points 2, 6, 7, 8, and 30 are clearly distinguished from the others.

These statistical analyses led to separating the alfalfa fields into populations based on biological criteria, either clean-cut and good stands or otherwise. Six points (1, 2, 3, 7, 8, and 30) were deleted from the latter population. The relationship between absolute and relative density estimates was based on the remaining 24 points, representing the population of clean-cut, good fields, which is the normal situation in Virginia. The statistical analysis between absolute and relative density (sweep) is as follows:

The LS analysis gives the estimated regression equation
Sweep = 0.025 + 0.0536 Absolute

Predictor	Coef	Stdev	t-ratio	p
Constant	0.0247	0.1115	0.22	0.827
absolute	0.053594	0.005202	10.30	0.000

```
s = 0.2732        R-sq = 82.8%        R-sq(adj) = 82.1%
```

Analysis of Variance

SOURCE	DF	SS	MS	F	p
Regression	1	7.9245	7.9245	106.15	0.000
Error	22	1.6423	0.0747		
Total	23	9.5668			

Unusual Observations

Obs.	absolute	sweep	Fit	Stdev.Fit	Residual	St.Resid
3	36.0	2.7600	1.9530	0.1064	0.8070	3.21R
21	43.4	2.4000	2.3512	0.1408	0.0488	0.21 X

```
R denotes an obs. with a large st. resid.
X denotes an obs. whose X value gives it large influence.
```

Note that the LS estimate of intercept and slope are 0.025 and 0.053, respectively. The analysis of the n = 24 points residuals begins with the scatterplot of the LS residuals versus x. The residuals for the 24 data points are in Figure 6.22.

Replicate 3 in the new dataset is "unusual." This is replicate 6 in the original dataset. The probability plot for the 24 data points is in Figure 6.22.

Again, old replicate 6 is an "unusual" point. The Huber M-estimates are, again, using k1 = 1.5 for the 24 data points.

```
# Fit the robust regression model using Huber's T norm
with t=1.5 huber_model <- rlm(y ~ x, data = df_cleaned,
psi = psi.huber, k = 1.5) # Extract weights
weights <- huber_model$w # Create a data frame for the
weights df_weights <- data.frame( Index =
1:length(weights), Weight = weights ) # Weighted index
plot for Huber M-estimation
ggplot(df_weights, aes(x = Index, y = Weight)) + geom_
point() + labs(title = 'Weight Index Plot for Huber
M-estimation', x = 'Index', y = 'Weight') + theme_mini-
mal() + theme(panel.grid.major = element_line(color =
"grey", size = 0.5), panel.grid.minor = element_
line(color = "grey", size = 0.25))
```

Huber M-estimation Summary

```
             Robust Linear Model Regression Results
==============================================================================
Dep. Variable:                  y   No. Observations:                     24
Model:                        RLM   Df Residuals:                         22
Method:                      IRLS   Df Model:                              1
Norm:                      HuberT
Scale Est.:                   mad
Cov Type:                      H1
No. Iterations:                13
==============================================================================
                coef    std err          z      P>|z|      [0.025      0.975]
------------------------------------------------------------------------------
const         0.0603      0.098      0.614      0.539      -0.132       0.253
x             0.0507      0.005     11.064      0.000       0.042       0.060
==============================================================================
```

Again, replicate number 3 (original number 6) is clearly detected as an "outlier" in Figure 6.23. Because no biological justification was found for this "outlier," it was not discarded from the dataset.

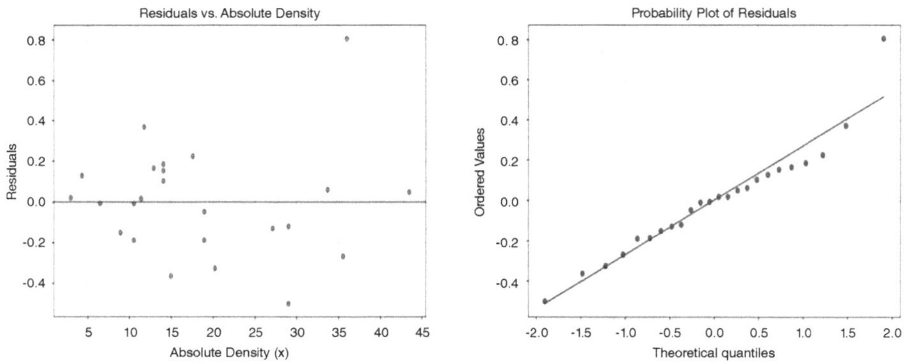

FIGURE 6.22
The residuals plot and Q-Q plot.

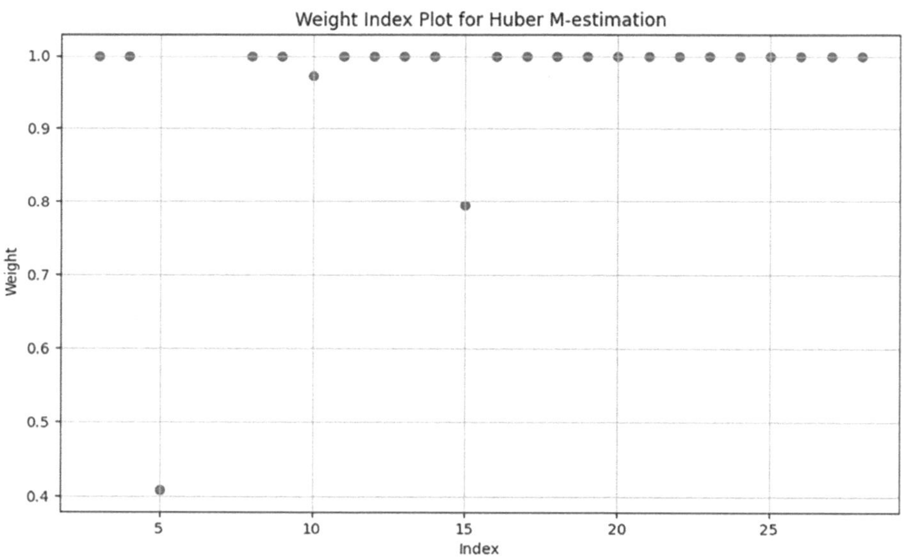

FIGURE 6.23
The weight index plot for Huber M-estimation.

The final analysis and conclusions concerning this data were based on the N = 24-point dataset and the least square analysis. The M-estimation procedure was used to identify unusual data points. In this example, biological justification was found that supported the setting aside of these data points. However, it is reassuring to note that had we decided not to set these points aside, the M-estimate of slope for the N = 30-point original dataset (0.043) was close to the value obtained by least squares on the reduced N = 24-point dataset (0.054). The M-estimates are easy to obtain using the R program. The residuals or weights can now be plotted and further analyzed, but we will not repeat this analysis here.

This example illustrates the three principles of outlier analysis: detection, accommodation, and deletion. Of course, detection refers to identifying the outliers in the dataset by using the weights obtained by IRLS. Accommodation refers to the ability of the M-estimation procedure to supply coefficients that are resistant to the outliers, even in the full presence of the outliers. Deletion refers to the notion that once the outliers are identified, a possible remedy is to remove them from the dataset and obtain the least squares estimates, estimates that are now unaffected by the outliers. Detection is an important first step in the analysis. The second step may be either accommodation or deletion, depending on the circumstances and goals of the analysis.

6.4 Nonparametric Regression

As we have seen, the SLR model can be useful in various applications. It does have its limitations, however. One problem is that the SLR model uses only one regressor variable. The response variable in real-world problems is often affected by more than one regressor. In the next chapter, the SLR model will be extended to handle any number of regressor variables through the multiple linear regression model.

A second problem with the SLR model is that it only allows for a straight line to be fitted to the x and y data. It hardly seems reasonable that a straight line adequately describes the relationship between x and y, except over a limited range of variables. Nonlinear regression methods are particularly well suited for fitting curves, other than the straight line, to the x and y data. Curves such as $y = \beta_0 + \beta_1 e^{\beta_2 x}$ or $y = \frac{1}{1+e^{-(\beta_0+\beta_1 x)}}$ are two examples of curves used to describe relationships between x and y found in such areas as biology, finance, and economics.

The SLR model and nonlinear models, such as those presented in the preceding paragraph, are examples of "parametric" models in that they can be expressed as equations involving parameters. Parametric models are extremely useful in many situations, especially where some underlying subject matter theory suggests the proper model form. For example, in physical chemistry, the log of vapor pressure is linearly related to the reciprocal of absolute temperature, suggesting a nonlinear relationship between vapor pressure and absolute temperature. However, it is not unusual for the researcher in any particular field to study phenomena where the "natural laws" of the physical sciences do not apply, leaving the researcher to grope for an adequate linear or nonlinear model to describe relationships between variables.

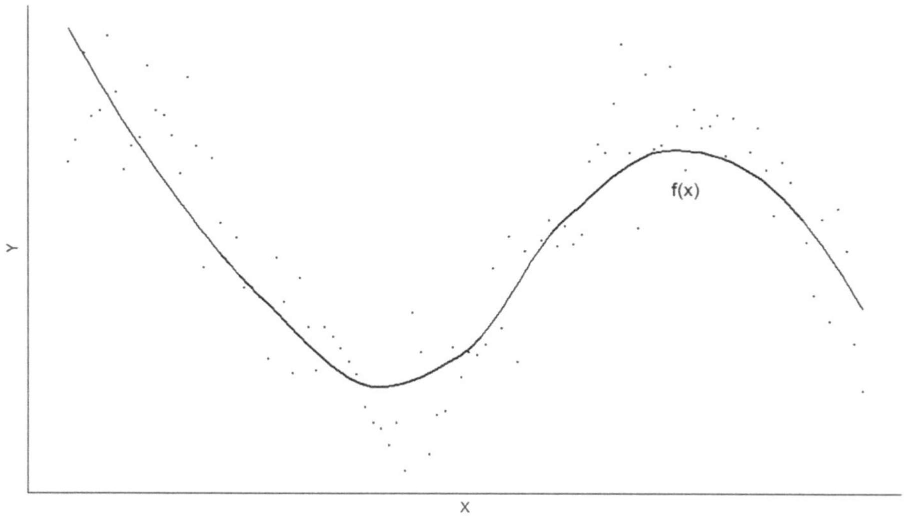

FIGURE 6.24
Scatterplot of y versus x, with the nonparametric function $f(x)$.

Additionally, in plotting the data (Figure 6.24), the researcher may see features in the scatterplot such as dips, valleys, peaks, or local wiggles that a parametric model of any type, linear or nonlinear, does not capture. In these situations, the researcher may still want a "function" to describe the relationship between x and y, although it may not be a parametric one. Nonparametric models are ones where the curve used to describe the x, y relationship cannot be expressed as a parametric function. In the figure 6.24, we see such a relationship between the two variables, x and y. The data reveals a strong upward trend for small values of x and a strong downward trend for large values of x. In between small and large values of x, the data "wiggles" through a pattern that does not appear to be caused by randomness. A parametric equation to describe this overall behavior may be difficult or impossible to find, yet the relationship seems to be well described by the solid line, denoted by $f(x)$ on the plot, a nonparametric function. The function $f(x)$ has no closed functional form, yet for a given value of x, say x_0, the nonparametric function value $f(x_0)$ may be found. There are several methods by which nonparametric functions can be obtained. We will review a few of them here.

6.4.1 Parametric versus Nonparametric Regression Models

The parametric regression approach is to specify a model of the form:

$$y_i = f(x_i; \beta) + \varepsilon_i,$$

where x_i is the value of the regressor variable for the *i*th subject and β is a *px*1 vector of unknown parameters. The error terms are usually assumed to have zero expectation with a constant variance of σ^2. The function f() is some specified form. It follows that the expected value of the response, y_i, is $f(x_i;\underline{\beta})$ with a variance of σ^2.

Some common examples are the SLR model, $y_i = \beta_0 + \beta_1 x_i + \varepsilon_i$, and the nonlinear model $y_i = \beta_0 + \beta_1 e^{\beta_2 x_i} + \beta_3 \sin(\beta_4 x_i) + \varepsilon_i$. The latter model may be flexible enough to fit the data represented in Figure 6.24, but the model is complex and likely offers little insight regarding the subject matter.

Another approach is to fit a nonparametric regression model of the form:

$$y_i = f(x_i) + \varepsilon_i,$$

where the function $f()$ is completely unspecified. Again, it is usually assumed that $E(\varepsilon_i) = 0$ and $Var(\varepsilon_i) = \sigma^2$, which implies that $E(y_i) = f(x_i)$ and $Var(y_i) = \sigma^2$.

Nonparametric regression techniques allow us to estimate $E(y)$ at any value of x. For example, at $x = x_0$, we can obtain $\hat{E}(y_0) = \hat{y}_0 = \hat{f}(x_0)$, where $\hat{f}(x_0)$ is the estimate of the function f at x_0. One can then plot $\hat{f}(x)$ versus x for many values of x to trace out the estimated curve $\hat{f}(x)$, such as the one displayed in Figure 6.24.

This nonparametric approach offers several advantages over the conventional parametric approach. One, the nonparametric approach is extremely flexible. We can essentially estimate any smooth curve without being restricted by a prespecified functional form. Hopefully, this will result in improved fits and predictions in those situations where the user is unable to accurately specify the proper functional form, thereby reducing the bias in fits and predictions resulting from model "misspecification."

A second advantage is that nonparametric regression offers an exploratory approach to evaluating a set of regression data. Nonparametric regression is a mechanism for determining if certain data characteristics are "real" or simply the result of noise in the data. This approach allows for new insights to be extracted from the system that gave rise to the data. For example, after examining the nonparametric fit, a parametric model may be suggested to the user.

On the negative side, nonparametric methods are nonconventional and, therefore, viewed somewhat skeptically. One reason for this is that the user has a wide variety of methods available, all requiring user interaction to determine any final estimated function. This subjectivity makes it very difficult for statistical theory to be applied to any given situation. In particular, this raises the question of how confidence statements or p-values are to be interpreted.

Some of the nonparametric techniques will be presented and reviewed very briefly. Two of the more common techniques, kernel regression and local linear regression, will be presented in some detail.

The basic philosophy of nonparametric regression is that an estimate of the fit at $x = x_0$, $\hat{f}(x_0)$, for any reasonably smooth function f(), should be based more on those responses associated with observed x values close to x_0 rather than on responses associated with observed x values far away from x_0. Consequently, if we think of $\hat{f}(x_0)$ as a weighted average of the n observed responses, $y_1,\ldots,y_n$ – that is, $\hat{f}(x_0) = \sum_{j=1}^{n} w_{0j} y_j$, where $0 \le w_{0j} \le 1$ – we need a method for determining the weights, $w_{01},\ldots,w_{0n}$, so that if x_i is "closer" to x_0 than another observed x value, say x_i, that w_{0i} is larger than w_{0i}. The problem is determining the assignment of the weights.

Although this concept is intuitive and basic to nonparametric regression, it appears to be different from the parametric approach discussed earlier in this chapter. Consider the SLR model and the estimation of the parameters using the least squares method. Here, we have $\hat{y}_0 = \hat{f}(x_0) = \hat{\beta}_0 + \hat{\beta}_1 x_0$. However, it will be shown in Chapter 7, that another expression for $\hat{y}_0$ is $\hat{y}_0 = \sum_{j=1}^{n} h_{0j} y_j$, where $h_{0j} = \dfrac{1}{n} + \dfrac{(x_0 - \bar{x})(x_j - \bar{x})}{S_{xx}}$. The h_{0j} satisfy $-1 \le h_{0j} \le 1$, with $\sum_{j=1}^{n} h_{0j} = 1$ when predictions are obtained at one of the n original data points, $x_1,\ldots x_n$. Thus, $\hat{y}_0$ in SLR can also be expressed as a weighted average with weights h_{0j}. For a fixed value of $(x_0 - \bar{x})$, we note that the term $(x_j - \bar{x})$ increases in magnitude as x_j moves away from $\bar{x}$. Thus, the weight given to a response at an observed x value close to x_0 is actually smaller than the weight given to a response at an observed x value farther away from x_0, just the opposite from the idea in nonparametric regression. If the true model is indeed the SLR model, the weighting scheme used by SLR is optimal; the best way to locate the straight line is to give the most weight to the extreme observations. On the other hand, this may be a ludicrous scheme if the SLR is incorrect.

To demonstrate some of these ideas, consider the dataset where ten observations are to be taken, one at each of the locations $x = -10, -3, -1.5, -1, -.5, .5, 1, 1.5, 3,$ and 10. Then the least squares prediction weights h_{0j}, based on the SLR model, associated with prediction y at locations $x_0 = 0, 3,$ and 6 are shown in Table 6.7.

Considering the least squares prediction at location $x_0 = 0$, the mean of the x's. Table 6.7 shows this prediction is the sample mean of all ten observations. Specifically, the remote observations at $x = -10$ and $x = 10$ are given the same weight as the nearby observations taken at $x = -.5$ and $x = .5$. For prediction at location $x_0 = 3$, the least squares method gives weight 0.233 to the observation taken at $x = 10$. This weight is substantially greater than the weight 0.140 given to the observation taken directly at location $x = 3$. The observation taken at $x = 10$ is given weight 0.367 in making the least squares prediction at $x_0 = 6$. This is in contrast to the much smaller weight (0.180) given to the observation at location $x = 3$, which is closer to $x_0 = 6$. Further, for this prediction, the weight given to the remote observation at $x = -10$,

TABLE 6.7

Least Squares Prediction Weights for Illustration

x	−10	−3	−1.5	−1	−0.5	0.5	1	1.5	3	10
$x_0 = 0$	.10	.10	.10	.10	.10	.10	.10	.10	.10	.10
$x_0 = 3$	−0.033	.060	.080	.087	.093	.107	.113	.120	.140	.233
$x_0 = 6$	−0.167	.020	.060	.073	.087	.113	.126	.140	.180	.367

namely −0.167, is nearly as large in magnitude as for $x = 3$. Moreover, the magnitude of the weight at $x = -10$, and therefore the importance placed on this in the least squares prediction, is larger than the weight given to any of the observations from $x = -3$ to $x = 1.5$. Such a weighting scheme is a direct ramification of the SLR model prescribed by the user. In general, the particular weighting scheme used for making predictions using the least squares method will be strongly tied to the form of the user's parametric model.

Suppose the user's proposed model is incorrect. In that case, as it would be if an SLR model is selected for the data depicted in Figure 1, it is hard to imagine how a response obtained for a small observed x could be helpful in predicting the mean response at a medium-size value of x, a value where the curve, $f(x)$, dips. Local information contained in the y's associated with observed medium x's will be more useful than y's associated with extreme values of x.

Two nonparametric regression methods, the moving mean (or moving median) method (Tukey, 1977) and the k nearest neighbor method (Hardle, 1992), are very simple to use but possess few other redeeming features and will not be mentioned further. Smoothing splines is a popular and powerful technique. Here, the estimates of $f(x_i)$, $i = 1,\ldots,n$ are found to minimize

$$S_\lambda(f) = \sum_{i=1}^{n}(y_i - f(x_i))^2 + \lambda \int (f''(x))^2\,dx,$$

where the quantity $\int (f''(x))^2\,dx$ is a "roughness penalty," with f'' denoting the second derivative of f, and the parameter λ is the "roughness parameter." Essentially, the roughness penalty measures the "roughness" of the resulting fitted curve.

For example, in Figure 6.25, the straight line on the left is smooth and has a roughness of zero, while the curve on the right is very bumpy and has a large roughness value.

The roughness parameter, λ, is chosen by the researcher, indicating the degree of smoothness desired in the estimated curve. If λ is picked as zero,

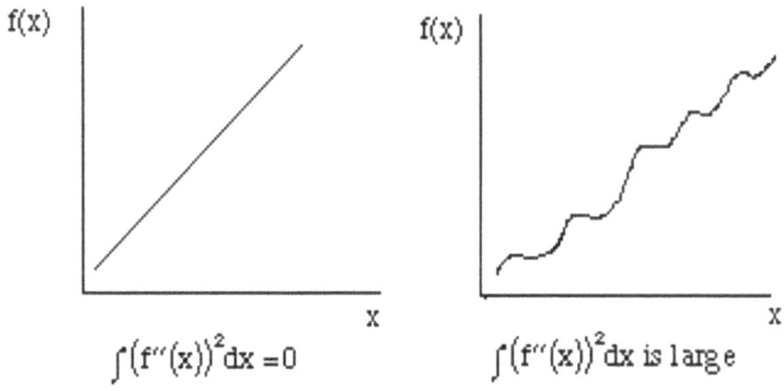

FIGURE 6.25
Sketch for two functions.

then $S_\lambda(f)$ will be minimized if $\hat{f}(x_i) = y_i$ and $S_\lambda(f) = 0$. If there is no penalty for a rough estimated curve, we allow $\hat{f}(x_i)$ to become as rough as it wants, and the method of smoothing splines will fit a curve that amounts to "connecting the dots" on the scatterplot.

On the other hand, if λ is made large, then the roughness penalty is large and $S_\lambda(f)$ will be minimized only when the estimated curve, $\hat{f}(x_i)$, is very smooth. As λ tends to be larger, the fitted curve tends to be smoother and, in the limit, $\hat{f}(x_i) = \bar{y}$, for all i. In this case, $\lambda \int (f''(x))^2 dx = 0$ and $S_\lambda(f) = \sum_{i=1}^{n} (y_i - \bar{y})^2$. Obviously, the trick is to pick λ as some middle value so that $\hat{f}(x_i)$ is a compromise between being too smooth and too rough. The interested reader is referred to Hardle (1992) for more information on smoothing splines.

Another nonparametric method is kernel regression. Several kernel regression procedures exist, but we will present one method due to Nadaraya (1964) and Watson (1964), referred to as the Nadaraya-Watson kernel regression procedure. This method yields an estimator of $f(x_0)$ that can be expressed as

$$\hat{y}_0^k = \hat{f}(x_0) = \sum_{j=1}^{n} h_{0j}^k y_j,$$

where the kernel weights or coefficients, h_{0j}^k, are defined as

$$h_{0j}^k = \frac{K\left(\frac{x_0 - x_j}{b}\right)}{\sum_{j=1}^{n} K\left(\frac{x_0 - x_j}{b}\right)},$$

for some $b > 0$. The function $\mathrm{K}(\mathrm{u})$ is called the kernel function, and the constant b is called the bandwidth. The kernel function, $\mathrm{K}(\mathrm{u})$ is a decreasing function of $|\mathrm{u}|$ and symmetric about zero. The so-called Gaussian kernel function, where $K(u) = \mathrm{e}^{-\mathrm{u}^2}$, a function proportional to the Gaussian (normal) probability density function, is a popular choice. Another example is the Epanechnikov kernel function, defined as $\mathrm{K}(\mathrm{u}) = .75(1-\mathrm{u}^2)^2$ for $-1 \le \mathrm{u} \le 1$. The Gaussian and Epanechnikov kernel functions are displayed in Figure 6.26, where the decreasing nature of these kernel functions is readily apparent.

Countless other examples of kernel functions exist, such as the uniform (on the interval $-1 \le u \le 1$) and the triangular (on the $-1 \le u \le 1$ interval). It has been shown that there is little difference in the estimated nonparametric function, $\hat{\mathrm{f}}(\mathrm{x})$, resulting from using these different kernel functions. One choice seems about as good as another. We will use the Gaussian kernel in all subsequent illustrations.

While the choice of Kernel function may not be crucial, the choice of the bandwidth b is crucial in obtaining an appropriate kernel fit to the data. The size of b determines the rate at which the kernel function decreases as $|\mathrm{u}|$ moves away from zero. For a small value of b, the kernel function concentrates its area around the point x_0, the point where the prediction is desired, leaving little area or weight, extending to neighboring x's. Consequently, for b close to zero, the kernel regression fit at the observed values of $\mathrm{x}, \mathrm{x}_\mathrm{i}$, is a weighted average, with the majority of the weight given to the observed response at x_i – that is, $\hat{f}(x_i) \approx y_i$. The resulting estimated kernel regression curve tends to be very "rough," described as undersmoothed or overfit, with possibly low prediction bias but with high prediction variability.

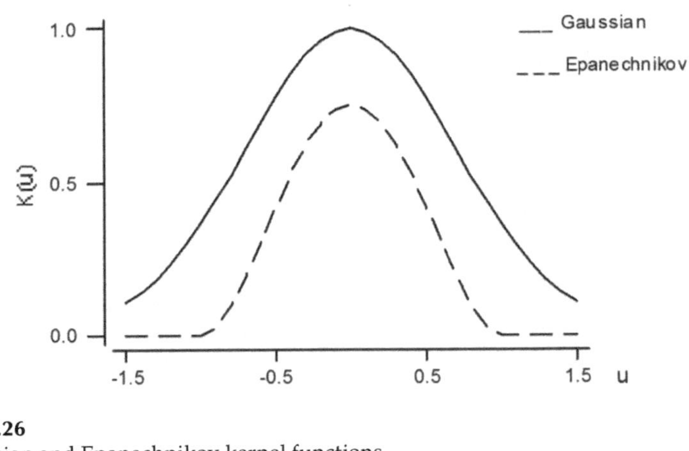

FIGURE 6.26
The Gaussian and Epanechnikov kernel functions.

On the other hand, large values of *b* concentrate the area of the kernel function over a much broader range than if *b* is small. Thus, in the weighted average, many of the *n* observed responses will receive some weight. And, as b tends to be very large, close to the range of the data, all points are given equal weight, and the weighted average becomes the mean of the responses – that is, $\hat{f}(x_i) \approx \bar{y}$, an over-smooth curve (underfit) with possibly high bias and almost certainly with low prediction variance. The shape of the Gaussian kernel for $b = .1, 1$, and 10 is shown in Figure 6.27.

Analogous to the choice of λ for smoothing splines, the choice of *b* is the researcher's responsibility. Without a priori knowledge of the appropriate value of *b*, the researcher must estimate b from the data. Thus, b is a parameter to be estimated in this nonparametric approach. A number of "data-driven" methods are available for the selection of *b*, all based on some "global" measure of the quality of fit of the resulting kernel regression function.

For example, one might be tempted to choose b so that the sum of squares of the residuals, $SSE = \sum_{i=1}^{n}\left(y_i - \hat{f}(x_i)\right)^2$ is minimized. Some reflection reveals, however, that this would be a very poor choice since we can make $SSE = 0$ by choosing $b = 0$ (recall $\hat{f}(x_i) \approx y_i$ if $b = 0$). The resulting fitted regression equation is too rough because it follows the variability inherent in the response data. Our goal is to fit a curve to the true pattern exhibited by the data (the true curve $f(x_i)$) and to ignore the variability of the responses (the σ^2). The hard part is to determine what a "pattern" is and what is variability.

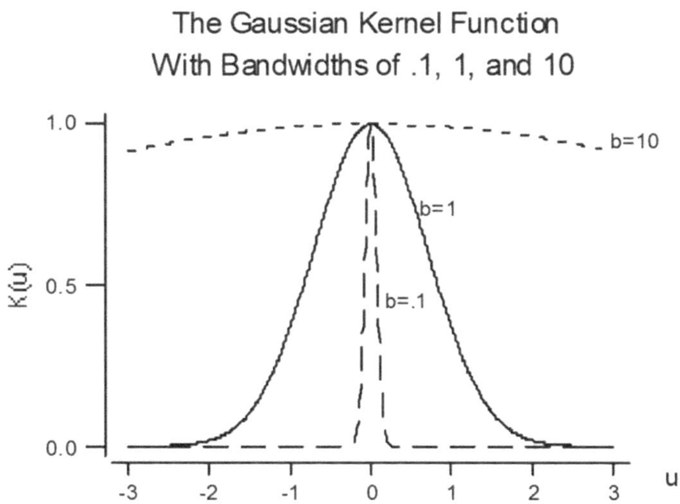

FIGURE 6.27
The Gaussian kernel function with different bandwidths.

Another possible choice is to choose b to minimize the mean squared error, defined as

$\mathrm{MSE} = \frac{\sum_{i=1}^{n}\left(y_i - \hat{f}(x_i)\right)^2}{n - df_{\mathrm{model}}}$, where $\mathrm{df}_{\mathrm{model}}$ reflects the "model degrees of freedom," those degrees of freedom associated with the fitted nonparametric function. The denominator in the MSE, analogous to the error degrees of freedom in a parametric model expression or $\mathrm{df}_{\mathrm{error}}$, can be thought of as a "penalty for overfitting" in that the model degrees of freedom range from 1 to *n*, with $\mathrm{df}_{\mathrm{model}} = 1$, if b is large (and $\hat{\mathrm{f}}(\mathrm{x}_\mathrm{i}) \approx \bar{\mathrm{y}}$ for all *i*), and $\mathrm{df}_{\mathrm{model}} = \mathrm{n}$, if b is close to zero (where $\hat{\mathrm{f}}(\mathrm{x}_\mathrm{i}) \approx \mathrm{y}_\mathrm{i}$ for all *i*). As *b* approaches zero, the numerator of MSE gets smaller, but so does the denominator. If the denominator gets smaller faster than the numerator, the MSE will become large for small values of b. Choosing b to minimize MSE will certainly result in a smoother fit than using the SSE as the measure of fit criteria. Empirical evidence suggests that the use of the MSE criteria still selects *b* "too small," resulting in an under-smooth kernel regression function.

Perhaps a better criteria is to choose b to minimize the penalized PRESS statistic, denoted PRESS* and defined as $\mathrm{PRESS}^* = \frac{\sum_{i=1}^{n}\left(y_i - \hat{y}_{i,-i}^{k}\right)^2}{n - df_{\mathrm{model}}} = \frac{\mathrm{PRESS}}{n - df_{\mathrm{model}}}$, where $\hat{y}_{i,-i}^{k}$ is the prediction of mean response at $\mathrm{x} = \mathrm{x}_\mathrm{i}$, obtained by kernel regression based on $\mathrm{n} - 1$ observations, with the *i*th data point, $(\mathrm{x}_\mathrm{i}, \mathrm{y}_\mathrm{i})$, removed. The numerator of PRESS* is commonly referred to in the regression literature as the "PRESS" statistic (see Myers, 1990). The PRESS statistic is typically used to compare models in terms of their "predictive" capabilities, where smaller values of the PRESS statistic are preferred. In the kernel regression literature, choosing *b* to minimize PRESS is called the method of "cross-validation."

In a manner similar to the MSE statistic, the denominator, $\mathrm{df}_{\mathrm{error}} = \mathrm{n} - \mathrm{df}_{\mathrm{model}}$, is used to penalize the kernel regression curve for being too rough (under-smoothed), as choosing *b* to minimize PRESS typically results in a *b* that is too small. One way to compute the model degrees of freedom, $\mathrm{df}_{\mathrm{model}}$, is $\mathrm{df}_{\mathrm{model}} = \sum_{\mathrm{i}=1}^{\mathrm{n}} \mathrm{h}_\mathrm{i}^\mathrm{k}$. For comparison purposes, in least squares simple linear regression, $\mathrm{h}_{\mathrm{ii}} = \frac{1}{\mathrm{n}} + \frac{(\mathrm{x}_\mathrm{i} - \bar{\mathrm{x}})^2}{\mathrm{S}_{\mathrm{xx}}}$, and it is easy to show that $\sum_{\mathrm{i}=1}^{\mathrm{n}} \mathrm{h}_{\mathrm{ii}} = 2$, the model degrees of freedom in the SLR model. In kernel regression, as *b* tends toward zero, the curve frequently becomes rougher, indicating greater flexibility of the curve and an increase in $\mathrm{df}_{\mathrm{model}}$. Thus, finding *b* to minimize the *PRESS** hopefully results in a smooth curve that provides an adequate fit to the data. Improving this fit by reducing PRESS* through reducing PRESS could only happen by using a smaller b, thereby using more $\mathrm{df}_{\mathrm{model}}$ and overfitting the model.

For a fixed value of bandwidth, the kernel fit at $x = x_0$ is expressed as $\hat{y}_0^k = \hat{f}(x_0) = \sum_{j=1}^{n} h_{0j}^k y_j$, a weighted average of the $y_1, \dots, y_n$. The $\text{var}(\hat{y}_0)$ can then easily be found as (considering the weights as fixed)

$$\text{var}(\hat{y}_0) = \sigma^2 \sum_{j=1}^{n} h_{0j}^{k^2}.$$

Consequently, the standard error of prediction at each value of x_0 is se $(\hat{y}_0) = s\sqrt{\sum_{i=1}^{n} h_{0j}^{k^2}}$, where $s = \sqrt{\text{MSE}}$. Approximate 95% confidence intervals may be obtained for $f(x_0)$ by using

$$\hat{y}_0 \pm t_{1-\frac{\alpha}{2}, df_{error}} \, se(\hat{y}_0).$$

6.4.2 The Tensile Data Application

All these concepts will be demonstrated in the following application. This real data is the "tensile data" taken in part from Montgomery and Peck (1992) as in Table 6.8. In this example, the response variable y is the tensile strength (in psi of paper), and the regressor variable x is the percentage of hardwood in the batch of pulp from which the paper was produced. This is the same dataset considered in Joglekar et al. (1989), who determined that the SLR model was inappropriate for this data.

We will examine this data, presented in the R output, using several different approaches. First, a "quadratic" model will be fit to the data and analyzed. Second, kernel regression will be applied to the data.

This R script performs the following tasks:

1. Fits a quadratic regression model to the tensile data and plots the fitted curve along with the observed data.
2. Plots the residuals from the quadratic regression to visualize the fit.
3. Finds the best bandwidth for kernel regression using leave-one-out cross-validation.
4. Fits a local linear regression (LLR) model to the data using the best bandwidth.
5. Calculates confidence intervals for the LLR fit.
6. Plots the observed data, the LLR fit, and the 95% confidence intervals.

TABLE 6.8

The Tensile Data

Number	1	2	3	4	5	6	7	8	9	10	11	12	13	14	15	16	17	18	19
Y	6.3	11.1	20	24	26.1	30	33.8	34	38.1	39.9	42	46.1	53.1	52	52.5	48	42.8	27.8	21.9
X	1	1.5	2	3	4	4.5	5	5.5	6	6.5	7	8	9	10	11	12	13	14	15

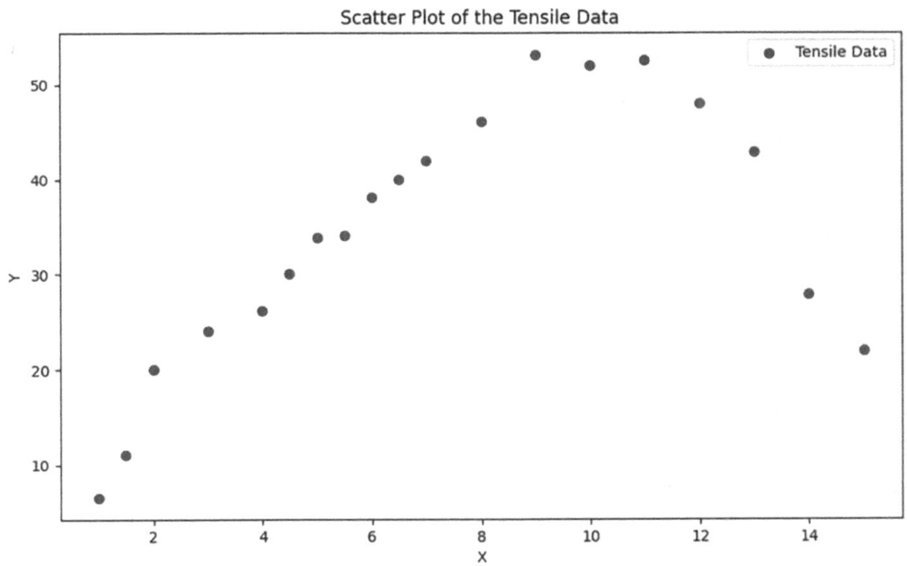

FIGURE 6.28
The tensile data.

Make sure to install the necessary packages **("MASS," "ggplot2," and "broom")** before running the script. The comprehensive script in Appendix A covers all the steps outlined in this application and produces the required plots. The first step is to plot the scatterplot for the tensile data, as shown in Figure 6.28.

A quadratic curve bends approximately the same way as the data bends in the plot above. The quadratic model is fit using the equation: $y = b_0 + b_1x + b_2x^2$.

The estimated regression equation is
$y = -6.67 + 11.8\ x + b_2 - 0.635\ x^2$.

Predictor	Coef	Stdev	t-ratio	p
Constant	-6.674	3.400	-1.96	0.067
x	11.764	1.003	11.73	0.000
xsq	-0.63455	0.06179	-10.27	0.000

```
s = 4.420        R-sq = 90.9%        R-sq(adj) = 89.7%
```

Analysis of Variance

SOURCE	DF	SS	MS	F	p
Regression	2	3104.2	1552.1	79.43	0.000
Error	16	312.6	19.5		
Total	18	3416.9			

SOURCE	DF	SEQ SS
x	1	1043.4
C3	1	2060.8

The curve numerically fits well, with an R^2 of 90.9%. But we need some plots to be sure. Plot the fit and 95% confidence bands. The graph menu was used to create Figure 6.29.

The fit is certainly poor throughout much of the *x* region, overestimating then underestimating in the middle portion of the curve. Plot the residuals versus x, as in Figure 6.30.

Let's try kernel regression on the tensile data. We might start with a guess since we don't know the bandwidth. We'll try $b = 10$. To perform kernel regression on the tensile data using a Gaussian kernel with a bandwidth of $b = 10$, we'll use the **"KernelReg"** class from the **"statsmodels"** library for this purpose.

```
The kernel variance is:
sigma2 155.318, and the error degrees of freedom is: dferror
  17.6597
```

We note that the MSE = 155.318 is far larger than the MSE = 19.5 obtained by the quadratic regression method. This suggests that the kernel fit is probably very poor. R was used to create Figure 6.31.

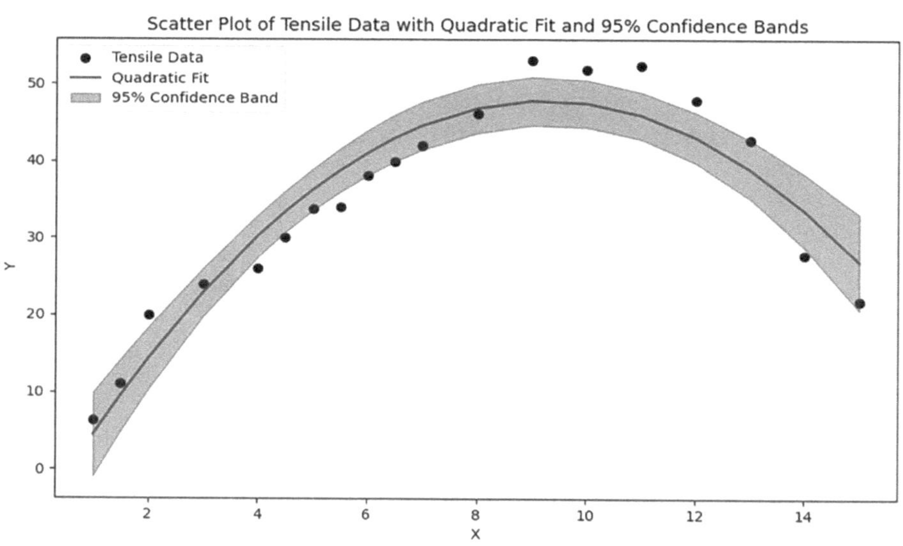

FIGURE 6.29
The quadratic fit with 95% confidence bands for the Tensile Data.

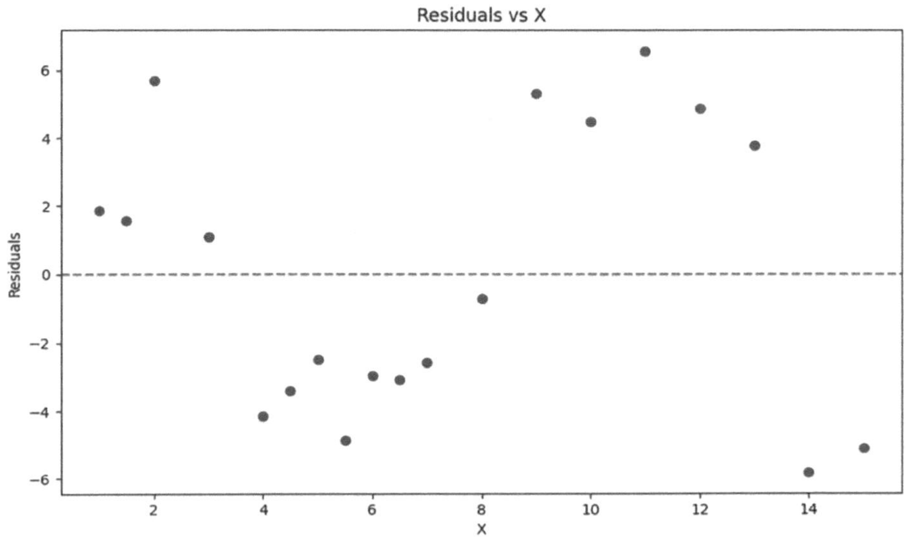

FIGURE 6.30
The residual plot for the quadratic model.

The kernel regression curve is very flat, indicating that a far too large bandwidth has been used. Note that the error df is 17.65, implying that the kernel fit uses only 1.35 df. The kernel fit with $b = 10$ fits a slightly more flexible curve than the sample mean. We must find a better value for the bandwidth. R was used to create the plot in Figure 6.32. The tensile data – press versus bandwidths from the candidate list as in Figure 6.32.

The plot reveals that the minimum press* occurs between 1 and 1.5. Let us update the candidate list and try again. Looks like $b = 1.09$ is best. We'll use this value in subsequent analyses.

Let's get the kernel fit at the xi data points. We found that the kernel variance is:

```
sigma2 6.85969, The error degrees of freedom is: dferror
  11.0282
```

This kernel fit the observed x's and can either be printed or plotted along with the 95% confidence bands. The graph menu was used to create Figure 6.33.

It is clear that this fit is much superior to the quadratic regression fit. The code generates both the fitted values and the associated 95% confidence bands for all entries in the specified column. The kernel fit is excellent, especially when compared to the quadratic fit, as seen in Figure 6.34.

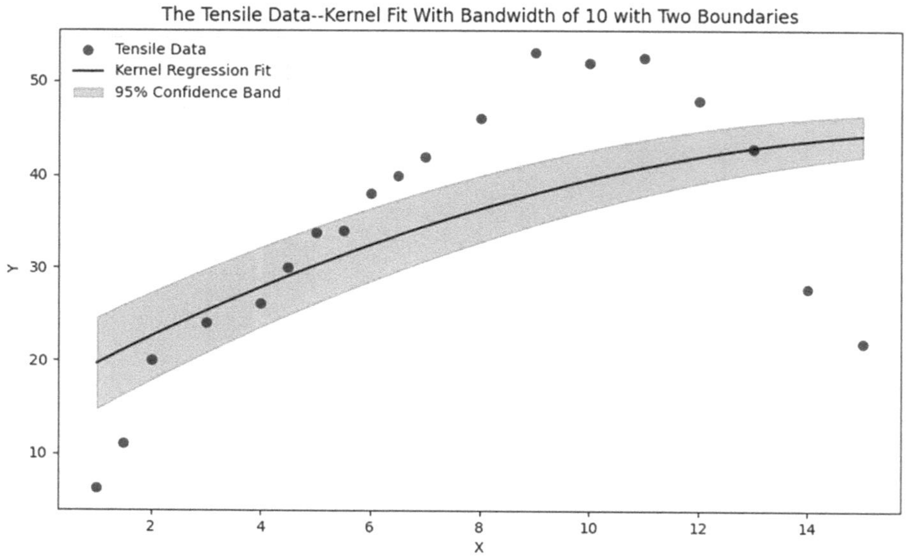

FIGURE 6.31
The tensile data–kernel fit with bandwidth of ten.

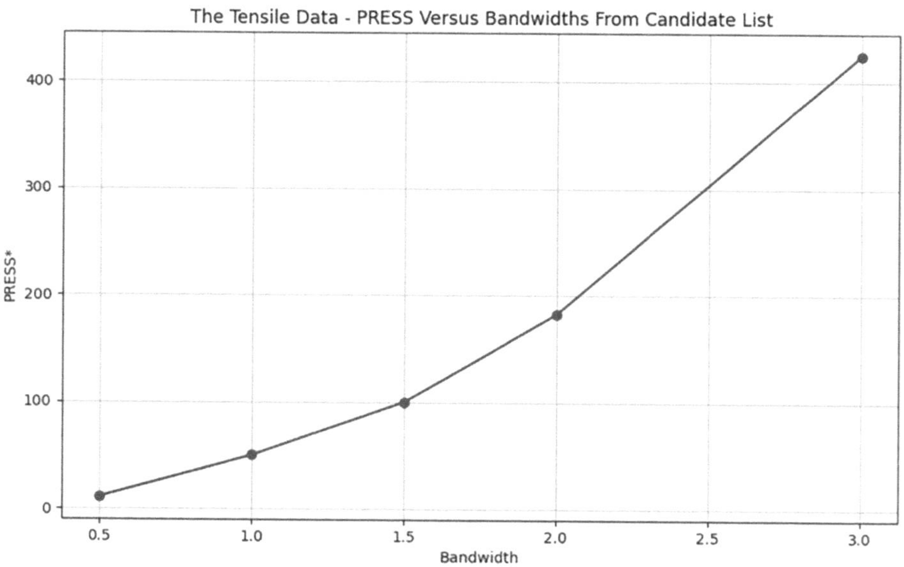

FIGURE 6.32
The tensile data – press versus bandwidths from the candidate list.

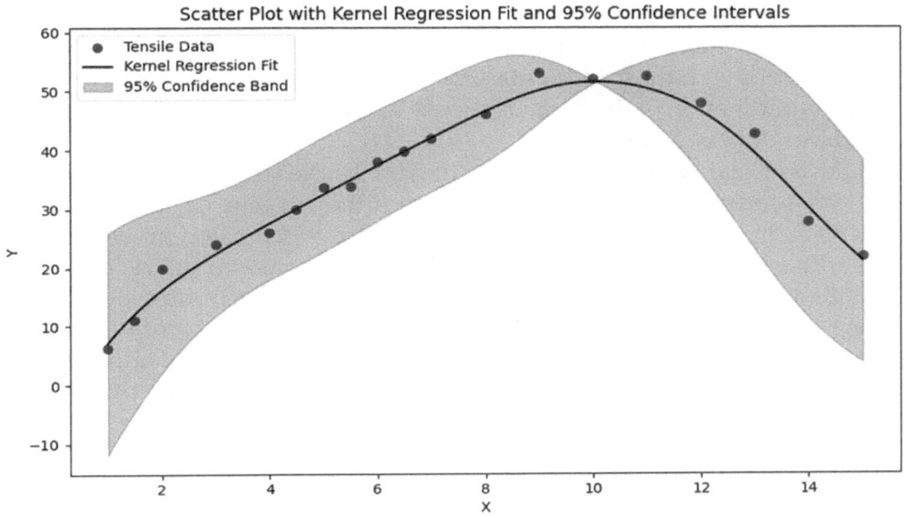

FIGURE 6.33
This kernel fits along with the 95% confidence bands.

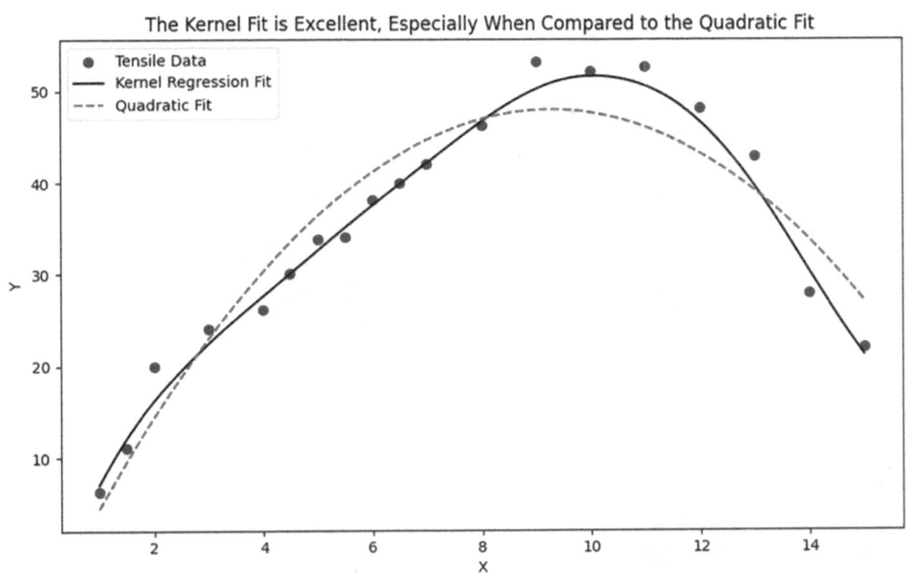

FIGURE 6.34
The tensile data, the quadratic fits, and the kernel fits.

Kernel regression is a very simple method for obtaining nonparametric regression fits that produce reasonably good results, as demonstrated in the above R script with the tensile data. One problem with kernel regression concerns the estimation of the mean response at the left and right boundaries of the curve – that is, at $x = x_{min}$ and $x = x_{max}$.

In the tensile data example, we note that the kernel regression curve (using $b = 1.09$) overestimates the response at both the left and right boundaries. This "bias in the boundaries" will always occur when using kernel regression with a symmetric kernel function because of the asymmetric nature of the data at the boundaries. This bias will be especially troublesome if the observed responses at the bounds follow either an increasing or decreasing trend. To see this, recall that when using kernel regression to obtain the fit at $x = x_0$, the kernel function assigns weights to the yi responses on the basis of how far their corresponding xi is from x_0. For values of $x = x_0$ toward the center of the x data, the kernel function assigns weights to y's on both sides of x_0.

However, at the left boundary, there are no x values to the left of x_{min}, and all weight must go to observations to the right of x_{min}. In the tensile data, we see that the response values to the right of x_{min} are all greater than the y at x_{min}, and consequently, the weighted average of these responses must be at least as large as this response at x_{min}, resulting in the overestimation at the left boundary. At the right boundary, all weight is given to responses at x values equal to and to the left of x_{max}.

Since these responses are all greater than the response at x_{max}, the kernel estimate at x_{max} will be at least as large as this response.

A nonparametric regression method that will remedy this annoying problem is to use LLR. In LLR, the estimated response is obtained at each $x = x_0$ by fitting a simple linear regression line at x_0 using the method of weighted least squares (WLS). The kernel weights, $h_{01}^k, h_{02}^k, \ldots, h_{0n}^k$, obtained using kernel regression, serve as the weights in the WLS formulas. LLR is the natural extension of kernel regression in the nonparametric regression environment, as the SLR model is the natural extension of the location model in the parametric model setting. Kernel regression may be viewed as the fitting of a "location parameter" at each value of $x = x_0$ by the WLS method.

The weights, $h_{01}^k, h_{02}^k, \ldots, h_{0n}^k$, are used to determine how much neighboring x values will contribute to the estimate of mean response (the "location") at x_0. A more flexible methodology is to fit a straight line in this "local" fashion, weighting the responses more heavily if they occur at x values closer to x_0 than at x values farther away. Thus, at any observed data point $x = x_i$, for $i = 1, 2, \ldots, n$ we find $\hat{y}_i^{LLR} = \hat{f}(x_i) = \hat{\beta}_{0i} + \hat{\beta}_{1i} x_i$, where $\hat{\beta}_{0i}$ and $\hat{\beta}_{1i}$ are obtained using WLS, with weights $h_{01}^k, h_{02}^k, \ldots, h_{0n}^k$ obtained using the same kernel function as in kernel regression. Using the WLS formulas introduced in Section 5.3, we have

$$\hat{\beta}_{0i} = \bar{y}_{wi} - \hat{\beta}_{1i} \bar{x}_{wi} \text{ and } \hat{\beta}_{1i} = \frac{S_{xw_iy}}{S_{xw_ix}},$$

$$\text{where } \bar{y}_{wi} = \frac{\sum_{j=1}^{n} w_{ij} y_j}{\sum_{j=1}^{n} w_{ij}}, \; \bar{x}_{wi} = \frac{\sum_{j=1}^{n} w_{ij} x_j}{\sum_{j=1}^{n} w_{ij}},$$

$$S_{xwy} = \sum_{j=1}^{n} \left(x_j - \bar{x}_{w_i}\right) w_{ij} \left(y_j - \bar{y}_{w_i}\right), \text{ and}$$

$S_{xw_ix} = \sum_{j=1}^{n} (x_j - \bar{x}_{w_i})^2 w_{ij}$. As stated above, the weights $w_{i1}, \ldots, w_{in}$ used at each $x = x_i$ are the kernel weights, $h_{01}^k, h_{02}^k, \ldots, h_{0n}^k$. To obtain the n fits at $x_1, \ldots, x_n$ requires n separate simple linear regressions using WLS.

Fits at non-observed data values, $x = x_i$, are obtained using the same formulas as above, with the subscript "*i*" replaced by " 0" and using the kernel weights $h_{01}^k, h_{02}^k, \ldots, h_{0n}^k$, the same weights used to obtain the kernel fit at $x = x_i$.

For those values of $x = x_i$ in the middle portion of the regression function, the kernel and local linear regression methods give very similar results. At the boundaries, however, the LLR fit is usually superior to kernel regression. To see this, imagine fitting the left boundary to the data, represented by the "o," in Figure 6.35. If kernel regression is used, the kernel function might assign nonzero weights to $x = x_{min}$ and the three x values to the right of x_{min}, as illustrated by the kernel function in the figure. The resulting fit at x_{min}, the weighted average of these four responses, denoted by the "*," is larger than the observed response at the left boundary.

On the other hand, if a simple linear regression line is fit, using WLS, through these four points, we see the fit at x_{min}, denoted by (*) is closer to the y at x_{min} and may be either above or below this response, depending on the neighbors' behavior, see Figure 6.35.

As with kernel regression, the LLR fit at any value of x can be expressed as a weighted average of the responses $y_1, \ldots, y_n$. That is,

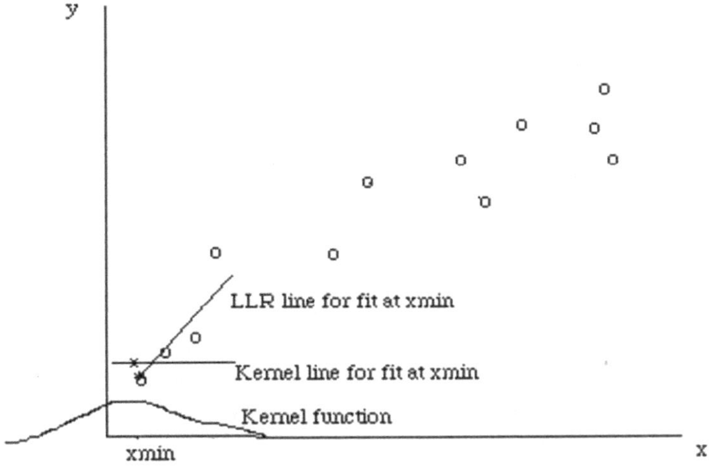

FIGURE 6.35
The kernel and LLR fits at the left boundary.

$$\hat{y}_0^{LLR} = \hat{f}(x_0) = \hat{\beta}_{00} + \hat{\beta}_{01}x_0 = \sum_{j=1}^{n} h_{0j}^{LLR} y_j,$$

where $h_{0j}^{LLR} = h_{0j}^k + \frac{(x_0 - \bar{x}_{w_0})\, h_{0j}^k\, (x_0 - \bar{x}_{w_0})}{S_{xw_0x}}$. This result provides a simple expression for the variance of $\hat{y}_0^{LLR}$, provided that the bandwidth is fixed (which, of course, it is not) as $var(\hat{y}_0^{LLR}) = \sigma^2 \sum_{j=1}^{n} (h_{0j}^{LLR})^2$. Analogous to the kernel regression case, an approximate 95% confidence interval for the mean response at $x = x_i$ may be obtained as

$$\hat{y}_0^{LLR} \pm t_{1-\frac{\alpha}{2}, n-df_{model}}\, se(\hat{y}_0^{LLR}),$$

where $df_{model} = \sum_{i=1}^{n} h_{ii}^{LLR}$ and $se(\hat{y}_0^{LLR}) = s\sqrt{\sum_{i=1}^{n} (h_{ii}^{LLR})^2}$.

To illustrate the LLR method using R, consider again the tensile data and the following main tasks:

1. Finding the best bandwidth to minimize the PRESS* statistic.
2. This bandwidth is used to obtain LLR fits at observed values.
3. Finding LLR fits for specified values and plotting the results with approximate 95% confidence bands.

To perform the LLR analysis and create the needed data, we can use R program. The R script for demonstrating how to implement LLR, calculate fits, and create plots is given in Appendix B.

```
Row   bandw    press*
  1       2   14.4140
  2       1   20.2375
  3       3   22.8465
```

It appears that the best bandwidth is between 1 and 2. Some further searching is used to find that the "best" bandwidth is b = 1.78 with a press* = 14.0692. We'll use this value in all LLR calculations. Obtain the LLR fits at the 19 observed data points for the tensile data. The bandwidth is 1.78 for this example, the "best" value. The LLR variance is sigma2 = 5.56203, and the error degrees of freedom is dferror 13.0823. Note that the variance estimate is 5.56 for LLR, compared to a variance estimate of 6.86 for kernel regression. Also, the error df is larger for LLR than for kernel regression. The LLR results may either be plotted as in Figure 6.36 or printed as in Table 6.9.

TABLE 6.9
The LLR Results

ROW	y	x	fits	uclm	lclm
1	6.3	1.0	7.5956	11.3149	3.8764
2	11.1	1.5	12.1196	14.9578	9.2813
3	20.0	2.0	15.9239	18.3103	13.5375
4	24.0	3.0	22.1566	24.1718	20.1414
5	26.1	4.0	27.5045	29.4582	25.5508
6	30.0	4.5	30.0210	31.9600	28.0820
7	33.8	5.0	32.5026	34.4027	30.6025
8	34.0	5.5	34.9607	36.8384	33.0831
9	38.1	6.0	37.3754	39.2880	35.4627
10	39.9	6.5	39.7330	41.7159	37.7502
11	42.0	7.0	42.0315	44.0821	39.9809
12	46.1	8.0	46.3088	48.4975	44.1200
13	53.1	9.0	49.5715	51.8967	47.2463
14	52.0	10.0	50.9675	53.3622	48.5728
15	52.5	11.0	49.8534	52.2629	47.4439
16	48.0	12.0	45.9398	48.3556	43.5240
17	42.8	13.0	39.1348	41.6033	36.6662
18	27.8	14.0	30.2465	33.0896	27.4034
19	21.9	15.0	21.0805	25.5889	16.5720

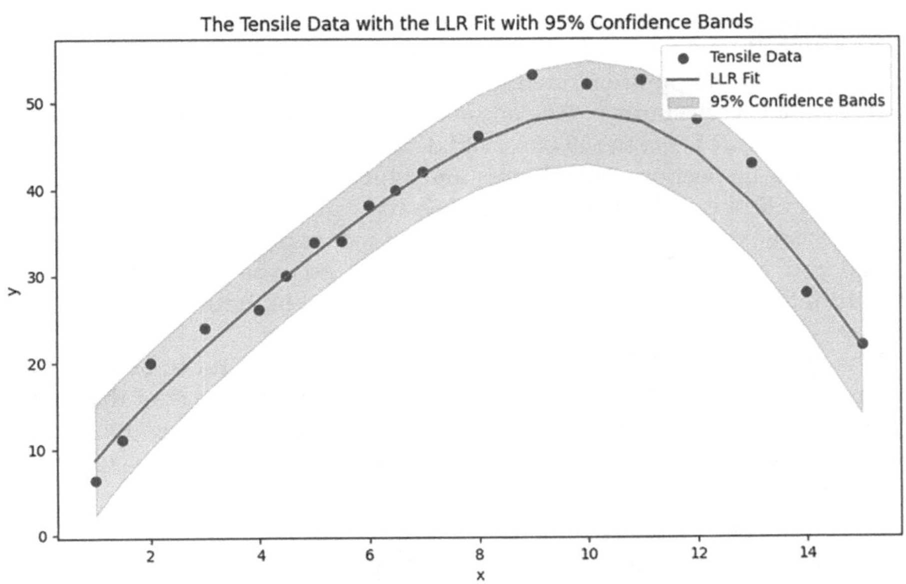

FIGURE 6.36
The tensile data with the LLR fit with 95% confidence bands.

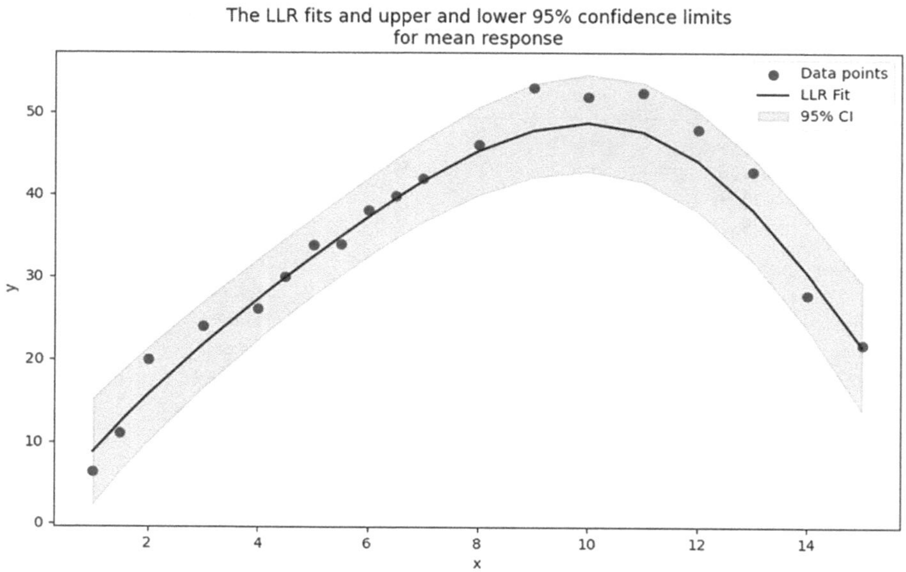

FIGURE 6.37
The LLR fits and 95% confidence band for the mean response.

Note that the fit is very good throughout the data range, with an excellent fit at the boundaries and much improvement over the kernel fit.

The LLR fits for the tensile data, and upper and lower 95% confidence limits for mean response are plotted in Figure 6.37.

This plot is very similar to the one above, where straight lines were used to connect the LLR fits at the 19 observed data points.

A careful inspection of the kernel and LLR fits to the tensile data reveals that the LLR fit is indeed better in the left and right boundaries, as expected. As the curves progress from small to larger values of x, they become nearly equal until around $x = 8$. There, the kernel fits tend to be larger than the LLR fits until the right boundary. Figure 6.38 shows that both the kernel and LLR fits are overlaid on the same graph.

The approximate 95% confidence intervals for LLR tend to be narrower than those for kernel regression throughout the entire range of the curve. Empirical studies by Mays (1995) and Mays and Birch (1996), among others, have shown that, as a general rule, LLR fits are preferred to kernel regression fits.

6.4.3 Smoothing the Bacteria Count Data Using R

The R program offers several built-in smoothing functions, including smoothing splines, kernel regression, LOESS, and the super smoother. Here,

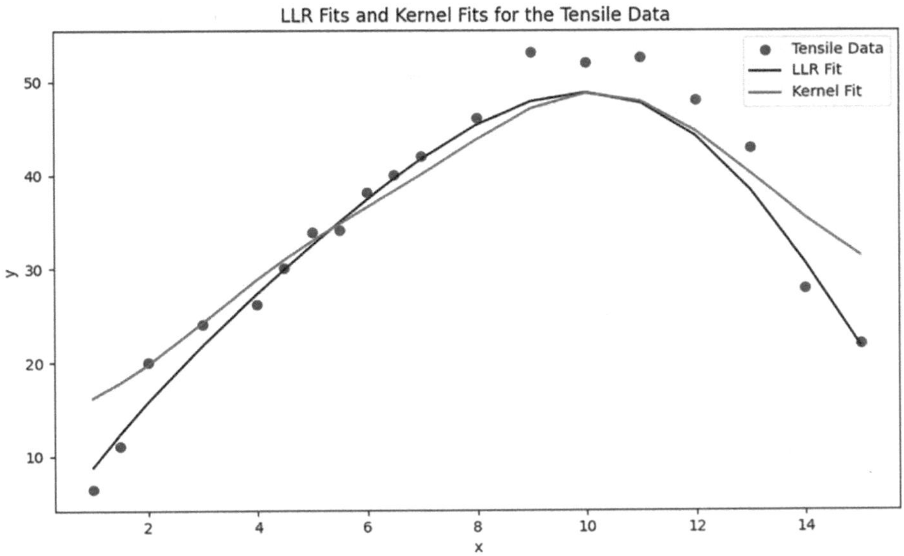

FIGURE 6.38
The kernel and LLR fits for the tensile data.

we will demonstrate the use of kernel smoothing and the super smoother function on the bacterial count data.

Environmental agencies often monitor the bacteria levels in lakes or streams to ensure water quality. In June 1989, management practices were altered to reduce bacteria levels in a specific stream in Virginia. This study aims to determine if the new management practices effectively reduced bacteria levels.

The following R commands produce a scatterplot of lnfc versus time, with both kernel smoothing (using a bandwidth of 85) and the super smoother lines overlaid. The data is provided.

It is interesting to determine whether or not the change in management practices successfully reduced bacteria levels in the stream. The data used for this study come from a station called station A. The variables of interest are lnfc, time, runoff, precipitation, and BMP; lnfc is the response variable referring to the log of the amount of bacteria in the water sample. Time is recorded in weeks after the initiation of the new management practice. The first sample was taken during week 1. The others are recorded as the number of weeks after the first week. Runoff is a measure of how well the stream flows. Precipitation is the amount of rain that has fallen within a certain time period before the sample was taken. Finally, BMP is a 0–1 variable, with 0 designating the time period for the old management practice and 1 representing the time period for the new management practice, going into effect at week 57. The data is contained on the dataset under the name "bacteria."

The following R script produces a scatterplot of lnfc versus week, with the kernel smooth (with a bandwidth of 85) and the super smoother lines overlaid on the plot. The data is contained in the data frame "fcount."

```
# Define kernel function
gaussian_kernel <- function(distances, bandwidth) {
exp(-0.5 * (distances / bandwidth)^2) / (bandwidth *
sqrt(2 * pi))}
# Perform kernel regression
kernel_regression <- function(x, y, bandwidth) {n
<- length(x)
  y_pred <- numeric(n)
  x_sorted_indices <- order(x)
  x_sorted <- x[x_sorted_indices]
  y_sorted <- y[x_sorted_indices]
    for (i in 1:n) { xi <- x_sorted[i]
      distances <- abs(x_sorted–xi)
      weights <- gaussian_kernel(distances, bandwidth)
      y_pred[i] <- sum(weights * y_sorted) / sum(weights) }
      return(list(x_sorted = x_sorted, y_pred = y_pred))}
# Super Smoother using LOWESS
super_smoother <- function(x, y, frac = 0.2) { smoothed
<- lowess(x, y, f = frac)
return(list(x = smoothed$x, y = smoothed$y))}
# Plotting function for the specified analysis
plot_smoothing <- function(x, y, y_kernel, y_super,
bandwidth) {
df <- data.frame(x = x, y = y, y_kernel = y_kernel,
y_super = y_super)
ggplot(df, aes(x = x)) + geom_point(aes(y = y), color =
'blue', label = 'Data points') + geom_line(aes(y = y_
kernel), color = 'black', linetype = 1, label =
sprintf('Kernel Smoother (bandwidth=%d)', bandwidth)) +
geom_line(aes(y = y_super), color = 'red', linetype = 2,
label = 'Super Smoother') + labs(x = 'Time', y = 'Log of
Fecal Coliform Concentration', title = 'Kernel and Super
Smoother for Bacteria Count Data') + theme_minimal() +
theme(legend.position = "best") }
# Example data (x and y should be your actual data)
x <- 1:100 # Replace with your actual data
y <- log1p(rnorm(100, 10, 5)) # Replace with your actual
data
# Set bandwidth
bandwidth <- 85
# Perform kernel regression
kernel_result <- kernel_regression(x, y, bandwidth)
```

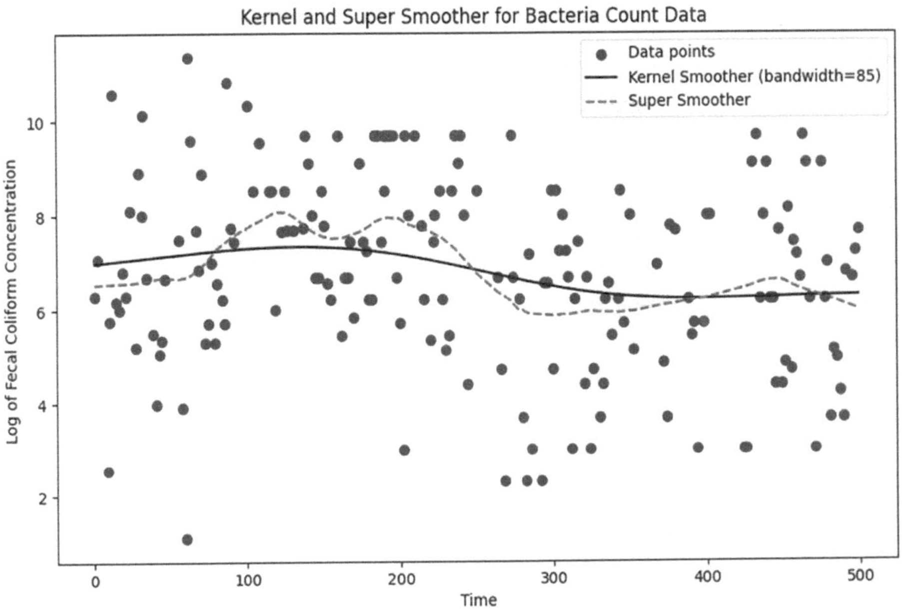

FIGURE 6.39
Kernel and smoothing the bacteria count data.

```
x_sorted <- kernel_result$x_sorted
y_pred_kernel <- kernel_result$y_pred
# Perform super smoother using LOWESS
super_result <- super_smoother(x, y)
x_super <- super_result$x
y_pred_super <- super_result$y
# Plot the results
plot_smoothing(x_sorted, y[order(x)], y_pred_kernel,
y_pred_super, bandwidth)
```

We see from Figure 6.39 that the two smoothers are nearly identical, both revealing the same up-down-up pattern in the data. The new management practice went into effect in week 57. We notice an actual increase in bacteria counts occurring about this time. In fact, the bacteria counts did not begin to drop until week 200, over two years after the new management system was put into place. Perhaps the new practice is ineffective, or other environmental conditions that influence bacteria count are changing. A multivariate analysis is required to evaluate this possibility. The nonparametric methods presented in this section can be extended to deal with more than one regressor, but we will not consider this situation here. The interested reader is referred to Fan and Gijbels (1996) for more on this topic.

6.5 Robust Nonparametric Regression

It should be clear from the previous section that the kernel and LLR regression techniques are not resistant to outliers. After all, the nonresistant least squares method is used to obtain the fit at any arbitrary data point x_0. Outliers in the response y can lead to distortion in the estimation of the mean response. These distortions tend to be local in nature, with their greatest impact in the neighborhood around the x value where the outlier occurred.

To illustrate this point, consider Figure 6.40, which contains a scatterplot of 25 data points generated by random sampling from the true mean function, a fourth-order polynomial $E(y|x) = f(x) = \frac{87}{18}x^4 - \frac{125}{18}x^3 + 2x^2$. The responses are $y_i = f(x_i) + \varepsilon_i$, with errors from a normal distribution with $\sigma^2 = 0.035$. The values for the regressor, x, are evenly spaced on the interval from 0 to 1. The response at x = 0.75 is actually −0.28 but was recorded by mistake as 0.28. An LLR fit (the dot-dash line) to the data, with a bandwidth of b = 0.15, is displayed in the figure along with the mean function (solid line) and a robust fit called "RLLR" (the dotted line).

It is readily observed that the LLR follows the mean function reasonably well except at the x locations from x = 0.5 to x = 1.0, where the single outlier exerts a distinct pull of the LLR fit away from the other data points. The

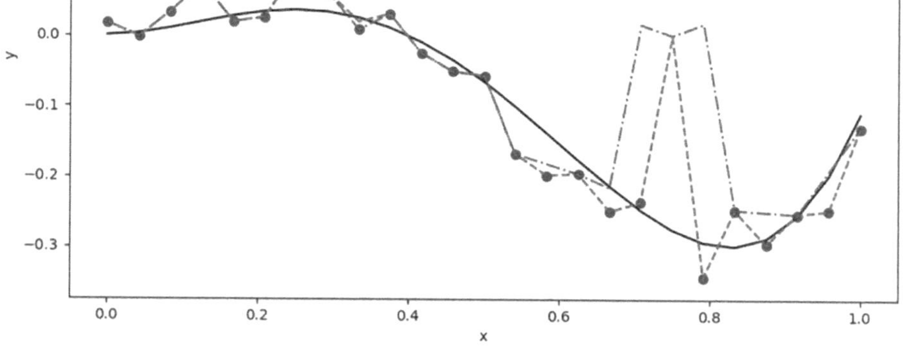

FIGURE 6.40
Illustrating the effects of an outlier on the LLR fit and the resistance of the RLLR fit.

RLLR fit follows the mean function very well throughout the entire curve and is resistant to the outlier. The RLLR method is the main topic of this section.

We have learned in Section 6.3 that the method of least squares can be made resistant to outliers through the use of WLS. The R program chooses the weights to down-weight the influence of outliers on the resulting intercept and slope estimates. The reader will recall that the weights are applied to the observed responses based on the rescaled residuals. LLR, on the other hand, can also be thought of as a WLS regression at each value of x_0 by weighing the observed regressor values through the kernel function. One way to "robustify" LLR is then to combine both of these weighting schemes in one, resulting in "robust local linear regression" (RLLR).

That is, using the notation of the previous section, the RLLR estimate of fit at any observed x location, x_i, may be written as

$$\hat{y}_i^{RLLR} = \hat{f}(x_i) = \hat{\beta}_{0i} + \hat{\beta}_{1i}x_i,$$

where $\hat{\beta}_{0i} = \overline{y}_{w_i} - \hat{\beta}_{1i}\overline{x}_{w_i}$ and $\hat{\beta}_{1i} = \dfrac{S_{xw_iy}}{S_{xw_ix}}$. The weights $w_{i1},\ldots,w_{in}$ used at each $x = x_i$ are simply the product of the "kernel weights," $h_{i1}^k,\ldots,h_{in}^k$, times the "M-estimate weights," $\dfrac{\Psi_{(r_i^*)}}{r_i^*},\ldots,\dfrac{\Psi_{(r_n^*)}}{r_n^*}$. That is, the n weights for RLLR at each $x = x_i$ may be written as

$$w_{ij} = h_{ij}^k \frac{\Psi_{(r_j^*)}}{r_j^*},$$

for $j = 1,2,\ldots,n$. Here, the Ψ-function could be chosen as either the Huber or bisquare functions given in Chapter 3. Fits at non-observed data values, $x = x_0$, are obtained using the same formulas as above, with the subscript "i" replaced by "0."

The algorithm to compute RLLR regression is outlined as follow.

1. For a fixed bandwidth value, b, obtain the LLR fits using WLS based on some appropriate kernel function. The weights $w_{i1},\ldots,w_{in}$ are the kernel weights $h_{i1}^k,\ldots,h_{in}^k$, for $i = 1,2,\ldots,n$. Obtain the n fits, $\hat{y}_i$. These are just the LLR fits as $\hat{y}_i = \hat{y}_i^{LLR}$ for all i.
2. Obtain the n residuals $r_1, r_2,\ldots,r_n$ as $r_i = y_i - \hat{y}_i$. Compute the rescaled mad, $\hat{s}$, and obtain the rescaled residuals as $r_i^* = \dfrac{r_i}{\hat{s}}$.

3. Obtain the robust nonparametric fits using WLS based on the robust kernel weights $w_{i1},\ldots,w_{in}$, where $w_{ij} = h_{ij}^{k}\left(\frac{\Psi_{(r_j^*)}}{r_j^*}\right)$, $j = 1,2,\ldots,n$ for each value of $i = 1,2,\ldots,n$. Obtain the n robust fits $\hat{y}_i^{robust}$ using WLS for simple linear regression.
4. If $\hat{y}_i \approx \hat{y}_i^{robust}$ for all i, quit. Compute the value of the "criteria function" for this bandwidth value. If, on the other hand, substantial differences exist between the $\hat{y}_i$ and $\hat{y}_i^{robust}$, replace the $\hat{y}_i$ by the $\hat{y}_i^{robust}$, and return to step 2.
5. Search through a range of bandwidths, returning to steps 1 through 4 for each value, using some appropriate search algorithm such as a binary search routine, finding that value of b where the criteria function is minimized. The fits at this value of b are the RLLR fits.

In LLR, the criteria function we used to determine the data-driven bandwidth was PRESS*, the penalized PRESS statistic. Recall that the PRESS statistic is a sum of squares of the PRESS residuals, $y_i - \hat{y}_{i,-i}$. On the other hand, one possible criteria function for RLLR uses a robust version of PRESS* where the quadratic loss function, associated with the non-robust LLR, is replaced with the ρ-function, associated with the robust weights calculated in step 3 of the algorithm. The reader will recall that the Ψ-function is found as $\Psi(r) = \frac{d\rho(r)}{dr}$. So, for example, the ρ-function associated with the Huber Ψ-function is

$$\rho(r) = \begin{cases} \frac{1}{2}r^2 & for|r| \le C_H \\ C_H|r| - \frac{1}{2}C_H^2 & for|r| > C_H \end{cases}.$$

The motivation for using this robust PRESS* criteria is the same motivation for using M-estimation over least squares in the first place. We don't want outliers to adversely affect the choice of bandwidth, and the robust counterpart of the quadratic loss function is the ρ-function.

Another possible criteria function is the weighted mean squared error, defined as

$$wmse(b) = \frac{\sum_{i=1}^{n}(\delta_i(y_i - \hat{y}_i^{RLLR}(b))^2}{\sum_{i=1}^{n}\delta_i - \sum_{i=1}^{n}h_{ii}^{RLLR}}.$$

Here, the δ_i, $i = 1, \ldots, n$, are the robust weights, $\delta_i = \frac{\Psi_{(r_i^*)}}{r_i^*}$, and w_{ii} is defined in step 3 of the RLLR algorithm given above. In the denominator, the $\sum_{i=1}^{n}\delta_i$ represents the "effective sample size." This quantity is n if all δ_i equal one and decreases as observations are down-weighted due to their degree of "outlierness." The second term in the denominator, $\sum_{i=1}^{n}h_{ii}^{RLLR}$, represents the "model degrees of freedom," a number that approaches two (because the RLLR fit would be just a robust line obtained by M-estimation involving two parameters) as b gets larger and approaches n as b approaches zero. The h_{ii}^{RLLR} play a role similar to the leverage values (h_{ii}) in SLR and are computed as $h_{ii}^{RLLR} = \frac{w_{ii}}{\sum_{j=1}^{n}w_{ij}} + \frac{(x_i - \overline{x}_{w_i})^2 w_{ii}}{s_{xw_ix}}$, for $i = 1, 2, \ldots, n$.

As with kernel regression and LLR, the RLLR fit at any value of x, say $x = x_i$, can be expressed as a weighted average of the responses, $y_i, \ldots, y_n$. That is, after some algebra, it can be shown that

$$\hat{y}_0^{RLLR} = \hat{f}(x_0) = \hat{\beta}_{0i} + \hat{\beta}_{1i}x_0 = \sum_{j=1}^{n}h_{0j}^{RLLR}y_j,$$

where $h_{0j}^{RLLR} = \frac{w_{0j}}{\sum_{j=1}^{n}w_{0j}} + \frac{\left(x_0 - \overline{x}_{w_0}\right)w_{0j}\left(x_0 - \overline{x}_{w_0}\right)}{s_{xw_0x}}$. This result provides a simple expression for the variance of $\hat{y}_0^{RLLR}$, provided that the bandwidth is fixed, as $var\left(\hat{y}_0^{RLLR}\right) = \sigma^2\sum_{j=1}^{n}(h_{0j}^{RLLR})^2$. Analogous to the LLR case, an approximate $(1-\alpha)100\%$ confidence interval for the mean response at $x = x_0$ may be obtained as

$$\hat{y}_0^{RLLR} \pm t_{1-\frac{\alpha}{2}, df_{error}}\, se\left(\hat{y}_0^{RLLR}\right),$$

where $df_{error} = \sum_{j=1}^{n}\delta_j - \sum_{j=1}^{n}h_{0j}^{RLLR}$ and $se\left(\hat{y}_0^{RLLR}\right) = \sqrt{wmse}\sqrt{\sum_{j=1}^{n}(h_{0j}^{RLLR})^2}$, where $wmse$ is the minimum value of $wmse(b)$.

Thus, the RLLR technique is both iterative (to achieve robustness) and requires a search routine (over b). Details of the theoretical and computational properties of the RLLR procedure may be found in Assaid and Birch (1998).

To illustrate the RLLR method, consider the simulated data displayed at the beginning of this section. The interested reader will find the R code in

Appendix B. An LLR analysis of this data revealed that b = 0.15 minimizes the PRESS* statistic.

The LLR fit, plotted in the figure at the beginning of this section, results in a nonresistant fit at x locations around the outlier that occurs at x = 0.75. The LLR fit results in an estimated variance of 0.01194 or an estimated standard deviation of 0.1092, severely overestimating the true value of $\sigma = 0.035$. Of course, this inflation of variance is due to the outlier.

To obtain the RLLR analysis in R, one can see that the best bandwidth is 0.01. Some further searching shows that the best bandwidth is b = 0.081 with a wmse of 0.00179. We'll use this value in all RLLR calculations. The RLLR variance is: sigma2 = 0.00179206, and the error degree of freedom is: dferror = 13.0452. The RLLR results may be either printed as in Table 6.10 or plotted as in Figure 6.41.

The bold line above indicates the outlier at x = 0.75 and indicates that the RLLR fit ignores the outlier. A scatterplot of the data with the RLLR fit and the 95% confidence bands for the mean response superimposed are plotted in Figure 6.41.

We note that the RLLR fit is in excellent agreement to the true mean function, as seen in the first figure of this section, and that the 95% confidence bands appear to be well behaved and reasonable estimates of mean response. We note from the RLLR program output that the variance estimate is 0.001792

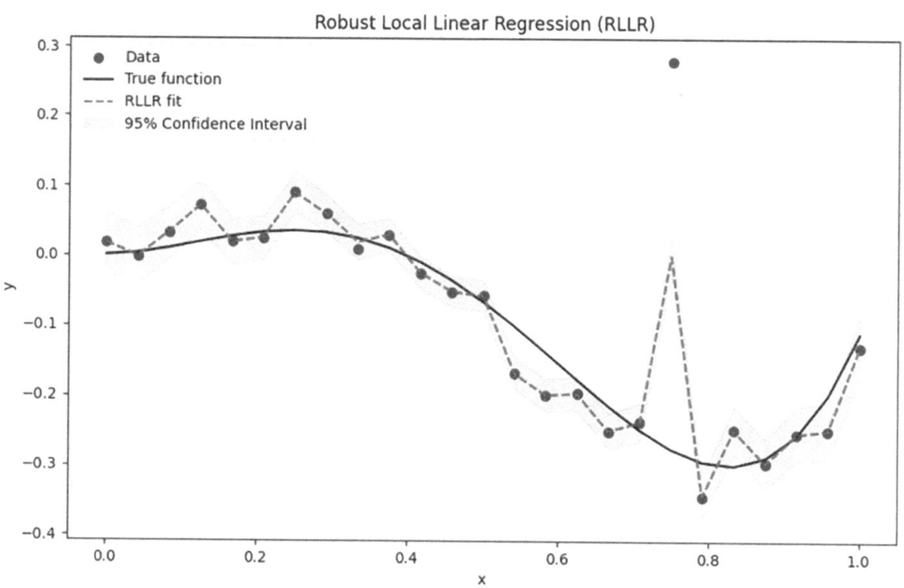

FIGURE 6.41
The RLLR fit with approximate 95% CI for mean response.

TABLE 6.10
The RLLR Results

Obs.	x	y	y-hat rllr	rlcl	rucl
1	−0.20	0.169953	0.166797	0.084791	0.248803
2	−0.15	0.089969	0.095792	0.043839	0.147745
3	−0.10	−0.009763	0.046222	0.002546	0.089897
4	−0.05	0.017323	0.016875	−0.028537	0.062287
5	0.00	−0.011822	0.009019	−0.036858	0.054895
6	0.05	0.010906	0.021245	−0.023615	0.066106
7	0.10	0.053513	0.043206	−0.001933	0.088345
8	0.15	0.063845	0.059670	0.015004	0.104337
9	0.20	0.099201	0.059038	0.015863	0.102213
10	0.25	0.037711	0.040063	−0.004608	0.084734
11	0.30	0.015440	0.010180	−0.034995	0.055355
12	0.35	−0.038935	−0.015165	−0.060017	0.029687
13	0.40	−0.036046	−0.026879	−0.071371	0.017613
14	0.45	−0.001718	−0.035413	−0.079225	0.008400
15	0.50	−0.063809	−0.053943	−0.098427	−0.009459
16	0.55	−0.064148	−0.083945	−0.128728	−0.039162
17	0.60	−0.126041	−0.122975	−0.168244	−0.077707
18	0.65	−0.168522	−0.163115	−0.211654	−0.114576
19	0.70	−0.212422	−0.191811	−0.241549	−0.142073
20	**0.75**	**0.280084**	**−0.217366**	**−0.264456**	**−0.170276**
21	0.80	−0.265186	−0.245738	−0.296898	−0.194578
22	0.85	−0.276653	−0.265194	−0.314032	−0.216355
23	0.90	−0.294506	−0.250949	−0.295437	−0.206462
24	0.95	−0.224276	−0.203659	−0.255075	−0.152242
25	1.00	−0.114371	−0.123315	−0.205274	−0.041357

or an estimate of standard deviation of 0.0438, much closer to the actual value $\sigma = 0.035$ than obtained by LLR.

Readers already acquainted with robust nonparametric regression will recognize the similarity of RLLR with the LOWESS method, introduced by Cleveland (1979). LOWESS stands for locally weighted scatterplot smoother. LOWESS works in much the same manner as the RLLR algorithm described above, with several exceptions. First, LOWESS uses the tricubed kernel function rather than the rescaled Gaussian kernel used in the previous section. The tricubed kernel function is defined as $K(u) = (1-u^3)^3$ for $-1 \le u \le 1$. Its shape is similar to the Gaussian kernel's, except it is identically zero outside the $[-1,1]$ interval.

Second, during the robust step of the algorithm, LOWESS uses the bisquare Ψ-function rather than the Huber.

Finally, rather than selecting a bandwidth, LOWESS achieves the scaling of the tricubed kernel function by always smoothing a fixed fraction, *f*, of the data, where $0 < f \leq 1$. Smoothers that work in this way are referred to as "nearest neighbor" smoothers because a fixed number of points, *r*, where $r = (f * n)$, nearest to the location where the fit is desired are assigned weights when estimating the fit at that location.

It is easily seen that LOWESS responds to *f* in the same way as RLLR responds to bandwidth *b*. As *f* is increased, the LOWESS fit becomes smoother, containing less detail than what exists in the data. As *f* approaches one in the limit, the LOWESS fit is tantamount to fitting a robust simple linear regression based on the bisquare function. As *f* gets smaller, the LOWESS becomes rougher, following more random observations in the data. Clearly, the proper value of *f* is crucial for LOWESS to effectively estimate the mean response function.

The R program has a built-in LOWESS function as a subcommand of the plot command. The essential arguments for LOWESS are the fraction *f*, referred to as the "degree of smoothing" in R, and the number of steps (*n* steps) of the IRLS algorithm. The default for *f* is 0.5, meaning that the nearest half of the data are used to obtain a fit at each data point.

The user is encouraged to alter *f* and monitor the resulting fit, adjusting *f* to obtain a fitted curve that appears to best represent the data. If the user chooses *n* steps equal to zero, then LOWESS is essentially LLR based on the tricubed kernel function. In this case, LOWESS would not be resistant to outliers. Otherwise, *n* steps represent the number of passes through steps 2 through 4 of the RLLR algorithm. It is recommended that *n* steps be set to at least two to allow for down-weighting of the outliers.

Before the LOWESS function is demonstrated in R, it should be pointed out that it does not minimize any criteria function and is not iterated to convergence when used as described by R. Assiad and Birch (2000) demonstrate that this is not the optimal way to use LOWESS.

Rather, R, when used in this fashion, provides the user with a valuable exploratory and diagnostic tool to describe the relationship between *y* and *x*.

LOWESS will be demonstrated using the simulated data illustrated above. To overlay the LOWESS fit with a scatterplot of the data as in Figure 6.42, we use the R code:

```
# Perform LOWESS fit
lowess_fit <- lowess(x, y, f = 0.2) # Adjust the smooth-
ing parameter f as needed
# Create a data frame for plotting
df <- data.frame( x = x, y = y, y_true = y_true,
lowess_x = lowess_fit$x, lowess_y = lowess_fit$y )
# Plotting LOWESS fit with scatterplot and true function
ggplot(df, aes(x = x)) + geom_point(aes(y = y), color =
'blue', label = 'Data') + geom_line(aes(y = y_true),
```

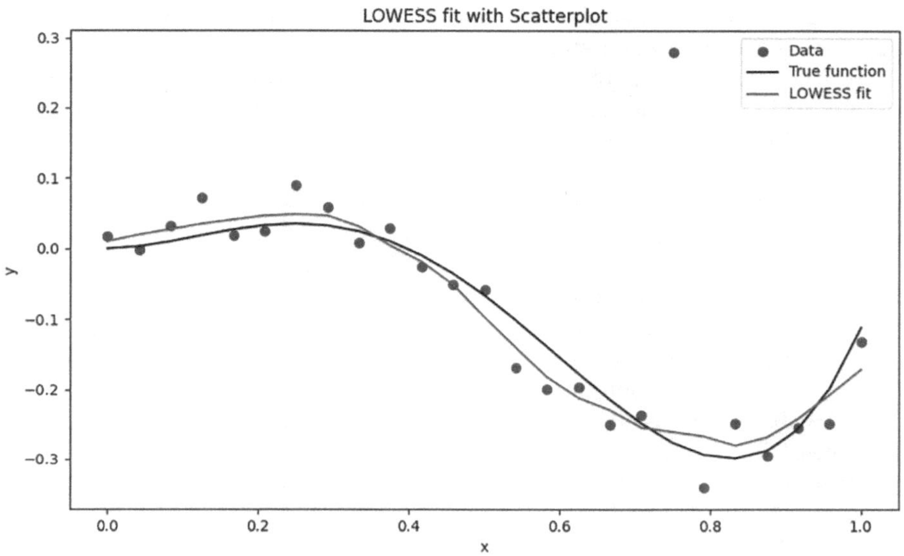

FIGURE 6.42
LOWESS fit to the simulated data.

```
color = 'black', linetype = '-', label = 'True func-
tion') + geom_line(aes(x = lowess_x, y = lowess_y),
color = 'green', linetype = '-', label = 'LOWESS fit') +
labs(title = 'LOWESS fit with Scatterplot', x = 'x', y =
'y') + theme_minimal() + theme(legend.position = "top")
```

We see the resulting LOWESS plot with $f = 0.3$.

```
# LOWESS fit
lowess_fit <- lowess(x, y, f = 0.3) # Adjust the smooth-
ing parameter f as needed
```

We obtain the following LOWESS fit. We see that the resulting LOWESS fit does a much better job of following the patterns expressed by the data while maintaining resistance to the outlier.

6.5.1 The Walleye Data Application

The large freshwater predatory walleye is considered by many anglers across North America as a highly desired species, mostly for its fine flesh quality. Many states engage in walleye stocking programs, usually at great monetary expense. Often, the survival of these stocked fish is very low, many not reaching a size to be of any value to anglers. The first winter after fall stocking is when a great deal of mortality occurs, seriously impacting the numbers reaching greater sizes.

The high mortality of small fish is attributed to insufficient energy reserves required to survive the winter period when food availability is nearly zero. The objective of one study (performed by Mike Cyterski, Department of Fisheries and Wildlife, Virginia Tech) was to determine if energy reserves in the fall period influence the overwinter survival of walleye fingerlings. In most animals, lipids (body fat) are the main form of energy storage, so the experimenter used lipid content of an individual fingerling as a surrogate measure of the amount of energy reserves.

One objective of the researcher was to examine whether commonly measured variables, such as the length and weight of an individual, could be used to predict lipid content.

The walleye data consists of eight variables measured on 54 individual walleyes. The variables are diet (fingerlings were fed one of four diets); season (measurements were made in the spring or the fall); weight and length; relative weight (wr) (computed as the weight relative to a length-specific standard for walleye); k (a measure of the health status of the fingerlings with higher values representing healthier fish); lipid % (percent lipid content (body fat) of fingerling). The walleye data is given on the book's website.

On R, built-in functions perform both LOWESS (LLR with the nearest neighbor smoothing using the tricubed kernel function) and LOWESS (RLLR with the nearest neighbor smoothing using the tricubed kernel function, just as described above for the R LOWESS function). We will demonstrate the LOWESS function to establish a relationship between the variables lipid and wr (the relative weight).

The R code is as follows:

```
# LOWESS fits with different smoothing fractions
fractions <- c(0.25, 0.40, 0.75)
styles <- c("solid", "dashed", "dotted")
# Create an empty data frame to store results
lowess_results <- data.frame()
# Perform LOWESS fits for each fraction and store the
results for (i in seq_along(fractions)) { lowess_fit
<- lowess(lipid, wr, f = fractions[i], iter = 3) df
<- data.frame( x = lowess_fit$x, y = lowess_fit$y, frac-
tion = fractions[i], style = styles[i] ) lowess_results
<- rbind(lowess_results, df) }
# Plot the results
ggplot() + geom_point(aes(x = lipid, y = wr), color =
'blue', label = 'Data') + geom_line(data = lowess_
results, aes(x = x, y = y, linetype = factor(fraction)),
color = 'black') + scale_linetype_manual(values =
styles) + labs(title = 'LOWESS fits with different
smoothing fractions', x = 'Lipid', y = 'WR') +
```

```
theme_minimal() + theme(legend.position = "top") +
guides(linetype = guide_legend(title = "Smoothing
Fraction"))
```

In the scatterplot of lipid versus wr, as shown in Figure 6.43, we see that lipid increases as wr increases, but at a decreasing rate. A SLR model would clearly be inappropriate.

Several outliers are also present in the data. We need to consider a robust local smoother to fit a nonparametric model that is robust to the outliers. Figure 6.43 shows the LOWESS smooth for three values of *f*, at 0.25, the wiggliest curve; at 0.40, the dotted curve, with some slight wiggles; and at 0.75, the very smooth dot-dash green curve. The curve with $f = 0.75$ appears to capture the general pattern in the data best. We see this curve in more detail in Figure 6.44.

```
# Best LOWESS fit
best_frac <- 0.75 best_lowess_fit <- lowess(lipid, wr,
f = best_frac, iter = 3)
# The result is a list with x and y components
best_lowess_fit
```

The decreasing nature of the relationship is clear, and it is apparent that the outlier in the lower right portion of the plot has no impact on the

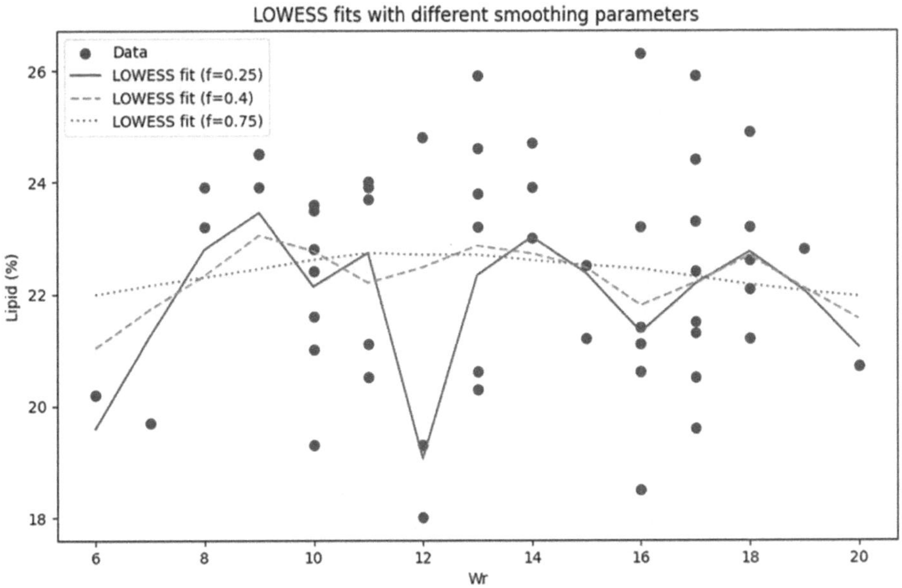

FIGURE 6.43
LOWESS fit with different smoothing parameters.

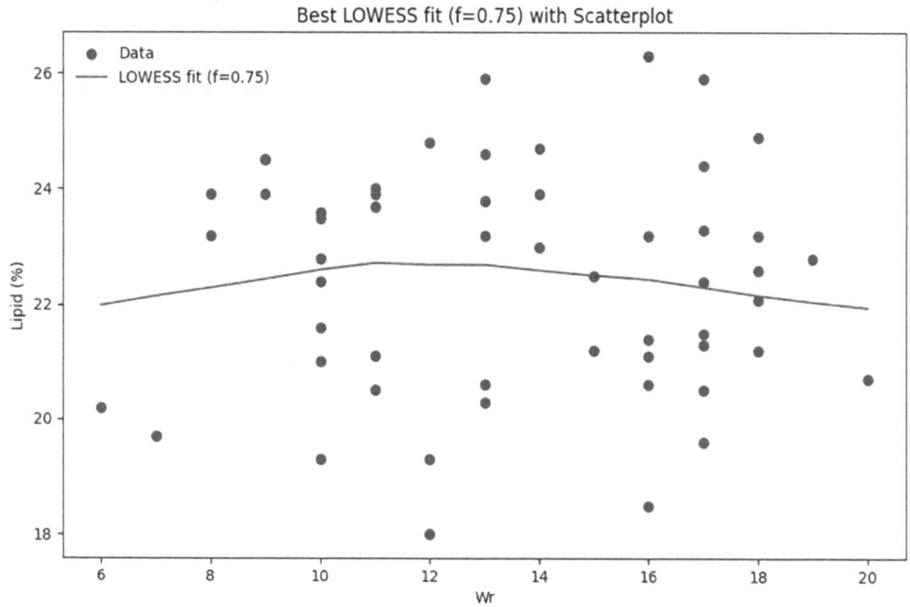

FIGURE 6.44
Best LOWESS fit with scatterplot.

LOWESS curve. From a biological viewpoint, the fact that the slope of the lipid % versus wr curve decreases as wr increases makes sense. After all, the body content of larger fish must consist of increasingly larger proportions of other tissue besides lipids, including bone, organs, and muscle.

Exercises

1) Consider the following hypothetical dataset (Table 6.11) involving the variables x and y.

   Use SPSS or R to do the following. Let y be the response variable.

   a) Obtain a scatterplot of this data.
   b) Obtain a complete LS analysis of the data. Comment on your results.
   c) Obtain the robust analysis using IRLS and the Huber psi function with a constant 1.5. Do not iterate the mad. Are outliers present in this dataset? If so, which observation? Compare the LS and robust analyses by comparing the fits, residuals, and weights. Use the potato leafhopper example in the text as a guide. Discuss your results.

TABLE 6.11

Hypothetical Data

Obs.	x	y
1	1.6	3.2
2	1.9	2.2
3	4.9	5.1
4	2.1	2.4
5	2.2	3.1
6	1.1	3.9
7	1.4	3.7
8	1.3	3.0
9	1.8	3.0
10	1.5	3.5

2) Consider the phosphorus data of Chapter 1. Let yield be the response and ferti be the regressor. Do the following analysis on your software package.

   a) Obtain a scatterplot of the data. Comment on your observations, especially those concerning the appropriateness of the SLR model and the constant variance assumption.

   b) Obtain a complete LS analysis of the data. Comment on your results.

   c) Can transformations of either the response and/or the regressor improve the SLR fit to this data? To find out, apply the Tukey method. If a transformation is suggested, repeat part b) of this question to the transformed data. Comment on your observations. Use the survival example of the text as a guide to your analysis.

   d) Repeat part c) of this question, except use the Box-Cox procedure instead of the Tukey method. What transformation on y is suggested by this method? Compare your results with those obtained in part c). What is your final model relating the response (or transformed response) to the regressor (or transformed regressor)?

3) Consider the Box-Cox method applied to the SLR model. Recall from Section 6.2 that the method only deals with transformations on the response variable, y. You are to generalize the Box-Cox method in SLR to consider power transformations on the y and x variables. That is, find λ_y and λ_x so that

   $$y^* = \beta_0 + \beta_1 x^* + \varepsilon,$$

   where $\varepsilon \sim N(0, \sigma^2)$.

a) Write the equation for $L(\lambda_y, \lambda_x)$, assuming fixed values for λ_y and λ_x.

b) Alter the Box-Cox programs to compute $L(\lambda_y, \lambda_x)$ for a series of pairs of values for λ_y and λ_x.

c) Apply your new program to the survival time data. Compare your results to the results obtained by the Tukey method.

4) For your data, consider your response variable and another variable, a single regressor. Please pick this single variable carefully; it should be linearly related to y. Otherwise, the next phase of your report will not be meaningful. Perform an SLR analysis of this data, mimicking the potato leafhopper example in your book. Your analysis should include:

a) The LS analysis, including appropriate plots.

b) The robust analysis includes both the Huber and bisquare estimates. You pick your own value of the tuning constants. Include appropriate plots (weight index plot, for example).

c) Discuss your results in terms of appropriateness of the LS and robust methods and any new insights into the research question gained by this analysis. Include all the parts a), b), and c) in your "report."

5) Consider the phosphorus data of Chapter 1. Let yield be the response and mehl-3p be the regressor. Do the following analysis on your software package.

a) Obtain a scatterplot of the data. Comment on your observations, especially those concerning the appropriateness of the SLR model and the constant variance assumption.

b) Perform kernel regression to this data. Your analysis should follow the text's tensile data example. Determine the proper bandwidth as explained in the text. Any insights gained by this analysis? Discuss.

c) Perform LLR regression to this data. Determine the proper bandwidth. Compare the LLR fit to the kernel fit. Which do you prefer and why? Any insights gained by this analysis? Discuss.

d) Perform robust smoothing to this data using RLLR and LOWESS (if you are using R) or using only LOWESS (if you are using some other software package). Compare your results to the LLR fit. Which do you prefer and why? Be sure you carefully choose your bandwidth (when using RLLR) and fraction f (when using LOWESS). Discuss your results.

6) Consider the LLR discussion in Section 6.4. Verify that h_{0j}^{LLR} may be written as

$$h_{0j}^{LLR} = h_{0j}^{k} + \frac{(x_0 - \bar{x}_{wo}) h_{0j}^{k} (x_0 - \bar{x}_{wo})}{s_{xw_0x}}$$

7) The following data represents the number of defective cars that came through the production line during the 30 eight-hour shifts for a typical two-week period at a certain plant in Detroit (the first shift lasts from midnight to 8 a.m.). The data is presented in order, starting from the first shift on Monday of the first week and continuing through the third shift on Friday of the second week (read across the row).

   9 7 6 8 6 1 5 5 6 7 5 6

   8 7 7 8 8 9 7 5 6 2 4 4

   6 5 4 7 5 6

   Do all the following analyses using your software package. Let y = number of defective cars, $x1$ = week classification (1 or 2), $x2$ = day classification (1 through 5), $x3$ = shift classification (1 through 3), and $x4$ = observation number (1 through 30).

   a) Obtain a scatterplot of y versus $x4$. Comment on your observations.
   b) Perform an LLR analysis of this data with y as the response and $x4$ as the regressor. Be sure to determine the proper bandwidth. Comment on the fit of this data.
   c) Perform a two-way factorial analysis with y as the response and $x2$ and $x3$ as the two factors. Summarize your results by describing the pattern that exists in this data and by the differences that exist between shifts and between days. What suggestions, if any, do you have for the manager of this plant?

8) Consider the bacteria data of this chapter. Let lnfc be the response variable and week be the regressor. Perform all the following analyses using your software package.

9) Consider the five (y, x) pairs of observations in Table 6.12.

   In all subsequent calculations, use the Epanechnikov kernel function defined as $K(u) = 0.75(1 - u^2)$ for $-1 \le u \le 1$, where $u = \dfrac{x_o - x_j}{b}$, the rescaled distance from x_0, the point where the fit is desired, to the

TABLE 6.12

The X and Y Data

Obs.	**1**	**2**	**3**	**4**	**5**
X	2	4	5	6	8
Y	2	4	8	6	5

data point x_j, for $j = 1, \ldots, n$. As usual, b is the bandwidth. Use $b = 2$ in all calculations.

a) Obtain a scatterplot of the data.

b) Find the kernel fit, $\hat{y}_5^K$, at $x = 5$ Do all calculations by hand.

c) Find the LLR fit, $\hat{y}_5^{LLR}$, at $x = 5$. Do all calculations by hand.

Multiple-Choice Questions

1. What does the simple linear regression (SLR) model assume about the relationship between the independent variable (x) and the dependent variable (y)?

   a) The relationship is exponential.
   b) The relationship is logarithmic.
   c) The relationship is linear.
   d) The relationship is quadratic.

   Answer: c)

2. In the SLR model, what does the term "linear" refer to?

   a) Linearity in the variables
   b) Linearity in the regression coefficients
   c) Linearity in the data points
   d) Linearity in the residuals

   Answer: b)

3. Which of the following is a common method used to estimate the coefficients in an SLR model?

   a) Maximum likelihood estimation
   b) Bayesian estimation
   c) Ordinary least squares
   d) Ridge regression

   Answer: c)

4. What is the primary objective of the ordinary least squares (OLS) method?

   a) To maximize the likelihood function
   b) To minimize the sum of squared residuals
   c) To minimize the sum of absolute residuals
   d) To maximize the correlation coefficient

   Answer: b)

5. Which of the following transformations might be considered if the scatterplot of y versus x reveals a strong nonlinear tendency?

   a) Box-Cox transformation

b) Logarithmic transformation
c) Exponential transformation
d) All of the above

Answer: d)

6. What is the purpose of applying the Box-Cox transformation in regression analysis?
   a) To reduce the number of predictors
   b) To stabilize variance and make the data more normally distributed
   c) To increase the correlation coefficient
   d) To reduce the number of observations

   Answer: b)

7. In the context of regression, what is an "outlier"?
   a) An observation that lies within the interquartile range
   b) An observation that deviates significantly from the other observations
   c) An observation that is the median value
   d) An obsservation that is equal to the mean

   Answer: b)

8. What is the effect of an outlier on the OLS estimates?
   a) Outliers have no effect on OLS estimates.
   b) Outliers can significantly skew OLS estimates.
   c) Outliers improve the accuracy of OLS estimates.
   d) Outliers reduce the variance of OLS estimates.

   Answer: b)

9. Which method can be used to mitigate the influence of outliers in regression analysis?
   a) Ordinary least squares
   b) Robust regression using Huber's T norm
   c) Maximum likelihood estimation
   d) Principal component analysis

   Answer: b)

10. In nonparametric regression, which of the following is a commonly used technique to fit the data?
    a) Ordinary least squares
    b) Kernel regression
    c) Ridge regression
    d) Lasso regression

    Answer: b)

Appendix A: R Code for the Tensile Application

```
# Load necessary libraries
library(ggplot2)
library(np)
library(stats)
# Tensile data
number <- 1:19
Y <- c(6.3, 11.1, 20, 24, 26.1, 30, 33.8, 34, 38.1,
39.9, 42, 46.1, 53.1, 52, 52.5, 48, 42.8, 27.8, 21.9)
X <- c(1, 1.5, 2, 3, 4, 4.5, 5, 5.5, 6, 6.5, 7, 8, 9,
10, 11, 12, 13, 14, 15)
# Convert to data frame
data <- data.frame(X = X, Y = Y)
# Candidate bandwidths
bandwidths <- c(1.0, 1.5, 2.0, 0.5, 3.0)
# Function to calculate PRESS statistic
calculate_press <- function(Y, y_pred) {
  sum((Y–y_pred) ^ 2)}
# Calculate PRESS for each bandwidth
press_values <- sapply(bandwidths, function(bw) {
  model <- npreg(Y ~ X, regtype = "ll", bw = bw)
  y_pred <- fitted(model)
  calculate_press(Y, y_pred)})
# Sort bandwidths and PRESS values
sorted_indices <- order(bandwidths)
bandwidths <- bandwidths[sorted_indices]
press_values <- press_values[sorted_indices]
# Best bandwidth (assumed given as 1.09 in original
code)
best_bandwidth <- 1.09
best_press <- min(press_values)
# Print the "best" bandwidth and PRESS statistic
cat(sprintf("Looks like b=%.2f is best. We'll use this
value in subsequent analyses.\n", best_bandwidth))
# Print the table of bandwidths and PRESS values
cat("\nRow bandw press*\n")
for (i in seq_along(bandwidths)) {cat(sprintf("%4d %.1f
%.4f\n", i, bandwidths[i], press_values[i]))}
# Plot PRESS versus bandwidths
ggplot(data.frame(bandwidths, press_values), aes(x =
bandwidths, y = press_values)) + geom_line() + geom_
point() + labs(title = "The Tensile Data–PRESS Versus
Bandwidths From Candidate List", x = "Bandwidth", y =
"PRESS*") + theme_minimal()
# Kernel regression with the best bandwidth
best_model <- npreg(Y ~ X, regtype = "ll", bw =
best_bandwidth)
x_fit <- seq(min(X), max(X), length.out = 100)
y_fit <- predict(best_model, newdata = data.frame(X =
x_fit))
```

```
# Calculate residuals
y_pred <- fitted(best_model)
residuals <- Y–y_pred
# Calculate kernel variance (sigma^2)
sigma2 <- var(residuals, ddof = 1)
# Calculate error degrees of freedom
n <- length(Y)
dferror <- n - 2 # Assuming 2 parameters for kernel
regression
# Print kernel variance and error degrees of freedom
cat(sprintf("The kernel variance is: sigma^2 = %.5f\n",
sigma2))
cat(sprintf("The error degrees of freedom is: dferror =
%.4f\n", dferror))
# Calculate 95% confidence intervals
y_std <- predict(best_model, newdata = data.frame(X =
x_fit), se.fit = TRUE)$se.fit
upper_bound <- y_fit + 1.96 * y_std
lower_bound <- y_fit - 1.96 * y_std
# Create data frame for fitted values and confidence
intervals
df_fit <- data.frame(X = x_fit, Y = y_fit, Upper =
upper_bound, Lower = lower_bound)
# Plot scatter plot with kernel regression fit and 95%
confidence intervals
ggplot() + geom_point(data = data, aes(x = X, y = Y),
color = 'blue') +
geom_line(data = df_fit, aes(x = X, y = Y), color =
'green') + geom_ribbon(data = df_fit, aes(x = X, ymin =
Lower, ymax = Upper), fill = 'green', alpha = 0.3) +
labs(title = 'The Tensile Data--Kernel Fit With Band-
width of 1.09 with Two Boundaries', x = 'X', y = 'Y') +
theme_minimal()
# Analysis note about the boundaries
cat('In the tensile data example, we note that the ker-
nel regression curve (using b=1.09) overestimates the
response at both the left and right boundaries.\n')
# Create figure for scatter plot with kernel regression
fit and the 95% confidence intervals
ggplot() + geom_point(data = data, aes(x = X, y = Y),
color = 'blue') +
geom_line(data = df_fit, aes(x = X, y = Y), color =
'green') +
geom_ribbon(data = df_fit, aes(x = X, ymin = Lower,
ymax = Upper), fill = 'green', alpha = 0.3) +
labs(title = 'Scatter Plot with Kernel Regression Fit
and 95% Confidence Intervals', x = 'X', y = 'Y') +
theme_minimal()
# Additional note about creating the required plots
```

```
cat('We create the required plots, including the scatter
plot with the kernel regression fit and the 95% confi-
dence intervals.\n')
# Quadratic fit for comparison
quadratic_model <- lm(Y ~ poly(X, 2))
y_quad_fit <- predict(quadratic_model, newdata = data.
frame(X = x_fit))
# Create data frame for quadratic fit
df_quad <- data.frame(X = x_fit, Y = y_quad_fit)
# Create figure comparing kernel fit and quadratic fit
ggplot() + geom_point(data = data, aes(x = X, y = Y),
color = 'blue') + geom_line(data = df_fit, aes(x = X,
y = Y), color = 'black') + geom_line(data = df_quad,
aes(x = X, y = Y), color = 'red', linetype = 'dashed') +
labs(title = 'The Kernel Fit is Excellent, Especially
When Compared to the Quadratic Fit', x = 'X', y = 'Y') +
theme_minimal()
```

Appendix B: R Code for the LLR and Kernel Regression

```
# Define the Gaussian kernel function
gaussian_kernel <- function(distances, bandwidth) {
exp(-0.5 * (distances / bandwidth)^2) / (bandwidth *
sqrt(2 * pi))}
# Perform local linear regression
local_linear_regression <- function(x, y, bandwidth) {
  n <- length(x)
  y_pred <- numeric(n)
  x_sorted_indices <- order(x)
  x_sorted <- x[x_sorted_indices]
  y_sorted <- y[x_sorted_indices]
    for (i in 1:n) { xi <- x_sorted[i]
    distances <- abs(x_sorted-xi)
    weights <- gaussian_kernel(distances, bandwidth)
    X <- cbind(1, x_sorted)
    W <- diag(weights)
    beta <- solve(t(X) %*% W %*% X) %*% t(X) %*% W %*%
    y_sorted
    y_pred[i] <- beta[1] + beta[2] * xi }
    list(x_sorted = x_sorted, y_pred = y_pred)}
# Perform kernel regression
kernel_regression <- function(x, y, bandwidth) {n
<- length(x)
  y_pred <- numeric(n)
  x_sorted_indices <- order(x)
  x_sorted <- x[x_sorted_indices]
  y_sorted <- y[x_sorted_indices]
  for (i in 1:n) { xi <- x_sorted[i]
    distances <- abs(x_sorted-xi)
    weights <- gaussian_kernel(distances, bandwidth)
    y_pred[i] <- sum(weights * y_sorted) / sum(weights) }
  list(x_sorted = x_sorted, y_pred = y_pred)}
# Compute confidence intervals
confidence_intervals <- function(x, y, y_pred, bandwidth,
alpha = 0.05) {n <- length(x)
  residuals <- y-y_pred
  se2 <- sum(residuals^2) / (n - 2)
  t_value <- qt(1 - alpha / 2, df = n - 2)
  distances <- as.matrix(dist(x))
  weights <- gaussian_kernel(distances, bandwidth)
  h <- rowSums(weights)
  se_fit <- sqrt(se2 * (1 / h))
  lower_bound <- y_pred-t_value * se_fit
  upper_bound <- y_pred + t_value * se_fit
  list(lower_bound = lower_bound, upper_bound =
  upper_bound)}
# Example data
x <- seq(0, 10, by = 0.1)
```

```
y <- sin(x) + rnorm(length(x), sd = 0.2)
# Set bandwidth
bandwidth <- 1
# Perform local linear regression
llr_result <- local_linear_regression(x, y, bandwidth)
x_sorted_llr <- llr_result$x_sorted
y_pred_llr <- llr_result$y_pred
# Perform kernel regression
kr_result <- kernel_regression(x, y, bandwidth)
x_sorted_kr <- kr_result$x_sorted
y_pred_kr <- kr_result$y_pred
# Compute confidence intervals
ci_result <- confidence_intervals(x_sorted_kr,
y[order(x)], y_pred_kr, bandwidth)
lower_bound <- ci_result$lower_bound
upper_bound <- ci_result$upper_bound
# Plot results
df <- data.frame(x = x_sorted_kr, y = y[order(x)], y_
pred_kr = y_pred_kr, lower_bound = lower_bound, upper_
bound = upper_bound)
ggplot(df, aes(x = x)) + geom_point(aes(y = y), color =
'blue') + geom_line(aes(y = y_pred_kr), color = 'green')
+ geom_ribbon(aes(ymin = lower_bound, ymax = upper_
bound), fill = 'green', alpha = 0.3) + labs(title =
'Kernel Regression with Confidence Intervals', x = 'X',
y = 'Y') + theme_minimal()
```

7

Multiple Linear Regression and the General Linear Model

Summary

This chapter will extend the results of the previous chapter to allow the regression model to deal with more than one regressor at a time, through the multiple linear regression (MLR) model. Even though it is assumed that the reader has already been exposed to multiple regression, we will present the basic material of this topic. Included are the estimation and testing of parameters, estimation of the response, either the mean or individual response, and some proper methods for determining the "best" model (commonly referred to as "model building"). Some useful data diagnostics will be presented and illustrated, such as the use of the "hat diagonals" and the "rstudent" statistic.

Various types of residuals will be discussed, as well as their role in model evaluation and interpretation. Section 7.2 deals with robust regression through the use of *m* estimators and bounded-influence estimators. Section 7.3 presents robust inference in the MLR model, and Section 7.4 gives applications of the robust methods to other examples of the general linear model. Each section will include a detailed illustration of the techniques using the SPSS and/or R program.

Learning Objectives

By the end of this chapter, readers should be able to:

- Extend understanding of regression to include multiple regressors using the MLR model.
- Learn parameter estimation, hypothesis testing, and model evaluation in MLR.
- Understand diagnostic tools like residual analysis, hat diagonals, and leverage for model assessment.
- Explore robust regression methods using *m*-estimators and bounded-influence estimators.
- Apply robust inference techniques to MLR and general linear models.
- Gain proficiency in statistical software (e.g., SPSS, R) for implementing MLR and robust regression techniques.
- Understand model-building strategies and criteria for selecting the "best" model.

DOI: 10.1201/9781003565321-7

- Recognize the impact of outliers and leverage points in regression analysis.

7.1 The Multiple Linear Regression Model

This overview of MLR assumes some familiarity with the material. In particular, the reader should be comfortable with the material on simple linear regression (SLR) in Section 6.1. The MLR model attempts to model one response variable, y, in terms of k regressor variables, $x_1,\ldots,x_k$. In the SLR model of Chapter 6, $k = 1$.

So, we see that MLR is an extension of the SLR model to handle more than one regressor variable. The MLR model is of the form:

$$y_i = \beta_0 + \beta_1 x_{1i} + \beta_2 x_{2i} + \ldots + \beta_k x_{ki} + \varepsilon_i,$$

for $i = 1, 2, \ldots n$ and where $\beta_0, \beta_1, \ldots, \beta_k$ are the regression coefficients or regression parameters. There are $k + 1 = p$ of these coefficients; they are unknown and must be estimated from the observed data. The ε_i in the MLR model represents the random error component of the model. These are conceptual terms only in that they cannot be observed. Consequently, assumptions must be made regarding them, as in the SLR model. On each of n experimental units, p variables have been measured, denoted by $(y_i, x_{1i}, \ldots, x_{ki})$ for the ith subject. Included among the k regressors may be functions of originally measured variables.

For example, in a chemical experiment, the response, y, is the amount of precipitate in the beaker after mixing together two chemicals, A and B. Perhaps x_1 is the amount of chemical A, x_2 is the amount of chemical B, x_3 is the amount of time these chemicals are mixed together, and x_4 is the temperature under which these chemicals are mixed.

It may be useful to the MLR model to consider the additional variables $x_5 = (\text{temperature})^2 = (x_3)^2, x_6 = (\text{time})^2 = (x_4)^2$, and $x_7 = (\text{time})$ (temperature) $= (x_3)(x_4)$. Variables x_1 through x_4 are the linear terms, x_5 and are the quadratic terms in the model, and x_7 is the interaction or cross-product term.

It is convenient to express the above MLR model in terms of matrices and vectors in the following way:

$$\begin{pmatrix} y_1 \\ y_2 \\ \vdots \\ y_n \end{pmatrix} = \begin{bmatrix} 1 & x_{11} & \cdots & x_{k1} \\ 1 & x_{12} & \cdots & x_{k2} \\ \vdots & \vdots & \cdots & \vdots \\ 1 & x_{1n} & \cdots & x_{kn} \end{bmatrix} \begin{pmatrix} \beta_0 \\ \beta_1 \\ \vdots \\ \beta_k \end{pmatrix} + \begin{pmatrix} \varepsilon_1 \\ \varepsilon_2 \\ \vdots \\ \varepsilon_n \end{pmatrix}$$

or

$$\underline{y} = X\underline{\beta} + \underline{\varepsilon},$$

where $\underline{y}$ is the $nx1$ vector of responses, X is the nxp model matrix consisting of a column of ones (for the intercept term) and the n observations of the k regressors, β is the $px1$ vector of unknown parameters, and $\underline{\varepsilon}$ is the $nx1$ vector of random errors. If we let $\underline{\beta}$ denote the vector of estimates of the parameters, then we can write the vector of estimated responses at the n observed data points as $\hat{\mathrm{y}} = \mathrm{X}\hat{\beta} = \left(\hat{\mathrm{y}}_1,\ldots,\hat{\mathrm{y}}_n\right)'$, and $\hat{y}_i$ represents the model's estimated response for the ith unit or subject.

The estimated response for an individual subject, say the ith subject, can be written as $\hat{y}_i = \underline{\acute{x}}_i\underline{\hat{\beta}}$, where $\acute{x} = (1x_{1i}\ x_{2i}\ x_{3i}\ldots\ \ldots\ x_{ki})'$ is the ith row of the X matrix. For estimation at an arbitrary new data point, say $\underline{x}_0{}' = (1 \quad \mathrm{x}_{10} \quad \mathrm{x}_{10} \quad \cdots \mathrm{x}_{k0})$, we have $\hat{\mathrm{y}}_0 = \underline{x}_0{}'\hat{\hat{\beta}}$.

7.1.1 Regression Coefficients: Estimation and Testing

As in the SLR model, we desire estimates of the coefficients so that the differences between the observed responses, $\underline{y}$, and the estimated responses, y, are "small." These differences are the residuals, collectively represented by the residual vector, $\underline{\underline{r}} = \underline{\mathrm{y}} - \hat{y}$. The size of $\underline{r}$ is often represented by its squared length or $\underline{\underline{r'}}\,\underline{\underline{r}} = \Sigma_{\mathrm{i=1}}^{\mathrm{n}}\mathrm{r}_{\mathrm{i}}^2$.

The method of least squares (LS) finds the estimates of coefficients so that $\Sigma_{i=1}^{n} r_i^2$ is minimized over all possible values of $\hat{\beta}$; the LS method is appropriate if $\mathrm{E}(\underline{\varepsilon}) = \underline{0}$ and $\mathrm{var}(\underline{\varepsilon}) = \sigma^2\mathrm{I}$, where I is a nxn identity matrix. These are the same assumptions made for the SLR model.

If we let $L = r'\underline{r}' = (\underline{y} - X\underline{\hat{\beta}})'(\underline{y} - X\underline{\hat{\beta}})'$, then we find the LS estimate, $\underline{\hat{\beta}}$, by calculating $\dfrac{\partial L}{\partial \hat{\beta}}$, setting this $p \times 1$ vector equal to $\underline{0}$ and solving for $\underline{\hat{\beta}}$.

We obtain

$$\frac{\partial L}{\partial \underline{\hat{\beta}}} = X'\left(\underline{y} - X\,\underline{\hat{\beta}}\,\right) \text{set} = \underline{0}$$

or

$$\mathrm{X'r} = 0,$$

the so-called normal equations. Solving for $\underline{\hat{\beta}}$, provided that the $pxp(X'X)$ matrix is of full rank p, resulting in the well-known least squares estimator,

$$\underline{\hat{\beta}} = (X'X)^{-1} X'\underline{y}.$$

It is interesting to note that the LS estimator not only results in the minimum value of the sum of squares of the residuals but also results in residuals that must satisfy the constraints imposed on them by the normal equations. That is, the latter expression of the normal equations implies that the residuals must satisfy the p constraints:

$$\sum_{i=1}^{n} r_i = 0$$

$$\sum_{i=1}^{n} x_{1i} r_i = 0$$

$$\vdots$$

$$\sum_{i=1}^{n} x_{ki} r_i = 0$$

Viewing the columns of the X matrix as p-dimensional vectors, we note that the inner product of each of these p-vectors with the residual vector is 0, indicating that each vector is orthogonal to the residual vector. Geometrically, the p columns of the X matrix "span" the estimation space, a p-dimensional subspace of n-space, and the residuals are in a space orthogonal to the estimation space. The residual vector is at a right angle to the estimation space. It can be shown that the LS estimator, $\underline{\beta}$, is an unbiased estimator of $\underline{\beta}$, the true parameters, and the variance of $\underline{\beta}$ is $\text{var}(\underline{\beta}) = \sigma^2 (X'X)^{-1}$, the pxp variance-covariance matrix of the LS estimator.

It is well known that if, in addition to the two assumptions listed above, for $\underline{\varepsilon}$, we also assume that $\underline{\varepsilon}$ are normally distributed (technically, $\underline{\varepsilon} \sim N(\underline{0}, \sigma^2 I)$), it follows that $\underline{\hat{\beta}} \sim N\left(\underline{\beta}, \sigma^2 (X'X)^{-1}\right)$. In this case, $\underline{\beta}$ is also the maximum likelihood estimator. The reader should be familiar with the ideas and notation used above. If not, it would be wise to consult a regression text, such as Myers (1990), for more details concerning the above topics and information on subsequent topics.

To help illustrate the ideas that have been presented so far, consider the SLR model, which has already been discussed in detail in Section 6.1. In the matrix notation, the model matrix X for the SLR model consists only of two columns, the column of 1's and the column of x data for the single regressor variable.

A worthwhile exercise for the interested reader is to work out the matrix calculations to show that the matrix formula for the computation of $\hat{\beta}$ results in the same formulas for $\hat{\beta}_0$ and $\hat{\beta}_1$ as given in Section 6.1. In addition, the distribution properties for the LS estimates of the intercept and slope for the SLR model given in Section 6.1 are simply those given above for $\underline{\beta}$. The reader will also recall partitioning the total sum of squares into two pieces, one due to unexplained variability and the other due to explained variability through the model.

This same identity holds for the MLR model as well. That is,

$$SS_{total} = SS_{error} + SS_{model},$$

where now SS_{model} refers to the amount of variability of the response variable accounted for by the MLR model. The degrees of freedom partition in an analogous manner, with $df_{total} = n-1 = df_{error} + df_{model} = (n-p)+(k)$. The coefficient of determination, R^2, is computed as $R^2 = \frac{SS_{model}}{SS_{total}}$, the same formula as in the SLR model, representing the fraction of total variability of y accounted for by the regressors through the MLR model. The estimate of the natural variability of the response variable is denoted by s^2 and is computed as $s^2 = \frac{SS_{error}}{df_{error}}$. The estimate of σ^2 is also referred to as the mean square error, or MS_{error}.

Of course, this is the same formula used in the SLR model, only now the SS_{error} and df_{error} are computed from the MLR model; s^2 or its square root, s, are used in all inferential procedures that involve $\underline{\hat{\beta}}$, as will soon be demonstrated.

A notation that we will find useful in this section is the "sum of squares due to regression" notation or "SSR" notation. In the SLR model containing the single variable x_2, for example, we could also write the model sum of squares due to x_2, SS_{model}, as $SSR(x_2)$.

In the MLR model with variables x_1 and x_2, we could write $SS_{model} = SSR(x_1, x_2)$, where it is now clear which two variables are in the model. If we consider building a two-variable MLR model by adding one variable at a time, we might be interested in the additional sum of squares regression due to adding the variable x_2 to a model containing the variable x_1. We would write $SSR(x_2 \mid x_1)$ to denote this additional sum of squares and compute this quantity as $SSR(x_2 \mid x_1) = SSR(x_1, x_2) - SSR(x_1)$. Where $SSR(x_2 \mid x_1)$ is referred to as the sequential sum of squares regression due to adding x_2 to the model already containing x_1.

Thus, we are comparing two models, one with only the variable x_1 and the other containing the two variables x_1 and x_2. If the variable x_2 has a significant impact on the model containing the variable x_1, then it should pick up a significant amount of regression sum of squares, and the quantity $SSR(x_2 \mid x_1)$ will be the measure of this amount. We note that $SSR(x_1, x_2) = SSR(x_1) + SSR(x_2 \mid x_1)$, a fact that is easy to demonstrate.

Of course, the SSR notation is completely general and can be used to compare any two models, models where one model is nested within the other. That is, one model (the "reduced" model) is contained in a second model (the "full" model).

For example, if we want to determine the contribution of three variables x_3, x_4, and x_5 to the model with two variables x_1 and x_2, we would obtain the SS_{model} for the two models, one containing only x_1 and x_2 (the reduced model) and one containing all five variables x_1, x_2, x_3, x_4, and x_5 (the full model). We see that the smaller model is nested within the larger model. The quantity represents the contribution of the three variables to the model containing the two variables. If this quantity is small, then the conclusion would be that these three variables collectively do not add much to the regression model containing the two variables.

On the other hand, a large quantity implies that these three variables will improve the reduced model, and the full model should be considered an improvement over the reduced one. In a more generic notation, we can rewrite the above SSR expression involving x_3, x_4, and x_5. As SSR("tested"|"reduced"), where "tested" refers to those variables that are being tested to determine whether they make a significant contribution in the presence of those variables in the "reduced" model. In our example, x_3, x_4, and x_5 are the variables being tested, and x_1 and x_2 are the variables in the reduced model. The significance of the contribution of the tested variables in the presence of the variables in the reduced model to the regression sum of squares can be made using the general F-test.

Thus, to test the hypotheses, H_0: The tested variables should be removed from the full model versus H_1: At least one of the tested variables should not be removed from the full model, we use the following general F-test, which can be written in the following equivalent forms:

$$F = \frac{\dfrac{SSR(tested \mid reduced)}{d}}{S^2_{full}} = \frac{\dfrac{SSR(full) - SSR(reduced)}{d}}{S^2_{full}} = \frac{\dfrac{SS_{error,\,reduced} - SS_{error,\,full}}{d}}{S^2_{full}},$$

where d is the number of variables being tested. Under H_0 being true, the above F-statistic follows an F-distribution with d and $df_{error,\,full}$ degrees of freedom for the numerator and denominator, respectively. Using the general F-test, any number of variables may be tested. If $d = 1$, then this F-test is called the partial F-test. The partial F-test is useful when individually evaluating each *k*-variable model. Then each *k* partial F-test measures the impact of the single tested variable in the presence of the other $k-1$ variables. Suppose the partial F is significant for the *j*th variable, for example. In that case, it indicates that the *j*th regressor variable is a useful regressor to a model already containing the other $k - 1$ regressors. Conversely, if the partial *F* is insignificant for the *j*th variable, it indicates that the *j*th regressor is not needed in the model already containing the other $k-1$ regressors. If $d = k$, the total number of regressors, the general F-test is called the "overall" or "model" F-test and is used to test the overall significance of the *k* regressor variables.

One certainly hopes that the null hypothesis that all *k* regressor coefficients equal zero is rejected. Otherwise, the *k*-variable regression model appears to lack any explanatory capability. Tests on individual variables may also be made using the familiar *t*-statistic. That is, to test the hypotheses

$$H_0 : \beta_j = 0 \text{ versus } H_1 : \beta_j \neq 0$$

in the presence of the other $k-1$ variables, the proper test statistic is

$$t_j = \frac{\hat{\beta}_j}{\text{se}\left(\hat{\beta}_j\right)}, \text{ for } j = 0, 1, \ldots, k,$$

where the *j* indicates that there is a *t*-statistic for each of the *p* parameters in the MLR model and where $\text{se}\left(\hat{\beta}_j\right)$ is the standard error of β_j and is computed as $\text{se}\left(\hat{\beta}_j\right) = s\sqrt{c_{jj}}$, where c_{jj} is the jth diagonal element of the $\left(X'X\right)^{-1}$ matrix.

Again, to further add to the extension of the SLR model to the MLR model, it is noted that the *t*-tests t_0 and t_1 for β_0 and β_1 in the SLR model obtained using the above formula are exactly those *t*-tests given in Section 6.1. As a further note, the *t*-statistic, t_j, for the *j*th regressor, is testing the same hypotheses as the partial F-test for β_j given earlier. In fact, it is well known that these two tests are equivalent tests in that one can show that $t_j^2 = F_j$, the partial F-test for the jth regressor, and that both tests have the same *p*-value, leading us to make the same decision regarding the usefulness of x_j in the presence of the other regressors.

7.1.2 Estimation of Response, Leverage, and the Hat Matrix

Using the matrix form of the MLR model, we see that the *n* estimated responses at the observed data points are expressed as the vector $\underline{y} = X\underline{\beta}$ and that an individual estimated response for the ith subject may be expressed as $y_i = \underline{x}_i{}'\beta$, where $\underline{x}_i{}'$ is the ith row of the *X* matrix. Estimation at any other setting of the regressor variables, say $\underline{x}_0{}' = \left(1 \quad x_{10} \quad x_{20} \quad \cdots \quad x_{k0}\right)$, is found by $y_0 = \underline{x}_0{}'\underline{\beta}$. Another useful representation of the estimated responses results by replacing $\hat{\beta} = (X'X)^{-1} X'\underline{y}$ in the above expression for $\underline{\hat{y}}$, obtaining $\underline{\hat{y}} = X(X'X)^{-1} X'\underline{y} = H\underline{y}$, where $H = X(X'X)^{-1} X'$, the so-called hat matrix due to its role in taking $\underline{\hat{y}}$ to $\underline{\underline{y}}$.

Readers with a linear model background will recognize H as a symmetric (meaning $H = H'$) and idempotent matrix (meaning $H = HH$). The elements of the nxn H matrix are of considerable interest in the study of modern regression methods.

We note that H can be written as

$$H = \begin{bmatrix} \underline{h}_1' \\ \underline{h}_2' \\ \vdots \\ \vdots \\ \underline{h}_n' \end{bmatrix},$$

where $\underline{h}'_i = (h_{i1}h_{i2}\ldots h_{in})$ represents the ith row of the matrix. With this notation, we see that $\hat{y}_i = \underline{h}'_i\underline{y} = \sum_{j=1}^{n} h_{ij} y_j$ can be written as a weighted sum of the observed responses with coefficients, the elements of the ith row of H. In particular, we note that the coefficient multiplying y_i is h_{ii}, the ith diagonal element of H. Thus, the role that y_i plays in the estimated response y_i is through the coefficient h_{ii}. These h_{ii} values also play a prominent role in other aspects of the regression analysis. It can be shown, see Myers (1990) for example, that we can compute h_{ii} as $h_{ii} = \underline{x}_i'(X'X)^{-1}\underline{x}_i$ and that these diagonal elements must satisfy the inequality $\frac{1}{n} \le h_{ii} < 1$ and the off-diagonal elements satisfy the inequality $-1 \le h_{ij} < 1$.

In addition, the n elements of any row of H sum to one. Thus, it is apparent that if h_{ii} is close to one, then from the above form for $\hat{y}_i$, it is clear that y_i plays a major role in the value of $\hat{y}_i$. It also can be shown that var $(\underline{\hat{y}}) = \sigma^2 H$, and, in particular, $\text{var}(\hat{y}_i) = \sigma^2 h_{ii}$, another occurrence of h_{ii}. Due to the restriction on the values taken by h_{ii}, it is clear that $\text{var}(\underline{\hat{y}}_i) < \text{var}(y_i) = \sigma^2$. That is, the estimated response at $\underline{x}_i'$ is more precise than the observed response there. This makes sense from an intuitive viewpoint. Each y_i is uncorrelated with the other responses. However, each $\hat{y}_i$ is estimated based on information contained in all n responses. Pooling this information together in obtaining $\hat{y}_i$ results in reduced variance when compared to y_i. The standard error of $\hat{y}_i$ is expressed as $se(\hat{y}_i) = s\sqrt{h_{ii}}$. If the normality assumption is reasonable, then a $(1-\alpha)\times 100\%$ confidence interval for the true mean response at $\underline{x}_i$ may be obtained as $\hat{y}_i \pm t_{1-\frac{\alpha}{2}, n-p}\, \text{se}(\hat{y}_i)$.

The h_{ii} can be used as a numerical diagnostic for the detection of "high-leverage points," those $\underline{x}_i$ values that are far removed from the center of the x data. The h_{ii}, often referred to as the "leverage values," can be thought of as a standardized distance measure, measuring the "distance" of $\underline{x}_i'$ from the center of the x data. Large values of h_{ii} indicate that $\underline{x}_i'$ is far from the center of the x data, $\overline{\underline{x}}' = (\overline{x}_1 \overline{x}_2 \ldots \overline{x}_k)'$, and is referred to as a high-leverage point. A typical rule of thumb, established through empirical studies, is that if $h_{ii} \geq 2\frac{p}{n}$, then $\underline{x}_i$ will be labeled as a high-leverage point.

High-leverage points are noteworthy in that they indicate a potential for the ith observation $(y_i, \underline{x}_i')$ to have a great influence on the regression analysis. Such points are called high-influence points. If an unusual response occurs at a high-leverage point, the joint impact of the high leverage of the regressors with the outlier in y can cause a significant distortion of all phases of the regression analysis. Coefficient estimates may be shifted either up or down, adversely affecting the fits at arbitrary x locations, resulting in inflated variances of coefficients and estimated responses and altering the values of test statistics. It is of great concern to the user of the MLR model to determine if the regression analysis has been so affected. And, if so, which points are the cause of these distortions?

Detection of high-influence points is usually accomplished through numerical diagnostics specifically designed to measure the impact of the point on some aspect of the regression analysis. The common philosophy in measuring the influence of the point $(y_i, \underline{x}_i')$ on the regression is to examine some aspect of the regression both with and without the ith point.

For example, the influence diagnostic, DFFITS_i, measures the standardized difference in the estimated responses (or fits) at $\underline{\mathbf{x}}_i'$ both with and without the point $(y_i, \underline{x}_i')$ in the dataset. In fact, the name DFFITS_i is an acronym for "difference in fits, standardized for the ith observation." That is, we compute

$$\text{DFFITS}_i = \frac{\hat{y}_i - \hat{y}_{i,-i}}{\text{se}(\hat{y}_i)} = \frac{\hat{y}_i - \hat{y}_{i,-i}}{s_{-i} \cdot \sqrt{h_{ii}}},$$

where the "-i" notation refers to the regression result obtained without the point $(y_i, \underline{x}_i')$. Thus, $\hat{y}_{i,-i}$ is the estimated response at $\underline{\underline{x}}_i'$ based on observations since $(y_i, \underline{x}_i')$ has been removed, and s_{-i} represents the square root of the mean squared error (MSE) obtained without the point.

The reader can see that if removal of the point $(y_i, \underline{x}_i')$ causes the regression analysis based on $n-1$ points to change much, that the numerator of DFFITS_i will be large. The denominator is used to scale the result so that the size of $DFFITS_i$ can be compared to some standard value, or "rule of thumb" value, applicable regardless of the units or scale of the original data.

For example, the rule of thumb for $DFFITS_i$ is to declare $(y_i, \underline{x}_i')$ as a possible high-influence point if $|DFFITS_i| > 2\sqrt{\frac{p}{n}}$. An often better way to determine high-influence points is to plot the diagnostic versus the observation number (or "index"). For example, plot $|DFFITS_i|$ versus *i*, the DFFITS index plot, and look for unusual behavior (look for "sore thumbs").

The reader, at this point, should be concerned that if a $DFFITS_i$, or some of the other influence diagnostics, is to be computed for each of the n observations in the dataset, that a great many regression analyses may need to be performed simply to obtain n values of these diagnostics.

For example, if $n = 100$, a single point removal would require 100 regressions, each with 99 observations have to be obtained in order to evaluate the 100 $DFFITS_i$. However, this is not the case because computational formulas exist that enable all influence diagnostics to be obtained with one regression model based on the full n observation dataset.

For this and many other issues relating to influential data diagnostics, the interested reader is referred to Ryan (1997) or Myers (1990).

7.1.3 Building MLR Models

One question of interest to most regression users is determining which variables from a collection of variables should be included in the model. It may be the case that the subject matter theory "suggests" the form of the MLR model. If this is the case, the user knows what model to evaluate. However, as is often the case, the user is unsure of the final form of the model and is trying to determine which MLR model "best" describes the data.

Of course, determining the best model must be based on appropriate numerical criteria. Possible criteria for selecting models are the following:

1. The R^2, a number that we want to be large.
2. The s^2, or MSE, a number that we want to be small.
3. The tests of significance on regression coefficients, the t_j (the *t*-test for the *j*th regressor) or F_{p_j} (the partial *F*-test for the jth regressor); we want to use significant variables in the model, so we want $|t_j|$ large, for all *j*, with small *p*-values.
4. Estimated coefficients should make sense in terms of their size and sign with respect to the subject matter. Signs opposite to what the subject matter theory suggests indicate that the model is misspecified.

These criteria should be evaluated across the competing models to determine which model is "best." Other criteria exist for comparing models, such as the PRESS statistic and the C_p statistic, but we will not discuss these options here.

TABLE 7.1

Breadwrapper Data

x1(F°) Sealing Temp	x2(F°) Cooling Temp.	x3 (weight in %) % Polyethylene	y (g / in) Strength
220	40	1.5	1,070
180	28	2.0	885
250	35	2.5	1,060
300	60	3.0	1,172
350	65	3.2	1,275
150	15	1.5	837
200	25	2.0	939
110	20	1.0	810
400	60	4.0	1,430

The interested reader is referred to Myers (1990) for further discussion on these and other selection criteria.

7.1.4 The Breadwrapper Example: Illustrating MLR on R

In the following example (Myers, 1976), it is desired to determine an empirical relationship between the seal strength (*s*) in grams per inch of a breadwrapper stock with the sealing temperature (*st*), the cooling bar temperature (*ct*), and the percent of polyethylene in the stock (*pe*). The combinations of the regressor variables were chosen in advance by the scientist. Nine experiment runs were made at the given settings with these settings, and the observed responses are given in Table 7.1.

Our analysis will consider the following four models: with *st*; with *st* and *ct*; with *st*, *ct*, and *pe*; and finally, with *st* and *pe*. We will use the R to highlight and annotate the output for each model. The following is the R session used to evaluate the four models.

```
> # Descriptive statistics, to get the descriptive statistics
  for each
> summary(data)
Sealing_Temp_F Cooling_Temp_F Polyethylene_Percent
  Strength_g_in
Min.   :110     Min.   :15.00   Min.   :1.0     Min.   : 810
1st Qu.:180     1st Qu.:25.00   1st Qu.:1.5     1st Qu.: 885
Median :220     Median :35.00   Median :2.0     Median :1060
Mean   :240     Mean   :38.67   Mean   :2.3     Mean   :1053
3rd Qu.:300     3rd Qu.:60.00   3rd Qu.:3.0     3rd Qu.:1172
Max.   :400     Max.   :65.00   Max.   :4.0     Max.   :1430
> # Scatterplot matrix
> ggpairs(data, columns = 1:4)
```

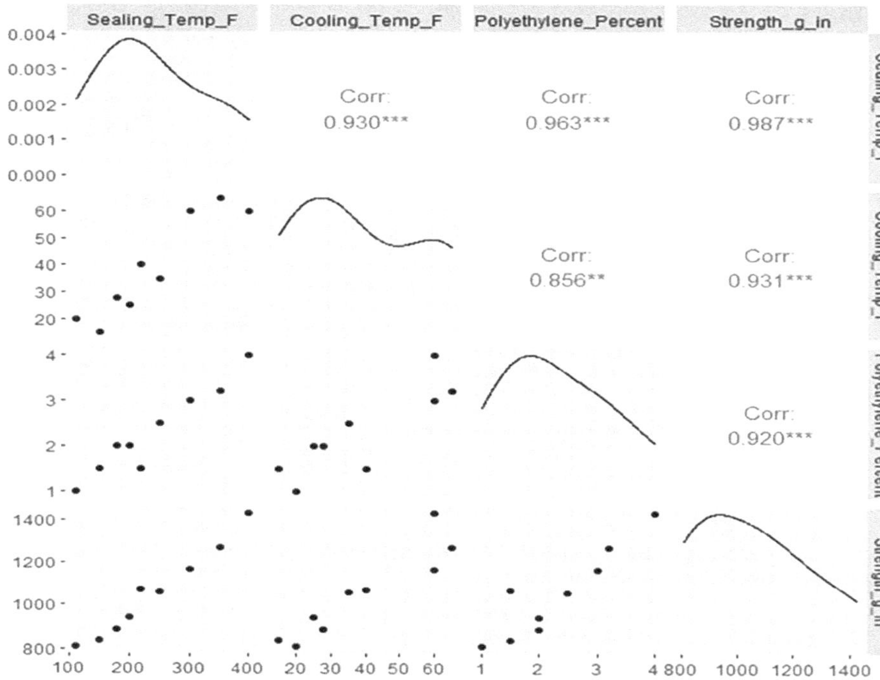

FIGURE 7.1
The scatterplot matrix for breadwrapper data.

The scatter plot matrix in Figure 7.1 shows that the plots of S versus ST, CT, and PE are seen in the bottom row of the graph. The plots of ST versus the other three variables are displayed in the fourth row, labeled ST, and so on.

The correlations between these variables as follow.

```
> # Correlation matrix
> cor(data)

                     Sealing_Temp_F Cooling_Temp_F Polyethylene_Percent
Sealing_Temp_F            1.0000000      0.9299615            0.9630364
Cooling_Temp_F            0.9299615      1.0000000            0.8555602
Polyethylene_Percent      0.9630364      0.8555602            1.0000000
Strength_g_in             0.9867798      0.9314304            0.9203235
                     Strength_g_in
Sealing_Temp_F           0.9867798
Cooling_Temp_F           0.9314304
Polyethylene_Percent     0.9203235
Strength_g_in            1.0000000
```

Notice the high correlations between the regressors and the response and between the regressors. Clearly, *st* is the best single variable in explaining *s*. However, other correlations are of interest. In particular, note that the correlation between *s* and *ct* is 0.93, indicating that *ct* would be a very useful variable in explaining *s*, if it were in the regression by itself.

Let's look at some different regression models. Try to determine which model is best. For each model, use the brief 3 output. This is the SLR using only *st*.

```
> # Simple linear regression: Strength ~ Sealing_Temp_F
> model1 <- lm(Strength_g_in ~ Sealing_Temp_F, data = data)
> summary(model1)
Call:
lm(formula = Strength_g_in ~ Sealing_Temp_F, data = data)
Residuals:
   Min     1Q Median    3Q   Max
-37.04 -19.51 -14.96 27.38 60.58
Coefficients:
               Estimate Std. Error t value Pr(>|t|)
(Intercept)    528.8444    34.7310   15.23 1.27e-06 ***
Sealing_Temp_F   2.1844     0.1356   16.11 8.64e-07 ***
---
Signif. codes: 0 '***' 0.001 '**' 0.01 '*' 0.05 '.' 0.1 ' ' 1
Residual standard error: 36.39 on 7 degrees of freedom
Multiple R-squared: 0.9737, Adjusted R-squared: 0.97
F-statistic: 259.5 on 1 and 7 DF, p-value: 8.635e-07
```

The regression equation is

$$S = 529 + 2.18\ ST$$

Obviously, *st* is an excellent variable in the model by itself, with a large R^2, and highly significant *t*- and F-statistics.

Now consider the *st*, *ct* model. Compare to the *st*-only model.

```
> # Multiple linear regression: Strength ~ Sealing_Temp_F +
  Cooling_Temp_F
> model2 <- lm(Strength_g_in ~ Sealing_Temp_F + Cooling_
  Temp_F, data = data)
> summary(model2)
Call:
lm(formula = Strength_g_in ~ Sealing_Temp_F + Cooling_Temp_F,
  data = data)
Residuals:
   Min     1Q Median    3Q   Max
-37.49 -23.86 -11.46 34.85 54.87
Coefficients:
               Estimate Std. Error t value Pr(>|t|)
(Intercept)    535.1755    38.0884  14.051 8.11e-06 ***
Sealing_Temp_F   1.9748     0.3876   5.095  0.00223 **
Cooling_Temp_F   1.1373     1.9558   0.582  0.58206
---
```

```
Signif. codes: 0 '***' 0.001 '**' 0.01 '*' 0.05 '.' 0.1 ' ' 1
Residual standard error: 38.24 on 6 degrees of freedom
Multiple R-squared: 0.9751, Adjusted R-squared: 0.9668
F-statistic: 117.7 on 2 and 6 DF, p-value: 1.537e-05
```

The regression equation is

$$S = 535 + 1.97\ ST + 1.14\ CT$$

Notice that the R^2 has increased by only a small amount over the value in the *st*-only model. Also, the MSE for this model has increased from 1,324 for the *st* model to 1,462 for the *st, ct* model. Obviously, adding *ct* has not helped to improve the *st*-only model. We further note that the *t*-statistic for the *st* variable is highly significant, with a *p*-value of 0.002, indicating that if *ct* is in the model, then the variable *st* makes a significant contribution to the model.

However, the *p*-value for *ct* is only 0.582, implying that *ct* is not making a worthwhile contribution to the model containing *st*. We see from the sequential sum of squares (SEQ SS) that if *st* has entered the model first, picking up 343,569 units of the sum of squares, when *ct* is added, it picks up only 494 additional units of the sum of squares. The small *t*-statistic for *ct* reflects this poor contribution. On the other hand, the estimation of the coefficient of *ct* (1.137) required the use of one degree of freedom, taken from the error degrees of freedom in the *st*-only model. Consequently, there is one less error df in the *st, ct* model and only 494 more sum of squares regression due to adding *ct*. The net result is that the MSE increases rather than decreases.

A visual comparison can be made between two models with at least one regressor in common by overlaying the two model fits on a scatterplot of the response versus one of the variables in common to both models. This plot has some special features if the models are in "full" and "reduced" form.

For example, if the model (ST, CT) is the full model and the model (ST) is the reduced model, we can compare the two models by plotting S versus ST with the two fits overlaid on the plot. We will call such a plot the "scatterfits" plot. The scatterfits plot for these two models is given in Figure 7.2.

A number of features are worth noting in Figure 7.2. First, true observations are labeled with their observation number (rather than a symbol such as "*") for ease of identification of those observations where the fit is exceptionally good or exceptionally poor. Second, both fits are displayed on the plot. The fits are connected with two different types of straight-line segments (solid and dashed lines, for example) to help distinguish the two fits. As a procedural issue, it is imperative that the entire dataset be sorted by the plotted regressor variable (ST in our example). The regressor is sorted, and all other variables, including the data labels, are carried along with the sort. If this operation is not performed, the plotted line segments may be a mass of wiggly lines, and it may be impossible to adequately compare them.

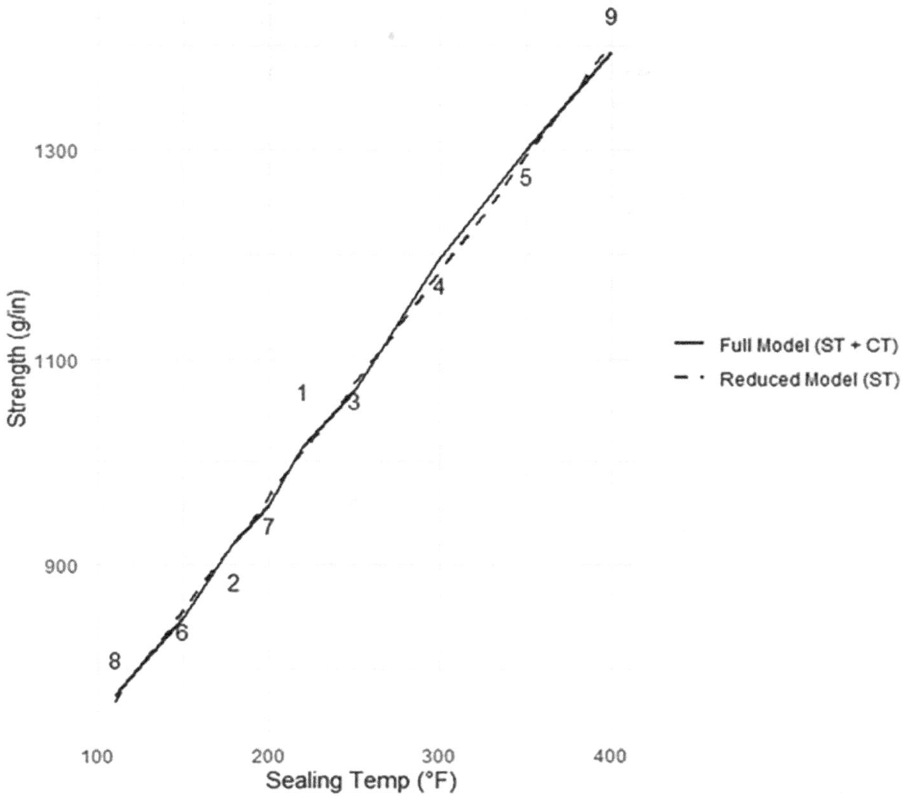

FIGURE 7.2
Strength versus sealing temp with full and reduced model fits.

The two fits can be compared by comparing the two lines. Obviously, in our example, there is very little difference between the two models since the two lines are nearly identical. Observations where the fit for either model is very good or very poor can now be easily identified by looking at the residuals, the vertical distances between the data labels, and the lines. In the example above, we see that observations 1, 8, and 9 do not fit well by either model.

If the models being compared are in full-reduced form and least squares is the estimation method, a numerical comparison of the two fits in the scatter-fits plot can easily be obtained by comparing either the fits or by comparing the residuals resulting from the two models.

The test statistic is simply the general F-test presented earlier in this section but represented in a slightly different form. In can be shown that the general F-test can be written as

$$F = \frac{\frac{SSR(\text{full}) - SSR(\text{reduced})}{d}}{s^2_{\text{ful}}} = \frac{\frac{SS_{\text{errorreduced}} - S_{\text{errorfull}}}{d}}{s^2_{\text{ful}}}$$

$$= \frac{\sum_{i=1}^{n}\left(\hat{y}_{i,full} - \hat{y}_{i,\text{re duced}}\right)^2}{d} + \frac{\sum_{i=1}^{n}\left(r_{i,\text{reduced}} - r_{i,full}\right)^2}{s^2_{\text{ful}}} = \frac{d}{s^2_{\text{ful}}}.$$

A statistical comparison between the models (ST, CT) and (ST) may be made by typing as follows:

```
> # Model comparison
> anova(model1, model2, model3, model4)

Analysis of Variance Table
Model 1: Strength_g_in ~ Sealing_Temp_F
Model 2: Strength_g_in ~ Sealing_Temp_F + Cooling_Temp_F
Model 3: Strength_g_in ~ Sealing_Temp_F + Polyethylene_Percent
Model 4: Strength_g_in ~ Sealing_Temp_F + Cooling_Temp_F +
  Polyethylene_Percent

  Res.Df    RSS Df Sum of Sq        F  Pr(>F)
1      7 9267.5
2      6 8773.0  1     494.5   0.5074  0.5081
3      6 4896.6  0    3876.5
4      5 4872.5  1      24.0   0.0247  0.8814
```

We note that this F-test has the same p-value as the t-test on CT, as the two tests are equivalent. Thus, we conclude that CT is not needed in the presence of ST.

For the following models, store the residuals in res1 and res2, and store the fits. Evaluate the *st*, *ct*, and *pe* model.

```
> # Model: Strength ~ Sealing_Temp_F + Cooling_Temp_F +
  Polyethylene_Percent
> model_full <- lm(Strength_g_in ~ Sealing_Temp_F + Cooling_
  Temp_F + Polyethylene_Percent, data = data)
> # Store residuals and fitted values
> data$RES1 <- residuals(model_full)
> data$FITS1 <- fitted.values(model_full)
> # Print summary of the model
> summary(model_full)
Call:
lm(formula = Strength_g_in ~ Sealing_Temp_F + Cooling_Temp_F +
Polyethylene_Percent, data = data)
Residuals:
     1       2      3     4       5       6       7      8      9
 5.504 -10.345 -6.902 1.369 -32.834 -14.220 -20.061 39.285 38.205
```

```
Coefficients:
                      Estimate Std. Error t value Pr(>|t|)
(Intercept)           523.5863    31.6296  16.554 1.47e-05 ***
Sealing_Temp_F          3.1447     0.6649   4.730   0.0052 **
Cooling_Temp_F         -0.2741     1.7456  -0.157   0.8814
Polyethylene_Percent -93.3054    46.6379  -2.001   0.1019
---
Signif. codes: 0 '***' 0.001 '**' 0.01 '*' 0.05 '.' 0.1 ' ' 1
Residual standard error: 31.22 on 5 degrees of freedom
Multiple R-squared: 0.9862, Adjusted R-squared: 0.9779
F-statistic: 119 on 3 and 5 DF, p-value: 4.543e-05
```

The regression equation is

$$S = 524 + 3.14\ ST - 0.27\ CT - 93.3\ PE$$

Some indicators for this model are impressive. The R^2 is large at 98.6, and the MSE has dropped to 975 due to the addition of *pe*. The *t*-statistic for *st* indicates that *st* is making a significant contribution to the model if *ct* and *pe* are already in the model. But if *st* and *pe* are in the model, the *t*-test for *ct* is highly insignificant, indicating it should be removed from the model. Surprisingly, *pe* is also an insignificant variable in the model containing *st* and *ct*, in spite of the fact that the addition of *pe* resulted in a major reduction of the MSE (from 1,462 to 975). It appears that *st* is a very strong variable in the model; *ct*, a weak variable in the model if *st* is already there; and *pe* is a moderately good variable. One is suspicious that if *ct* is removed from these three variables, the resulting model of *st* with *pe* will give us the "best" overall results.

The scatter fits plot comparing the models (ST) with (ST, CT, PE) is shown in Figure 7.3.

The model with (ST, CT, PE) does a much better job of fitting observations 1 and 2. However, this difference is not significant, as the general F (the reader may verify) gives a value of 2.25 and a *p*-value of 0.200. We will evaluate the ST, PE model. Is this a better model than (ST, CT, PE)?

```
> # Print summary of the reduced model
> summary(model_reduced)
Call:
lm(formula = Strength_g_in ~ Sealing_Temp_F + Polyethylene_
  Percent, data = data)
Coefficients:
                      Estimate Std. Error t value Pr(>|t|)
(Intercept)           525.2306    27.3128  19.230 1.28e-06 ***
Sealing_Temp_F          3.0653     0.3952   7.756 0.000242 ***
Polyethylene_Percent -90.3455    39.0382  -2.314 0.059916 .
---
Signif. codes: 0 '***' 0.001 '**' 0.01 '*' 0.05 '.' 0.1 ' ' 1
```

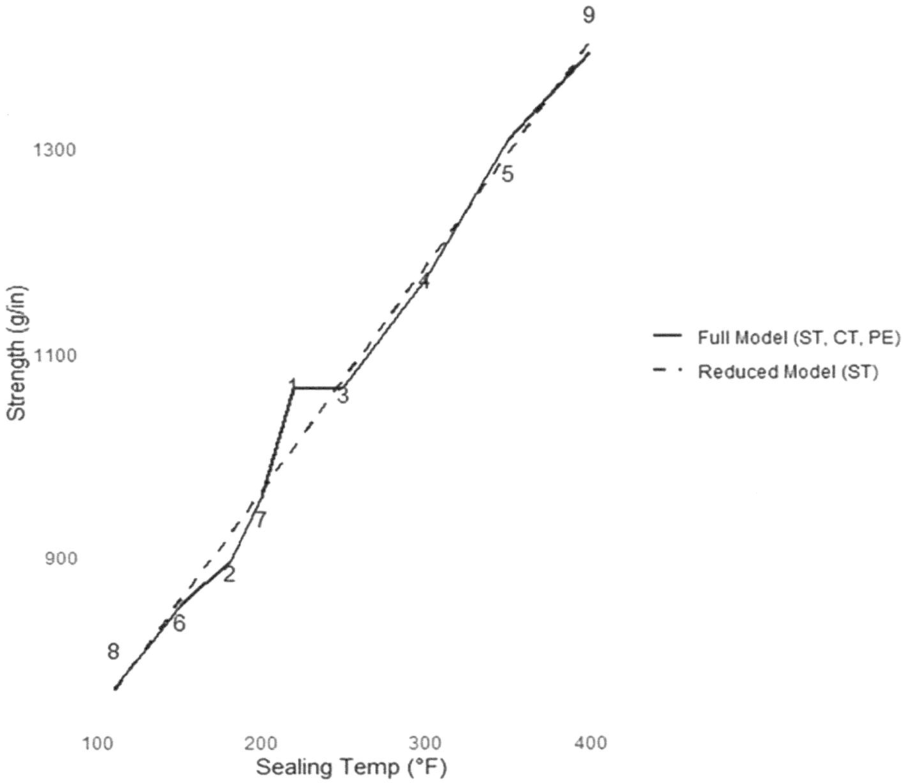

FIGURE 7.3
Strength versus sealing temp with full and reduced model fits.

```
Residual standard error: 28.57 on 6 degrees of freedom
Multiple R-squared: 0.9861, Adjusted R-squared: 0.9815
F-statistic: 213.2 on 2 and 6 DF, p-value: 2.673e-06
```

The regression equation is

$$S = 525 + 3.07\ \mathrm{ST} - 90.3\ \mathrm{PE}$$

We see that this *st*, *pe* model also has a 98.6 R^2 but has a smaller MSE of 816. We see that *pe* is significant only at the 0.06 level (instead of the usual 0.05) but that the addition of *pe* has greatly lowered the MSE (from 1,362 for the *st*-only model). The scatterfits plot comparing (ST, CT, PE) to (ST, PE) is shown in Figure 7.4.

We note that the two fits are nearly identical. It is clear the regressor CT is not needed in the (ST, PE) model.

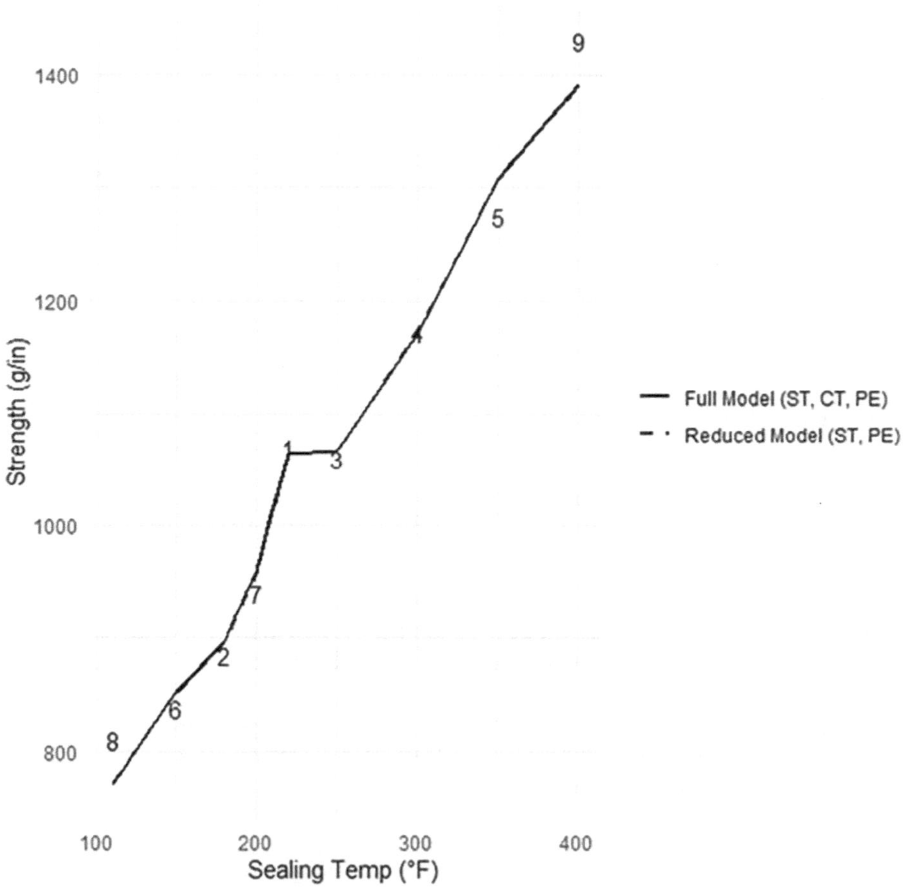

FIGURE 7.4
Strength versus sealing temp comparing full and reduced models.

Next, plot the residuals versus the fits for the (ST, CT, PE) and the (ST, PE) models.

```
# Residuals vs Fitted Values Plot for the Full Model (ST, CT, PE)
ggplot(data, aes(x = FITS1, y = RES1)) + geom_point(color =
  "black") +
  geom_hline(yintercept = 0, linetype = "dashed", color =
  "black") +
  labs( title = "Residuals vs Fitted Values for Full Model
  (ST, CT, PE)",
    x = "Fitted Values",
    y = "Residuals"
  ) + theme_minimal()
```

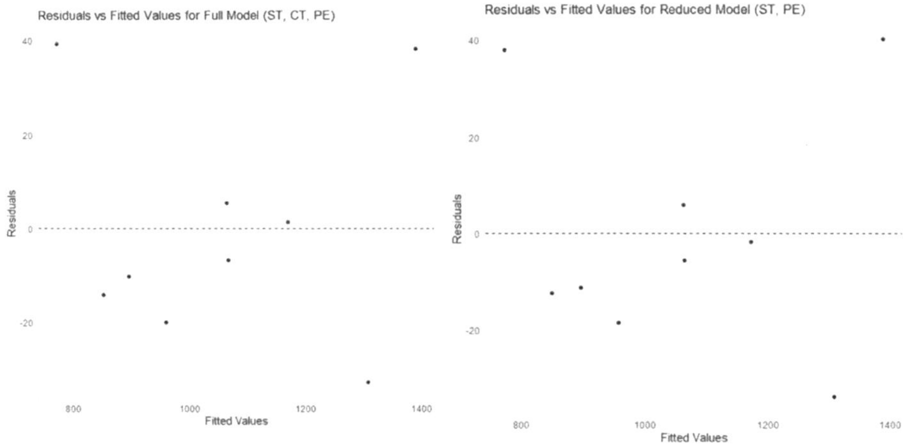

FIGURE 7.5
Residuals versus the fits for the (ST, CT, PE) and the (ST, PE) models.

Figure 7.5 shows that there is very little difference between these two plots. Apparently, little is given up by taking out the "CT" variable. Notice the similarities between the residuals. The shorter model is preferred over the longer one.

Our final model uses only ST and PE. Now that we have found which variables to use in the model, let's examine this model in more detail. We'll get the hii and dffits for this model.

```
> # Model: Strength ~ Sealing_Temp_F + Polyethylene_Percent
> model_selected <- lm(Strength_g_in ~ Sealing_Temp_F +
  Polyethylene_Percent, data = data)
> # Calculate leverage values (hii)
> data$hii <- hatvalues(model_selected)
> # Calculate DFFITS values
> data$dffits <- dffits(model_selected)
> # Summary of the selected model
> summary(model_selected)

Call:
lm(formula = Strength_g_in ~ Sealing_Temp_F +
  Polyethylene_Percent,
data = data)
Coefficients:
                      Estimate Std. Error t value Pr(>|t|)
(Intercept)           525.2306    27.3128  19.230 1.28e-06 ***
Sealing_Temp_F          3.0653     0.3952   7.756 0.000242 ***
Polyethylene_Percent  -90.3455    39.0382  -2.314 0.059916 .
---
```

```
Signif. codes: 0 '***' 0.001 '**' 0.01 '*' 0.05 '.' 0.1 ' ' 1
Residual standard error: 28.57 on 6 degrees of freedom
Multiple R-squared: 0.9861, Adjusted R-squared: 0.9815
F-statistic: 213.2 on 2 and 6 DF, p-value: 2.673e-06
```

The regression equation is

$$S = 525 + 3.07\ ST - 90.3\ PE$$

Figure 7.6 shows that the first observation is a high-leverage point. Let's plot the |DFFITSi| to check for high influence.

One can see from Figure 7.7 that observation 9 is certainly an unusual data point in that the change in fit is over 3 standard error units different when observation 9 is present than when it is absent. Referring back to the

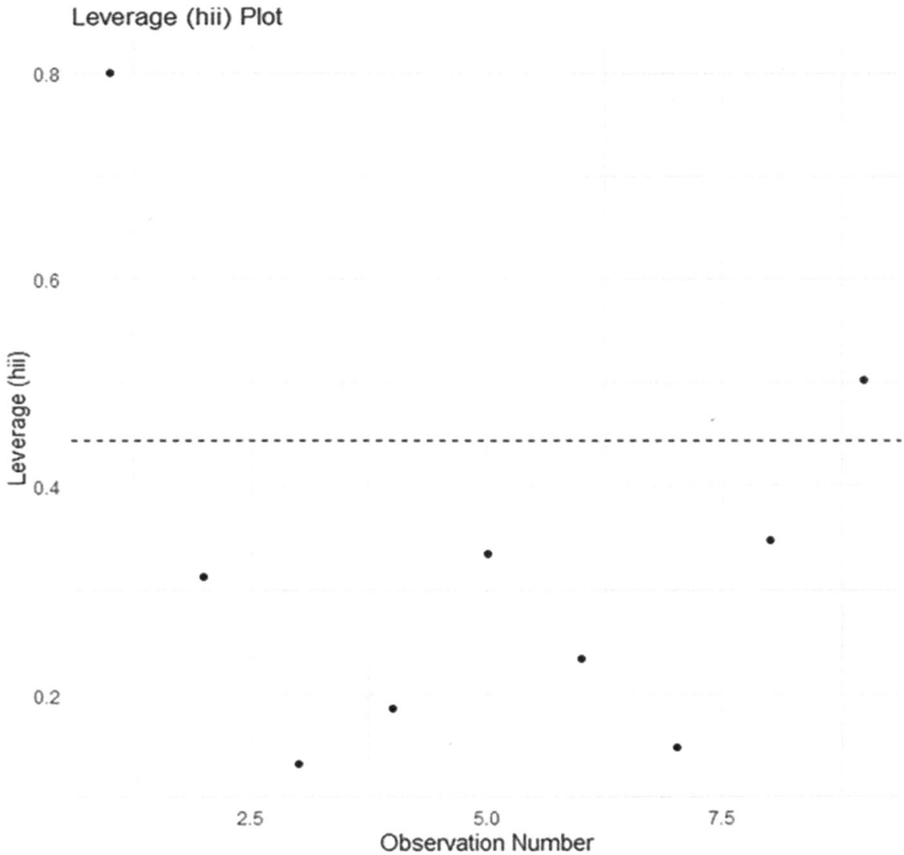

FIGURE 7.6
The leverage index plot for the ST, PE models.

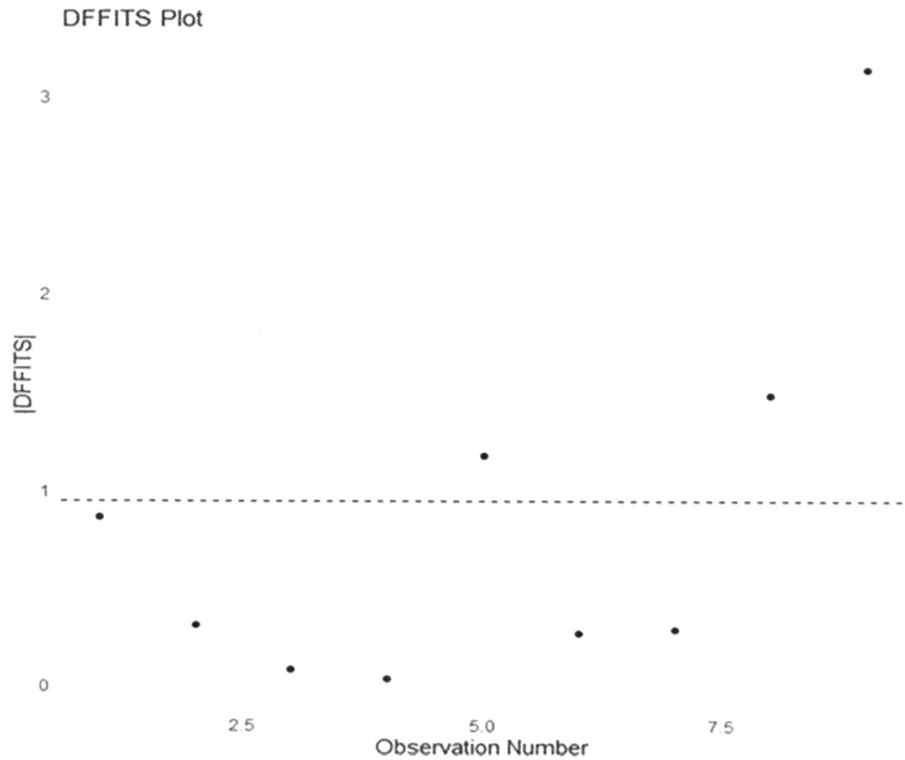

Figure 7.7

The |DFFITSi| plot for the ST, PE models.

scatterplots of *s* versus *st* and *s* versus *pe*, we see that observation 9 is indeed unusual in that both the *st* and *pe* values are very large compared to the other values of these variables, and the response at observation 9 must be much larger than predicted by the *st*, *pe* model. To see this, we note that from the scatterplot of the residuals versus the fits for the *st*, *pe* model, the residual for observation 9 is 40, so the *st*, *pe* model underestimates the response at observation 9 by 40 units. This fit is reduced by 3.2 standard error units when observation 9 is removed from the dataset.

7.2 Robust Multiple Regression

7.2.1 M-Estimation

Let us now extend the concepts of robust estimation, which have been developed so far for the location and SLR models, to the MLR model as well as the general linear model (GLM). The reader will recall that the GLM expresses the $nx1$ vector of responses, $\underline{y}$, as

$$y = X\underline{\beta} + \underline{\varepsilon},$$

where X is the nxp matrix of constants which may or may not be of full rank, β is the $px1$ vector of unknown parameters that must be estimated, and $\underline{\varepsilon}$ is the $nx1$ vector of random error terms. Our assumption is that $E(\underline{\varepsilon}) = \underline{0}$ and $\text{var}(\underline{\varepsilon}) = \sigma^2 I_n$, where I_n is the nxn identity matrix.

Of course, we recognize that the GLM can be used to represent the MLR situation and other forms of linear models, such as analysis of variance (ANOVA) and analysis of covariance (ANCOVA) models.

In Section 7.4, we will consider those ANOVA models that have been reparameterized into full-rank models by the use of indicator variables. The primary emphasis in this section will be on MLR models.

For a quick review, recall the M-estimation technique applied to the location model in Chapter 3. This was our first application of robust estimation applied to the most elementary form of the GLM. Our location model, written in the altered form, is

$$y_i = \theta + \varepsilon_i,$$

where $\varepsilon_i \sim f \subset F$, the family of symmetric, heavy-tailed distributions. Then is that value which minimizes

$$\sum \rho\left(\frac{y_i - \theta}{\hat{s}}\right)$$

or, equivalently, solves

$$\sum \psi\left(\frac{y_i - \theta}{\hat{s}}\right) = 0,$$

where $\psi(u) = \partial\rho(u)/\partial u$ and $\hat{s}$ is an estimate of the scale of the y-data, typically taken as the median absolute deviation, or mad. Recall that we find

$$\hat{\theta}_m = \frac{\sum w_i y_i}{\sum w_i},$$

where $w_i = \dfrac{\psi(r_i^*)}{r_i^*}, r_i^* = \dfrac{y_i - \hat{\theta}_m}{\hat{s}}$ by using the iteratively reweighted least squares (IRLS) algorithm.

It has previously been demonstrated that outliers also affect estimation and testing for the GLM. Accordingly, we will consider the following "altered glm" as $y = X\beta + \underline{\varepsilon}$, where $\underline{\varepsilon} \sim f \subset F$, the family of heavy-tailed distributions, with $\text{var}(\underline{\varepsilon}) = \sigma^2 I_n$, as in the usual GLM.

Let's consider the parallel development of the M-estimator with that of the LS estimator. For the LS estimator, $\hat{\beta}$ is chosen so that $\sum(y_i - \underline{x}_i\underline{\beta}')^2$ is minimized, where $\underline{x}_i'$ is the ith row of the X matrix. Taking the derivative of this expression with respect to $\underline{\beta}$ results in the $px1$ linear system of equations,

$$\sum \underline{x}_i\left(y_i - \underline{x}_i'\,\underline{\beta}\right) = \underline{0}$$

or, equivalently,

$$X'\left(\underline{y} - X\underline{\beta}\right) = 0.$$

The resulting least squares estimator is

$$\hat{\beta} = (X'X)^{-1}X'\underline{y}.$$

Robust regression based on M-estimators proceeds in a similar fashion as the LS estimator with the exception that the quadratic loss function is replaced by the ρ function. This function is exactly the same as that used in our discussion of M-estimation in the location and SLR models. Specifically, $\hat{\beta}_m$ is chosen to minimize $\sum\rho\left(\dfrac{y_i - \underline{x}_i'\underline{\beta}}{\hat{s}}\right)$, where $\hat{s}$ is an estimate of the scale of the y data. Taking derivatives with respect to $\underline{\beta}$ results in

$$\sum_{i=1}^{n}\underline{x}_i\psi\left(\frac{y_i - \underline{x}_i'\underline{\beta}}{\hat{s}}\right) = \underline{0}$$

or, equivalently,

$$X'\psi\left(\underline{r}^*\right) = \underline{0},$$

where $\underline{r}^*$ is the $nx1$ vector of rescaled residuals of the form $r_i^* = \left(y_i - \underline{x}_i\underline{\beta}\right)/\hat{s}, i = 1,\ldots,n$ and $\psi\left(\underline{r}^*\right) = \left[\psi\left(r_1^*\right)\ldots\psi\left(r_n^*\right)\right]$. This $px1$ system of equations is nonlinear and cannot be solved in closed form. We can solve them by a number of different numerical methods, including the Newton-Raphson (NR) technique or the IRLS method presented in Chapters 4 and 6. We will discuss only the IRLS method here.

For a thorough presentation of the NR and other methods, the reader is referred to Staudte and Sheather (1990). Motivation for the IRLS algorithm proceeds as in the case of the location and SLR models. We desire to solve

$$\sum_{i=1}^{n} \underline{x}_i \psi\left(\frac{y_i - \underline{x}_i' \underline{\beta}}{\hat{s}}\right) = \underline{0}$$

or, equivalently,

$$\sum_{i=1}^{n} \underline{x}_i \psi(r_i^*) \left(\frac{(y_i - \underline{x}_i' \underline{\beta})/\hat{s}}{(y_i - \underline{x}_i' \underline{\beta})/\hat{s}}\right) = \underline{0},$$

which implies

$$\sum \underline{x}_i w_i \left((y_i - \underline{x}_i' \underline{\beta})/\hat{s}\right) = \underline{0}'$$

the weighted normal equations. In matrix notation, we have

$$X'W(\underline{y} - X\underline{\beta}) = \underline{0},$$

where W is a nxn diagonal matrix with diagonal elements. The weighted solution is

$$\hat{\underline{B}}_{\mathrm{M}} = (X'WX)^{-1} X'Wy.$$

Of course, this is not the real solution since $r_i^* = (y_i - \underline{x}_i' \underline{\beta})/\hat{s}$ is a function of the unknown. In addition, we need an estimate of s, the scale of the response data. One easy way to handle the scale problem is to use $s = \mathrm{mad} = \mathrm{med}|r_i - \mathrm{med}(r_i)|/.6745$, the rescaled mad about the median.

The IRLS algorithm used to solve the weighted normal equations can be written as follows:

1) Compute $\hat{\beta}_0$; for example, let $\hat{\beta}_0 = \hat{\beta}_{LS}$, or one of the high breakdown bound estimators, such as $\underline{\beta}_{\mathrm{LMS}}, \hat{\beta}_{\mathrm{LTS}}$, or $\hat{\beta}_{\mathrm{LAV}}$.
2) Compute $r_i = y_i - \underline{x}'_i \hat{\beta}_0, i = 1, \ldots, n$ and $\hat{s} = \mathrm{mad} = \mathrm{med}|r_i - \mathrm{med}(r_i)|/.6745$. Compute $w_i = \psi(r_i^*)/r_i^*, i = 1, \ldots, n$, where $r_i^* = (y_i - \underline{x}_i \hat{\beta}_0)/\hat{s}$, for some appropriately chosen ψ function.
3) Compute $\underline{\beta}_1 = (X'WX)^{-1} X'Wy$.
4) If $\hat{\beta}_1 \approx \hat{\beta}_0$, quit; $\hat{\beta}_1$ is the M-estimator, $\hat{\beta}_M$. Otherwise, replace $\hat{\beta}_0$ by $\hat{\beta}_1$, and return to step 2.

One advantage of this IRLS algorithm over the NR algorithm is that it can be used by most existing statistical software. All one needs is a weighted LS algorithm with input of the proper weights, or one can use the usual *LS* computations, replacing $\underline{y}$ by $\underline{y}^* = W^{\frac{1}{2}}\underline{y}$ and X by $X^* = W^{\frac{1}{2}}X$. Then compute the LS estimator using $\underline{y}^*$ and X^*, resulting in

$$\underline{\beta}^*_{LS} = \left(X^{*\prime}X\right)^{-1}X^{*\prime}\underline{y}^* = \left(X'WX\right)^{-1}X'W\underline{y} = \hat{\beta}_M.$$

Of course, the procedure still must be iterated.

Starting values are very important in M-estimation for MLr. If severe outliers or high-influence points are present, then setting $\hat{\beta}_0 = \hat{\beta}_{LS}$ results in very poor initial guesses of the parameters. In this case, the IRLS algorithm may not converge or may converge to some undesirable solution. Estimators that are highly resistant to severe outliers are much preferred, such as the $\hat{\beta}_{LMS}, \hat{\beta}_{LTS}$, or $\hat{\beta}_{LAV}$.

The least median of squares (LMS) and least trimmed squares (LTS) estimator, $\hat{\beta}_{LMS}$, for the MLR model are presented in Section 7.3.

If $k = 1$, we have the SLR model, and the one obtained by the IRLS algorithm reduces to the weighted LS estimates given in Section 6.3. That is,

$$\hat{\beta}_{0w} = \bar{y}_w - \hat{\beta}_{1w}\bar{x}_w$$
$$\hat{\beta}_{1w} = \frac{S_{xwy}}{S_{xwx}},$$

where $\bar{y}_w = \frac{\sum w_i y_i}{\sum w_i}, \bar{x}_w = \frac{\sum w_i x_i}{\sum w_i}, S_{xwy} = \sum(x_i - \bar{x}_w)v_i(y_i - \bar{y}_w)$, and $S_{xwx} = \sum(x_i - \bar{x}_w)^2 w_i$. As in the SLR model, the weights are used as a diagnostic for outliers, where low weight implies that the point $(y_i, \underline{x}_i')$ is outside the pattern expressed by the remainder of the data. The robust method will now be illustrated with the stackloss data.

7.2.2 The Stackloss Example

Let's consider the stackloss data, first introduced by Brownlee (1960), and analyzed extensively by Daniel and Wood (1960). The data in Table 7.2 represents 21 successive days of operation of a plant oxidizing ammonia to nitric acid. Regressor x_1 is the flow of air to the plant, x_2 is the temperature of the cooling water entering the countercurrent nitric oxide absorption tower, and x_3 is the concentration of nitric acid in the absorbing liquid (coded for ease of hand calculation by subtracting 50 and then multiplying by 10). The response, *y*, is ten times the percentage of the ingoing ammonia that is lost as unabsorbed

TABLE 7.2

Data from the Operation of a Plant for the Oxidation of Ammonia to Nitric Acid

Obs. #	stkloss	airflow	temp	acidcon
1	42	80	27	89
2	37	80	27	88
3	37	75	25	90
4	28	62	24	87
5	18	62	22	87
6	18	62	23	87
7	19	62	24	93
8	20	62	24	93
9	15	58	23	87
10	14	58	18	80
11	14	58	18	89
12	13	58	17	88
13	11	58	18	82
14	12	58	19	93
15	8	50	18	89
16	7	50	18	86
17	8	50	19	72
18	8	50	19	79
19	9	50	20	80
20	15	56	20	82
21	15	70	20	91

nitric oxide and is an indirect measure of the yield of nitric acid. The data is available on the book's website. Careful analysis of data by Daniel and Wood (1960) and by Andrews (1974), in one of the earliest papers on robust regression, discovered that four observations should be considered outliers.

We will use robust regression to see if we can find these outliers. We will use the R code, the same used for the SLR model in Chapter 6, but now set up for the MLR model.

We'll first obtain scatterplots of stackloss versus each of the regressors.

From Figure 7.8, we see a strong linear trend with airflow, a strong bow-shaped trend with temperature, and a very weak trend is observed for acid concentration, with several distinct groupings of points. We note that in all the plots, the points labeled 1, 2, 3, and 4 are distinct from the other points.

The MLR model using all three regressors is as follows:

```
> # Fit the multiple linear regression model
> mlr_model <- lm(stkloss ~ airflow + temp + acidcon, data =
  stackloss_data)
> # Display the summary of the model
```

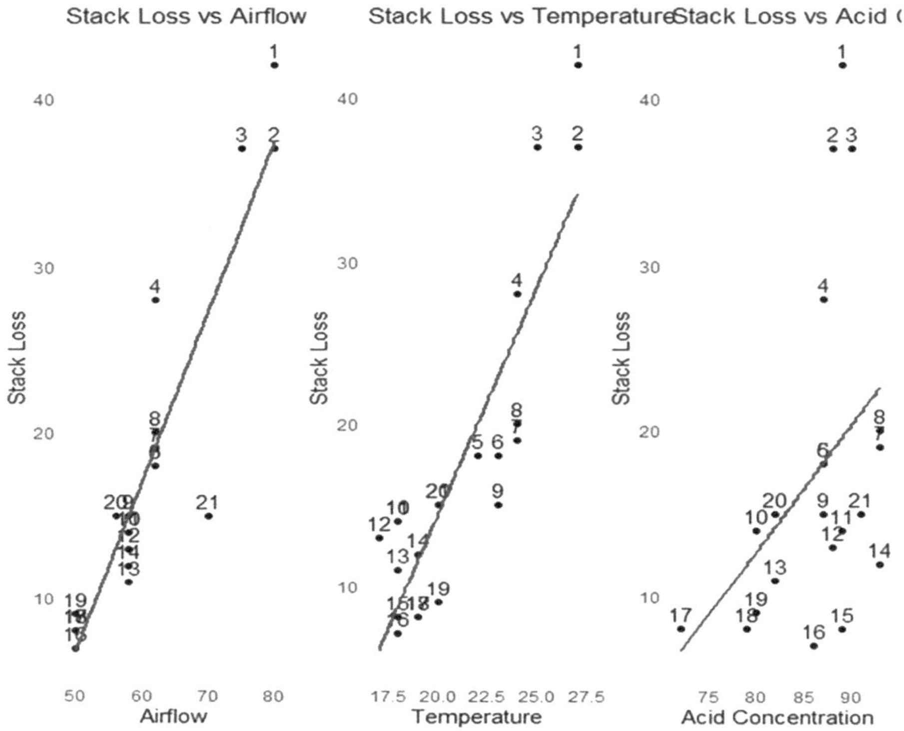

FIGURE 7.8
Scatterplot for stackloss versus airflow, temperature, and acid concentration.

```
> summary(mlr_model)
Call:
lm(formula = stkloss ~ airflow + temp + acidcon, data =
  stackloss_data)
Coefficients:
             Estimate Std. Error t value Pr(>|t|)
(Intercept) -39.9197    11.8960  -3.356  0.00375 **
airflow       0.7156     0.1349   5.307  5.8e-05 ***
temp          1.2953     0.3680   3.520  0.00263 **
acidcon      -0.1521     0.1563  -0.973  0.34405
---
Signif. codes: 0 '***' 0.001 '**' 0.01 '*' 0.05 '.' 0.1 ' ' 1
Residual standard error: 3.243 on 17 degrees of freedom
Multiple R-squared: 0.9136, Adjusted R-squared: 0.8983
F-statistic: 59.9 on 3 and 17 DF, p-value: 3.016e-09
```

The regression equation is

$$\text{stkloss} = -39.9 + 0.716\,\text{airflow} + 1.30\,\text{temp} - 0.152\,\text{acidcon}$$

To check for the outlier observations:

```
> # Calculate the standardized residuals
> stackloss_data$std_residuals <- rstandard(mlr_model)
> outliers <- stackloss_data[abs(stackloss_data$std_residuals)
  > 2, ]
> cat("Potential outliers based on large standardized
  residuals:\n")
Potential outliers based on large standardized residuals:
> print(outliers)
obs stkloss airflow temp acidcon fitted_values leverage dfits
  residuals
21  15     70   20   91    22.23771 0.2845335 -2.100296 -7.237713
    std_residuals
      -2.63822
```

We note a fairly good fit with an R^2 of 91.4. However, the regressor acid is insignificant at the 0.344 level, and the standardized residual has detected observation 21 as being large. We examine the hii, the dffits, and the residuals, as follows:

```
> # Cutoff for leverage: 2*(p/n)
> leverage_cutoff <- 2 * p / n
> high_leverage_points <- stackloss_data[stackloss_data$hii >
  leverage_cutoff, ]
> # Identify outliers based on DFFITS (rule of thumb:
  abs(dffits) > 2/sqrt(n))
> outliers <- stackloss_data[abs(stackloss_data$dffits) >
  dffits_cutoff, ]
> # Print the results
> cat("\nHigh leverage points (hii > 2(p/n)):\n")
High leverage points (hii > 2(p/n)):
> print(high_leverage_points[, c("obs", "hii")])
   obs       hii
17  17 0.4121235
> cat("\nOutliers based on large DFFITS (|DFFITS| > 2/
  sqrt(n)):\n")
Outliers based on large DFFITS (|DFFITS| > 2/sqrt(n)):
> print(outliers[, c("obs", "dffits", "olsres")])
   obs     dffits    olsres
1    1  0.7947205  3.234637
2    2 -0.4813230 -1.917485
3    3  0.7441582  4.555533
4    4  0.7878844  5.697774
7    7 -0.4376723 -2.389491
12  12  0.5091767  2.779460
17  17 -0.5020211 -1.519951
21  21 -2.1002964 -7.237713
```

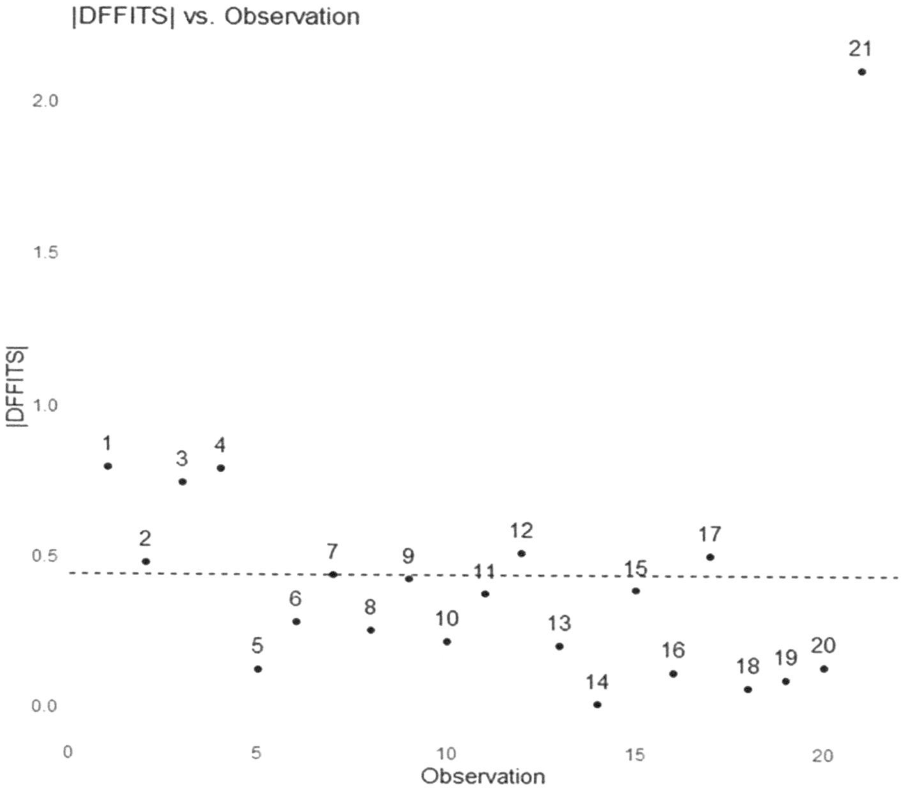

FIGURE 7.9
The |DFFITSi| plot versus observations.

These diagnostics, as in Figure 7.9, reveal that observation 17 is a high-leverage point since *hii* is 0.41 and exceeds the cutoff point of $2(p/n) = 2(4/21) = 0.38$. Observation 21 is an outlier since it has a very large dffits when compared to the other dffits (as revealed in Figure 7.9), though its leverage is relatively small. Observation 21 also has the largest residual.

The scatterfits plot of the full three-regressor model versus the SLR with airflow only is shown in Figure 7.10.

Figure 7.10 reveals that the three-regressor model provides a far superior fit to the data than using SLR with airflow only. Additionally, the figure reveals that either model very poorly fits observations 1, 3, 4, and 21.

```
The robust analysis is as follows:
Set the starting values for the huber estimator using ch=1.00.
> # Fit the robust regression model using Huber's estimator
> huber_model <- rlm(stkloss ~ airflow + temp + acidcon,
  data = stackloss_data, psi = psi.huber, init = "ls", scale.
  est = "Huber", k2 = 1.0)
```

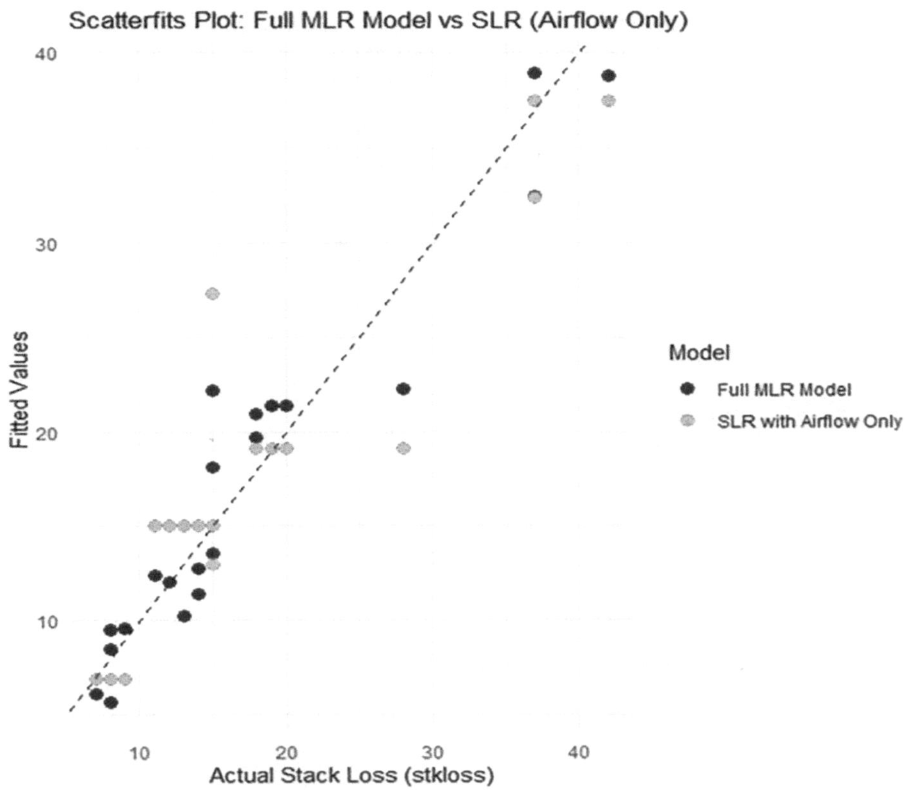

FIGURE 7.10
The scatterfits plot of the full three-regressor model versus the SLR with airflow.

```
> # Display the Huber m coefficients
> cat("The Huber m coefficients are:\n")
The Huber m coefficients are:
> print(coef(huber_model))
(Intercept)    airflow       temp    acidcon
-41.1615121 0.8073399 1.0214138 -0.1336725
> # Display the number of iterations required for convergence
> cat("\nConvergence achieved in", huber_model$converged,
  "iterations.\n")
Convergence achieved in TRUE iterations.
> # Calculate and display the weights, robust residuals, and
  OLS residuals
> stackloss_data$weights <- huber_model$w
> stackloss_data$robres <- residuals(huber_model, type =
  "response")
> stackloss_data$olsres <- residuals(lm(stkloss ~ airflow +
  temp + acidcon, data = stackloss_data))
> # Display the results
```

```
> cat("\nRow Weights Robres OLSres\n")
      Row   Weights    Robres    OLSres
       Row   Weights    Robres    OLSres
         1   0.40476   4.30263   3.23464
         2   1.00000  -0.80672  -1.91748
         3   0.34534   5.04159   4.55553
         4   0.23966   7.26306   5.69777
         5   1.00000  -1.26670  -1.71165
         6   0.86891  -2.00182  -3.00694
         7   1.00000  -1.08087  -2.38949
         8   1.00000  -0.08087  -1.38949
         9   1.00000  -1.67431  -3.14438
        10   1.00000   0.23587   1.26719
        11   1.00000   1.21998   2.63630
        12   1.00000   0.84575   2.77946
        13   0.68381  -2.54544  -1.42856
        14   1.00000  -1.07776  -0.05050
        15   0.92775   1.87498   2.36142
        16   1.00000   0.54695   0.90505
        17   1.00000  -0.71900  -1.51995
        18   1.00000   0.04641  -0.45509
        19   1.00000   0.42064  -0.59826
        20   1.00000   1.64807   1.41215
        21   0.19306  -9.01408  -7.23771
```

Four observations, 1, 3, 4, and 21, are given weight substantially less than one; see the bold lines. Note that the robust residuals are smaller than the LS residuals at 14 of the remaining 17 observations. The residuals are displayed in the box plot in Figure 7.11.

The majority of the robust residuals, denoted by group 2 are noticeably smaller than the LS residuals.

The bisquare function can also be used to obtain estimates of parameters. Recall that the bisquare function has the ability to give zero weight to outliers. We use the tuning constant of 3.5 as follows:

```
> # Fit the robust regression model using the bisquare (MM)
  estimator with tuning constant 3.5
> bisquare_model <- rlm(stkloss ~ airflow + temp + acidcon,
  data = stackloss_data, psi = psi.bisquare, init = start_
  coeffs, maxit = 20, acc = 0.001)
> # Display the Bisquare m coefficients
> cat("The Bisquare m coefficients are:\n")
The Bisquare m coefficients are:
> print(coef(bisquare_model))
(Intercept)     airflow        temp     acidcon
-42.2849116   0.9274800   0.6508544  -0.1123196
```

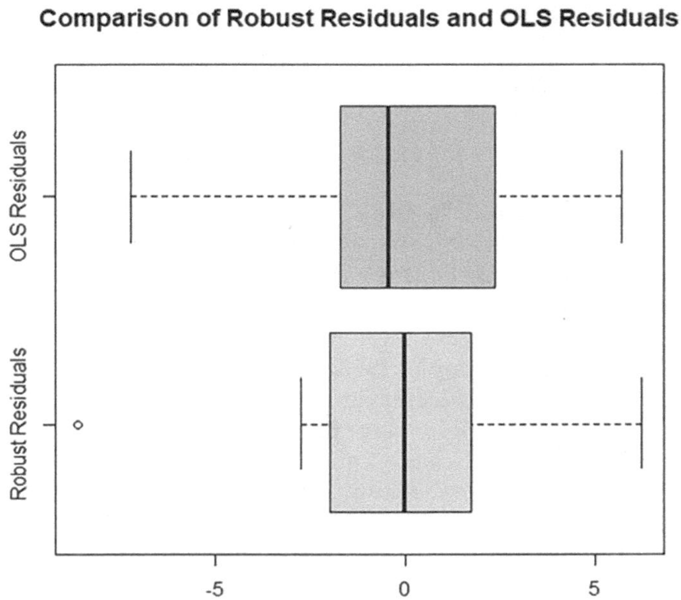

FIGURE 7.11
The side-by-side box plots for the residuals.

Notice the M-estimates are considerably different from the LS estimates, with 61% relative change in the coefficient for the third regressor. The weights make it clear which data points are considered outliers.

Row	Weights	Robres(Bisquare)	OLSres
1	**0.000000**	**6.06609**	**3.23464**
2	0.910723	0.99212	-1.91748
3	**0.000000**	**6.26829**	**4.55553**
4	**0.000000**	**8.25526**	**5.69777**
5	0.950626	-0.73309	-1.71165
6	0.862262	-1.23891	-3.00694
7	0.991622	-0.30089	-2.38949
8	0.954937	0.69911	-1.38949
9	0.918594	-0.94566	-3.14438
10	0.999596	0.06566	1.26719
11	0.950719	0.73143	2.63630
12	0.997500	0.16329	2.77946
13	0.407893	-2.78639	-1.42856
14	0.807059	-1.47849	-0.05050
15	0.844471	1.31793	2.36142

```
16 0.999117     0.09601     0.90505
17 0.981672    -0.44548    -1.51995
18 0.999498     0.07235    -0.45509
19 0.962033     0.64050    -0.59826
20 0.706970     1.84857     1.41215
21 0.000000    -9.01202    -7.23771
```

A box plot of the residuals in Figure 7.12 clearly reveals the superior fit of the robust method based on the bisquare function compared to the LS fit.

The weight index plot in Figure 7.13 (a) shows which observations have been declared as outliers.

Other methods developed in Chapters 1 and 2 may be used to compare and evaluate this model. For example, the normal probability plot of the LS residuals and the robust residuals are in Figure 7.13 (b) and (c). As can be observed in the two plots the probability plot for the LS residuals is not as well behaved in the center of the data (between –1 and 1) as the robust residuals.

The four outliers are clearly distinguished from the other 17 data points, indicating the high quality of fit at these points.

A scatterfits plot is shown in Figure 7.14 where the two fits are the LS and the robust fit based on the bisquare function.

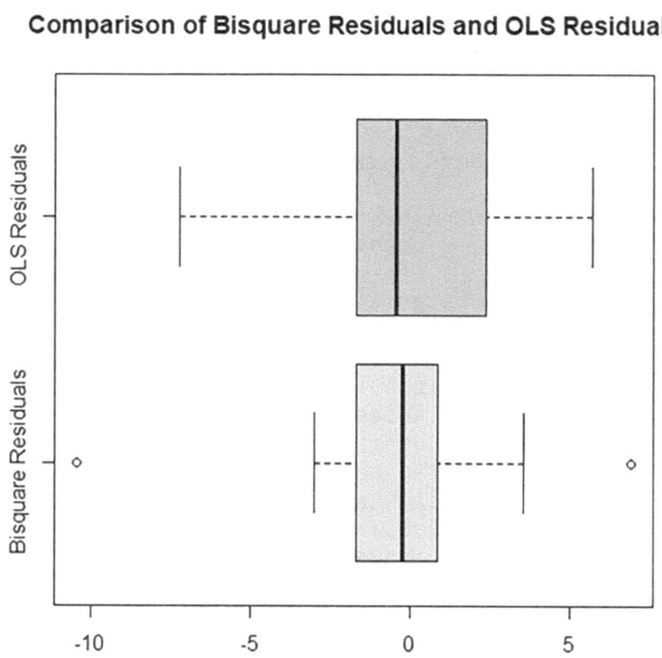

FIGURE 7.12
The side-by-side box plots for the residuals from the bisquare and LS methods.

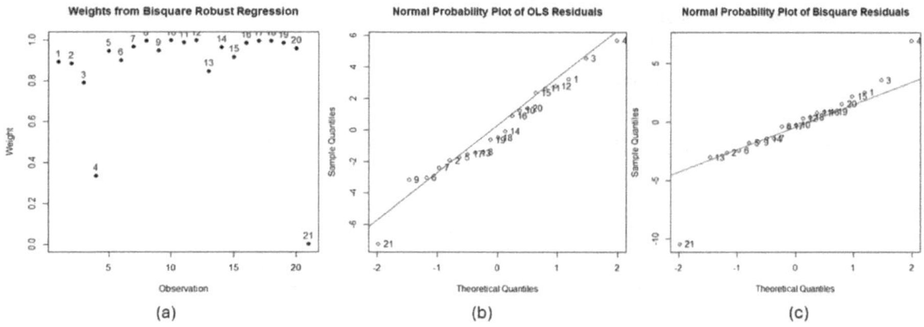

FIGURE 7.13
(a) The weight index; (b) Normal probability plot of the LS residuals; (c) Normal probability plot of the robust residuals.

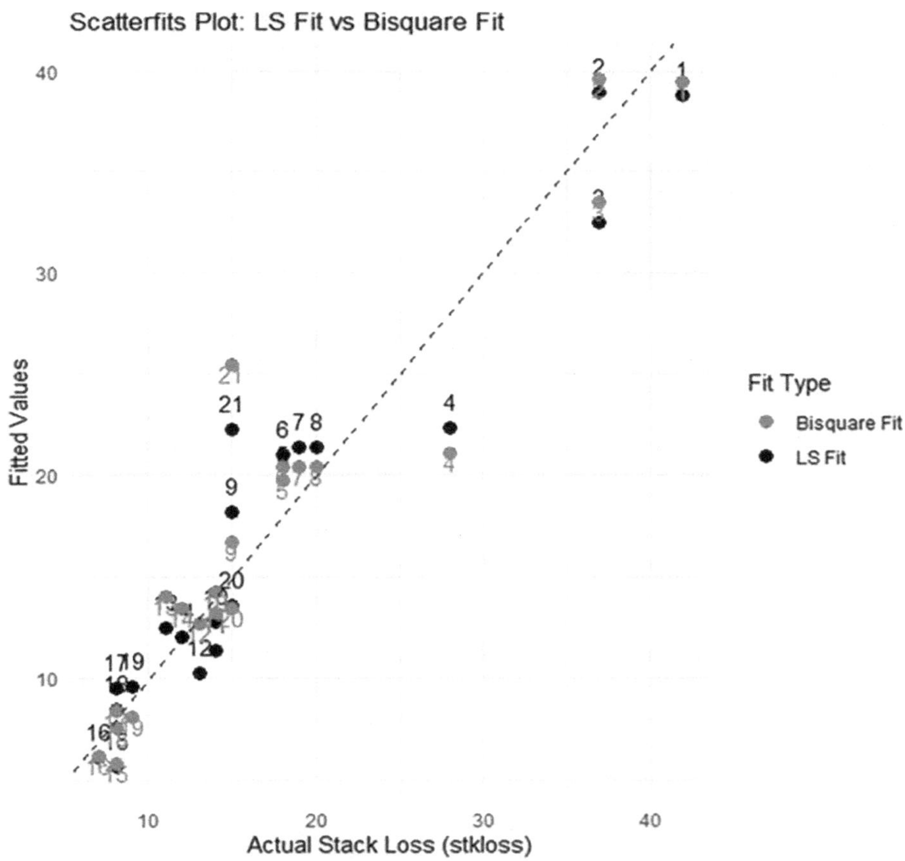

FIGURE 7.14
Scatterfits from the LS and the robust fit based on the bisquare function.

We see that the fits at observations 1, 3, 4, and 21 are closer to the data labels for the LS fit and that the fits are closer to the data labels for the majority of the other points for the robust fit.

Now that these four outliers have been identified, one might consider redoing the LS analysis with these four points removed. On R, type as follows:

```
# Exclude outliers (observations 1, 3, 4, and 21) and refit the
  OLS model subset_data <- stackloss_data[-c(1, 3, 4, 21), ]
# Refit the OLS model without outliers
ols_model_no_outliers <- lm(stkloss ~ airflow + temp + acid-
  con, data = subset_data)
# Display the summary of the new OLS model
summary(ols_model_no_outliers)
Coefficients:
             Estimate Std. Error t value Pr(>|t|)
(Intercept) -37.65246    4.73205  -7.957 2.37e-06 ***
airflow       0.79769    0.06744  11.828 2.48e-08 ***
temp          0.57734    0.16597   3.479  0.00408 **
acidcon      -0.06706    0.06160  -1.089  0.29611
---
Signif. codes: 0 '***' 0.001 '**' 0.01 '*' 0.05 '.' 0.1 ' ' 1
Residual standard error: 1.253 on 13 degrees of freedom
Multiple R-squared: 0.975, Adjusted R-squared: 0.9692
F-statistic: 169 on 3 and 13 DF, p-value: 1.159e-10
```

The regression equation is

$$\text{stkloss} = -37.7 + 0.798\ \text{airflow} + 0.577\ \text{temp} - 0.0671\ \text{acidcon}$$

```
Unusual Observations
Obs.   C12      C11    Fit Stdev.Fit Residual  St.Resid
  10  58.0   11.000 13.506     0.552   -2.506    -2.23R
R denotes an obs. with a large st. resid.
```

The fit is certainly much better with an R^2 of 97.5% (compared to 91.4% for LS based on all 21 observations) and an MSE of 1.57 (compared to 10.52 for the original dataset).

A comparison of the three sets of coefficients: LS using $n = 21$; robust using the bisquare, $n = 21$; and the LS using $n = 17$ is shown in Table 7.3.

Table 7.3 The Coefficients from the LS, Robust, and LS Methods

Coefficients	LS n=21,	Robust n=21,	LS n=17
Intercept	−39.92	−37.00	−37.65
airflow	0.72	0.82	0.80
temp	1.30	0.51	0.58
acidcon	−0.15	−0.07	−0.07

It is seen that the coefficients based on the robust method with $n = 21$ are nearly equal to those from the LS method with $n = 17$, with both sets of coefficients being much different than those obtained using LS with all $n = 21$ observations.

7.2.3 Bounded-Influence Regression

A very interesting problem arises when performing robust regression on data where outliers occur at high-leverage points in the *X*-space. To illustrate, consider Figure 7.15.

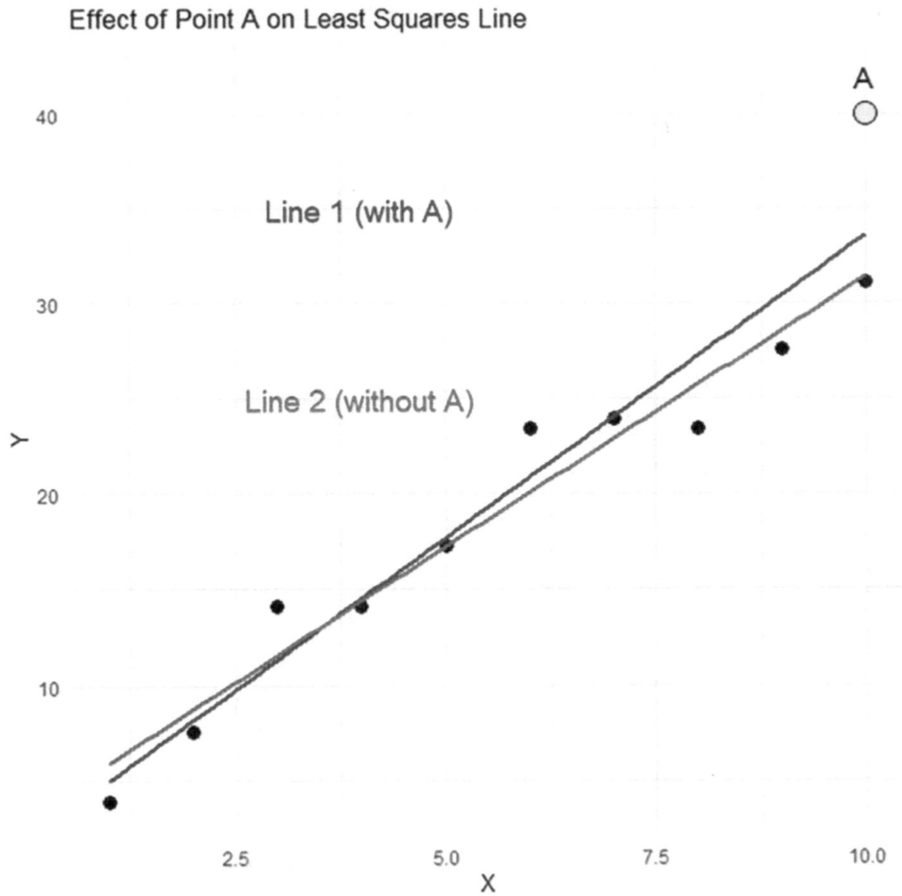

FIGURE 7.15
Scatterplot illustrating the effect of circled point A on the least squares line. The LS line with point A included in the dataset is line blue, while the LS line without point A included in the dataset is line red.

Obviously, the point A is a high-influence point (hip). We know that the LS line will be drawn toward point A. The resulting residual there, r_A, will be very small. Now, M-estimation works by down-weighting outliers, observations that have large residuals. Thus, M-estimation cannot down-weight point A, and the resulting M-estimation fit will be the same as the LS fit, which is undesirable. M-estimation fails when hips cause distortion in the coefficients. One solution to this problem is to use bounded-influence regression. Here, the name refers to "bounding" the influence that the point x_i' has in the X-space.

One form of bounded-influence (BI) estimators is the solution to the following set of equations where $r_i^* = (y_i - \underline{x}_i'\hat{\beta})/\hat{s}$, u(), and v() are appropriate functions applied to x_i'. A related form of the BI-estimator, called the Schweppe form (see p. 258 of Staudte and Sheather, 1990), is written as

$$\sum_{i=1}^{n} u(\underline{x}_i)\,\psi\left(\frac{r_i^*}{u(\underline{x}_i)}\right)\underline{x}_i = \underline{0}.$$

That is, $v(\underline{x}_i) = 1/u(\underline{x}_i)$, where $u(\underline{x}_i)$ is chosen so that the effect of a large x_i' is reduced if (y_i, x_i') is a hip.

One choice of $u(\underline{x}_i)$ is $u(\underline{x}_i) = \sqrt{1-h_{ii}}$, where $h_{ii} = x_i'(X'X)^{-1}\underline{x}_i$. This is an obvious choice since h_{ii} is the leverage value and is large if x_i' is far from the center of the X data. Then the argument of the ψ function becomes $\psi\left(\frac{r_i}{\hat{s}\sqrt{1-h_{ii}}}\right)$. It is recognized that $\frac{r_i}{\hat{s}\sqrt{1-h_{ii}}}$ is approximately equal to the influence diagnostic rstudent_i. We know that rstudent_i will get big if r_i is of medium size and h_{ii} is large, say greater than 0.5. But, if h_{ii} is very large (indicating that x_i' is a high-leverage point) and r_i very small, as is the case in Figure 7.15, then the *rstudent*$_i$ may not be large. So, choosing this form of $u(\underline{x}_i)$ does not enlarge r_i enough to result in down-weighting the hip. A better choice for $u(\underline{x}_i)$ is to set which results in $\frac{r_i}{\hat{s}\sqrt{1-h_{ii}}}\frac{\sqrt{h_{ii}}}{\sqrt{1-h_{ii}}}$ as the argument of the ψ function. The reader may recognize this argument as similar in form to the influence diagnostic *DFFITS*$_i$ except that, traditionally, *DFFITS*$_i$ uses $\hat{s}$ equal to the root MSE from the regression model with the point (y_i, x_i') removed from the dataset. That is, *DFFITS*$_i$, following the notation of Myers (1992), uses $\hat{s} = s_{-i}$. The quantity $\frac{1-h_{ii}}{\sqrt{h_{ii}}}$, appearing in the argument of the ψ function, can be called the "leverage inflation factor" since it is large if h_{ii} is large. In fact, it is easily seen that $\frac{1-h_{ii}}{\sqrt{h_{ii}}}$ tends to infinity as h_{ii} approaches one.

Thus, this choice of $u(\underline{x}_i)$ results in a large argument of the ψ function if the outlier (in the *Y*-space) occurs at a hip (in the *X*-space) with a large h_{ii} value. With this adjustment, we must solve the system of equations

$$\sum_{i=1}^{n} u(\underline{x}_i)\,\psi\left(\frac{r_i^*}{u(\underline{x}_i)}\right)\underline{x}_i = \underline{0},$$

with $u(\underline{x}_i) = \frac{1-h_{ii}}{\sqrt{h_{ii}}}$. The solution is obtained exactly as in the M-estimation case using the IRLS algorithm. This is seen by rewriting the previous equations as

$$\sum_{i=1}^{n} \frac{\psi(''DFFITS'')}{''DFFITS''}\,\underline{x}_i\left(\frac{y_i - \underline{x}_i{}'\underline{\beta}}{\hat{s}}\right) = \underline{0},$$

where $''DFFITS'' = \frac{r_i}{\hat{s}\sqrt{1-h_{ii}}}\frac{\sqrt{h_{ii}}}{\sqrt{1-h_{ii}}}$. The solution can then be written as

$\underline{\hat{\beta}}_{BI} = (X'WX)^{-1}X'W\underline{y}$,

a weighted LS estimator as before, but with $w_i = \frac{\psi(''DFFITS'')}{''DFFITS''}$. Thus, as the IRLS algorithm is iterated, we need only change the rescaling of r_i^* from $r_i^* = \frac{r_i}{\hat{s}}$, used in M-estimation, to $r_i^* = \frac{r_i}{\hat{s}\pi_i}$, where $\pi_i = \frac{1-h_{ii}}{\sqrt{h_{ii}}}$ for BI-estimation.

Again, the starting value is very important. BI-estimation will fail if there are several hips clustered together. This is due to BI-estimators' low breakdown point (see Chapter 6 on robust concepts for a discussion on the breakdown point). It can be shown (see Rousseeuw and Leroy, 1987) that the breakdown point for BI-estimators cannot exceed $1/(p+1)$, where p is the number of parameters to be estimated in the general linear model. This results in 1/2 for the location model, 1/3 for SLR, and small fractions for MLR as the number of regressors in the model increases. Therefore, $\underline{\hat{\beta}}_0 = \underline{\hat{\beta}}_{LMS}$ or $\underline{\hat{\beta}}_0 = \underline{\hat{\beta}}_{LTS}$ is recommended.

An additional point concerns the tuning constant used to determine the scale of the weight function. In BI-estimation, the rescaled residuals are $r_i^* = \frac{r_i}{\hat{s}\pi_i}$ with $\pi_i = \frac{1-h_{ii}}{\sqrt{h_{ii}}}$. We know from M-estimation, where $r_i^* = \frac{r_i}{\hat{s}}$, that proper tuning constants for the Huber are c_H, generally between 1 and 1.5 and, for the bisquare, c_H, generally between 3.5 and 4.7 (4.685 achieves 95% efficiency against LS for the location model when the errors follow a normal curve). Since large values of c_B are those exceeding $2p/n$, it follows by simple algebra that a large value of π_i is one exceeding $\frac{\sqrt{2pn}}{n-2p}$. Consequently, we use the tuning constant $c\frac{\sqrt{2pn}}{n-2p}$ for BI-estimation, where c is c_H when using the Huber weight function and c_B when using the bisquare weight function.

7.2.4 The High-Leverage Data

The need for a bounded-influence estimator is illustrated in the following example, called the "high-leverage data." The data are presented in Table 7.4.

Not too careful inspection of this data reveals that observation 3 is very different from the others.

We will analyze this data on R, using ordinary least squares, M-estimation, and BI-estimation. The R codes for the LS, Huber, and bisquare weight functions, respectively.

```
#Least Squares (LS) Analysis
ols_model <- lm(y ~ x, data = data)
ols_summary <- summary(ols_model)
> print(ols_summary)
Call:
lm(formula = y ~ x, data = data)
Coefficients:
            Estimate Std. Error t value Pr(>|t|)
(Intercept)  2.4428     0.4864   5.023 0.00102 **
x            0.4380     0.2180   2.009 0.07938 .
---
Signif. codes: 0 '***' 0.001 '**' 0.01 '*' 0.05 '.' 0.1 ' ' 1
Residual standard error: 0.7089 on 8 degrees of freedom
Multiple R-squared: 0.3354, Adjusted R-squared: 0.2523
```

Table 7.4 The High-Leverage Data

Obs.	y	x
1	3.2	1.6
2	2.2	1.9
3	5.1	4.9
4	2.4	2.1
5	3.1	2.2
6	3.9	1.1
7	3.7	1.4
8	3.0	1.3
9	3.0	1.8
10	3.5	1.5

```
F-statistic: 4.037 on 1 and 8 DF, p-value: 0.07938
```

The regression equation is

$$\mathbf{y = 2.44 + 0.438\ x}$$

```
Unusual Observations
Obs.     x       y     Fit    Stdev.Fit     Residual  St.Resid
   3 4.90  5.100  4.589        0.675         0.511    2.35RX
R denotes an obs. with a large st. resid. X denotes an obs.
  whose X value gives it large influence.
```

We see that observation 3 has been detected as an "outlier" by the standardized residual and as a high-leverage point, given that h3 is large.

```
> # Print OLS results
> print(ols_results)
Obs  y   x Std_Residual  Fitted_Values  Leverage  Residuals   DFFITS
 1    1 3.2 1.6   0.08455141 3.143570 0.1136536  0.05642965  0.028334056
 2    2 2.2 1.9 -1.59894380 3.274962 0.1006051 -1.07496218 -0.606433015
 3    3 5.1 4.9   2.35419571 4.588880 0.9062027  0.51111952 12.349210343
 4    4 2.4 2.1 -1.43234969 3.362557 0.1013616 -0.96255673 -0.521846903
 5    5 3.1 2.2 -0.45669317 3.406354 0.1045764 -0.30635401 -0.147933824
 6    6 3.9 1.1  1.51325025 2.924584 0.1732224  0.97541604  0.766913842
 7    7 3.7 1.4  0.97501093 3.055976 0.1318079  0.64402421  0.378570348
 8    8 3.0 1.3 -0.01856531 3.012179 0.1437216 -0.01217852 -0.007114905
 9    9 3.0 1.8 -0.34431526 3.231165 0.1030635 -0.23116490 -0.109995267
10   10 3.5 1.5  0.60245018 3.099773 0.1217852  0.40022693  0.214785057
```

This data clearly reveals that observation 3 is very different from the others. Will M-estimation detect this hip and give a negative estimate for the slope?

```
> print(m_summary_huber)
Call: rlm(formula = y ~ x, data = data, psi = psi.huber)
Coefficients:
            Value  Std. Error  t value
(Intercept) 2.4545 0.6447      3.8073
x           0.4494 0.2889      1.5552
Residual standard error: 0.6043 on 8 degrees of freedom
```

The IRLS algorithm converges in four iterations to coefficients nearly exactly equal to those obtained by LS. Print the weights and residuals.

Row	y	x	weight	robres	olsres
1	3.2	1.6	1.00000	0.02839	0.05643
2	2.2	1.9	0.74945	-1.10605	-1.07496
3	**5.1**	**4.9**	**1.00000**	**0.44949**	**0.51112**
4	2.4	2.1	0.83255	-0.99568	-0.96256
5	3.1	2.2	1.00000	-0.34050	-0.30635
6	3.9	1.1	0.86897	0.95247	0.97542

```
      7   3.7   1.4   1.00000    0.61802    0.64402
      8   3.0   1.3   1.00000   -0.03716   -0.01218
      9   3.0   1.8   1.00000   -0.26124   -0.23116
     10   3.5   1.5   1.00000    0.37321    0.40023
```

Three observations (2, 4, and 6) receive less than one weight, while observation 3 receives full weight. Observation 3 obviously is not detected as unusual by the M-estimator using the Huber weight function. What if the bisquare weight function is obtained (use Huber to start).

```
> cat("Bisquare M Estimation Coefficients:\n")
Bisquare M Estimation Coefficients:
> print(coef(m_model_bisquare))
(Intercept)           x
  2.4346103   0.4536331
> print(bisquare_results)
    Obs   y   x Weight_Bisquare Residuals_Bisquare OLS_Residuals
      1 3.2 1.6    0.9996204    0.03957678 0.05642965
      2 2.2 1.9    0.7301046   -1.09651315 -1.07496218
      3 5.1 4.9    0.9531281 0.44258759 0.51111952
      4 2.4 2.1    0.7779673 -0.98723977 -0.96255673
      5 3.1 2.2    0.9734009 -0.33260307 -0.30635401
      6 3.9 1.1    0.7866717 0.96639332 0.97541604
      7 3.7 1.4    0.9061249 0.63030340 0.64402421
      8 3.0 1.3    0.9998569 -0.02433330 -0.01217852
      9 3.0 1.8    0.9847900 -0.25114984 -0.23116490
     10 3.5 1.5    0.9644439 0.38494009 0.40022693
```

No outliers are detected, and observation 3 is not distinguished from the others. We see that the estimation for this dataset has failed to work properly.

Now, try bounded-influence estimation. First, use the Huber weight function.

```
The Huber BI coefficients are:
bihuber
3.03860 0.06971
```

The bi coefficients are slightly different from either the LS or M-estimates, but the slope is still positive and not negative as it should be. Look at the weights to see why.

```
> print(comparison_results)
Obs y   x Weight_Huber Residuals_Huber Weight_Bisquare Residuals_Bisquare
 1 3.2 1.6   1.0000000     0.02651048     0.9996204      0.03957678
 2 2.2 1.9   0.7333388    -1.10829841     0.7301046     -1.09651315
 3 5.1 4.9   1.0000000     0.44361267     0.9531281      0.44258759
 4 2.4 2.1   0.8142520    -0.99817101     0.7779673     -0.98723977
 5 3.1 2.2   1.0000000    -0.34310731     0.9734009     -0.33260307
 6 3.9 1.1   0.8544207     0.95119196     0.7866717      0.96639332
 7 3.7 1.4   1.0000000     0.61638307     0.9061249      0.63030340
 8 3.0 1.3   1.0000000    -0.03868063     0.9998569     -0.02433330
```

```
 9 3.0 1.8   1.0000000    -0.26336212      0.9847900     -0.25114984
10 3.5 1.5   1.0000000     0.37144677      0.9644439      0.38494009
   OLS_Residuals
1      0.05642965
2     -1.07496218
3      0.51111952
4     -0.96255673
5     -0.30635401
6      0.97541604
7      0.64402421
8     -0.01217852
9     -0.23116490
10     0.40022693
```

We see the problem. The third observation, while given a much smaller weight than the other observations, is still given a nonzero weight—and the influence of this point is so great that using it in a small amount results in distorted coefficients. This perfectly illustrates the fault with the Huber weight function; it cannot give zero weight to any point.

The bisquare weight function is needed here to eliminate the effect of observation 3.

```
The Bisquare m coefficients are:
bisquare
  4.89649 -1.07926
> # Print the results
> print(results)
    Obs y   x   Weight_Bisquare Residuals_Bisquare OLS_Residuals
    1   3.2 1.6     0.9996204           0.03958         0.05643
    2   2.2 1.9     0.7301046          -1.09651        -1.07496
    3   5.1 4.9     0.0000000           0.00000         0.51112
    4   2.4 2.1     0.7779673          -0.98724        -0.96256
    5   3.1 2.2     0.9734009          -0.33260        -0.30635
    6   3.9 1.1     0.7866717           0.96639         0.97542
    7   3.7 1.4     0.9061249           0.63030         0.64402
    8   3.0 1.3     0.9998569          -0.02433        -0.01218
    9   3.0 1.8     0.9847900          -0.25115        -0.23116
    10  3.5 1.5     0.9644439           0.38494         0.40023
```

We see that the third observation is properly given zero weight and that the residuals are smaller at seven of the ten observations.

The scatterplot of the data with the bisquare and OLS fits is in Figure 7.16.

From Figure 7.16, it is obvious that the bi (Bisquare) estimate is clearly superior to the fits obtained by LS, M-estimation, or BI-estimation using the Huber weight function.

This example serves to illustrate that one cannot blindly use these estimation techniques. In some cases, one method works much better than another method. Least squares estimation works well when there are no outliers in the dataset. M-estimation works well in the presence of outliers that occur at low- to moderate-leverage points. BI-estimation is

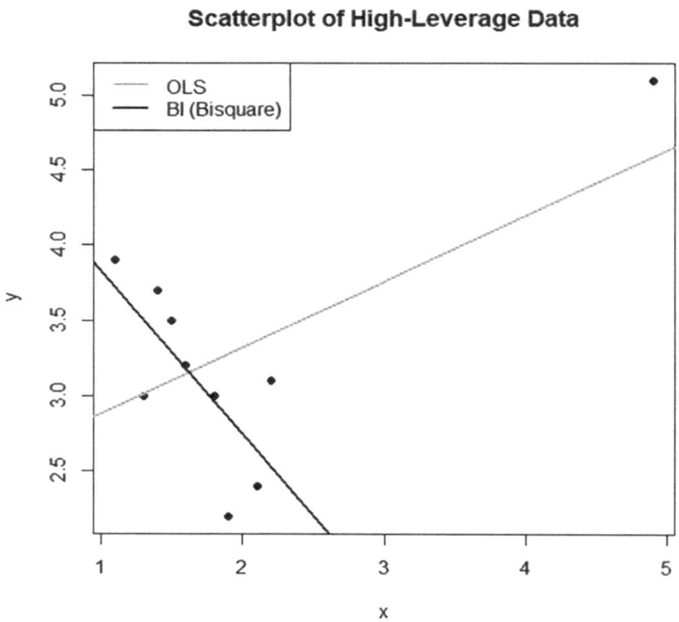

FIGURE 7.16
Scatterplot of the high-leverage data.

designed for those situations where the outliers occur at high-influence points (hips). One way to determine which method to use is to first obtain all three types of estimates, using both the Huber and bisquare weight functions for the m and bi procedures. If all sets of estimates are similar, then outliers must not be present in the data and one would feel comfortable using the LS estimates. If the LS and BI-estimates are nearly equal but different from the M-estimates, then there are outliers in the dataset occurring at low- to moderate-leverage points, and the M-estimation results should be used.

On the other hand, if the LS and M-estimates are nearly equal but different from the BI-estimates, then there are outliers occurring at high-leverage points in the data, and the BI-estimation results should be used.

The next section will discuss high breakdown methods (LMS and LTS) for the MLR situation.

7.3 High Breakdown Point Estimators in Regression

In Section 4.3, the estimators LMS and LTS were introduced for the location model. The estimators were developed by Rousseeuw (1990) as high breakdown point estimators, with asymptotic breakdown points of 50%, for use not so much in location but for regression.

The reader will recall that due to the lack of efficiency of these estimators, their primary use is to supply starting values to other robust procedures such as the M- or BI-estimators. The LMS estimator, $\underline{\hat{\beta}}_{\text{LMS}}$, in the MLR model is defined as the value of $\hat{\beta}^*$ that minimizes $\text{med}_i\left(r_i^2\right)$, where $r_i = y_i - x_i'\hat{\beta}^*$. Actually, the term "median" is not technically correct for the MLR model in that LMS actually minimizes $\left(r^2\right)_{(h)}$, the hth ordered squared residual, where $h = [n/2] + [(p+1)/2]$ and p is the number of parameters to be estimated for the model. In the location model, $p = 1$, and the LMS estimator is indeed the median squared residual. Otherwise, the use of p is an adjustment for the fact that the residuals are correlated random variables. So, if $n = 26$ and $p = 4$, then $h = 15$, whereas the median would be found as the average of the 13th and 14th ordered squared residuals. The breakdown point for the LMS estimator approaches 50% as n gets large.

The LTS estimator, $\underline{\hat{\beta}}_{\text{LTS}}$, on the other hand, finds the value of $\underline{\hat{Q}}^*$ that minimizes $\sum_{i=1}^{h}\left(r^2\right)_{(i)}$, the sum of the first h squared residuals, where $h = [n(1-\alpha)] + [\alpha(p+1)]$ for $0 \le \alpha \le .5$, where α represents the trimming fraction. Larger values of have higher breakdown points, approaching 50% when α is 0.5. For reasons to be discussed shortly, it may be desirable to use some other value of, say 0.25. The breakdown point would then only approach 25%, but better estimates may result. As in the LMS and LTS estimates for the location model, there is no closed-form solution to either objective function, and solutions to both are obtained algorithmically. The estimates may be "exact" or "approximate," depending on the algorithm used to obtain the solution. All methods rely on evaluating many values of $\underline{\hat{\beta}}^*$.

One common way to obtain the solution is to draw repeated subsamples of size p from the data using sampling without replacement. This subset of data is called the "elemental set of size p" because it represents the smallest subset of the original data for which the p parameters may be estimated. The method of least squares is used to obtain the estimates, labeled as $\hat{\beta}^*$. Of course, $\hat{\beta}^*$ exists so long as the resulting X matrix (the model matrix) is of rank p. If X is not of full rank, the sample is discarded and replaced by another elemental set. The objective function is computed for this value of $\underline{\hat{\beta}}^*$, either $\left(r^2\right)_{(h)}$ for LMS or $\sum_{i=1}^{h}\left(r^2\right)_{(i)}$ for LTS. This process is repeated a large number of times, say 500 times, to ensure with a high probability that at least one of the elemental sets will contain a set of p "clean" (uncontaminated by outliers) observations. The estimate that results in the minimum value of the objective function is the final LMS and/or LTS estimate. The repeated sampling method gives an approximate estimate rather than the exact one.

Another way in which the elemental sets may be obtained is to examine all $C_p^n = \binom{n}{p}$ combinations of p observations. C_p^n represents all possible elemental sets of size p that can be obtained from the data. In this way, the algorithm would search through all possible values of $\hat{\beta}^*$ that arise from the elemental sets. The reader can see that C_p^n can quickly become a very large number. For example, in fitting an SLR model, with $p = 2$, and $n \leq 32$, then C_p^n is ≤ 496. If n is only 50, then there are 1,225 possible elemental sets of size 2, and if $n = 100$, the number is 4,950. Clearly, examining all elemental sets is superior to examining only a subset of them since it allows more values of $\hat{\beta}^*$ to be calculated. Computationally, there may be too many combinations to allow all to be examined. Obtaining the estimates in this manner also results in an approximate solution to either objective function.

The exact solutions are what we want but are harder to obtain. In the SLR model, the exact solution for LMS is found by examining all C_2^n elemental sets of size two, fitting the slope, $\hat{\beta}_1^*$, to each pair of points, computing the n residuals $r_i^* = y_i - \hat{\beta}_1^* x_i$, and finding the LMS intercept, estimate via the exact LMS method for the location model as given in Section 4.3. That is, $\hat{\beta}_0^*$ is the midpoint of the shortest interval containing h residuals. Each pair of estimates $\left(\hat{\beta}_0^* \ \hat{\beta}_1^*\right)$ gives residuals of the form $r_i = y_i - \hat{\beta}_0^* - \hat{\beta}_1^* x_i$. The exact LMS estimates are those values of $\left(\hat{\beta}_0^*, \hat{\beta}_1^*\right)$, that minimize $\left(r^2\right)_{(h)}$. The exact LTS solution for the SLR model may be obtained in the same fashion except the exact LTS method for the location model is used to obtain the intercept for each C_2^n elemental set. Exact fits for the MLR model may be obtained for the LMS estimator using "Chebyshev fits," a method we will not discuss further here. The interested reader may refer to books by Rousseeuw and Leroy (1987) and Ryan (1997) for more details concerning these computational problems and other issues as well.

The R code has been written to obtain the LMS and LTS estimates in the SLR model. The macros compute the approximate solutions described by using all C_2^n elemental sets when $n \leq 32$ and by using 500 elemental sets of size two, obtained by subsampling when $n > 32$.

To illustrate the use of the R code and the robust properties of the LMS and LTS estimators, consider again the high-leverage data of Section 7.2. The reader will recall that $n = 10$ and one *hip* exists at $x = 4.9$ with a response thereof. See Section 7.1 for a scatterplot of the data with the bi fit line based on the bisquare function. The LS and M-estimated fits all incorrectly followed the hip, resulting in a positive slope instead of the desired negative slope. To examine these fits to the high-leverage data, use this code:

```
> lms_coef <- lms_estimator(x, y)
> cat("LMS Coefficients:\nIntercept:", lms_coef[1], "\
  nSlope:", lms_coef[2], "\n")
```

```
      LMS Coefficients:
      Intercept: 5.76
      Slope: -1.6
> # Calculate LTS estimator with alpha = 0.5
> lts_coef <- lts_estimator(x, y, alpha = 0.5)
> cat("LTS Coefficients:\nIntercept:", lts_coef[1], "\
  nSlope:", lts_coef[2], "\n")
      LTS Coefficients:
      Intercept: 5.76
      Slope: -1.6
```

The LMS and LTS coefficients, LMS fits at observed x values, and LMS residuals are obtained. The LMS and LTS estimates are very nearly identical for this example. This is not always the case. We note that these estimates match closely the bi-estimates based on the bisquare function of 4.90 for the intercept and –1.08 for the slope. Figure 7.17 compares the LMS, LTS, and bi (based on the bisquare) fits with the LS fit.

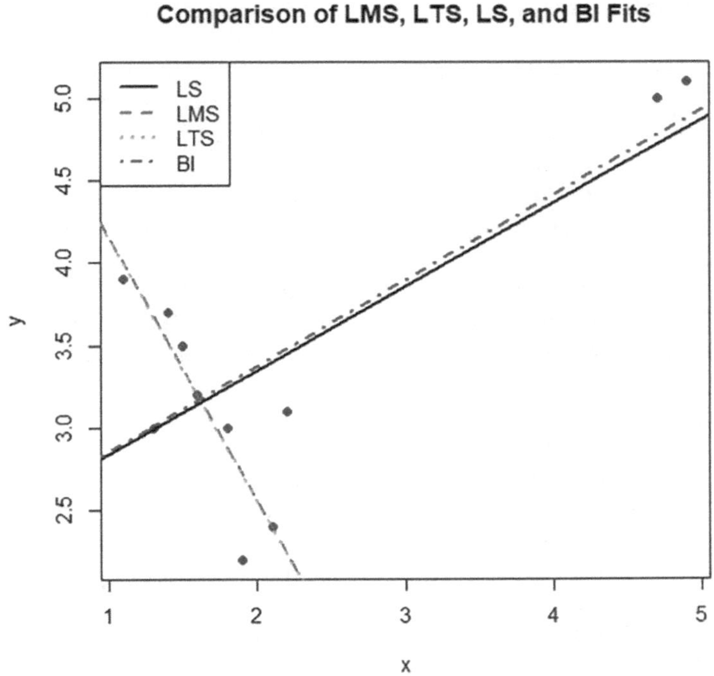

FIGURE 7.17
The high-leverage data comparing the LS, BI, LMS, and LTS fits.

The LMS and LTS estimators work well on this example. However, the situation represented by the high-leverage data hardly justifies the use of a 50% breakdown point estimate; after all, only 10% of the observations are anomalous. To see one advantage of these high breakdown estimators, consider adding an 11th point to the dataset at $y = 5.0$. Now there are two points instead of one in the upper-left corner of the scatterplot. Two points out of 11 (18%) are against the trend established by the other 9 observations. It is apparent that LS fits will be drawn toward the two anomalous points, as will the M-estimates based on the Huber and bisquare psi functions. The BI-estimates, using the LS estimates as starting values, will also fail to detect the correct pattern, even if the BI-estimates are based on the bisquare function. The pull of the two hips together is powerful enough to distort the bi fit. The LMS and LTS fits, on the other hand, find the correct trend, resulting in a negative slope. In fact, the estimates are exactly the same as in the ten-point high-leverage dataset. The high breakdown point estimators have ignored the two hips, as we would hope. The scatterplot of the data containing several of the fits is given in the plot Figure 7.17. The bisquare estimate cannot overcome the combined effects of the poor starting values and the pull of the two hips. However, if we alter the bi algorithm to start with the resistant LMS or LTS estimates, we will get the same coefficients obtained using the ten-point dataset. This result occurs whether we use the Huber or bisquare functions.

In other words, the Huber estimates continue to be pulled toward the outliers, while the bisquare estimates continue to ignore the outliers. This example illustrates one of the intended uses of the LMS and/or LTS estimators, as starting values to the more efficient procedures such as BI-estimation. The net effect is to obtain the resistance of the high breakdown point (but low efficiency) estimators while achieving the higher efficiency of the BI-estimators.

Unfortunately, this approach may also have problems. We recall from Section 4.3 that the LMS estimator, and to a lesser extent the LTS estimator, suffers from numerical problems. The LMS estimator is extremely sensitive to local "wiggling" of the data, resulting is an unstable estimator. A similar type of numerical instability occurs in the regression case.

Consider the altered high-leverage data seen in the scatterplot in Figure 7.18. It can be observed that several data points – namely, observations 6, 9, 10, and 11 (the extra point) – have been "wiggled" toward the LS line fit to the original data. There are now three hips in the dataset: two severe, observations 10 and 11; and a mild hip, observation 9. It is observed that the collection of six observations 2, 5, 6, 9, 10, and 11 all lie on a very strong positive linear trend. In fact, the fit to this linear trend is stronger that the best linear fit to any six points of the nine points that follow the downward trend represented by points 1, 3, 4, 5, 2, 6, 7, 8, and 9. Hence, the best $h = 6$ residuals used by the LMS and LTS algorithms belong to the upward, false trend rather than the correct downward trend.

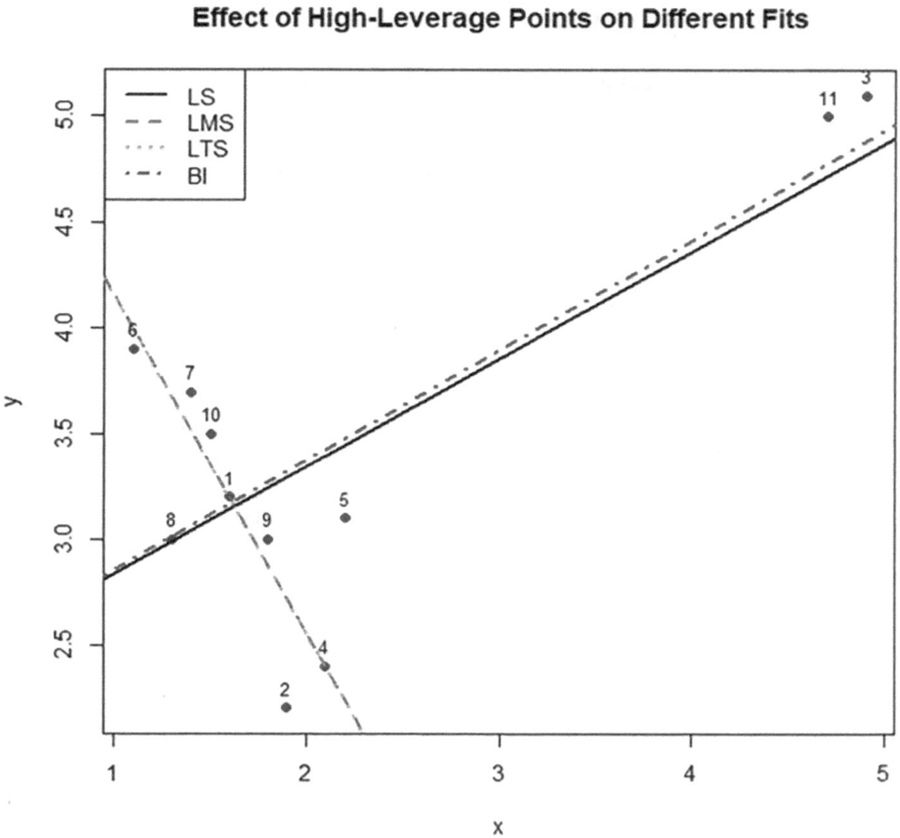

FIGURE 7.18
Effect of high-leverage points.

This example shows that the LMS and LTS estimators have failed, with only 3/11 = 27% of the data as outliers. Obviously, the 50% breakdown point does us little good if the "bad" data happens to lie on a strong linear trend represented by as few as h observations. The problem lies in the fact that requiring a 50% breakdown may mean a fit to the wrong 50% of the data. In our example, the proper trend is represented by 8/11 (73%) of the data, as observation 9 is not really along the trend. One remedy to this problem might be to use the LTS estimator with a smaller value of α, say $\alpha = 0.25$, rather than 50%. The loss of 25% in the breakdown point is compensated by the fact that the dominant trend must represent about 75% of the data. It has been speculated (Ryan, 1997) that the resulting LTS fit would then be similar to the LS fit to h data points, with the other n-h points removed.

We can check this assertion on the altered 11-point high-leverage dataset by setting $\alpha = 0.25$. Here, $h = 9$ when $\alpha = 0.25$. Does the LTS fit match the LS fit to the nine data point exhibiting the downward trend, observations numbered one through nine? The resulting LTS fit to all 11 points, and the LS fit to the nine points is shown in Figure 7.18. It is seen that the LTS fit does not match the LS fit in this example. Of course, the LMS and LTS estimators may be applied to MLR models as well.

Our example serves to illustrate several features of the LMS and LTS estimators in the MLR model. The first is that they are resistant to large hips that may tend to distort other estimators such as the LS, M-, and BI-estimators. Second, they tend to be subject to the same type of numerical instability as LMS and LTS in SLR models. That is, since the exact solutions are so often very time consuming to obtain if even achieved, we must rely on the approximate solutions obtained by the resampling of elemental sets. Even with a large number of elemental sets, we saw that the LMS and LTS results varied with each new set of elemental sets. Sometimes this variation is substantial. As mentioned earlier in this section, the LMS and LTS estimators may be used as starting values for the more efficient forms of robust estimators such as the M- or BI-estimators.

The R code presented in this text that performs M-estimation could easily be altered to allow the LTS (or LMS) estimates as starting values. Using LTS as a starting value for BI-estimation is a different matter, however. The reader will recall from our discussion of the BI-estimators that they work by determining which points are high-influence points by computing the leverage values, h_{ii}, and $\pi_i = \frac{1-h_{ii}}{\sqrt{h_{ii}}}$. If we start the Huber BI-estimates with the LS estimates, the h_{ii} and, consequently, the π_i are readily available as a result of the LS calculations. But the h_{ii} themselves are not resistant to outliers among the regressors (the high-leverage points, *hlps*) and could be distorted if several high-influence points or clusters of *hlps* exist in the data. If the h_{ii} are not truly reflective of the unusual data points, the π_i are incorrect, and the IRLS algorithm would assign incorrect weights to the observations, perhaps giving too much weight to the anomalous data points. Unfortunately, high breakdown versions of leverage values, h_{ii}, are not so easily obtained when using the LTS estimators. One method, developed by Rousseeuw and van Zomeren (1990), obtains high breakdown estimates of leverage based on the minimum volume ellipsoid estimators (MVE). These robust measures of leverage and the LTS estimates can then be used in high breakdown–high efficiency estimators, often referred to as generalized M-estimators.

The interested reader should see Simpson et al. (1992) for a presentation of such a procedure. These methods are often successful at detecting clusters of *hips* and providing robust estimates of coefficients.

7.4 Robust Inference in MLR

It is well known that outliers and high-influence points can severely distort the usual least squares estimates of the parameters. As well, they can lead to incorrect inferences on these parameters when traditional test statistics, such as F and *t*-tests, are employed. This is especially disconcerting when inferential methods are utilized in the regression model-building process. For example, the presence of outliers can lead to an erroneous set of variables in the model when using variable selection procedures such as forward, backward, and stepwise regression.

In fact, the status of a regressor variable or group of variables in terms of their significance in the model can be solely determined by one or more anomalous cases. Borrowing terminology from the regression (multicollinearity) literature, an outlier(s) or high-influence point(s) that causes significance for a variable(s) that would be insignificant without that point is termed a significance inducing point, whereas an anomalous point that causes insignificance for a variable that would be significant without that point is termed a significance masking point.

In this section, weighted least squares is suggested as an appropriate robust method for inference as well as for estimation. The weights are obtained from the IRLS algorithm used in computing either the M- or bounded-influence estimator. The resulting weights can also be utilized to establish a robust analysis of the variance table and other related statistical measures. Consider the usual MLR model expressed in vector-matrix notation as

$$\underline{y} = X\underline{\beta} + \underline{\varepsilon},$$

as described in Section 7.1. The usual LS estimator of $\underline{\beta}$ is $\hat{\underline{\beta}} = (X'X)^{-1} X'\underline{y}$. Tests of hypotheses on linear combinations of parameters of the form $H_0 : G\beta = g$, where G is a dxp matrix of rank d, and g is a $dx1$ vector of chosen constants, are performed using the test statistic

$$F = \frac{(G\hat{\underline{\beta}} - \underline{g})'\left(G\left((X'X)^{-1}\right)G'\right)^{-1}\left(G\hat{\underline{\beta}} - \underline{g}\right)}{ds^2},$$

Where s^2 is the usual MSE term. Under the assumption that the ε's follow a normal distribution, the null distribution of F is an F-distribution with d and $n - p$ degrees of freedom. In addition, $\hat{\beta}$ is also the maximum likelihood estimator and F is the uniformly most powerful unbiased (UMPU) test. The optimality of this F-test can be seriously compromised when outliers and high-influence points are present in the data.

Many inferential methods based on BI-estimators have also been developed in recent years; see, for example, Huber (1973), Schrader and McKean (1977), Schrader and Hettmansperger (1980), Sen (1982), Ronchetti (1982), and Markatou and Hettmansperger (1990). Perhaps one problem with these proposed methods from a practitioner's viewpoint is that while some methods are mathematically elegant, they do not provide a unified theory comparable to the one that exists for least squares analysis.

In an extension of the distribution theory for *M*-estimators in the location model, we know that for M- and BI-estimates in the GLM that (Huber, 1973)

$$\left(\hat{\beta}_{\mathrm{M}}\left(\text{or } \hat{\beta}_{\mathrm{BI}}\right)-\beta\right) \rightarrow \mathrm{N}\left(0, \mathrm{V}^{2}\left(\mathrm{X}'\mathrm{X}\right)^{-1}\right)$$

Where $V^2 = \dfrac{s^2 E\left(\psi^2\right)}{\left(\mathrm{E}\left(\psi'\right)^2\right)}$. It seems reasonable to use a *t*-like statistic to test

$H_0 : \beta_j = 0$ as $\mathrm{t}_{\mathrm{j}} = \dfrac{\hat{\beta}_{Mj}}{se(\hat{\beta}_{Mj})} = \dfrac{\hat{\beta}_{Mj}}{\sqrt{c_{jj}}}$ (or use $\hat{\beta}_{\mathrm{BI}_j}$), where c_{jj} is the jth diagonal element of $\mathrm{V}^2\left(\mathrm{X}'\mathrm{X}\right)^{-1}$. However, the small-sample properties of such a procedure are questionable. Another approach is to mimic what we did in the location model in Section 3.10. That is, at the convergence of the IRLS algorithm, we have $\underline{\beta} = \left(X'WX\right)^{-1} X'W\underline{y}$, where $\hat{\beta}$ is either $\hat{\beta}_M$ or $\hat{\beta}_{BI}$. Now, it is a fact that if $\text{var}\left(\underline{\mathrm{y}}\right) = \sigma^2\Sigma$, then $\text{var}\left(\underline{\hat{\beta}}_w\right) = \sigma^2\left(X'WX\right)^{-1}$, where $\underline{\hat{\beta}}_w$ is the weighted least squares estimator and the matrix $\mathrm{W} = \Sigma^{-1}$. If we simply combine the asymptotic distribution theory for the M- or BI-estimators with the fact that we obtained the solution as a weighted (iterated) least squares estimator, the following distribution "theory" is suggested:

$$\hat{\beta}_{\mathrm{M}}\left(\text{or} \hat{\beta}_{\mathrm{BI}}\right) \sim \mathrm{N}\left(\underline{\beta}, \mathrm{V}^2\left(\mathrm{X}'\mathrm{W}\mathrm{X}\right)^{-1}\right)$$

for small samples.

Using these ideas, an inferential procedure developed by Birch and Agard (1991) provides a unified approach to estimation and inference. They advocate the use of the previous general F-test with $\hat{\beta}$ replaced by either $\hat{\beta}_{\mathrm{M}}$ when using M-estimation or $\hat{\beta}_{\mathrm{BI}}$ when using BI-estimation, $\left(\mathrm{X}'\mathrm{X}\right)^{-1}$ replaced by $\left(\mathrm{X}'\mathrm{W}\mathrm{X}\right)^{-1}$, and s^2 replaced by a robust estimate of variability $\mathrm{s}_{\mathrm{B}}^2 = \hat{\mathrm{V}}^2 = \hat{\mathrm{s}}^2\left(\dfrac{\mathrm{n}^2}{\mathrm{n}-\mathrm{p}}\right)\sum_{\mathrm{i}=1}^{\mathrm{n}}\psi^2\left(\mathrm{r}_{\mathrm{i}}^*\right)/\left(\sum_{\mathrm{i}=1}^{\mathrm{n}}\psi'\left(\mathrm{r}_{\mathrm{i}}^*\right)\right)^2$, where $r_i^* = \dfrac{r_i}{\hat{\mathrm{s}}\pi_i}$ $\hat{\mathrm{s}}$ is usually the rescaled mad, and $\pi_i = 1$ for M-estimation or $\pi_{\mathrm{i}} = \dfrac{1-\mathrm{h}_{\mathrm{ii}}}{\sqrt{\mathrm{h}_{\mathrm{ii}}}}$ for BI-estimation, as discussed earlier in this chapter.

TABLE 7.5

The Robust ANOVA

Source	df	SS
model	k	$\sum w_i(\tilde{y}_i - \overline{y_w})^2 = SS_{regw}$
error	$n-p$	$\sum w_i(y_i - \tilde{y}_i)^2 = SS_{error}$
total	$n-1$	$\sum w_i(y_i - y_w)^2 = SS_{totalw}$

Essentially, they suggest that the variance-covariance matrix of $\underline{\beta}_M$ or $\underline{\beta}_{BI}$ be estimated by the more robust $s_B^2(X'WX)^{-1}$, where W is the weight matrix obtained at the convergence of the IRLS algorithm. Agard (1991) and Birch and Agard (1991) show that this form of the F-test, termed the F-test based on robust weights and denoted by F_w, provides for both robustness of validity (accurate α levels under a true null hypothesis) and robustness of sensitivity (high power under the alternative hypothesis) when the errors are from contaminated normal distributions. Their work supports and extends the results obtained by Birch and Myers (1982) and Birch (1983) for the simple analysis of the covariance model with two groups and one regressor. They also demonstrate that the test F_w compares favorably with the other robust tests, with greater power under certain circumstances. Thus, through the IRLS algorithm, the weights that are used for estimation are also utilized for testing.

In addition, the weights can be employed to construct a robust analysis of the variance table of the form in Table 7.5, where $\tilde{y}_i$ is the ith estimated response based on either $\hat{\beta}_M$ or $\hat{\beta}_{BI}$ and $\overline{y}_w$ is the weighted mean of the y's, the M- or BI-estimator of location.

The overall F-test, or model F-test, based on robust weights is simply the weighted mean square regression divided by s_B^2. We do not advocate the use of the weighted MSE since it has previously been demonstrated (Birch, 1983) that this estimator of σ^2 provides unstable test results.

Performing MLR using robust weights provides an approach with both intuitive appeal and a unified methodology. The IRLS procedure identifies the anomalous observations and accommodates their presence by assigning them low weights compared to the other observations. Furthermore, the weighted analysis of the variance table provides an interpretation consistent with that used in standard least squares analysis. Thus, with few exceptions, the method requires no new statistical concepts.

Other aspects of the traditional least squares analysis are also applicable in the weighted least squares context. For example, to test $H_0: \beta_j = 0$, one can use F_w, the partial F-test based on robust weights, with G equal to a $1xp$ vector with a one in the position of the jth coefficient and zeros elsewhere. Equivalently, one could compute the t-test based on robust weights as

$$t_{w_j} = \frac{\hat{\beta}_{M_j}}{\text{se}\left(\hat{\beta}_{M_j}\right)} \text{ or } t_{w_j} = \frac{\hat{\beta}_{BI_j}}{\text{se}\left(\hat{\beta}_{BI_j}\right)},$$

where $\text{se}\left(\beta_{M_j}\right) = s_B \cdot \sqrt{c_{jj}}$, for example, and c_{jj} is the jth diagonal element of $\left(X'WX\right)^{-1}$. And, as in the usual least squares setting, it can be shown that $t^2_{w_j} = F_{Wj_j}$.

Thus, a set of k t-tests, along with their p-values, could also be included with the ANOVA table, along with either $\hat{\beta}_M$ or $\hat{\beta}_{BI}$ and their standard errors. An R^2-type value, a "robust R^2" can be calculated by computing $R^2_w = \frac{SS_{\text{regw}}}{SS_{\text{totalw}}}$.

Robust inference using this procedure may be obtained on R as in the coming example.

7.4.1 The Potato Leafhopper Example

To illustrate the usefulness of the M-estimation procedure, the Huber weight function was used to identify the outliers. The example is taken in part from Birch and Fleisher (1982). As part of a developing alfalfa integrated pest management program in Virginia, the relationship between relative and absolute density estimates of adult potato leafhopper (*Empoasca fabae*) was deemed valuable. Regression analyses were needed to relate the mean field counted sweep-net catch from a given field (y variable) to the mean absolute density estimate from that field (x variable). The data appears in Table 7.6. Sweep-net counts had been obtained with pendular sweeps of a net. Absolute density estimates had been obtained from suction sampling using a special vacuum cleaner.

Further details concerning the sampling procedure may be found in Birch and Fleischer (1982). Each replicate resulted in a single x, y data point and consisted of a unique field or a field sampled on different dates, as in Table 7.6. Where the y variable was the sweep-net catch from a given field and the x variable was the mean absolute density estimated from the field.

The M-estimation procedure, using the Huber weight function, lead to the identification of seven outliers. The analysis of the full 30-point dataset continues.

```
> # Ordinary Least Squares (OLS) Regression
> ls_fit <- lm(sweep_density ~ absolute_density, data = data)
> summary(ls_fit)
Call:
lm(formula = sweep_density ~ absolute_density, data = data)
Coefficients:
              Estimate   Std. Error  t value  Pr(>|t|)
(Intercept)   0.715236     0.213569    3.349   0.00233 **
```

TABLE 7.6

Observations for Mean Absolute Density Estimates (x) and Mean Field Counted Sweep-Net Catch (y) of Adult Potato Leafhoppers in Alfalfa

Repl-cate	Harvest method	Stand quality	absolute density	sweep density
1	Strip	Good	39.96	1.55
2	Strip	Good	35.65	3.88
3	Strip	Good	56.43	2.74
4	Clean	Good	27.13	1.35
5	Clean	Good	33.66	1.89
6	Clean	Good	35.98	2.76
7	Clean	Poor	71.93	4.36
8	Strip	Good	54.45	3.45
9	Clean	Good	14.02	0.88
10	Clean	Good	12.85	0.88
11	Clean	Good	11.68	1.02
12	Clean	Good	17.52	1.19
13	Clean	Good	14.00	0.96
14	Clean	Good	14.00	0.93
15	Clean	Good	29.00	1.46
16	Clean	Good	29.00	1.08
17	Clean	Good	20.15	0.78
18	Clean	Good	14.89	0.46
19	Clean	Good	10.51	0.58
20	Clean	Good	10.51	0.40
21	Clean	Good	18.88	0.85
22	Clean	Good	18.88	0.99
23	Clean	Good	2.92	0.20
24	Clean	Good	11.34	0.65
25	Clean	Good	6.42	0.36
26	Clean	Good	43.41	2.40
27	Clean	Good	35.48	1.66
28	Clean	Good	8.91	0.35
29	Clean	Good	4.21	0.38
30	Clean	Poor	158.40	2.66

```
absolute_density 0.025103   0.005216    4.813  4.62e-05 ***
Signif. codes: 0 '***' 0.001 '**' 0.01 '*' 0.05 '.' 0.1 ' ' 1
Residual standard error: 0.8333 on 28 degrees of freedom
Multiple R-squared: 0.4528, Adjusted R-squared: 0.4332
F-statistic: 23.17 on 1 and 28 DF, p-value: 4.623e-05
> # Huber's M-estimator using rlm
> huber_fit <- rlm(sweep_density ~ absolute_density, data =
  data, psi = psi.huber, k = 1.5)
> summary(huber_fit)
Call: rlm(formula = sweep_density ~ absolute_density, data =
  data, psi = psi.huber, k = 1.5)
```

```
Coefficients:
                     Value  Std. Error  t value
(Intercept)         0.2232      0.0857   2.6051
absolute_density   0.0430      0.0021  20.5385
Residual standard error: 0.2954 on 28 degrees of freedom
> print(coefficients_comparison)
                 Row        ols     hubers
(Intercept)        1 0.71523557 0.22323887
absolute_density  2 0.02510286 0.04298102
> print("Robust ANOVA Table:")
[1] "Robust ANOVA Table:"
> print(robust_anova_table)
 Row  df        ss       ms        last
1  1   1  14.67075  14.67075  0.2227099
2  2  28  27.61515        NA  421.8289501
3  3  29  42.28590        NA          NA
> print("T-test based on Robust Weights:")
[1] "T-test based on Robust Weights:"
> print("T-test based on Robust Weights (Approximate):")
[1] "T-test based on Robust Weights (Approximate):"
> print(robust_t_test)
                 Row      hubers          se      ttest      pvaluet
(Intercept)        1 0.22323887 0.085693982  2.605071 9.185534e-03
absolute_density  2 0.04298102 0.002092707 20.538475 9.757885e-94
```

The output from the R lists the LS and *m* coefficients. It is clear that the outliers have greatly impacted the LS values. The robust ANOVA table shows an R^2 of 0.74, much higher than the LS value of 0.45, and much closer to the R^2 value of 0.83 for the 24-point dataset which results when six (of the seven outliers) are removed from the original dataset. We also see that the *t*-test based on robust weights is 20.53 and is highly significant (at $p < 0.0000001$), much more so than the LS test based on the 30 observations, and in agreement with the LS *t*-test based on the 24-point dataset. It can be observed that the overall F-test (based on robust weights) is the square of the *t*-test on the slope term, analogous to the LS result.

7.4.2 The Stackloss Data

The stackloss data of Section 7.2 consists of three regressors and 24 observations. The analysis on this data revealed four outliers, observations 1, 3, 4, and 21, with M-estimates based on the bisquare weight function as the preferred results. Robust inference on these M-estimates can be obtained on R as follows:

```
> print("OLS and Huber's M-Estimator Coefficients:")
[1] "OLS and Huber's M-Estimator Coefficients:"
> print(coefficients_comparison)
```

```
                    Row          OLS        Huber
(Intercept) Intercept -39.9196744 -41.1718411
airflow       Airflow   0.7156402   0.8133042
temp             Temp   1.2952861   0.9994335
acidcon       AcidCon  -0.1521225  -0.1324047
> print(robust_t_test)
          Row Huber_Coefficients          SE    T_Value      P_Value
(Intercept) Intercept -41.1718411 10.8742350 -3.7861828 1.529792e-04
airflow  Airflow        0.8133042  0.1232750  6.5974765 4.182150e-11
temp        Temp        0.9994335  0.3364142  2.9708420 2.969845e-03
acidcon  AcidCon       -0.1324047  0.1428698 -0.9267512 3.540557e-01
```

It is noted by comparing these results to the LS results that the robust standard errors for the robust estimates are about 50% of those for the LS results. The robust R^2 is 0.981, indicating that the fit to the remaining 17 observations is very strong. Of course, this value agrees with the least square results for $n = 17$ where the R^2 is 97.5%. Note that the first two regressors are highly significant, while the third is highly insignificant and probably should be removed from the model.

7.4.3 The High-Leverage Data

Recall the high-leverage data of Section 7.2 with ten observations, including one high-influence point. BI-estimation using the bisquare weight function was needed to enable the estimate for the slope to be negative, indicative of the trend in the bulk of the data. Continuing with the analysis of this data, we obtain the robust ANOVA table.

```
# Ordinary Least Squares (OLS) Regression
ols_fit <- lm(y ~ x, data = high_leverage_data)
# Robust Regression using Bisquare (Biweight) Estimator
bisquare_fit <- rlm(y ~ x, data = high_leverage_data,
  psi = psi.bisquare, c = 4.685)
#Huber's M-estimator with tuning constant 1.345
huber_fit <- rlm(y ~ x, data = high_leverage_data, psi =
  psi.huber, k = 1.345)
# Extract coefficients for OLS and Bisquare regression
ols_coef <- coef(ols_fit)
bisquare_coef <- coef(bisquare_fit)
# Display OLS and Bisquare coefficients side by side
coefficients_comparison <- data.frame(
Row = c("Intercept", "Slope"),OLS = ols_coef, Bisquare =
  bisquare_coef)
> print(coefficients_comparison)
                   Row       OLS  Bisquare
(Intercept) Intercept 2.4428139 2.4346103
x               Slope 0.4379728 0.4536331
**********************************************
```

```
> # Calculate the robust F-test
> robust_f_test <- bisquare_t_value[2]^2
> # Compile the ANOVA results into a table
> robust_anova_table <- data.frame(
+ Row = 1:3, + df = c(1, 8, 9),+ ss = c(ss_total-ss_
  resid_bisquare, ss_resid_bisquare, ss_total),+ ms =
  c(ss_total-ss_resid_bisquare, ss_resid_bisquare / 8,
  NA),+ last = c(robust_r_squared, robust_f_test,
  bisquare_p_value[2])+ )
> print(robust_anova_table)
  Row df        ss       ms       last
1   1  1 2.020896 2.020896 0.33408762
2   2  8 4.028104 0.503513 3.35396001
3   3  9 6.049000       NA 0.06704355
> print(robust_t_test)
                  Row Bisquare_Coefficients SE  T_Value      P_Value
(Intercept) Intercept 2.4346103 0.5526539 4.405307 0.0000105634
x               Slope 0.4536331 0.2476999 1.831382 0.0670435470
```

The robust ANOVA table indicates an R^2 of 51%, clearly higher than the 33% obtained by the LS analysis of all ten observations. The slope is highly significant at the 0.006 level.

Notice the similarity with the bi results and the results obtained using LS on the remaining nine data points, with observation 3 set aside.

```
# Regression without observation 3
high_leverage_data_omit3 <- high_leverage_data[-3,]
# Perform OLS regression on remaining 9 data points
ols_fit_omit3 <- lm(y ~ x, data =
  high_leverage_data_omit3)
summary(ols_fit_omit3)
# Display the results of the regression after omitting
  observation 3
print(summary(ols_fit_omit3))
Coefficients:
            Estimate Std. Error t value Pr(>|t|)
(Intercept)  4.8768     0.6770   7.203 0.000177 ***
x           -1.0665     0.4001  -2.666 0.032202 *
---
Signif. codes: 0 '***' 0.001 '**' 0.01 '*' 0.05 '.'
0.1 ' ' 1
Residual standard error: 0.4201 on 7 degrees of freedom
Multiple R-squared: 0.5037, Adjusted R-squared: 0.4329
F-statistic: 7.106 on 1 and 7 DF, p-value: 0.0322
```

The regression equation is

$$\mathbf{y = 4.88 - 1.07\ x}$$

7.5 Robust General Linear Model

7.5.1 Applications of Robust Estimation and Inference in the General Linear Model

In the previous section, we concentrated on the application of the *M*- and BI-estimation and inferential methods applied to the MLR model. In this section, we want to demonstrate these techniques applied to other examples of the GLM. Specifically, we will consider comparing population means for the two sample problems with both dependent (paired data) and independent samples and the *k*-sample problem based on independent samples. For each of these three situations, the classical methods will be compared to two alternative techniques, one based on nonparametric methods using the rank transformation and the other based on *M*-estimation. The final example will illustrate estimation and inference applied to the simple analysis of the covariance model. All applications will be demonstrated on R as in Chapters 3 and 6.

7.5.2 The Paired-Data Problem: Comparing Two Dependent Samples

7.5.2.1 The Paired t-Test Based on Robust Weights

The reader will recall the "paired-data" problem, where n independent pairs of observations are obtained through either a "matched-pair" or "before-after" experiment. In the matched-pair experiment, n pairs of experimental subjects are matched based on the similarity of important characteristics thought to be sensitive to the two treatments. One subject of each pair is randomly assigned to one of the treatments, while the other member of the pair is assigned to the other treatment. Another experimental design that leads to the same model and analysis is the "before-after" experiment, where the same subject is measured twice, once before the application of the treatment and again after the application of the treatment. In either design, the difference between the elements of the pair represents the "treatment effect." These n differences are then analyzed as a single sample of n independent observations in the location model.

Our model for the paired-data experiment, where we obtain two dependent samples each of size n, is that $\mathrm{y}_{1\mathrm{i}}, \mathrm{i} = 1,\ldots,\mathrm{n}$ represents n random observations from population 1 where treatment 1 has been applied and, similarly $y_{2i}, i = 1,\ldots,n$ represents n randomly obtained observations from population 2, where treatment 2 has been applied. The classical approach to this problem is to assume $\mathrm{y}_{1\mathrm{i}} \sim \mathrm{N}\left(\mu_1, \sigma_1^2\right)$ and $\mathrm{y}_{2\mathrm{i}} \sim \mathrm{N}\left(\mu_2, \sigma_2^2\right)$ for $\mathrm{i} = 1,\ldots,\mathrm{n}$, where the observations y_{1i} and y_{2i} occur in n pairs. That is, there exists a nonzero covariance between $\mathrm{y}_{1\mathrm{i}}$ and $\mathrm{y}_{1\mathrm{i}}$ for $\mathrm{i} = 1,\ldots,\mathrm{n}$. This implies that the pair (y_{1i}, y_{1i}) follows a bivariate normal distribution with the above means and variances and with covariance σ_{12}.

The experimenter is frequently interested in comparing the means, μ_1 and μ_2. This may be accomplished through tests of hypotheses of the form:$H_0 : \mu_1 = \mu_2$ versus one of the alternative hypotheses of the form $H_1 : \mu_1 \neq \mu_{2'}$ or $H_1 : \mu_1 < \mu_{2'}$ or $\mathrm{H}_1 : \mu_1 > \mu_2$.

One way to handle this problem is to remove the dependency between observations of each pair by computing the difference $d_i = y_{1i} - y_{2i}$. Under the assumptions of the model, we have $d_i \sim N\left(\mu_d, \sigma_d^2\right)$ where $\mu_d = \mu_1 - \mu_2$ and $\sigma_d^2 = \sigma_1^2 + \sigma_2^2 - 2\sigma_{12}$ The hypotheses of interest may then be rewritten in terms of $\mu_{d'}$; for example, $\mathrm{H}_0 : \mu_\mathrm{d} = 0$ versus $\mathrm{H}_1 : \mu_\mathrm{d} \neq 0$ would indicate a two-sided alternative in terms of μ_d. The analysis proceeds by computing the n differences, $d_i = y_{1i} - y_{2i'}$, summarizing this single sample of data by obtaining $\bar{d}$, the sample mean of the d's, and the sample standard deviation of the d's. Any of these hypotheses may be tested by computing the single sample t-test as

$$\mathrm{t}_{\text{paired}} = \frac{\bar{\mathrm{d}}}{\mathrm{s_d} / \sqrt{\mathrm{n}}}.$$

This t-test is usually referred to as the "paired t-test." The sampling distribution of this test, under a true null hypothesis and the above model assumptions, is that of a t-distribution with $n-1$ degrees of freedom. We will demonstrate the use of this procedure shortly.

The exploratory data analysis in the paired-data problem is performed on the n d's. If the dot plot or box plot on the d's reveals non-normal characteristics, the normality assumption is questionable. Of course, outliers among the d's cause problems for the paired test just as they did for the t-test in the location and regression models, with the primary effect being a loss of power of the test. Several alternative approaches exist for dealing with the problem of non-normal data, including non-normality due to outliers. Two methods will be considered here.

The first alternative is an adjustment to the popular nonparametric procedure referred to as the "Wilcoxon signed-rank test." A description of this method may be found in any nonparametric text (e.g., Conover, 1971). The procedure is as follows: Consider the n differences $\mathrm{d_i} = \mathrm{y_{1i}} - \mathrm{y_{2i}}$, and assign ranks of 1 to n to the smallest through largest ordered $|\mathrm{d_i}|$, respectively.

At this stage, any $|\mathrm{d_i}|$ equal to zero must be dropped from the analysis since they contribute no information toward the decision, and the sample size is reduced by the number of dropped values. Additionally, if any nonzero $|\mathrm{d_i}|$ are tied, we use the "tied rank method," which assigns to each tied $|d_i|$ the average of the ranks associated with the tied values. We then attach the sign of the d_i to the rank of d_i, that is, we obtain $R_i = \mathrm{sign}(d_i) * \mathrm{rank}(|d_i|), i = 1, 2, \ n$. These R_i are referred to as the "signed ranks." At this point, the $\mathrm{R_i}$ may be used to compute the Wilcoxon signed-rank test. However, another approach is to compute an approximation to this test by considering the signed ranks as a transformed version of the original data and using the paired *t*-test using the signed ranks.

That is, compute $\bar{R}$; the sample mean of the signed ranks; and s_R, the sample standard deviation of the signed ranks, and use the test

$$t_{\text{pairedrank}} = \frac{\bar{R}}{s_R / \sqrt{n}},$$

the paired *t*-test based on ranks. The approximate sampling distribution under the null hypothesis of this test is a *t*-distribution with "$n-1$" degrees of freedom, where "n" refers to the sample size after the $|d_i|$'s that equal zero, if any, are dropped. This procedure will be demonstrated on R.

The final alternative method will be based on using the M-estimate. Once the $d_i = y_{1i} - y_{2i}$ are obtained, we may use the estimation procedure for the location model presented in Chapter 3, specifically Sections 3.8 and 3.10. That is, we obtain the M-estimate and its standard error, $\hat{\theta}_m$ and $se\left(\hat{\theta}_m\right)$, respectively, based on the d_i's. The paired *t*-test based on robust weights, t_{pairedm}, then is used to test any of the three possible alternative hypotheses. This method will be demonstrated in subsection 7.5.3.

7.5.3 The Insulin Data

Koopmans (1987, p. 354) describes an experiment to study the insulin-producing capabilities of the (beta) islets of Langerhans cells in the pancreas. There is thought to be a difference between this capability between normal and obese individuals. A study on a special strain of rats was made to provide information regarding this hypothesis. Insulin production data was obtained over several weeks for both lean and obese rats. The data in Table 7.7 represents the amount of insulin produced from obese rat cell cultures at the start of the experiment (week 0) and one week later (week 1).

TABLE 7.7

The Insulin Data

Obs.	week0	week1
1	31.2	18.4
2	72.0	37.2
3	31.2	24.0
4	28.2	20.0
5	26.4	20.6
6	40.2	32.2
7	27.2	23.0
8	33.4	22.2
9	17.6	7.8

Is there a statistically significant difference between the amount of insulin produced during this one week for obese rat cell cultures? Since the data from week 0 and week 1 are measured in the same cell culture, the samples are dependent and paired by rat.

The analysis will be performed on R using the three methods. Unfortunately, due to carelessness on the part of the researcher, the week 0 and week 1 values for the second rat (observation 2) were reversed and recorded as week 0 of 37.2 and week 1 of 72.0. Note that the data entry error goes undetected at this point.

```
# Box plot of differences

boxplot(insulin_data$diff, main = "Boxplot of Differences",
xlab = "Differences", horizontal = TRUE)
```

It is clear from Figure 7.19 that one difference is an outlier. Again, the outlier is due to a recording error that is not corrected.

```
Descriptive Statistics for Differences:
> print(desc_stats)
   Min. 1st Qu.  Median    Mean 3rd Qu.    Max.
  -34.8    5.8      8.0    3.6       9.8   12.8
```

Note that the sample mean is greater than zero (mean = 3.6). The *t*-test will be insignificant because the outlier has shifted the sample mean to the left

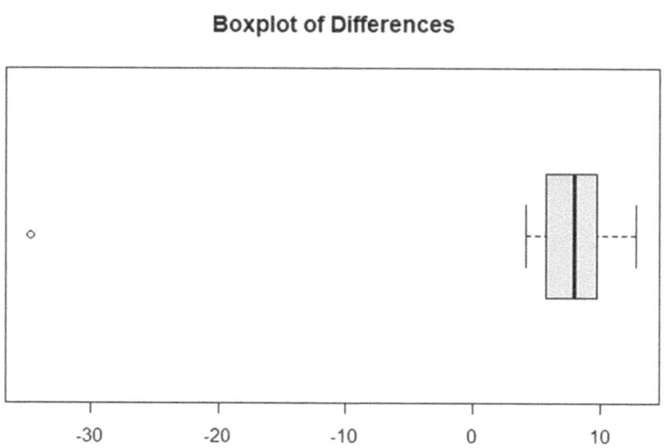

FIGURE 7.19
The box plot of the differences for the insulin data.

and inflated the standard error, and the 95% confidence interval on the difference in the population means.

```
> # Paired t-test
> t_test_result <- t.test(diff)
> print(t_test_result)
        One Sample t-test
data: diff
t = 0.73782, df = 8, p-value = 0.4817
alternative hypothesis: true mean is not equal to 0
95 percent confidence interval: -7.651473 14.851473
sample estimates:
mean of x 3.6
```

The *p*-value on the *t*-test is 0.48, a nonsignificant result suggesting that there is no difference between the average insulin production from week 0 to week 1 for cells from obese rats. This result is in agreement with the 95% confidence interval, which covers zero, indicating no difference between the two groups. To apply the *t*-test based on ranks, obtain the signed ranks as follows.

```
> # Wilcoxon signed-rank test
> wilcox_test_result <- wilcox.test(diff, exact = TRUE)
> print(wilcox_test_result)
        Wilcoxon signed rank exact test
data: diff
V = 36, p-value = 0.1289
alternative hypothesis: true location is not equal to 0
```

The result is not significant at 0.128. Use the M-estimation method. Try the Huber weight function with a constant of 1.5. Obtain the 95% confidence interval.

```
> cat("Huber M-estimator mean:", huber_est$mu, "\n")
Huber M-estimator mean: 7.77143
> cat("Huber M-estimator 95% CI:", ci_lower, "-", ci_upper, "\n")
Huber M-estimator 95% CI: 0.2498898 - 15.29297
> cat("Huber M-estimator p-value:", p_value, "\n")
Huber M-estimator p-value: 0.0443641
```

We see that the estimate of the average difference is 7.8 based on the M-estimate, compared to 3.4 for the LS method. The standard error is now 1.43 compared to 4.88 for LS. The 95% confidence interval for the population difference is 4.5–11.1, much narrower than the classical result and indicating a significant difference from zero. The weight index plot in Figure 7.20 reveals that the second observation receives a weight of zero. It should be noted that

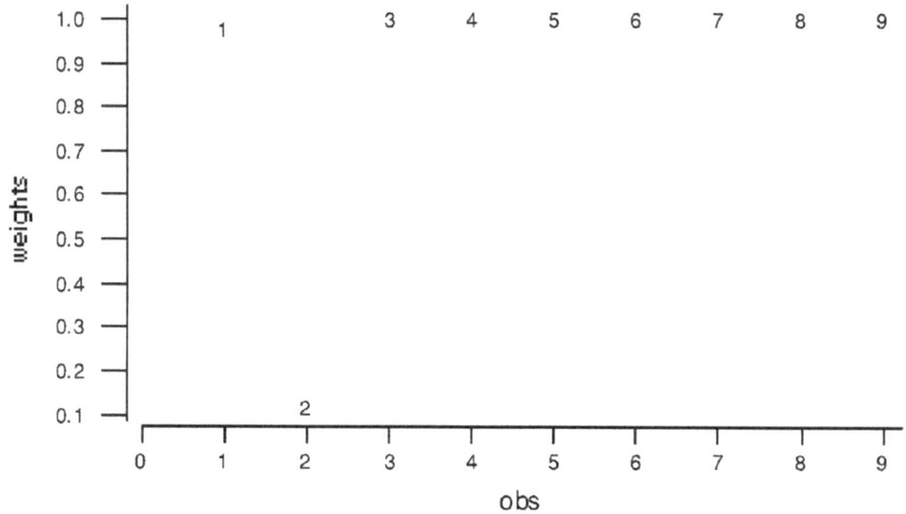

FIGURE 7.20
Weights using the Huber function for the insulin data.

had the second observation been read correctly, the difference would still have been an outlier.

The paired *t*-test based on ranks and the paired *t*-test based on robust weights would both have resulted in a highly significant test. However, the M-estimate based procedure would have resulted in a smaller *p*-value and a much narrower 95% confidence interval.

7.5.4 Comparing Two Independent Samples

7.5.4.1 The Two-Sample t-Test Based on Robust Weights

Suppose that the two samples are independent and it is desired to compare the population means across the two groups or treatments. The assumptions for the classical model analysis are to consider the n_1 observations from the first treatment, y_{1i}, $i = 1,\ldots,n_1$, and the n_2 observations from the second treatment, y_{2i}, $i = 1,\ldots,n_2$, as independent observations from normal populations with means μ_1 and μ_2 and variances σ_1^2 and σ_2^2. An additional assumption is that $\sigma_1^2 = \sigma_2^2 = \sigma^2$, where σ^2 represents the common variance in both populations. This is the homogeneity of variance assumption (homoscedasticity). Thus, our two-sample model is

$$y_{ij} = \mu_i + \varepsilon_{ij},$$

where $i = 1, 2,\ j = 1, \ldots, n_i$, and $\varepsilon_{ij} \sim N(0, \sigma^2)$.

Under these assumptions, the best test statistic for testing hypotheses of the form $H_0 : \mu_1 - \mu_2 = \delta_0$ versus either of the three alternative hypotheses $H_1 : \mu_1 - \mu_2 \neq \delta_0$, or $H_1 : \mu_1 - \mu_2 < \delta_0$, or $H_1 : \mu_1 - \mu_2 > \delta_0$ is the ever-popular two-sample *t*-test, based on common variance, or

$$t = \frac{\bar{y}_1 - \bar{y}_2 - \delta_0}{s_p \sqrt{\frac{1}{n_1} + \frac{1}{n_2}}},$$

where the pooled estimate of variance is $s_p^2 = \dfrac{(n_1 - 1)s_1^2 + (n_2 - 1)s_2^2}{(n_1 - 1) + (n_2 - 1)}$. The null distribution t is that of a *t*-distribution with $n_1 + n_2 - 2$ degrees of freedom.

Similar to the location and regression models, the *t*-test is extremely sensitive to outliers occurring in either or both samples, and many attempts have been made to obtain robust versions of this procedure.

One nonparametric method is the Wilcoxon rank sum test, which offers protection against not only outliers but other non-normal characteristics as well. The procedure to be demonstrated here is the approximation to the rank sum test. Both procedures begin in a similar fashion: pool the data from the two samples into one large sample, keeping track of each observation's treatment number, either a 1 or a 2. Assign ranks of 1 to $n_1 + n_2$ from the smallest through largest observations, respectively.

At this point, the Wilcoxon rank sum test considers the test statistic as the sum of the ranks associated with the group with the smallest sample size and compares this sum to special tables appropriate for this test statistic.

The approximation method proceeds as follows: considering the ranks as new data for each treatment, compute the sample means and variances for each treatment to obtain $\bar{R}_1, \bar{R}_2$ $s_{R_1}^2$, and $s_{R_2}^2$, respectively. Compute as the test statistic

$$t_R = \frac{\bar{R}_1 - \bar{R}_2 - \delta_0}{s_{R_p} \sqrt{\frac{1}{n_1} + \frac{1}{n_2}}}$$

the two-sample *t*-test based on ranks. Here, $s_{R_p}^2 = \dfrac{(n_1 - 1)s_{R_1}^2 + (n_2 - 1)s_{R_2}^2}{(n_1 - 1) + (n_2 - 1)}$ is the pooled estimate of variance based on the ranks assigned to the data. The null distribution of t_R is approximated by the with $n_1 + n_2 - 2$ degrees of freedom. It is easy to see that this test is in exactly the same format as the two-sample *t*-test with the exception that the ranks of the data are used rather than the raw data.

The third alternative is the two-sample t-test based on robust weights. The M-estimate is obtained for each sample along with their standard errors. This information is then combined exactly as in the two tests to construct the t-test based on robust weights as

$$t_w = \frac{\hat{\theta}_{m1} - \hat{\theta}_{m2} - \delta_0}{se\left(\hat{\theta}_{m1} - \hat{\theta}_{m2}\right)} = \frac{\hat{\theta}_{m1} - \hat{\theta}_{m2} - \delta_0}{\hat{V}_p\sqrt{\dfrac{1}{\sum_{i=1}^{n_1} W_{1i}} + \dfrac{1}{\sum_{i=1}^{n_2} W_{2i}}}}$$

where $\hat{\theta}_{mi};i=1,2$ are the M-estimates for the two treatments and $\hat{V}_p$is the robust pooled estimate of variance computed as

$$\hat{V}_p^2 = \left(\frac{(n_1+n_2)^2}{n_1+n_2-2}\right)\frac{\hat{s}^2\sum_{i,j}\psi^2\left(r_{ij}'\right)}{\left(\sum_{i,j}\psi'\left(r_{ij}'\right)\right)^2},$$

with $r_{ij}' = \dfrac{y_{ij} - \hat{\theta}_{m_i}}{\hat{s}}, i=1,2$, and $\hat{s}$ is the rescaled mad based on all n_1+n_2 residuals, and $w_{11},\ldots,w_{1n_1}$ and $w_{21},\ldots,w_{2n_2}$ are the weights at the convergence of the IRLS algorithm. The approximate null distribution of t_m is a t with n_1+n_2-2 degrees of freedom.

Perhaps the easiest way to perform the two-sample t-test based on robust weights is to rewrite the two-sample model as an MLR model using an indicator variable. In this case, the existing estimation MLR programs used in Section 6.2 can be used to compare μ_1 to μ_2.

To illustrate this approach, we let the indicator variable z be defined as zero if the observation is from treatment 1 and as one if the observation is from treatment 2. In this definition of z, the treatment receiving the "zero" is called the "reference treatment" and the treatment receiving the "one" is referred to as the "indicated treatment." The reparameterized means model may now be written as

$$y_{ij} = \beta_0 + \beta_1 z_{ij} + \varepsilon_{ij},$$

where $\varepsilon_{ij} \sim N\left(0,\sigma^2\right)$. To relate the two sample means model to the SLR model, simply consider the two values of z. If an observation is from treatment 1, then the SLR model states that $E\left(y_{1j}\right) = \beta_0$. But the means model already states that $E\left(y_{1j}\right) = \mu_1$, so it follows that $\beta_0 = \mu_1$. For an observation from treatment 2, the SLR model states that $E\left(y_{2j}\right) = \beta_0 + \beta_1(1) = \beta_0 + \beta_1$. Since the means model states that $E\left(y_{2j}\right) = \mu_2$, it follows that $\beta_0 + \beta_1 = \mu_2$ or, rewriting, $\beta_1 = \mu_2 - \mu_1$. Thus, β_1 is a parameter that measures the difference between the two population means. Consequently, an estimate of β_1 is the estimate of the difference in the means from the indicated to the reference population. The LS estimate of β_1 will be $\bar{y}_2 - \bar{y}_1$, and the M-estimate will be $\theta_{m_1} - \theta_{m_2}$.

Testing hypotheses of interest involving $\mu_2 - \mu_1$ is equivalent to testing hypotheses on β_1. For example, to test $H_0 : \mu_2 - \mu_1 = 0$ versus $H_1 : \mu_2 - \mu_1 \neq 0$ in the means model is equivalent to testing $H_0 : \beta_1 = 0$ versus $H_1 : \beta_1 \neq 0$ in the SLR model using indicator variables. In fact, when using *M*-estimation to test the two-sided alternative, the two-sample test based on robust weights is equivalent to the model F-test based on robust weights since both tests have the same *p*-value.

7.5.5 The Mill Data Revisited

Each of the previous three techniques will be demonstrated on the mill data, first introduced in Section 5.1.1. Recall that there are five mills to be compared for the amount of wastage of material that results from their production processes. The one-way ANOVA analysis from Chapter 5 using least squares failed to detect any significant differences between the average amount of wastage across the five mills. This was due to many outliers that exist in the dataset.

On the other hand, the nonparametric method, the approximate Kruskal-Wallis method, found some differences between average wastage across the mills; specifically, there was a significant difference between mills 1 and 4. Suppose mills 1 and 4 represent our two treatments, respectively. We obtain the following analysis on R.

```
# Data Preparation

mill_data <- data.frame(

  mill1 = c(0.12, 1.01, 0.15, -0.20, -0.30, -0.07, 0.32, 0.27,
-0.32, -0.17, 0.24, 0.03, 0.35, -0.08, 1.94, 0.28, 1.30, 4.27, 0.14,
0.30, 0.24, 0.13),

  mill4 = c(1.15, 1.02, 0.38, 0.83, 0.66, 1.02, 0.88, 0.27, 0.51, 1.12,
0.59, 1.30, 0.68, 1.45, 0.52, 0.73, 0.71, 0.34, 0.07, NA, NA, NA))

# Remove rows with NA values in mill4

mill_data <- mill_data[!is.na(mill_data$mill4), ] # Remove rows with
NA in mill4

# Create Horizontal Box plot for Mill 1 and Mill 4

boxplot(mill_data$mill1, mill_data$mill4,
        names = c("Mill 1", "Mill 4"),main = "Horizontal Boxplot of
Wastage for Mill 1 and Mill 4", xlab = "Wastage", horizontal = TRUE,
        bty = "n") # 'bty = "n"' removes the box around the plot
```

Box plots in Figure 7.21 clearly reveal outliers in mill 1 and a distinct shift in mill 4 (labeled as mill 2) values of wastage.

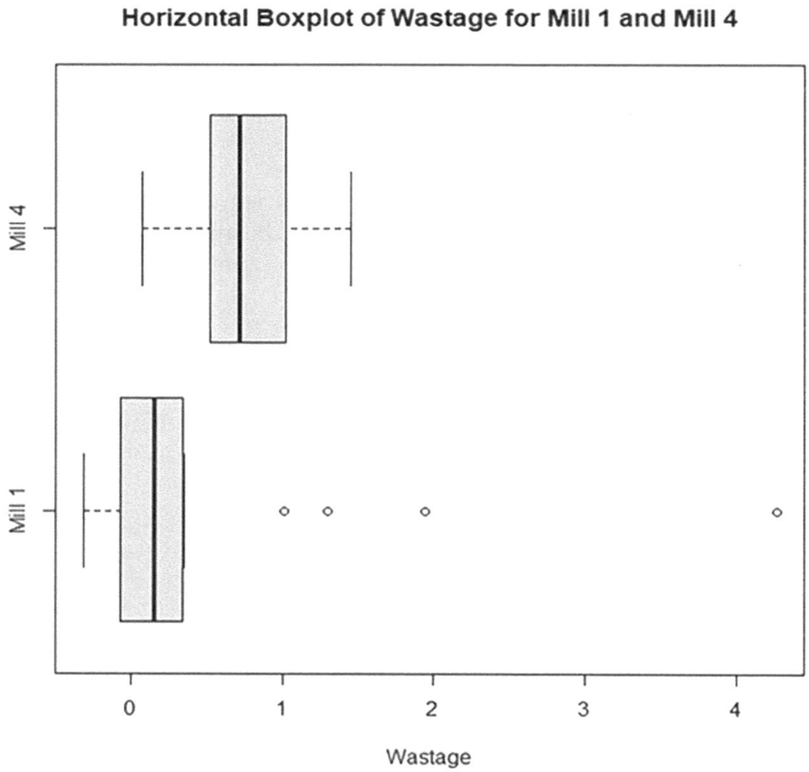

FIGURE 7.21
Box plot for mill 1 and mill 4.

```
> # Two-sample t-test (pooled)
> t_test_result <- t.test(wastage ~ mill, data = mill_data_
  long, var.equal = TRUE)
> print(t_test_result)
        Two Sample t-test
data: wastage by mill
t = -0.99716, df = 36, p-value = 0.3253
alternative hypothesis: true difference in means between group
mill1 and group mill4 is not equal to 0
95 percent confidence interval: -0.7904022 0.2693495
sample estimates:
mean in group mill1 mean in group mill4
          0.4884211           0.7489474
```

As suspected, the classical method cannot detect a significant difference between the two mills in terms of average wastage.

The *t*-test based on ranks is obtained by first transforming the raw data by the rank transformation and then using the two-sample *t*-test.

```
data: rank_wastage by mill
t = -3.262, df = 36, p-value = 0.002425
alternative hypothesis: true difference in means between
group mill1 and group mill4 is not equal to 0
95 percent confidence interval:
  -16.985590 -3.961779
sample estimates:
mean in group mill1 mean in group mill4
          14.26316            24.73684
```

This test is highly significant at the $p = 0.0005$ level. We obtain the two-sample *t*-test based on robust weights or the equivalent F-test based on robust weights as follows.

```
> # Robust Regression (Huber M-Estimator)
> # Create indicator variable for robust regression
> mill_data_long <- mill_data_long %>%
+ mutate(indicator = ifelse(mill == "mill1", 1, 0))
> # OLS regression
> ols_model <- lm(wastage ~ indicator, data = mill_data_long)
> summary(ols_model)
Call:
lm(formula = wastage ~ indicator, data = mill_data_long)
Coefficients:
            Estimate Std. Error t value Pr(>|t|)
(Intercept) 0.7489    0.1847    4.054   0.000257 ***
indicator  -0.2605    0.2613   -0.997   0.325342
---
Signif. codes: 0 '***' 0.001 '**' 0.01 '*' 0.05 '.' 0.1 ' ' 1

Residual standard error: 0.8053 on 36 degrees of freedom
Multiple R-squared: 0.02688, Adjusted R-squared: -0.0001532
F-statistic: 0.9943 on 1 and 36 DF, p-value: 0.3253
```

The regression equation is

```
wastage = 0.749 - 0.297 mils
```

Note that the slope coefficient is just the LS estimate of the difference in population means of mill 4 from mill 1. The *t*-test and *p*-value for the slope are just the two-sample t-test as seen earlier. Also, note that the t-test when squared is the overall F-test. Both have the same p-value.

To get the M-estimation analysis, type as follows:

```
> # Display the results with the p-value included
> cat("Robust Regression (Huber M-Estimator) Results:\n")
Robust Regression (Huber M-Estimator) Results:
Coefficients:
            Value Std. Error t value p-value
(Intercept) 0.7466     0.0971  7. 6922   0.0000
indicator  -0.5532     0.1373  -4.0301   0.0003
```

The first portion of the output compares the OLS estimates (the sample mean for mill 4 and the difference in sample means of mill 1 and mill 4) to the M-estimates based on the Huber weight function with a constant of 1.5. We see that there is little difference between the OLS and the M-estimates of the mean for mill 4 since the data in mill 4 is well behaved. But there is a considerable difference between the estimate of the difference in the means, with the M-estimate being much larger in magnitude. The t-test based on robust weights is given in bold. Note that the result is highly significant, at a p-value of 0.0001. The overall F-test based on robust weights (in bold) is the square of the two-sample t-tests based on robust weights.

A nice comparison of the two fits is obtained by a scatterplot of wastage versus z in Figure 7.22. The slope of the estimated line is a visual representation of the difference between the two estimates of the means.

The pull of the outliers at $z = 1$ (at mill 1) is readily apparent. Note that the bottom line represents the robust analysis.

7.5.6 Robust One-Way ANOVA

7.5.6.1 The F-Test Based on Robust Weights and a Robust Multiple Comparison Method

As a third model, consider the one-way ANOVA model presented in Chapter 5. The classical analysis based on the LS method was presented in Section 5.1. One alternative to the classical method is the approximation of the Kruskal-Wallis test, which simply uses rank-transformed data in the classical formulas. The third approach to be illustrated here uses M-estimation and the GLM. The method will require that the usual means model be rewritten in terms of $k-1$ indicator variables, where k represents the number of treatments, with $k \geq 2$.

We will illustrate the procedure using the mill data and all five mills. We will then need four indicator variables, z_1, z_2, z_3, and z_4. Let mill 1 be the reference mill with all indicators set to zero; let z_1 indicate mill 2 (i.e., if an observation is from mill 2 let $z_1 = 1$ and if an observation is from some other

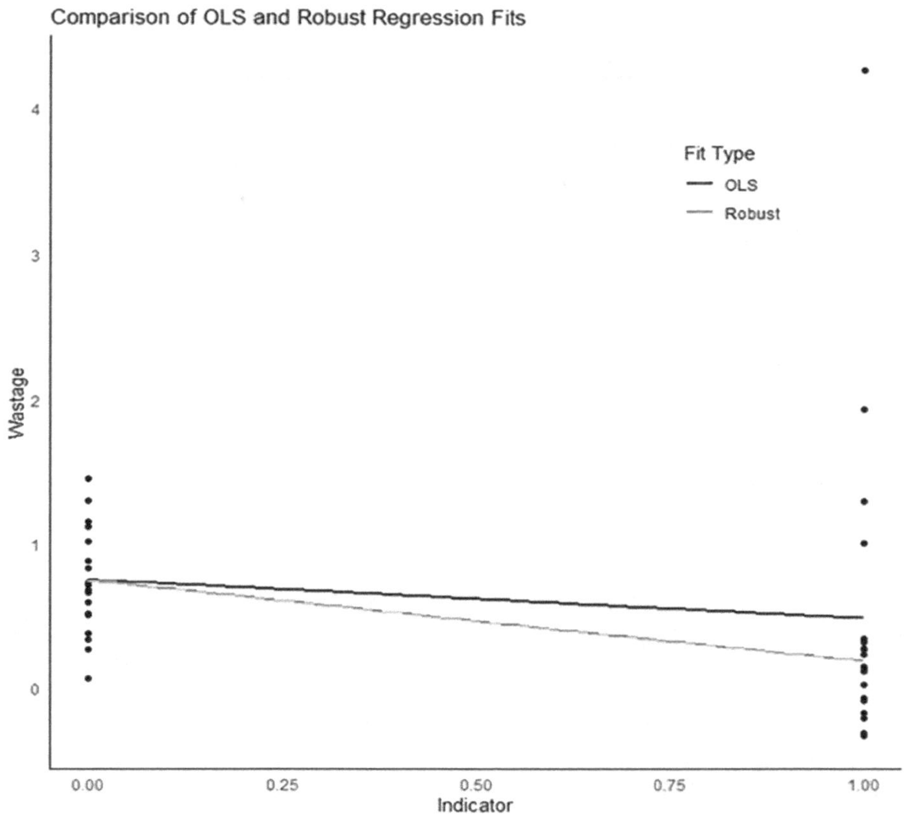

FIGURE 7.22
Scatterplot of the mill 1 and mill 4 data with the LS and M-estimate fits.

mill, let $Z_1 = 0$), let z_2 indicate mill 3, z_3 indicate mill 4, and z_4 indicate mill 5. Let our means model be written as a four-variable MLR as

$$y_i = \beta_0 + z_1\beta_1 + z_2\beta_2 + z_3\beta_3 + z_4\beta_4 + \varepsilon_i,$$

where $i = 1,\ldots,n$, where $n = \sum_{i=1}^{k} n_j$ with $\varepsilon_{ij} \sim N\left(0,\sigma^2\right)$. To establish the equivalence between the means model and the MLR, set up the following equivalence table.

We see that the intercept term in the MLR model represents the population mean for the reference population, mill 1 in this example. The coefficient on z_1, the indicator of mill 2, represents the difference in population means between the indicated mill and the reference mill. The other three coefficients in the MLR model have a similar interpretation.

TABLE 7.8

The MLR and One-Way ANOVA

Treatment	z_1, z_2, z_3, z_4	$E(y_i)$ MLR	$E(y_{ij})$ 1-way		Equivalence
mill 1	0 0 0 0	β_0	$= \mu_1$		$\beta_0 = \mu_1$
mill 2	1 0 0 0	$\beta_0 + \beta_1$	$= \mu_2$	$\Rightarrow$	$\beta_1 = \mu_2 - \mu_1$
mill 3	0 1 0 0	$\beta_0 + \beta_2$	$= \mu_3$	$\Rightarrow$	$\beta_2 = \mu_3 - \mu_1$
mill 4	0 0 1 0	$\beta_0 + \beta_3$	$= \mu_4$	$\Rightarrow$	$\beta_3 = \mu_4 - \mu_1$
mill 5	0 0 0 1	$\beta_0 + \beta_4$	$= \mu_5$	$\Rightarrow$	$\beta_4 = \mu_5 - \mu_1$

Analogous to the two-sample model, it follows that the estimates of the betas in the MLR model match the corresponding estimates on the means model. Thus, for example, $\hat{\beta}_0 = \hat{\mu}_1$. And, if OLS is used to obtain the estimates, we will have $\hat{\beta}_0 = \hat{\mu}_1 = \overline{y}_1$; if M-estimation is used, we will have $\hat{\beta}_0 = \hat{\mu}_1 = \hat{\theta}_{m_1}$. Similarly, using OLS estimates, we have $\hat{\beta}_1 = \hat{\mu}_2 - \hat{\mu}_2 = \overline{y}_2 - \overline{y}_1$ and using m -estimation, we have $\hat{\beta}_1 = \hat{\mu}_2 - \hat{\mu}_1 = \hat{\theta}_{m_2} - \hat{\theta}_{m_1}$. The other three estimates have similar interpretations. We see that estimates of the slope terms compare the four indicated sample mean estimates to the reference sample mean estimate.

The overall F-test based on robust weights obtained from the robust ANOVA table will be a test of the hypotheses: H_0: all μ_i's equal $\Leftrightarrow \mathrm{H}_0 : \beta_1 = \beta_2 = \beta_3 = \beta_4 = 0$ versus H_1: not H_0. If it is desired to obtain an ordering of the population means based on the sample means, then multiple pairwise tests will have to be performed. For LS analysis, the protected Fisher's least significant difference (LSD) is a useful procedure. For M-estimation, the robust version of Fisher's multiple comparison method is easy to obtain. One simply uses the overall F-test based on robust weights as the protection test. If this test is significant at the predetermined level, say 0.05, the pairwise comparisons are made using the two-sample t-test based on robust weights, discussed above or, equivalently, the F-test based on robust weights, with both tests using the pooled variance estimate based on all treatments. By considering the matrix G used in that test as a $1xk$ vector with a "1" and a "–1" in the appropriate positions, all comparisons needed to complete the robust version of Fisher's method may be obtained. This technique will be illustrated in the next example.

7.5.7 The Mill Data Again

We will demonstrate this robust method by referring again to the mill data using all five mills. The 95 observed wastage values are placed in $c1$ with the

$c2$ through $c5$ containing the four indicator variables, Z_1, Z_2, Z_3 and Z_4. The indicator variables are entered through the "set" command.

On R, type as follows:

```
> #Fit the OLS regression model
> ols_model <- lm(wastage ~ z1 + z2 + z3 + z4, data = mill_data)
> summary(ols_model)
lm(formula = wastage ~ z1 + z2 + z3 + z4, data = mill_data)
Residuals:
    Min       1Q    Median      3Q      Max
-1.9677  -0.3173  -0.0889  0.2859  3.8177
Coefficients:
             Estimate Std. Error t value  Pr(>|t|)
(Intercept)  0.45227     0.14626   3.092  0.00265 **
z1           0.29182     0.20685   1.411  0.16175
z2           0.07404     0.21486   0.345  0.73119
z3           0.29667     0.21486   1.381  0.17076
z4           0.58542     0.23999   2.439  0.01668 *
---
Signif. codes: 0 '***' 0.001 '**' 0.01 '*' 0.05 '.' 0.1 ' ' 1
Residual standard error: 0.686 on 90 degrees of freedom
Multiple R-squared:  0.07492,  Adjusted R-squared:  0.0338
F-statistic: 1.822 on 4 and 90 DF,  p-value:  0.1314
```

The regression equation is

```
wastage = 0.452 + 0.431 z2 + 0.031 z3 + 0.297 z4 + 0.585 z5
```

This results match exactly the LS one-way ANOVA results obtained in Section 4.1. That is, the overall F is highly insignificant, indicating that there are no differences in average wastage across the five mills. Careful inspection of the coefficients shows that the intercept is the sample average for mill 1, the coefficient for $z2$ is the difference in sample means between mill 2 and mill 1, the coefficient for $z3$ is the difference in sample means between mill 3 and mill 1, and so on. As expected, each is associated with an insignificant t-statistic. The conclusion is that, in terms of average wastage, the five mills are indistinguishable.

The robust analysis using the Huber weight function is as follows.

```
> # Robust regression using the Huber M-estimator
> huber_model <- rlm(wastage ~ z1 + z2 + z3 + z4, data =
  mill_data, psi = psi.huber)
> huber_summary <- summary(huber_model)
> print(huber_summary)
Call: rlm(formula = wastage ~ z1 + z2 + z3 + z4, data = mill_data,
    psi = psi.huber)
Residuals:
    Min          1Q      Median        3Q      Max
-2.09321   -0.26212   -0.03713   0.27449 4.06964
```

```
Coefficients:
            Value  Std. Error t value
(Intercept) 0.2004 0.0973     2.0597
z1          0.5635 0.1376     4.0963
z2          0.3331 0.1429     2.3313
z3          0.5468 0.1429     3.8263
z4          0.9628 0.1596     6.0323
Residual standard error: 0.4045 on 90 degrees of freedom
> print(anova_table)
      df        ss        ms    F_value     p_value
     1  4  8.701792 2.1754479 4.453001  0.002494351
     2 90 43.968172 0.4885352       NA           NA
     3 94 45.787899        NA       NA           NA
> # Run the robust pairwise comparison for mill 3 vs mill 5
> contrast_m3_m5 <- c(0, 0, 1, 0, -1)
> f_test_m3_m5 <- robust_f_test(contrast_m3_m5, robust_coefs,
  inv_xpwX, ms_residual, df_residual)
> cat("F test comparing mill 3 to mill 5: F = ", f_test_m3_
  m5[1], ", p-value = ", f_test_m3_m5[2], "\n")
F test comparing mill 3 to mill 5: F = 6.265245, p-value =
  0.01411722
> inv_xpwX <- solve(crossprod(model.matrix(huber_model)))
> # Contrast matrix example to compare mill 3 to mill 2
> contrast_m3_m2 <- c(0, 0, 1, -1, 0)
> f_test_m3_m2 <- robust_f_test(contrast_m3_m2, robust_coefs,
  inv_xpwX, ms_residual, df_residual)
> cat("F test comparing mill 3 to mill 2: F = ", f_test_m3_
  m2[1], ", p-value = ", f_test_m3_m2[2], "\n")
F test comparing mill 3 to mill 2: F = 0.8875485, p-value =
  0.3486645
> # Compare mill 3 to mill 5
> contrast_m3_m5 <- c(0, 0, 1, 0, -1)
> f_test_m3_m5 <- robust_f_test(contrast_m3_m5, robust_coefs,
  inv_xpwX, ms_residual, df_residual)
> cat("F test comparing mill 3 to mill 5: F = ", f_test_m3_
  m5[1], ", p-value = ", f_test_m3_m5[2], "\n")
F test comparing mill 3 to mill 5: F = 6.265245, p-value =
  0.01411722
# Perform OLS and Robust Regression (Huber M-Estimator)
> # OLS regression
> ols_model <- lm(wastage ~ z1 + z2 + z3 + z4, data =
  mill_data)
> # Robust regression using the Huber M-estimator
> huber_model <- rlm(wastage ~ z1 + z2 + z3 + z4, data =
  mill_data, psi = psi.huber)
> # Extract coefficients from both models
> ols_coefs <- coef(ols_model)
> huber_coefs <- coef(huber_model)
> # Combine the coefficients into a matrix for comparison
```

```
> comparison_matrix <- cbind(OLS = ols_coefs, Hubers =
  huber_coefs)
> # Print the comparison table in the desired format
> cat(" Row        ols       hubers\n\n")
  Row         ols      hubers
  1       0.452273   0.200359
  2       0.291818   0.563530
  3       0.074043   0.333134
  4       0.296675   0.546774
  5       0.585420   0.962848
> # Print the ANOVA table in the desired format
The Analysis of Variance Table is:
  Row   df        ss         ms        last
  1     4     8.7018     2.17545   4.45300
  2     90    43.9682    0.48854   0.48854
  3     94    45.7879    0.48854   0.00249
> # Print the Coefficients with t-values and p-values
  Row      hubers          se      ttest     pvaluet
  1      0.200359  0.097277  2.05967   0.042319
  2      0.563530  0.137571  4.09630   0.000092
  3      0.333134  0.142898  2.33127   0.021972
  4      0.546774  0.142898  3.82632   0.000240
  5      0.962848  0.159615  6.03232   0.000000
```

We note that the overall F-test based on robust weights is highly significant with a *p*-value of 0.00029. We also see from the coefficients that mill 2, mill 4, and mill 5 are significantly different in mean wastage from mill 1 due to having *p*-values all smaller than 0.05. Since there are five treatments, an ordering of the five means requires 5(4)/2 = 10 comparisons, of which four are already made through the coefficients of *z*2 through *z*4. A quick scan of the coefficients reveals that the ordering of the M-estimates of mean wastage are (in order of smallest to largest values) as follows:

Mill 1 Mill 3 Mill 2 Mill 4 Mill 5

Since mill 2 through mill 5 have already been compared to mill 1, we need only to make the other six comparisons.

```
> # Perform the robust F-test for mill 3 vs mill 2
> contrast_m3_m2 <- c(0, 1, -1, 0, 0) # contrast for comparing
  mill 3 to mill 2
> f_test_m3_m2 <- robust_f_test(contrast_m3_m2, robust_coefs,
  inv_xpwX, ms_residual, df_residual)
> cat("The F test based on robust weights comparing mill 3 to
  mill 2:\n")
The F test based on robust weights comparing mill 3 to mill 2:
frobust = 1.107765
pvaluef = 0.2953861
> #Perform the robust F-test for mill 3 vs mill 5
```

```
> contrast_m3_m5 <- c(0, 0, 1, 0, -1) # contrast for comparing
  mill 3 to mill 5
> f_test_m3_m5 <- robust_f_test(contrast_m3_m5, robust_coefs,
  inv_xpwX, ms_residual, df_residual)
The F test based on robust weights comparing mill 3 to mill 5:
> cat("frobust =", f_test_m3_m5[1], "\n")
frobust = 6.265245
pvaluef = 0.01411722
> # Grouping interpretation
> cat("Based on the robust F-tests, the following groupings
  are identified:\n")
Based on the robust F-tests, the following groupings are
  identified:
> cat("{mill 1, mill 3} < {mill 3, mill 2, mill 4} < {mill 4,
  mill 5}\n")
{mill 1, mill 3} < {mill 3, mill 2, mill 4} < {mill 4, mill 5}
```

Note that this represents one more significant difference than found by the F procedure based on ranks appearing in Section 4.2.

7.5.8 An Example of Analysis of Covariance: A Comparison of Regression Lines Based on Robust Weights

The final GLM that will be demonstrated is an example of the analysis of the covariance (ANCOVA) model. This model may be thought of as one that combines both the means model (the ANOVA model) with the MLR model. That is, in general, k treatment means are to be compared, just as in the ANOVA model, but the response is also a linear function of one or more regressors, commonly called "covariates" in the ANCOVA model. Consequently, the model is a combination of the ANOVA and MLR models.

For example, the simple ANCOVA model, the primary subject of this illustration, compares the response across two treatments with one covariate. We write the model as

$$y_{ij} = \mu_i + \beta_1 x_{ij} + \varepsilon_{ij} = \beta_{0i} + \beta_1 x_{ij} + \varepsilon_{ij},$$

where $i = 1, 2$ and $j = 1, \ldots, n_i$. The first model on the right-hand side expresses the response as a combination of the means and SLR models, while the second model on the right shows that the first is equivalent to comparing two SLR lines with a common slope. One hypothesis of interest in the ANCOVA model is to compare the means, μ_1 to μ_2, after adjustment for the covariate, x. This is the so-called test of the "adjusted means." We see from the second model that this is tantamount to comparing the two intercepts for the SLR lines.

The general ANCOVA model requires that the slope be the same for each covariate across all k treatments. The assumption of equal slope for the case of $k = 2$ and one covariate may be tested in two equivalent ways. The first is to fit each SLR separately, allowing for unequal slopes, β_{11} and β_{12}, for each of the two groups. If the two slopes are to be compared via hypothesis testing, a test on the contrast $\beta_{11} - \beta_{12}$ may be used with a "pooled" estimate of σ^2 required in the denominator of the test. Similarly, the two intercepts may be compared by using a *t*-test on the contrast.

Another more convenient approach is to reparameterize the SLR model with unequal slopes to a MLR model using indicator variables. Defining

$$z_i = \begin{cases} 0 & \text{if ith observation is from treatment1} \\ 1 & \text{if ith observation is from treatment2}' \end{cases}$$

we can rewrite the two SLR models as a single MLR model

$$y_i = \beta_0 + \beta_1 x_i + \beta_2 z_i + \beta_3 xz_i + \varepsilon_i,$$

where $xz_i = (x_i)(z_i)$, the product of the covariate and the indicator variable and $i = 1, \ldots, n_1 + n_2$. Using a similar argument that was used when reparameterizing the one-way ANOVA model using indicator variables, we may establish the following equivalence relationship between the two SLRs and MLR model:

From Table 7.9, it follows that the parameters for the two models are related as

$$\beta_0 = \beta_{01}$$

$$\beta_1 = \beta_{11}$$

$$\beta_2 = \beta_{02} - \beta_{01}$$

and

$$\beta_3 = \beta_{12} - \beta_{11}$$

TABLE 7.9

The Equivalence between the MLR and Two SLRs

Treatment	z_i	MLR	SLRs	Equivalence
1	0	$\beta_0 + \beta_1 x_i$	$\beta_{01} + \beta_{11} x_{1j} \Rightarrow$	$\Rightarrow \beta_0 = \beta_{01}$ and $\beta_1 = \beta_{11}$
2	1	$\beta_0 + \beta_1 x_i + \beta_2(1) + \beta_3 x_i(1)$ $= (\beta_0 + \beta_2) + (\beta_1 + \beta_3)x_i$	$\beta_{02} + \beta_{12} x_{2j} \Rightarrow$	$(\beta_0 + \beta_2) = \beta_{02}$ and $(\beta_1 + \beta_3) = \beta_{12}$

We see that the terms of the MLR model that do not involve the indicator variable, β_0 and β_1, represent the intercept and slope for treatment 1, the reference treatment. On the other hand, the terms of the MLR model involving the indicator variable, β_2 and β_3, represent the differences in intercepts and slopes between the indicated treatment and the reference treatment. Estimates of the parameters in the MLR model, either the LS or M-estimates, are then simply based on the corresponding intercept and slope estimates in the individual SLR models. Tests of hypotheses that compare the two separate SLRs can now be made in terms of the appropriate parameters of the MLR model.

For example, the test of common slopes or "test of parallelism,"

$$\mathrm{H}_0 : \beta_{11} = \beta_{12} \Leftrightarrow \mathrm{H}_0 : \beta_{12} - \beta_{11} = 0 \Leftrightarrow \mathrm{H}_0 : \beta_3 = 0$$

may be performed by testing $\beta_3 = 0$ using the *t*-test, where the *t*-test in the MLR is the correct test statistic for testing this hypothesis. If this test fails to reject the null hypothesis, it is appropriate to assume a common slope – that is, adopt the ANCOVA model in further analysis.

Using the indicator variable approach as presented above, we can write the simple ANCOVA model as an MLR model:

$$y_i = \beta_0 + \beta_1 x_i + \beta_2 z_i + \varepsilon_i,$$

where $\mathrm{i} = 1,\ldots,\mathrm{n}_1 + \mathrm{n}_2$. This model represents two SLR lines with a common slope, β_1; an intercept of β_0 for the reference treatment; and a difference in intercepts between the indicated treatment and the reference treatment of β_2. The two null hypotheses of interest are $H_0 : \beta_1 = 0$, testing that the covariate is linearly associated with the response, and $H_0 : \mu_1 = \mu_2 \Leftrightarrow H_0 : \beta_{01} = \beta_{02} \Leftrightarrow H_0 : \beta_2 = 0$, the test of equality of adjusted means. Each hypothesis may be tested using the appropriate *t*-test from the MLR model. Estimates of parameters for either of the models, the MLR with four parameters (two intercepts and two slopes) and the MLR with three parameters (two intercepts and one slope), may be obtained using the method of LS or, if outliers are suspected, the method of *M*- or BI-estimation.

For a discussion of the properties of using *M*-estimation and the test based on robust weights for the simple ANCOVA, see Birch and Myers (1982) and Birch (1983). These procedures will be illustrated in the following example.

7.5.9 The Magnesium Data

The amount of magnesium uptake is measured at several levels of time. It is anticipated that the two treatments used may result in different regression equations. The data in Table 7.10, obtained from the Environmental Protection Agency, are as follows.

TABLE 7.10

Amount of Magnesium Uptake over Time Obtained by Two Treatments

Obs.	Mgtrt1	Mgtrt2	time
1	2.17	2.03	1
2	2.15	2.04	1
3	2.16	2.01	1
4	2.15	2.08	1
5	2.08	1.88	2
6	2.02	1.86	2
7	2.06	1.96	2
8	2.04	1.91	2
9	1.97	1.77	3
10	1.96	1.60	3
11	1.98	1.78	3
12	1.96	1.80	3
13	1.89	1.57	4
14	1.87	1.47	4
15	1.88	1.59	4
16	1.89	1.56	4
17	1.81	1.52	5
18	1.72	1.43	5
19	1.81	1.48	5
20	1.75	1.52	5
21	1.66	1.53	6
22	1.64	1.40	6
23	1.61	1.98	6
24	1.59	1.45	6

We want to compare the two SLRs fit to each treatment for equal slopes, and if the slopes are equal, we want to check for equality of adjusted means. The two MLR models involving indicator variables will be used to accomplish this.

We will compare the LS method with M-estimation. On R, type as follows:

```
# Plot magnesium uptake vs. time with LS lines and outliers
  highlighted
ggplot(magnesium_long, aes(x = time, y = Uptake, color =
  Treatment)) +
  geom_point(aes(shape = as.factor(outlier))) + # Highlight
  outliers
  geom_smooth(method = "lm", se = FALSE) + # Add LS regression
  lines
  labs( title = "Magnesium Uptake Over Time by Treatment",x =
  "Time",y = "Magnesium Uptake" ) + scale_shape_manual
  (values = c("FALSE" = 16, "TRUE" = 8)) + theme_minimal() +
  theme(legend.position = "top")
```

Figure 7.23 shows that the lines look parallel, with different intercepts, and one outlier is apparent.

```
> # Linear regression with time, z, and time*z
> ols_model <- lm(magnes ~ time + z + interaction, data =
  magnesium_data)
> summary(ols_model)
Call:
lm(formula = magnes ~ time + z + interaction, data =
  magnesium_data)
Coefficients:
              Estimate Std. Error t value   Pr(>|t|)
(Intercept)  2.26717    0.04706   48.178   < 2e-16 ***
time        -0.10229    0.01208   -8.465  8.76e-11 ***
z           -0.18117    0.06655 -2.722     0.00926 **
```

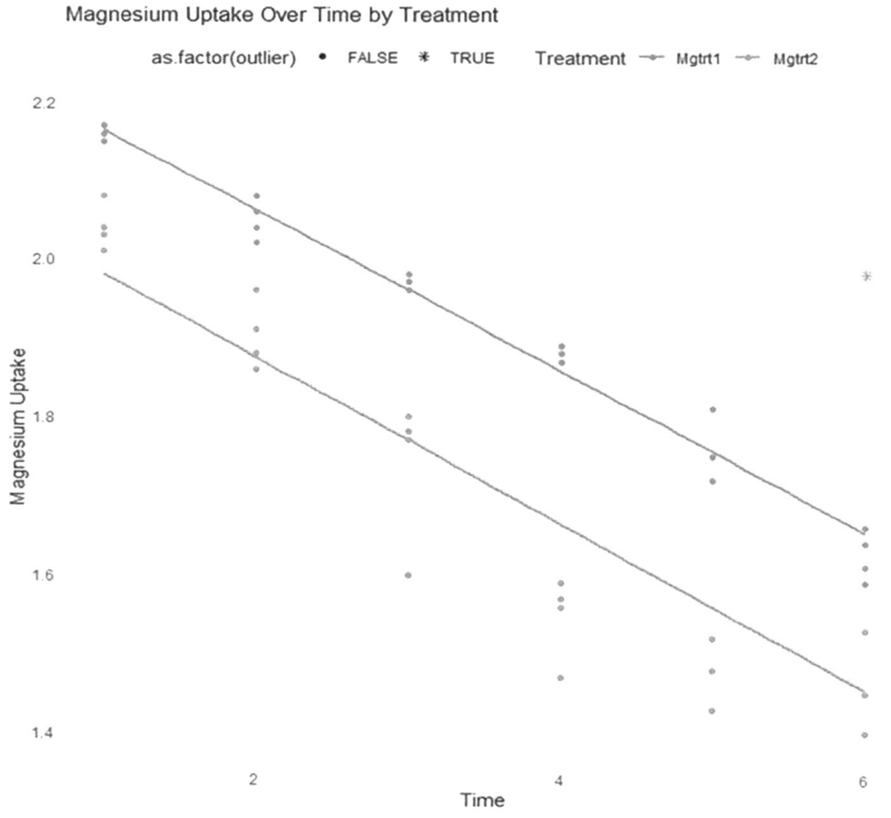

FIGURE 7.23
Magnesium uptake versus time for treatments 1 and 2.

```
interaction -0.00300 0.01709 -0.176 0.86145
---
Signif. codes: 0 '***' 0.001 '**' 0.01 '*' 0.05 '.' 0.1 ' ' 1
Residual standard error: 0.1011 on 44 degrees of freedom
Multiple R-squared: 0.8125, Adjusted R-squared: 0.7997
F-statistic: 63.57 on 3 and 44 DF, p-value: 4.926e-16
> # Analysis of Variance (ANOVA)
> anova(ols_model)
Analysis of Variance Table
Response: magnes
             Df  Sum Sq  Mean Sq  F value   Pr(>F)
time          1  1.50801 1.50801 147.5440 1.196e-15 ***
z             1  0.44083 0.44083  43.1313 4.936e-08 ***
interaction   1  0.00031 0.00031   0.0308 0.8614
Residuals    44  0.44971 0.01022
---
Signif. codes: 0 '***' 0.001 '**' 0.01 '*' 0.05 '.' 0.1 ' ' 1
> # Identify unusual observations
> unusual_obs <- which(abs(std_residuals) > 2)
> # Display the unusual observations
> magnesium_data[unusual_obs, ]
        magnes time z interaction
        47 1.98 6 1 6
```

The test of parallelism is accomplished by the *t*-test on time*z. It is highly insignificant, indicating that the two slopes for the separate SLRs are equal. To compare the adjusted means use the simple ANCOVA model. This is the model with only time and z.

Has the outlier distorted these conclusions? Run the Huber M-estimate to find out.

```
> # Robust regression using Huber M-estimator
> huber_model <- rlm(magnes ~ time + z + interaction, data =
  magnesium_data, psi = psi.huber)
> summary(huber_model)
Call: rlm(formula = magnes ~ time + z + interaction, data =
  magnesium_data,
  psi = psi.huber)
Coefficients:
             Value    Std.  Error t value
(Intercept)  2.2659    0.0216    105.0928
time        -0.1019    0.0055    -18.4030
z           -0.1190    0.0305     -3.9029
interaction -0.0251    0.0078     -3.2005
Residual standard error: 0.03096 on 44 degrees of freedom
> print(comparison)
               Row        OLS         Hubers
(Intercept)     1   2.2671667     2.26586181
time            2  -0.1022857    -0.10188385
```

```
z                3   -0.1811667    -0.11900474
interaction      4   -0.0030000    -0.02505846
> # Analysis of Variance for the Huber model
> anova_huber <- anova(huber_model)
> anova_huber
Analysis of Variance Table
Response: magnes
            Df  Sum Sq Mean Sq F value Pr(>F)
time         1 1.41720 1.41720
z            1 0.41590 0.41590
interaction  1 0.01822 0.01822
Residuals 0.48791
> # Interpret and compare the slopes and intercepts
> cat("OLS Intercept and Slopes:\n", ols_coefs, "\n")
OLS Intercept and Slopes:
 2.267167 -0.1022857 -0.1811667 -0.003
> cat("Huber Intercept and Slopes:\n", huber_coefs, "\n")
Huber Intercept and Slopes:
 2.265862 -0.1018839 -0.1190047 -0.02505846
```

We see the robust estimate of the difference in the slopes has increased about tenfold. The outlier actually made the LS slopes come closer together. The robust test of parallelism is now highly significant (in bold).

```
# Extract the weights from the robust regression model
weights <- huber_model$w
# Create an index plot of the weights with data labels
plot( weights, type = "p", # Use 'p' for points
  main = "Weight Index Plot for the Magnesium Data",
  xlab = "obs", ylab = "Weights", ylim = c(0, 1), xlim = c(1,
length(weights)),
  pch = 19 # Use solid circles for the points)
```

The robust analysis leads to the conclusion that there are differences between the two treatments in terms of magnesium uptake, with treatment 2 having both different slopes and intercept from treatment 1. It also appears from Figure 7.24, by examining the weights, that treatment 2 is not nearly as well behaved as treatment 1 since so many of treatment 2 observations receive low weight. The SLR model for treatment 2 appears to provide a very poor fit to the data.

These four examples show that the robust methods introduced in earlier chapters may easily be extended to other models within the GLM framework. The reader is encouraged to consider other models and examples and to reexamine them using the techniques presented here.

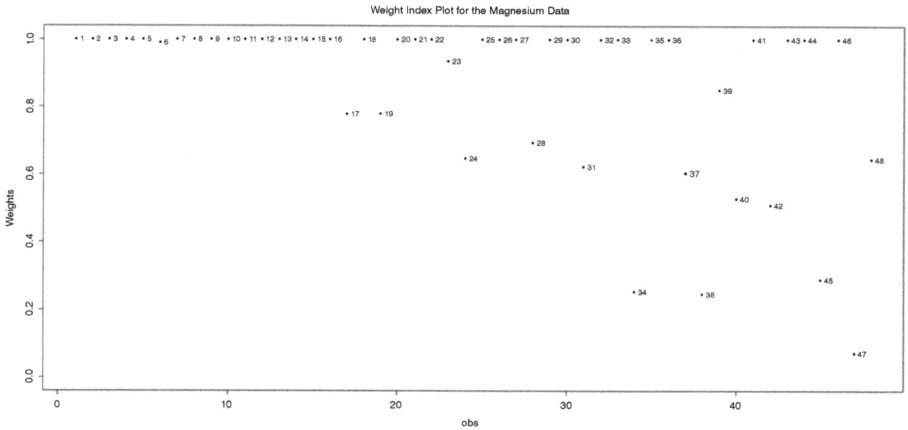

Figure 7.24 Weight index plot for magnesium data; observations 1–24 are for treatment 1.

7.6 The Bootstrap Method Applied to the Regression Problem

The bootstrap method, discussed in Section 4.5, was presented as a methodology for evaluating estimators of the location parameter. The bootstrap technique was especially helpful in obtaining estimates of standard errors of estimators and confidence intervals for the location parameter based on estimators where analytic formulas were not available. This situation occurred in the location model for such estimators as the median, M-, LMS, and LTS estimators, where only asymptotic formulas were available. A similar problem exists for regression models such as SLR and MLR and the nonparametric regression models considered in this text.

In the SLR model, for example, a proposed method for dealing with outliers is M-estimation, a procedure where the inferential techniques of testing hypotheses and creating confidence intervals are based on asymptotic theory, a theory that may not be applicable for finite sample problems. Fortunately, it is easy to extend the bootstrap method to the regression models. Indeed, the bootstrap method can be easily extended to any other model type such as ANOVA models, for example. The reader will recall that the bootstrap method mimics the Monte Carlo method by resampling with the replacement of the observed data to obtain the bootstrap sample. The resampling process is repeated B times, giving rise to B bootstrap samples. The estimation procedure is applied to each bootstrap sample, and the results are summarized in such a way to enable computation of the bootstrap standard error of the estimator and/or the bootstrap confidence interval for the parameter. The bootstrap method will be illustrated in this section for the SLR model.

The bootstrap method applied to the simple linear regression model, where "simple" refers to the one regression case, must reflect the design that was used to obtain the original data. There are two cases to consider, either the random or fixed regressor situation. In Chapter 6, we assumed that the regressor, x, was fixed. That is, the values taken by x are set and fixed by the researcher as opposed to being random variables as in the random case. Determining which is the case for any regression problem can be answered by considering the concept of repeated sampling. If new data were to be obtained by the same sampling scheme that was used to obtain the original data, would the values of the regressor remain the same? If so, then we have the fixed sample case. For the sake of completeness, the bootstrap method for both cases will be presented, explored, and illustrated here.

The bootstrap paradigm for the random case will be discussed first since this case more closely resembles the bootstrap method presented in Section 4.5 for the location model. Consider the n observed pairs $(y_i, x_i), i = 1,\ldots,n$. Select n new pairs of observations, the bootstrap sample, from the n original pairs, $(y_i, x_i), i = 1,\ldots,n$, using sampling with replacement. Compute the estimates of the intercept (β_0) and slope (β_1) based on this bootstrap sample. The estimates may be obtained by any available estimation procedure such as least squares, M-estimation, or LMS. Repeat this process B times, obtaining upon completion of the process, the B pairs of estimates of the intercept and slope, respectively, as $\left(\hat{\beta}^*_{0i}, \hat{\beta}^*_{1i}\right), i = 1,\ldots, B$.

Summarize these results by computing the bootstrap mean estimate of intercept and slope and bootstrap variance estimate of intercept and slope for the estimation procedure as

$$\bar{\hat{\beta}}^*_j = \frac{\sum_{i=1}^{B} \hat{\beta}^*_{ji}}{B} \text{ and } s^2\left(\hat{\beta}^*_j\right) = \frac{\sum_{i=1}^{B}\left(\hat{\beta}^*_{ji} - \bar{\hat{\beta}}^*_j\right)^2}{B-1}, \text{ for } j = 0,1 \cdot$$

The bootstrap standard error for the intercept and slope for the estimation procedure is simply $s\left(\hat{\beta}^*_j\right) = \sqrt{s^2\left(\hat{\beta}^*_j\right)}$, for $j = 0,1$. To obtain $(1-\alpha)100\%$ confidence intervals for the intercept and the slope, the bootstrap percentile method can be applied in the same manner as discussed in Section 4.5.

One fact to keep in mind when determining the proper bootstrap procedure is that the bootstrap method must mimic the sampling design that led to the original observations. This philosophy has already been illustrated for the random regressor case, where it was assumed that the original response and the regressor were observed as random pairs from some theoretical population of (y,x) pairs. Hence, the bootstrap method must resample from the original pairs of observations.

For the fixed regressor situation, the responses at the fixed x values are assumed to be random observations about the true regression line. Consequently, in the fixed regressor case, the bootstrap method must be adjusted to account for the fact that every bootstrap sample must contain the same n regressor values, $x_1,\ldots,x_n$. This is accomplished by using the estimation procedure (estimation, for example) to obtain the n fits, $\hat{y}_i$, and n residuals $r_i = y_i - \hat{y}_i$, for $i = 1,\ldots,n$, based on the original data.

Given these n residuals, $r_1,\ldots,r_n$, select n new residuals, $r_1^*,\ldots,r_n^*$, called the "bootstrap residuals," from $r_1,\ldots,r_n$ using sampling with replacement. Compute the bootstrap responses, as $y_i^* = \hat{y}_i + r_i^*$, for $i = 1,\ldots,n$. Consider the n pairs (y_i^*, x_i) as the bootstrap sample.

Compute the estimates of the intercept and slope, $(\hat{\beta}_0^*, \hat{\beta}_1^*)$, based on this bootstrap sample for the given estimation procedure. Repeat this process B times. The B pairs of bootstrap estimates, $(\hat{\beta}_{0i}^*, \hat{\beta}_{1i}^*)$, may then be used to obtain the bootstrap means, bootstrap estimates of standard errors, and confidence intervals by exactly the same procedure as used for the random regression case described above. In the bootstrap literature, this bootstrap method is called fixed resampling or residual resampling.

It is easy to perform the bootstrap method for regression using R. The R script will be provided in later parts of this chapter and on the book's website. Each code is written for the SLR model only, obtaining bootstrap information for the LS and M-estimator based on the Huber weight function with a constant 1.5. The reader, after examining the code, could make adjustments to them to accommodate MLR and other estimators, if desired.

The method will be illustrated here using the potato leafhopper data introduced in Section 6.3. The use of these R codes, along with an interpretation of the bootstrap results will now be demonstrated. Within the potato leafhopper worksheet, the response and the regressor are saved. Under the assumption that the regressor is fixed, we will first obtain results using the fixed resampling scheme, followed by the random scheme. On R, type as follows:

```
# Step 1: Define the data (using an example similar to potato
leafhopper data)
set.seed(123)
x <- 1:10
y <- 0.5 + 0.3 * x + rnorm(10, 0, 0.2) # Example data
data <- data.frame(x, y)
# Step 2: Bootstrap function for fixed resampling
bootstrap_fixed <- function(data, indices) {d <- data[indices, ]
  # Resample the data with replacement
  fit <- lm(y ~ x, data = d) # Fit the linear model
  return(coef(fit)) # Return the coefficients}
# Step 3: Bootstrap function for random resampling
bootstrap_random <- function(data, indices) {
  x <- data$x # Fixed x
residuals <- data$y—predict(lm(y ~ x, data = data)) # Residu-
  als from the original model
  y_boot <- predict(lm(y ~ x, data = data)) +
  residuals[indices] # Resample residuals
  fit <- lm(y_boot ~ x) # Fit the linear model to the boot-
  strap sample
  return(coef(fit)) # Return the coefficients}
# Step 4: Run bootstrap for fixed and random resampling
set.seed(123)
n_boot <- 500 # Number of bootstrap samples
# Fixed resampling
fixed_boot <- boot(data, bootstrap_fixed, R = n_boot)
# Random resampling
random_boot <- boot(data, bootstrap_random, R = n_boot)
# Step 5: Extract bootstrap estimates
fixed_intercepts <- fixed_boot$t[, 1]
fixed_slopes <- fixed_boot$t[, 2]
random_intercepts <- random_boot$t[, 1]
random_slopes <- random_boot$t[, 2]
# Step 6: Summarize the bootstrap results
summary_fixed <- data.frame(
  Method = rep("Fixed Resampling", 2),
  Parameter = c("Intercept", "Slope"),
  Mean = c(mean(fixed_intercepts), mean(fixed_slopes)),
  Median = c(median(fixed_intercepts), median(fixed_slopes)),
  SD = c(sd(fixed_intercepts), sd(fixed_slopes)))
```

Descriptive Statistics

```
> summary_results <- rbind(summary_fixed, summary_random)
> print(summary_results)
         Method Parameter      Mean    Median         SD
1  Fixed Resampling Intercept 0.6124099  0.5888603 0.14407919
2 Fixed Resampling      Slope 0.2837055  0.2843485 0.02337017
3 Random Resampling Intercept 0.6063981 0.5969859  0.12449227
4 Random Resampling     Slope 0.2832514 0.2830856  0.02013742
```

In the results, the two rows summarize the bootstrap results for the fixed sampling case, while the next two rows contain the fixed sampling results.

The histograms of these columns in Figure 7.25 (a) and (b) depict the estimates' bootstrap distributions.

```
# Histogram for intercept
ggplot(bootstrap_results, aes(x = Intercept, fill = Method)) +

  geom_histogram(binwidth = 0.02, alpha = 0.7, position =
"identity") + facet_wrap(~Method, scales = "free_y") +
labs(title = "Histogram of Bootstrap Intercepts", x =
"Bootstrap Intercept Estimates",y = "Frequency") +
theme_minimal()
# Histogram for slope
ggplot(bootstrap_results, aes(x = Slope, fill = Method)) +
  geom_histogram(binwidth = 0.005, alpha = 0.7, position =
  "identity") +

  facet_wrap(~Method, scales = "free_y") + labs(title = "Histogram
of Bootstrap Slopes", x = "Bootstrap Slope Estimates", y = "Fre-
quency") +

  theme_minimal()
```

The histogram reveals very different results depending on the nature of the resampling mechanism. The histogram for random resampling, for either the LS or M-estimate, shows clear bimodality of two nearly normal shapes, suggesting non-normal distributions of the estimators. On the other hand,

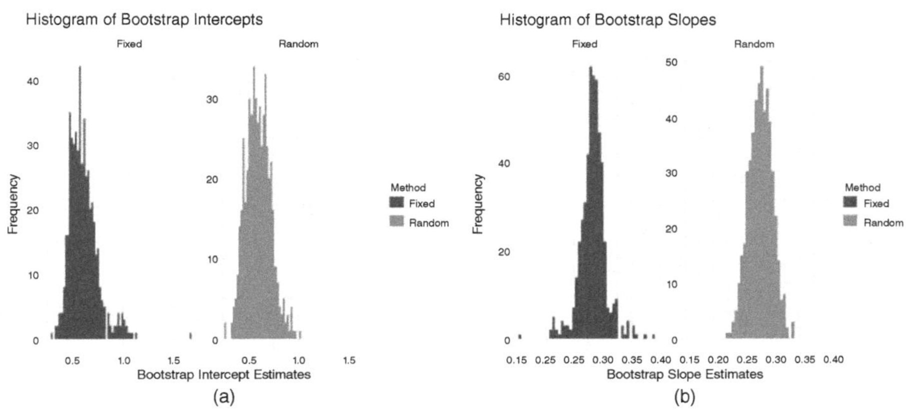

FIGURE 7.25
(a) Histogram of bootstrap intercepts; (b) Histogram of bootstrap slopes.

for fixed resampling, for either the LS or M-estimate, the histograms are distinctly symmetric and bell shaped, suggesting normal distributions of the estimators. For the fixed resampling case, we note that the histogram is centered around the bootstrap means of the estimates. Finally, the bimodal nature of the histogram for the random resampling case reflects the contamination of the raw data with outliers.

The analytical results for the potato leafhopper data, presented in Section 7.3, are compared with those obtained via the bootstrap as follow.

```
# Define the bootstrap functions
bootstrap_fixed <- function(data, indices) {d <- data[indices, ]
  # Resample the data with replacement
  fit <- lm(y ~ x, data = d) # Fit the linear model
  return(coef(fit)) # Return the coefficients}
bootstrap_random <- function(data, indices) {x <- data$x #
  Fixed x
  residuals <- data$y—predict(lm(y ~ x, data = data)) # Resid-
  uals from the original model
  y_boot <- predict(lm(y ~ x, data = data)) +
  residuals[indices] # Resample residuals
  fit <- lm(y_boot ~ x) # Fit the linear model to the boot-
  strap sample
  return(coef(fit)) # Return the coefficients}
# Run bootstrap for fixed and random resampling
set.seed(123)
n_boot <- 500 # Number of bootstrap samples
fixed_boot <- boot(data, bootstrap_fixed, R = n_boot)
random_boot <- boot(data, bootstrap_random, R = n_boot)
# Extract bootstrap estimates
fixed_intercepts <- fixed_boot$t[, 1]
fixed_slopes <- fixed_boot$t[, 2]
random_intercepts <- random_boot$t[, 1]
random_slopes <- random_boot$t[, 2]
# Analytical Estimates (using LS and M-estimation)
ls_fit <- lm(y ~ x, data = data)
ls_se <- summary(ls_fit)$coefficients[, 2]
m_fit <- rlm(y ~ x, data = data, psi = psi.huber, k = 1.5)
m_se <- summary(m_fit)$coefficients[, 2]
# Combine all results into a single table

comparison_table <- data.frame(Estimate_Type = c("LS inter-
  cept", "LS slope", "M intercept", "M slope"), Analytical_
  Estimate = c(coef(ls_fit)[1], coef(ls_fit)[2], coef(m_fit)
  [1], coef(m_fit)[2]), Analytical_SE = c(ls_se[1], ls_se[2],
  m_se[1], m_se[2]), Bootstrap_Random_Estimate =
  c(mean(random_intercepts), mean(random_slopes), mean(random_
  intercepts), mean(random_slopes)), Bootstrap_Random_SE =
  c(sd(random_intercepts), sd(random_slopes), sd(random_inter-
  cepts), sd(random_slopes)), Bootstrap_Fixed_Estimate =
```

```
  c(mean(fixed_intercepts), mean(fixed_slopes), mean(fixed_
  intercepts), mean(fixed_slopes)), Bootstrap_Fixed_SE =
  c(sd(fixed_intercepts), sd(fixed_slopes), sd(fixed_inter-
  cepts), sd(fixed_slopes)))
# Print the comparison table
print(comparison_table)
  Estimate_Type Analytical_Estimate Analytical_SE
  Bootstrap_Random_Estimate
1  LS intercept    0.6050935    0.13345532    0.6063981
2      LS slope    0.2836058    0.02150827    0.2832514
3   M intercept    0.5914447    0.15345491    0.6063981
4       M slope    0.2837245    0.02473150    0.2832514
  Bootstrap_Random_SE Bootstrap_Fixed_Estimate
  Bootstrap_Fixed_SE
1           0.12449227      0.6050935          0
2           0.02013742      0.2836058          0
3           0.12449227      0.6050935          0
4           0.02013742      0.2836058          0
```

We note the strong agreement between the bootstrap results obtained through fixed resampling and those obtained through the analytic method. Even the standard errors are very close. However, the bootstrap results obtained through random resampling are very different from the analytic results for both the estimates and the standard errors. These results are not too surprising since all analytic formulas are based on the assumption that the regressors are fixed. Our results raise two issues. One is that the analytic formulas presented in the previous section for the M-estimators seem to provide an adequate estimate of the true standard deviation of the estimators, at least under the assumption of fixed regressors. The second issue is that there are large discrepancies between the results of a regression problem depending on whether the regressors are assumed to be fixed or random.

The majority of regression presentations, including those in this text, assume that the regressors are fixed. This assumption may not be appropriate for many practical regression problems. However, the fixed regressor case is much easier to handle analytically. The bootstrap method allows the differences between the two cases to be handled via the computer, where the difficult analytical issues may be ignored.

Confidence intervals for slope and intercept may also be obtained in our example by applying the bootstrap percentile method. The bootstrap estimates are simply sorted, and the 95% confidence intervals are estimated by taking the average of the 12th and 13th smallest and largest bootstrap estimates (since (0.025)(500) = 12.5) as the lower confidence limit and the upper confidence limit, respectively. The 95% confidence limits for the intercept and slope are given for the random and fixed bootstrap and analytical methods. The analytical results are obtained by the usual method of $\hat{\beta}_j \pm t_{1\frac{a}{2},n-2}\, se\left(\hat{\beta}_j\right)$, $j = 0$ and 1 for each estimation procedure.

```
# Analytical 95% Confidence Intervals
analytical_ci <- data.frame(Parameter = c("LS intercept", "LS
  slope", "M intercept", "M slope"), Lower = c(coef(ls_fit)[1]
  - 1.96 * ls_se[1], coef(ls_fit)[2] - 1.96 * ls_se[2],coef(m_
  fit)[1] - 1.96 * m_se[1], coef(m_fit)[2] - 1.96 * m_se[2]),
  Upper = c(coef(ls_fit)[1] + 1.96 * ls_se[1], coef(ls_fit)[2]
  + 1.96 * ls_se[2],coef(m_fit)[1] + 1.96 * m_se[1], coef(m_
  fit)[2] + 1.96 * m_se[2]))

# Bootstrap 95% Confidence Intervals (using the percentile
  method)
bootstrap_ci <- function(bootstrap_estimates) {lower_limit
  <- apply(bootstrap_estimates, 2, function(x) quantile(x,
  probs = 0.025))
  upper_limit <- apply(bootstrap_estimates, 2, function(x)
  quantile(x, probs = 0.975))
  return(data.frame(Lower = lower_limit, Upper =
  upper_limit))}
fixed_ci <- bootstrap_ci(fixed_boot$t)
random_ci <- bootstrap_ci(random_boot$t)

# Combine confidence intervals into a single table
comparison_ci <- data.frame( Parameter = c("LS intercept", "LS
  slope", "M intercept", "M slope"), Analytical_Lower = ana-
  lytical_ci$Lower, Analytical_Upper = analytical_ci$Upper,
  Bootstrap_Random_Lower = random_ci$Lower, Bootstrap_Random_
  Upper = random_ci$Upper, Bootstrap_Fixed_Lower = fixed_
  ci$Lower, Bootstrap_Fixed_Upper = fixed_ci$Upper)
> # Print the comparison table for confidence intervals
> print(comparison_ci)
Parameter Analytical_Lower Analytical_Upper
  Bootstrap_Random_Lower
1 LS intercept    0.3435210    0.8666659   0.3891009
2     LS slope    0.2414495    0.3257620   0.2440469
3  M intercept    0.2906730    0.8922163   0.3891009
4      M slope    0.2352508    0.3321983   0.2440469
Bootstrap_Random_Upper Bootstrap_Fixed_Lower Bootstrap_Fixed_Upper
1       0.8717915     0.6050935     0.6050935
2       0.3222252     0.2836058     0.2836058
3       0.8717915     0.6050935     0.6050935
4       0.3222252     0.2836058     0.2836058
```

Similar results are seen here as observed in the previous outputs. The bootstrap results obtained by fixed resampling match closely those obtained via analytic formulas, while those results obtained via random resampling vary greatly from the analytic results. Indeed, the randomly obtained bootstrap intervals tend to be much wider than those obtained through the formulas or the fixed resampling scheme.

We are left in somewhat of a quandary. What results do we use to make our inferences, those from the random scheme or those resulting from fixed resampling? It seems logical to use the results that mimic how the original data was obtained. In the leafhopper example, the sampling scheme was clearly random, indicating that the random resampling results were appropriate.

Differences between the bootstrap results obtained via the fixed and random schemes remind us that the assumption of fixed regressors should not be taken lightly. Certainly, there are instances where the regressors are truly fixed, in designed experiments, for example. Dose-response experiments – where a fixed amount of some substance, the "dose," is given to experimental units – is a commonly designed experiment where assuming fixed regressors is appropriate. Replication of the experiment will result in exactly the same values for the regressors. On the other hand, in observational studies, common in such areas as economics, biology, and the social sciences, the regressors are often random, and replications of the studies may lead to very different values of the regressors. Assuming fixed regressors and performing the regression analysis based on this assumption, as is often done, may be misleading.

7.7 Semiparametric Regression: Model-Robust Regression

Consider the basic regression problem of explaining the response of a variable y by using one or more regressor variables, according to a model of the form. The term is a random error from the process, often assumed to have a mean of zero and variance. Of interest is when the researcher has some knowledge as to the form of f, but there are regions in the data that do not conform to this proposed underlying structure. Of particular interest is dealing with small datasets, which are common in environmental and biological settings. Also, this section emphasizes the "single regressor model," with all terms in the model defined to be polynomial expressions of this single regressor $\left(x, x^2, x^3, \ldots\right)$.

For illustration, consider once again the tensile data of Section 6.4, which is plotted in Figure 7.26. The response y is the tensile strength (in psi) of paper, and the regressor x is the percentage of hardwood in the batch of pulp from which the paper was produced (Montgomery and Peck, 1992). This data exhibits a strong quadratic structure but contains an interesting peak that deviates from the quadratic trend.

The classical parametric regression approach is to assume a known linear form for f, rewrite the model as $\underline{y} = X\underline{\beta} + \underline{\varepsilon}$, and estimate the unknown parameters in $\underline{\beta}$ from the data. Here X is an n by model matrix of k regressors augmented with a column of ones. Ordinary least squares (OLS) minimizes the sum of squared residuals to obtain the fitted values

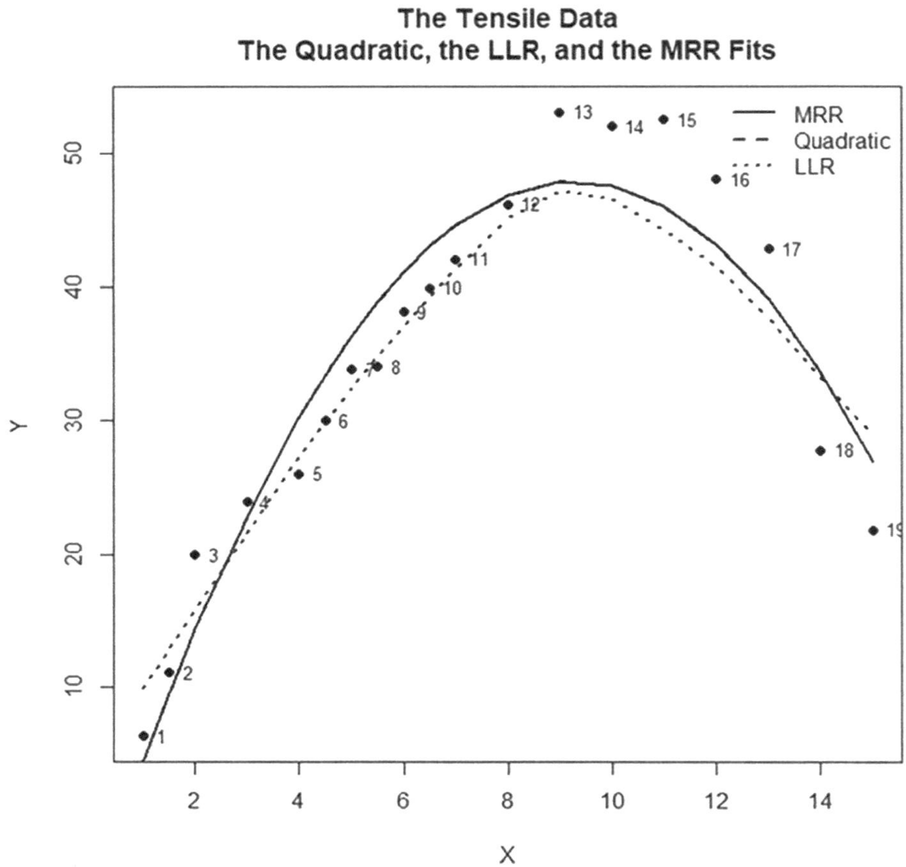

FIGURE 7.26
The quadratic LS and the MRR fits for the tensile data.

as $\hat{y}_{LS} = X\hat{\beta}_{LS} = X(X'X)^{-1}X'y = H^{LS}y$, where H^{LS} is the LS "hat" matrix of Section 7.1 (Myers, 1990). The fit for y_i at location x_i may be expressed as, a weighted sum (for $j = 1,\ldots,n$) of the n observations y_j, where $h_{i1}^{LS}, h_{i2}^{LS}, \ldots, h_{in}^{LS}$ are the elements of the ith row of H^{LS}. The weighting scheme is a direct consequence of the prescribed model. If the model is misspecified, even over only portions of the data, then inferences from this technique will be misleading. Note the inadequacy of the quadratic LS fit at the peak of the data in Figure 7.26.

An alternative approach is to use nonparametric regression, as discussed in Section 6.4, which is most useful in situations where the function f is considered to be unknown and the user has no parametric specification as to its form. Kernel and local linear regression (LLR) are two nonparametric forms presented in Section 6.4. The prediction of $f(x_0)$ at location x_0 is a weighted sum of y observations, with the weights being a decreasing function of the distances of the respective regressor locations. Since nonparametric techniques rely totally on the data itself and have no underlying stability of a specified functional form, they may fit irregular patterns in the data and may be more variable than parametric fits.

The LLR fit to the tensile data is also displayed in the figure above. Note how the LLR fit in the figure fits closely to the data. Subsequent results establish that this is a fit high in variance. Thus, a dilemma is created for the common problem where there is partial knowledge about the underlying model. A parametric model would explain most of the data but would be inadequate where the data deviates from the chosen model, whereas a nonparametric fit would ignore the important information that the researcher has about the underlying structure.

Also, it is possible that even with the presence of important deviations from the underlying model, a lack of fit test (see Myers, 1990) would conclude that the parametric fit is adequate, resulting in misleading inferences. One possible solution to this dilemma involves a semiparametric approach where the parametric method is combined with the nonparametric method. The goal is to obtain a simple procedure that maintains robustness to a misspecified model while adequately fitting the data.

One semiparametric method, termed model-robust regression or MRR (Mays and Birch, 2000, and Mays, Birch, and Starnes, 2001), yields the third fitted curve in the figure. This fit maintains the smooth fit of LS but uses a nonparametric fit to the residuals to also capture the peak in the data. Clearly, the MRR fit is better than that for LS (sum of squares error = 78.07 vs. 312.64 for LS). The improvement over the LLR fit is not as obvious from the graph, because the improvement is in variance. However, the cross-validation criterion PRESS* (see Section 6.4) yields a value of 10.85 for MRR and 14.07 for LLR.

Mays and Birch (2000) establish the advantageous bias and variance properties of MRR and several other semiparametric methods through the development of a theoretical MSE criterion. Following the presentation of the general concept of MRR, the method will be illustrated using the tensile strength data.

MRR is motivated by the need to improve upon the shortcomings of LS (the parametric procedure used here) with possible high bias in fit resulting from model misspecification and LLR (the nonparametric procedure used) with high variance of fit resulting from possible improper choice of bandwidth. In MRR, two separate fits are combined to obtain the final fit. The parametric portion of MRR is a parametric fit, say $\hat{\mathrm{y}}^{\mathrm{LS}}$, to the raw data. The nonparametric portion of MRR comes from a fit to the residuals $\left(\underline{r} = \underline{y} - \underline{\hat{y}}^{\mathrm{LS}}\right)$ from the parametric fit. The nonparametric fit (using LLR) may be expressed as

$\hat{\underline{r}} = H^{LLR}\underline{r}$, where H^{LR} is the local linear hat matrix for fitting the residuals. The final MRR fit is obtained by adding a portion of the residual fit back to the original parametric fit as: $\hat{y}^{MRR} = \hat{y}^{LS} + \lambda\hat{\underline{r}}$, where $\lambda \in [0,1]$.

The parameter λ, termed the mixing parameter, plays an important role in adjusting the fit obtained through the parametric model (using $\hat{y}^{LS}$ here).

In general, one would expect λ to increase from 0 to 1 as model misspecification increases. To see this, consider the case where the parametric model is correct. One would expect the resulting residuals to contain little more than random variation. A smooth fit to the residuals (using LLR here) will essentially be a straight line with a slope of zero. If the value of λ is zero, or close to zero, the LS fits will be unadjusted through MRR. The MRR fits will maintain the no-bias, low-variance properties of the LS fits. On the other hand, if the model is misspecified (perhaps due to leaving out needed polynomial terms in the model or from using a linear model when a nonlinear model is correct), then the residuals contain useful model information concerning the shape of $f(x_i)$ as well as randomness.

A nonparametric fit to these residuals will recover much of this structure. Perhaps all or a large portion of this smoothed structure should be added to $\hat{y}^{LS}$ to improve the final MRR fit. The amount added to $\hat{y}^{LS}$ is determined by the mixing parameter λ. If the amount of model misspecification is large, then λ will be one or close to one. The benefit of the MRR fit is a greatly reduced bias in fit with a slight increase in variance in fit, resulting in improved MSE of fit properties. If the parametric model is only moderately misspecified, then we may want to add only a fraction, say $\lambda = 0.5$, of the smoothed residuals to $\hat{y}^{LS}$, again resulting in improved MSE of fit properties. Using $\lambda = 1$, in this case, may result in a slight improvement in the bias of fit for MRR but with too much variance of fit, resulting in less-than-optimal MSE of fit properties.

Obviously, the choice of λ is crucial to the success of the MRR method. Data-driven methods, such as those used to determine the bandwidth in nonparametric regression, are suggested. One such method is to choose λ to minimize $\text{PRESS}^*(\lambda)$, where

$$\text{PRESS}^*(\lambda) = \frac{\sum_{i=1}^{n}\left(y_i - \hat{y}_{i,-i}^{MRR}(\lambda)\right)^2}{n - df_{model}}.$$

Here df_{model} is an appropriate model's degrees of freedom of the MRR fit and $\hat{y}_{i,-i}^{MRR}$ is the MRR fit at x_i obtained with (x_i, y_i) removed from the dataset. In terms of hat matrices, we can express the MRR fit, $\underline{\hat{y}}^{MRR}$, at the *n* data points as

$$\underline{\hat{y}}^{MRR} = H^{LS}\underline{y} + \lambda H^{LLR}\underline{r} = \left[H^{LS} + \lambda H^{LLR}\left(I - H^{LS}\right)\right]\underline{y} = H^{MRR}\underline{y},$$

where H^{MRR} is the "hat" matrix associated with the MRR fit. A fit at an individual data point, x_i, may be expressed as

$$\hat{y}_i^{MRR} = \underline{\underline{h}}_i^{MRR}\underline{y} = \sum_{j=1}^{n} h_{ij}^{MRR} y_j,$$

where $\underline{\underline{h}}_i^{MRR}$ is the ith row of H^{MRR}.

For a fixed value of b and λ, one can easily obtain an expression for the variance of the MRR fit at any arbitrary x value, say x_0, as

$$\mathrm{var}\left(\hat{y}_0^{MRR}\right) = \sigma^2 \sum_{j=1}^{n}\left(h_{0j}^{MRR}\right)^2.$$

Analogous to the LLR and robust local linear regression (RLLR) situations, an approximate $(1-\alpha)100\%$ confidence interval for the mean response at $x = x_0$ may be obtained as

$$\hat{y}_0^{MMR} \pm t_{1-\frac{a}{2}, df_{error}}\, se\left(\hat{y}_0^{MMR}\right)$$

where $df_{error} = n - \sum_{i=1}^{n} h_{ii}^{MRR} = n - df_{model}$ and $se\left(\hat{y}_0^{MRR}\right) = s\sqrt{\sum_{j=1}^{n}\left(h_{0j}^{MRR}\right)^2}$ with $s^2 = \dfrac{\sum_{i=1}^{n}\left(y_i - \hat{y}_i^{MRR}\right)^2}{df_{error}}$.

Work by Mays, Birch, and Starnes (2001) indicates that the use of MRR in the single regressor case offers certain advantages over the use of either the parametric or nonparametric approaches. If the user can correctly specify the model with a parametric form, then obviously that model should be used. If the error distribution can also be correctly specified, statistical theory dictates the proper method for estimation and inference.

For example, consider the SLR model with normally distributed errors, in the usual situation. The theory states that the LS estimates are unbiased with minimum variance among all unbiased estimators. The resulting LS fits are unbiased estimates for the true mean response at an arbitrary value of the regressor with minimum variance. Thus, the LS fit at any value of the regressor has minimum MSE. The MRR fit, which incorporates the user's correct model and some amount of the LLR smoothing to the residuals can do no better than the LS fit. Simulation studies conducted by Mays and Birch (2000) show that the smoothed residuals are nearly equal to zero, and the MRR method properly chooses a very small value of the mixing parameter. The resulting MRR fit is nearly the same as the LS fit, with slightly larger MSE values.

On the other hand, if the user's model is misspecified by a moderate amount, the MRR fit has better MSE of fit properties that are superior to either the LS fit or the LLR fit. If the user's model is greatly misspecified, providing little useful information with which to reduce the residuals, the MRR fit is only slightly worse than the LLR fit, both being superior to the LS fit. Thus, it is argued by Mays and Birch that the MRR method is the proper method when the user has some idea of the proper model but that the model is inadequate over some range of the regressor.

To illustrate the MRR method, consider once again the tensile data. The quadratic model has been specified by the user as an inadequate model over the range of the regressor, as seen in the scatterplot at the beginning of this section.

The residuals from this fit can be smoothed by LLR using R, as introduced in Section 6.4. To obtain the quadratic fit, type as follows:

```
# Create the quadratic term
tensile_data$x_squared <- tensile_data$x^2
# Fit the quadratic regression model
quadratic_model <- lm(y ~ x + x_squared, data = tensile_data)
  > summary(quadratic_model)
Call:
lm(formula = y ~ x + x_squared, data = tensile_data)
Residuals:
      Min         1Q  Median      3Q      Max
  -5.8503    -3.2482 -0.7267 4.1350    6.5506
Coefficients:
             Estimate Std. Error t value Pr(>|t|)
(Intercept)  -6.67419  3.39971  -1.963   0.0673 .
x            11.76401  1.00278  11.731 2.85e-09 ***
x_squared    -0.63455  0.06179 -10.270 1.89e-08 ***
---
Signif. codes: 0 '***' 0.001 '**' 0.01 '*' 0.05 '.' 0.1 ' ' 1
Residual standard error: 4.42 on 16 degrees of freedom
Multiple R-squared: 0.9085, Adjusted R-squared: 0.8971
F-statistic: 79.43 on 2 and 16 DF, p-value: 4.912e-09
The regression equation is
y = - 6.67 + 11.8 x - 0.635 x_squared
> anova(quadratic_model)
Analysis of Variance Table
Response: y
          Df  Sum Sq Mean Sq F value    Pr(>F)
x          1 1043.43 1043.43   53.40   1.758e-06 ***
x_squared  1 2060.82 2060.82  105.47   1.894e-08 ***
Residuals 16 312.64    19.54
---
Signif. codes: 0 '***' 0.001 '**' 0.01 '*' 0.05 '.' 0.1 ' ' 1
```

It is observed that all terms are significant and that our three-parameter model has an MSE of 19.5. A plot of the data with the quadratic fit is given earlier in the section and a plot of the residuals is given in Figure 7.27. The residual plot clearly reveals a strong sinusoidal tendency.

In order to perform MRR regression on R, we must first smooth the residuals from the user's model with LLR. Find the proper bandwidth. For example, see below.

```
> # Display optimal bandwidth and sigma2
> print(paste("Optimal Bandwidth:", optimal_bandwidth))
[1] "Optimal Bandwidth: 1.5"
> print(paste("Estimated Variance (sigma^2):", sigma2))
[1] "Estimated Variance (sigma^2): 7.22915753301257"
> print(paste("Degrees of Freedom:", df_error))
[1] "Degrees of Freedom: 17"
```

Figure 7.27 shows the LLR fit to the residuals. The structure in the residuals is obvious.

The question now remains as to what fraction of this LLR fit should be added back to the quadratic fit. To determine this, we need to run the R code, with a candidate list of values of lambda. Since the structure is so strong, it is clear that a large portion of the LLR fit is needed. Therefore, only large values of lambda should be considered, such as 0.5, 0.6, 0.7, 0.8, 0.9, 0.94, 0.95, 0.96, 0.99, and 1. The "best" lambda in the candidate list is lamb = 0.94, and the press* statistic at the "best" lambda is 10.8492.

```
# Create the quadratic term
tensile_data$x_squared <- tensile_data$x^2
# Fit the quadratic regression model
quadratic_model <- lm(y ~ x + x_squared, data = tensile_data)
tensile_data$residuals <- residuals(quadratic_model)
# Apply LLR to the residuals with a specified bandwidth (2.45)
bandwidth <- 2.45
llr_model <- locfit(residuals ~ lp(x, deg = 1), data = ten-
  sile_data, alpha = bandwidth)
tensile_data$llr_fit <- predict(llr_model, newdata =
  tensile_data$x)
# Use the optimal lambda (0.939) to calculate the MRR fit
optimal_lambda <- 0.939
tensile_data$mrr_fit <- (1 - optimal_lambda) *
  predict(quadratic_model) + optimal_lambda *
  tensile_data$llr_fit
# Use the provided variance for the MRR fit
sigma2_mrr <- 6.04203
df_error_mrr <- 14.3354
# Calculate the 95% confidence bands
conf_band <- qt(0.975, df_error_mrr) * sqrt(sigma2_mrr)
tensile_data$upper_band <- tensile_data$mrr_fit + conf_band
tensile_data$lower_band <- tensile_data$mrr_fit—conf_band
```

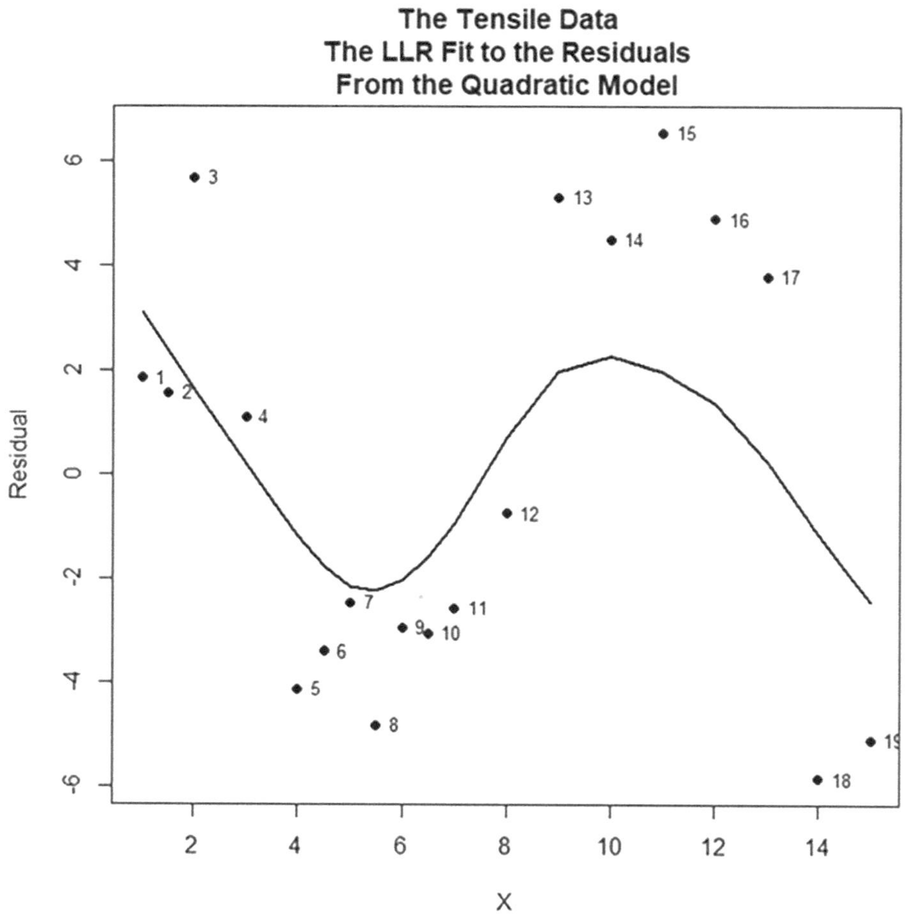

FIGURE 7.27
The LLR fits for the residuals from quadratic model.

```
# Plot the MRR fit with 95% confidence bands
plot(tensile_data$x, tensile_data$y, xlab = "x", ylab = "y",
  pch = 16, col = "black", main = "The Tensile Data\nThe MRR
  Fit With 95% Confidence Bands")
# Add the MRR fit line
lines(tensile_data$x, tensile_data$mrr_fit, col = "black",
  lwd = 2)
# Add the confidence bands
lines(tensile_data$x, tensile_data$upper_band, col = "black",
  lty = 2, lwd = 2)
lines(tensile_data$x, tensile_data$lower_band, col = "black",
  lty = 2, lwd = 2)
# Add the original quadratic fit for comparison (dashed line)
lines(tensile_data$x, predict(quadratic_model),col =
  "black",lty = 3,lwd = 2)
```

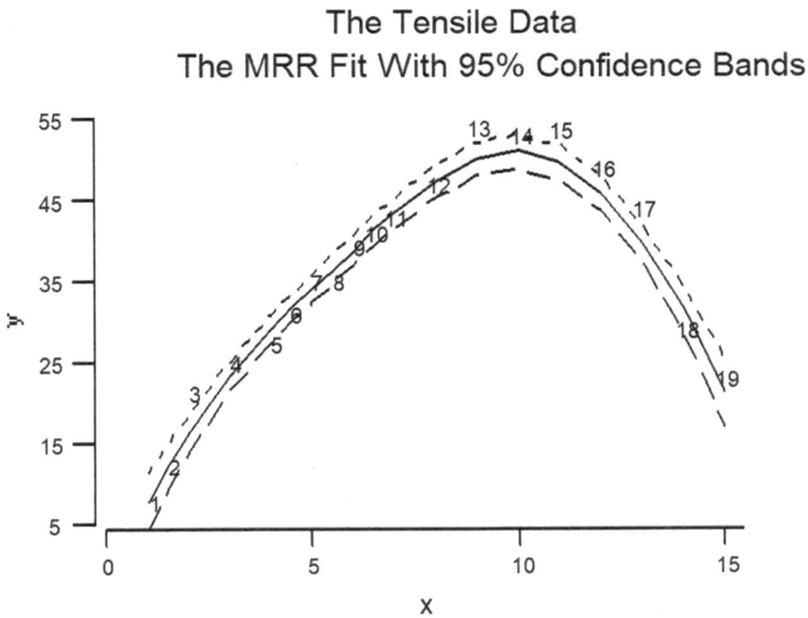

FIGURE 7.28
The MRR fit with 95% confidence bands.

```
# Add point labels for clarity
text(tensile_data$x, tensile_data$y, labels = seq_
  along(tensile_data$y), pos = 4, cex = 0.8)
```

The MRR estimate of variance (sigma squared) is 6.04, compared to 19.5 for the quadratic model, and 5.56 for the LLR fit. Figure 7.28 shows the MRR fit with the 95% confidence bands.

The reader will note the similarity between the LLR and the MRR fits for the tensile data. As noted above, the estimate of σ^2 is 6.04 for MRR, slightly larger than the estimate of 5.56 for LLR. The "model degrees of freedom" is 4.66 ($n - df_{error}$) for MRR, slightly less than the 5.92 required for LLR. This difference is due to the MRR's use of the user's quadratic model to form the bulk of the fit to the raw data. MRR does have a smaller PRESS* statistic of 10.85, compared to 14.07 for LLR and 29.93 for the user's quadratic model.

Mays, Birch, and Starnes (2001) demonstrate, using a variety of curve shapes and designs (configurations of the x values) that MRR is superior to LLR in situations similar to those represented by the tensile data – that is, situations with moderate model misspecification, uniformly spaced x values, and sample sizes of approximately 20. The MSE of fit, when averaged over the range of the x values is approximately 1.21 times greater for LLR than MRR in this situation. The techniques presented here have been adapted successfully to the quantal bioassay setting (Nottingham and Birch, 1999), the non-normal error case (Assaid and Birch, 2000), and the nonconstant variance situation (Robinson and Birch, 2000). Other forms of semiparametric regression methods exist.

Though other methods will not be presented here, the interested reader is encouraged to consult Mays, Birch, and Starnes (2000); Hastie and Tibshirani (1990); Speckman (1988); Abdel-Salam, A. S. G., Birch, J. B., & Jensen, W. A. (2013); Williams, J. D., Birch, J. B., & Abdel-Salam, A. S. G. (2015); Waterman, M. J., Birch, J. B., & Abdel-Salam, A. S. G. (2015), and Siddiqui, Z., & Abdel-Salam, A. S. G. (2019) for more information and references.

Exercises

1) **Pro Football Data**: If one listens to the experts of the game of pro football, one would hear contradictory viewpoints on what aspect of the game makes for a successful season (specifically, the number of victories). There is an old adage in basically all sports, at any level of competition, that says that defense wins championships, and professional football is no exception. This comes from the viewpoint that a team cannot lose if the opponent is held scoreless (for the most part). However, proponents of the offensive side of the ball say that while a team cannot lose if the opponent is held scoreless, the same team cannot win if they do not score themselves. How can anyone doubt a team's chances of victory if they are an offensive juggernaut? Finally, there are those who say that when push comes to shove, it is the turnovers that can make or break a game, thus make or break a season. Turnovers are the key to the thrill of victory or the agony of defeat. Thus, the research aims to find out which side, if any or if all, plays a significant role in a team 5-season total of wins by looking at some pertinent offensive and defensive variables. This experiment is an observational study of the 1995 season of the National Football League (season-ending approximately the second week in December). Each team plays a total of 16 games in a regular season, with 8 games against opponents within the team's division, 4 games against teams in a division in the opposite conference (i.e., Dallas (NFC East) played four teams from the AFC West this year), and the remaining 4 games are determined by the NFL based on the team's previous year's success. Thus, not all possible games, between any two teams, can happen in one season. The sample size is 30, the number of professional football teams in the NFL. The variables of interest are as follows:

Team: The team in the NFL. This variable is not used statistically but just as an indicator of the corresponding data.

Wins: The number of victories for the team during the 1995 season.

ORush: The total number of rushing yards the offense accumulated during the 1995 season.

OPass: The total number of passing yards the offense accumulated during the 1995 season.

DRush: The total number of rushing yards the defense allowed during the 1995 season.

DPass: The total number of passing yards the defense allowed during the 1995 season.

TAway: The total number of takeaways the defense created during the 1995 season.

GAway: The total number of giveaways the offense had during the 1995 season.

PF: The total number of points scored by the team.

PA: The total number of points allowed by the team.

The data is found on the website in the file football.xlsx.

a) Consider the **pro football data** of Chapter 7. Let wins be the response variable, and let the other eight variables be the regressors in an MLR model. Our goal is to perform an MLR analysis to determine which regressors have a significant impact on the number of wins achieved. Do all the analyses using your software package.
b) Obtain scatterplots of the response versus each regressor. Comment.
c) Obtain the correlations between the response and each regressor. Using these correlations and the scatterplots in part a), which variables appear to affect the response?
d) Obtain the classical MLR analysis. Also, obtain the residuals, leverage values, and DFFITS. Graphically, display the leverage values and the DFFITS. Comment on your results. Which regressors appear to be significant, and which are insignificant? Do high-influence points exist in the dataset? If so, why?
e) Consider the backward elimination variable selection technique. The goal of using this method is to select some subset of the full regressor dataset that results, in some sense, in the best model. The method begins with all regressors in the MLR model. A variable is eliminated if it is insignificant based on the t-statistic for that variable. This may be determined by comparing the p-value for the t-statistic with some benchmark value, often selected to be 0.10. If at least one regressor is insignificant, based on the p-value for the t-statistic, the most insignificant variable is removed from the model. The MLR analysis is then obtained for the MLR model with one less regressor. Note that the technique eliminates only one variable at a time. This process continues until all variables in the MLR model are significant. At this point, the regressors in the MLR model represent the final subset of variables chosen to appear in the MLR model. In the classical application of the method, all analysis is performed using the least squares method. Apply the

backward elimination technique to this dataset to find a final model. What are the final variables in the model, and which have been eliminated?

f) Describe the final model. Does it make sense in light of the subject matter?

g) Perform a robust analysis of this data for the full set of regressors using M- and BI-estimation for both the Huber and bisquare psi functions. You need to determine the proper values of the tuning parameters. What are your observations? Are outliers and/or high-influence points present in this dataset? Using the weights to justify your answer.

h) Compare your results obtained in d) to those obtained in c). Use the analysis of the stackloss data as a guide. You should consider coefficient values for the classical and robust analysis. Use graphical displays to support your conclusions. You should include such displays as the weight index plots, probability plots of residuals, and box plots of residuals.

Use the t-test based on robust weights and the backward elimination algorithm to perform a robust backward elimination analysis of this data. What is your final model? Compare your results to those obtained in d).

i) Summarize the results of all your analyses into a statement about your final MLR model for this data.

2) **Longley Data**: The infamous Longley data (J. W. Longley. 1967. "An appraisal of least-squares programs from the point of view of the user," JASA, 62, 819–841.) is often used to demonstrate the problems of using multiple regression when the regressors are highly correlated. The goal was to model the number of people employed (NPeoE) in terms of regressors:

Label: Comment

GNP-IPD: GNP implicit price deflator with 1954 set to 100 GNP

UNE: Number of eligible workers who are unemployed

Armed: Number serving in the armed forces

NI Pop: Size of noninstitutionalized population >=14 years of age

The data is contained on the website in file longley.xlsx. Consider the **longley data** of Chapter 7. Let NPeoE be the response variable, and let the other five variables (exclude year) be the regressors. Use your software package to perform all the following analyses. Do parts a) through g) of question 1.

3) Consider the **snowfall data** of Chapter 1. Let snowfall be the response variable and the other three variables be the regressors. Use your software package to perform all the following analyses. Do parts a) through g) of question 1.

4) Consider the **phosphorus data** of Chapter 1. Let yield be the response variable and fertilize, mehl-1-p, and tissue-p be the three regressors. Use your software package to perform all the following analyses. Do parts a) through g) of question 1.
5) Consider the **bolts data** of Chapter 5. Let t20bolt be the response variable and speed1, total, speed2, number2, and sens be the regressors. Use your software package to perform all the following analyses. Do parts a) through g) of question 1.
6) Consider the **growing season data** of Chapter 2. Let the season be the response and the other four variables be the regressors. Use your software package to perform all the following analyses. Do parts a) through g) of question 1.
7) Ten observations were generated using the SLR model with $\beta_0 = 10$ and $\beta_1 = -2$. The errors are independent normal errors with $\sigma^2 = 0.25$. An outlier was added as the 11th observation. The data is found on the website in the file slr.xlsx. Use your software package to perform all the following analyses.
   a) Obtain a scatterplot of the data. Where is the outlier?
   b) Obtain the LS, M-, and BI-estimates using the Huber and bisquare psi functions with tuning constants 1.345 and 4.685, respectively.
   c) Compare the estimated coefficients for these methods. Which method gives coefficients closest to the true values?
   d) Alter the dataset by replacing observation 6 with the new observation Y = 15.23 and X = 4.89. Repeat parts a) through c) above.
   e) Consider the dataset in part d) above. Add a 12th data point at Y = 13.21 and X = 5.11. Repeat parts a) through c) above.
   f) Consider the result of your analysis of all three datasets. What conclusions can you now draw concerning the estimation techniques?
8) For your data, consider your response variable and several other variables, multiple regressors. Please pick these regressors variables carefully so that a meaningful MLR model may be developed for your response variable. Perform the MLR analysis of this data, mimicking the breadwrapper example. Your analysis should include the following:
   a) The LS analysis, including appropriate plots.
   b) The robust analysis, using both the M- and BI-estimates (using the Huber and bisquare weight functions). You pick your own value of the tuning constants. Include appropriate plots (weight index plot, for example). Which analysis, m or bi, is most appropriate for your data? Why?
   c) Discuss your results in terms of appropriateness of the LS and robust methods and any new insights into the research question gained by this analysis. Include all the parts a), b), and c) in your "report."

9) **Deer Browse Data**: A study was undertaken at Virginia Tech (*Evaluation of Prescribed Burning in Relation to Available Deer Browse*, 1964) to determine if fire can be used as a viable management tool to increase the amount of forage available to deer during the critical months in late winter and early spring. Calcium is a required element for plants and animals. The amount taken up and stored in the plant is closely correlated to the amount present in the soil. It was hypothesized that a fire may thus affect the amount available to the deer. A large tract of land in the Fishburn Forest was selected for a prescribed burn. Soil samples were taken from 12 plots of equal area just prior to the burn on May 1964 and analyzed for calcium. On July 16, 1964, postburn calcium levels were analyzed from the same plots. These values were in kilograms per plot. The data is taken from Walpole and Myers (1985). The data is stored on the website under the name deer_browse_data.xlsx.

10) Consider the deer browse data of Chapter 7. Compute the difference between preburn and postburn for each plot. The alternative hypothesis of interest is that the mean difference between the calcium level present in the soil before the burn exceeds the mean calcium level present in the soil after the burn. Before proceeding with your analysis, change the postburn score for the 12th plot from 9 to 90 to represent a recording error in entering the data. Do all calculations using your software package.

    a) Perform the appropriate analysis of these differences. Use the insulin data example of the text as a guide. Is the normality assumption of the differences reasonable?

    b) To test the hypothesis, use the four tests, the classical paired t-test, the paired t-test based on ranks, the Wilcoxon signed-rank test, and the paired t-test based on robust weights (use Huber's psi with a constant of 1.5). Compare your results using the p-value.

    c) Correct the error, changing the 90 back to 9. Redo the paired t-test and the paired t-test based on robust weights. Comment on any differences that exist between your results and the results obtained in part b).

11) **Tool-Life Data:** Suppose a mechanical engineer wishes to relate the effective life of a cutting tool (y) used on a lathe to the lathe speed in rpm (x) and the type of cutting tool used (z). Twenty observations on tool life and lathe speed are obtained, ten observations on each tool type. The data is on the website under toollife.xlsx. Data is taken from Montgomery and Peck (1982). Consider the tool-life data of Chapter 7. Let life be the response variable, speed be the regressor variable, and type be the two-valued categorical variable. Do the following analyses using your software package.

a) Obtain a scatterplot of life versus speed, where the points are labeled by type. Use the magnesium data as a guide. Discuss the appropriateness of the normality assumption and the common variance assumption. Do the slopes appear to be equal? The intercepts?

b) Perform the classical ANCOVA analysis. Test the equality of slopes. If the slopes are equal, test for equality of intercepts under a common slopes model.

c) Perform the robust ANCOVA analysis. Perform the test of equality of slopes. If the slopes are equal, test for equality of intercepts under a common slopes model. Compare your results to those obtained in part b) of this question. Discuss your final conclusion regarding tool life across the two tool types.

12) Consider the crab data of Chapter 7. Let APC be the response variable, and use the indicator variables to represent the four categories of the variable hour. Use the robust inference method of Section 7.4 to assess the role of hor on APC. Compare your results to those obtained in Section 7.6 based on the bootstrap. Discuss your results.

13) Consider the mill data of Chapter 7. Repeat the analysis presented in Section 7.5 except use the LTS method to obtain estimates of the differences between mills rather than M-estimates. Use the bootstrap method, using residual resampling, to assess the statistical significance of your results. Use the crab data analysis of Section 7.6 as a guide. Discuss your results.

14) For parts a) through c), consider the MLR model where we have $y = X\underline{\varepsilon} + \underline{\varepsilon}$. We know that the M-estimator, $\hat{\beta}_M$, is the solution to the altered normal equations, where $r^* = \dfrac{r}{\hat{s}}$.

a) Show that the iteration step required to solve the above equations using the Newton-Raphson (NR) method may be written as $\hat{\beta}_1 = \hat{\beta}_0 + \hat{s}\left(X'\langle\psi'(\underline{r}^*)\rangle X\right)^{-1} X'\psi(\underline{r}^*)$.

b) Give a step-by-step description of the NR algorithm required to obtain the solution to the altered normal equations. That is, write the algorithm (in English, not in computer code) required to obtain $\hat{\beta}_M$ using the NR method.

15) Suppose that M-estimation is used to estimate β_0 and β_1 in the simple linear regression model. You are interested in testing the hypotheses:

$H_0 : \beta_1 = 0$ versus $H_1 : \beta_1 \neq 0$

Give the formula for the test based on robust weights, t_{w_1}, used to test these hypotheses. Your answer should follow the robust inference section of Chapter 7. Be sure to write out all the formulas used in your test based on robust weights. Be sure to define all the terms used in your formulas.

16) Consider a two-way factorial with factors A and B, both with two levels. One model used to represent this situation is

$$y_{ijk} = \mu_{ijk} + \varepsilon_{ijk}$$

where $i = 1,2; j = 1,2;$ and $k = 1,\ldots,n_{ij}$, with the usual constraints applied to the effects. The first model is the means model, and the second is the effects model. Or the model may be written in MLR form as

$$y_i = \beta_0 + \beta_1 z_{Ai} + \beta_2 z_{Bi} + \beta_3 z_{ABi} + \varepsilon_i,$$

where $i = 1,\ldots,n$, with $n = \sum_{i,j} n_{ij}$ and z_A indicating level 2 of factor A, z_B indicating level 2 of factor B, and $z_{AB} = z_A z_B$.

a) Write out the equivalence table that relates to the β_j from the means model to the μ_{ij} from the MLR model.
b) Consider the following hypothesis tests on parameters for the effects model:

$$H_0\text{: all } \gamma_{ij} = 0 \text{ versus } H_1\text{: all } \gamma_{ij} \neq 0$$
$$H_0\text{: all } \alpha_i = 0 \text{ versus } H_1\text{: all } \alpha_i \neq 0$$
$$H_0\text{: all } \beta_j = 0 \text{ versus } H_1\text{: all } \beta_j \neq 0$$

Write the equivalent hypotheses using the parameters of the MLR model.

c) The robust general F-test is to be used to make the three tests using the parameters of the MLR model, which are the hypotheses that you wrote in part b) above. What is the G matrix necessary to test each of these three pairs of hypotheses?
d) Use the means model for this question. Consider the data in Table 7.11.

Using Huber's psi function with a constant 1.5, find the robust M-estimates of the parameters of the means model. Use only one iteration of IRLS. Do the calculations by hand.

TABLE 7.11

Two Factors Data

		Factor B	
		0	**1**
Factor A	0	1	3
		2	4
		3	11
	1	2	4
		3	5
		4	6

Multiple-Choice Questions

1. What does the term "multicollinearity" refer to in the context of multiple regression analysis?
   a) The correlation between the error terms
   b) The correlation between the independent variables
   c) The correlation between the dependent and independent variables
   d) None of the above

   Answer: b)
2. Which of the following is an assumption of the classical linear regression model?
   a) Heteroscedasticity
   b) Homoscedasticity
   c) Autocorrelation
   d) Multicollinearity

   Answer: b)
3. In regression analysis, what does the R^2 value indicate?
   a) Total variation in the dependent variable
   b) The proportion of variance in the dependent variable that can be predicted from the independent variables
   c) The residual error of the model
   d) The coefficient of the regression line

   Answer: b)
4. What is the purpose of using an F-test in multiple regression analysis?
   a) To test the individual coefficients of the regression model
   b) To test the overall significance of the regression model
   c) To check the goodness of fit of the model
   d) To compare two competing models

   Answer: b)
5. Which of the following statements about outliers in regression analysis is true?
   a) Outliers do not influence the regression coefficients.
   b) Outliers can affect both the slope and intercept of the regression model.
   c) Outliers are typically ignored in regression analysis.
   d) Outliers increase the R^2 value.

   Answer: b)

6. Which of the following is a method to assess the fit of a regression model?
   a) Increasing the sample size
   b) Calculating the residual standard error
   c) Adding more predictors
   d) Reducing the significance level

   Answer: b)

7. What is the main advantage of adding more predictors to a regression model?
   a) It always reduces the bias of the model.
   b) It necessarily improves the accuracy of predictions.
   c) It can potentially explain more variance in the dependent variable.
   d) It simplifies the model interpretation.

   Answer: c)

8. In the context of model diagnostics, what does leverage indicate?
   a) The error term is constant.
   b) The influence of a single observation on the overall model's predictions.
   c) The degree of correlation between predictors.
   d) The accuracy of the regression coefficients.

   Answer: b)

9. How does adding interaction terms in a regression model affect the model?
   a) It simplifies the interpretation.
   b) It accounts for the effect of one variable on another.
   c) It reduces the collinearity between variables.
   d) It decreases the model's degrees of freedom.

   Answer: b)

10. In regression analysis, what does the term "residuals" refer to?
    a) The difference between observed and predicted values of the dependent variable
    b) The predicted values of the dependent variable
    c) The independent variables in the model
    d) The coefficients of the independent variables

    Answer: a)

8

Analysis of Non-normal Data: The Generalized Linear Model

Summary

In this chapter, we will examine the analysis of non-normal data. Examples include those cases where the response variable is discrete. For example, in many practical problems, the value of the response variable is one of only two possible outcomes, commonly labeled as "success" and "failure." This two-valued discrete random variable is often coded for convenience, one for success and zero for failure. Clearly, the distribution of such a random variable is not the normal distribution. The Bernoulli distribution is often used to model this type of data. In another common situation, the response variable represents the number of occurrences of some outcome of interest. For example, we may be interested in the number of red blood cells in a sample of blood, the number of bubbles in a pane of glass, or the number of eggs in a bald eagle's nest. Such "count" data is non-normal, and the Poisson distribution is frequently used to model this situation.

Many other examples exist where theory suggests a particular distribution other than the normal. If the hypothesized distribution is a member of a certain class of distributions, namely the exponential family of distributions, there is a general method of modeling the response. This method, referred to as the generalized linear model or "GLM" will be the topic of this chapter. The first section will present the analysis of $0-1$ data; the second section will discuss count data. The third section will present the general case. The fourth section will use the general theory to give examples of the location, two-sample, and k-sample problem. The R program will be used in each section to illustrate the numerical and graphical analysis.

Learning Objectives

By the end of this chapter, readers should be able to:

- Understand the concept and application of the GLMs for analyzing non-normal data.
- Perform the quantal regression models for binary response variables, such as logistic and probit regression.
- Learn to model count data using Poisson regression and related techniques.
- Gain proficiency in estimating model parameters using maximum likelihood methods.

DOI: 10.1201/9781003565321-8

- Apply diagnostic and model evaluation tools, including pseudo R^2, classification accuracy, and likelihood ratio tests.
- Use statistical software (e.g., R) to implement GLM analyses and interpret results effectively.
- Develop skills in hypothesis testing within the GLM framework, including Wald tests and generalized likelihood ratio tests.

8.1 The Quantal Regression Model

In this section, we consider the following very common situation where the response variable, y, is a $0-1$ variable rather than a measurement variable. For example, $y=1$ if the subject "responds," and $y=0$ if the subject fails to respond. Thus, y could be just about any two-valued outcome. In medical situations, $y=1$ might indicate that the subject is sick, on medication, has lung cancer, or smokes cigarettes. A $y=0$ would indicate that the subject is not sick, or is not on medication, or does not have lung cancer, or does not smoke cigarettes. In an industrial application, a $y=1$ might indicate that a component is defective, and $y=0$ indicates that the component is not defective. In assessing voting preferences, $y=1$ might indicate a vote for the Democratic candidate, whereas $y=0$ is a vote for some candidate other than the Democrat. There are a countless number of possibilities.

The data in our quantal regression setting is of two basic types. In the first case, we observe

$$\begin{bmatrix} y_1 & x_{11} & \cdots & x_{k1} \\ y_2 & x_{12} & \cdots & x_{k2} \\ \vdots & \vdots & \cdots & \vdots \\ y_n & x_{1n} & \cdots & x_{kn} \end{bmatrix},$$

where y_i is a $0-1$ variable. This is exactly the standard multiple linear regression (MLR) setup, as presented in Section 7.1, except y is not a measurement variable. This is called the "ungrouped" data case.

In the second case, we observe data of the form

$$\begin{bmatrix} n_1 & r_1 & x_{11} & \cdots & x_{k1} \\ n_2 & r_2 & x_{12} & \cdots & x_{k2} \\ \vdots & \vdots & \vdots & \cdots & \vdots \\ n_s & r_s & x_{1s} & \cdots & x_{ks} \end{bmatrix},$$

where n_i is the number of subjects observed at $(x_{1i} \quad x_{2i} \quad \cdots \quad x_{ki})$ and r_i is the number responding at $(x_{1i} \quad x_{2i} \quad \cdots \quad x_{ki})$ (the number of y' equal to 1).

This is called the "grouped" data case. Of course, the grouped case is just the ungrouped case with n_i repetitions at each setting of the k regressors and r_i is the number of ones. There are distinct settings of the k regressors. The presentation in this chapter will emphasize the ungrouped data case.

8.1.1 Possible Models

It is clear that the random variable y is a Bernoulli random variable. The essential parameter of interest for a Bernoulli random variable is $\text{P} = \text{Prob}(\text{y} = 1)$, the probability that the subject will respond. In our current regression setting, this probability is modeled as a function of the regressor information and unknown parameters. That is, our model is $\text{Prob}(y_i = 1 \mid \underline{x}_i) = P(\underline{x}_i, \underline{\beta}) = P$, where $P()$ is a function of $\underline{x}_i$ and $\underline{\beta}$ and P_i is a convenient way to represent $p(\underline{x}_i, \underline{\beta})$. The possible forms of $P()$ will be presented shortly. Similarly, $\text{Prob}(y_i = 0 | \underline{x}_i) = 1 - P(\underline{x}_i, \underline{\beta}) = 1 - P_i = Q_i$. It is well known that for a Bernoulli random variable that $E(y_i) = P_i$ and $\text{Var}(y_i) = P_i Q_i$. One possible model for this new situation is the standard MLR model. That is, $\text{y}_\text{i} = \underline{x}_i{}'\underline{\beta} + \varepsilon_i$. The unknown parameters, $\underline{\beta}$, could then be estimated using the method of least squares, as usual. If this model is used, it must be that ε_i is a discrete random variable equal to $-\underline{x}_i\underline{\beta}$ when $y_i = 0$ (found by setting the left-hand side of the MLR model equal to zero and solving for ε_i) with probability Q_i and equal to $1 - \underline{\text{x}}'_\text{i}\, \underline{\beta}$ with probability P_i. Thus, ε_i is a two-valued random variable (a "shifted Bernoulli") with $E(\varepsilon_i) = P_i - \underline{x}_i{}'\underline{\beta}$ (so that $\text{E}(\text{y}_\text{i}) = \text{P}_\text{i}$) and $\text{Var}(\varepsilon_\text{i}) = \text{P}_\text{i}\text{Q}_\text{i}$. Consequently, the variance of ε_i is not constant as it must be for the LS estimator to be appropriate. In addition, another problem with this model is that if $\hat{\beta} = (\text{X}'\text{X})^{-1}\text{X}'y$ and $\hat{y}_i = \underline{x}'_i\, \hat{\beta} = \hat{P}_i$ is used as the estimate of $E(y_i) = P_i$, there is no assurance that $0 \le \hat{P}_i \le 1$. Clearly, an estimated probability of response of -0.221 is nonsense.

A better model that will eliminate the problem of estimated probabilities exceeding one or being less than zero is the model

$$\text{Prob}(y_\text{i}\underline{\text{x}}'_\text{i}) = P(\underline{x}'_i\, \underline{\beta}) = P_i = F(\underline{x}'_i\, \underline{\beta}),$$

where $F()$ is some appropriately chosen, smooth cumulative distribution function (CDF) that is a function of $\underline{x}_i$ and β through the linear combination $\underline{x}'_i\, \underline{\beta}$, called the "linear predictor." That is, a subject supplies the researcher with some $\underline{x}_i$ information; the linear predictor, $\underline{x}'_i\, \underline{\beta}$, is computed; and $F(\underline{x}_i\underline{\beta})$, the probability that the subject will respond, given this setting of the regressors, is found. Graphically, this concept is represented in Figure 8.1, where the lines with the arrows indicate that the probability is obtained as a function of the linear predictor.

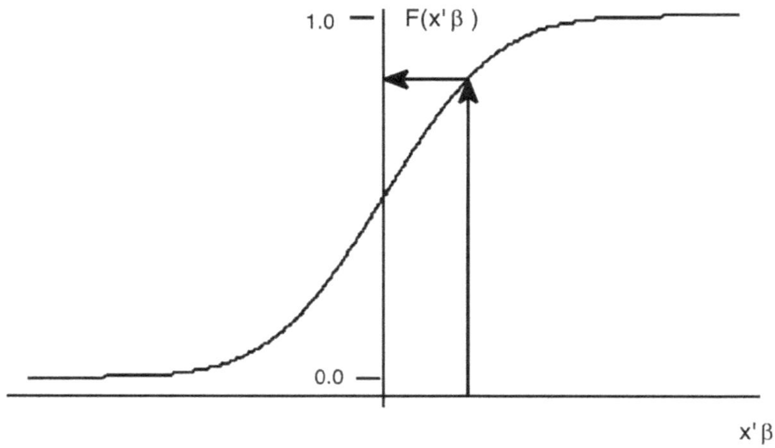

FIGURE 8.1
The CDF versus the linear combination of $\underline{x}_i' \ \underline{\beta}$.

Some goals in quantal regression are to build models, test hypotheses of interest, estimate response probabilities, and evaluate the model's quality. Each of these goals will be addressed in this section.

8.1.2 Estimation of $\underline{\beta}$

First, consider the estimation of $\underline{\beta}$. The method of least squares (LS) could be used to obtain estimates of $\underline{\beta}$, where $\hat{\beta}$ is found so that $\sum_{i=1}^{n}\left(y_i - F\left(\underline{x}_i\underline{\beta}^*\right)\right)^2$ is minimized over all possible values of, $\underline{\beta}^*$. However, recall that $\text{Var}(y_i) = P_iQ_i$ and is, in general, not constant across observations. This lack of constant variance violates one of the assumptions of LS, and consequently, the LS technique would not be the optimal way of obtaining the estimates of $\underline{\beta}$.

Consider instead the maximum likelihood (ML) estimator of $\underline{\beta}$. The reader will recall from Section 2.2 that the basic technique of the ML estimation is to first obtain the probability function of the random variable, y_i (either the density or mass function), and obtain the joint probability function of the n observations, $y_1,\ldots,y_n$. The likelihood function is defined as the joint probability function of $y_1,\ldots,y_n$, viewed as a function of the unknown parameters. The ML estimator is the value of the estimator that maximizes the likelihood function or, equivalently, maximizes the log of the likelihood function.

In the quantal regression problem, the probability mass function (PMF) for the discrete random variable, y_i is needed. Since y_i is a Bernoulli random variable, the PMF may be written as

$$\text{pmf}_i = F\left(\underline{x}_i' \ \underline{\beta}\right)^{y_i}\left(1 - F\left(\underline{x}_i' \ \underline{\beta}\right)\right)^{1-y_i}.$$

We note that if $y_i = 1$, then $pmf_i = P_i = F(\underline{x}_i'\underline{\beta})$, as it should, and if $y_i = 0$, then $pmf_i = 1 - P_i = 1 - F(\underline{x}_i' \_)$, as it should.

The likelihood function for $y_1, \ldots, y_n$ is then

$$L(\underline{\beta}, \underline{x}_i) = \prod_{i=1}^{n} F(\underline{x}_i' \underline{\beta}')^{y_i} \left(1 - F(\underline{x}_i' \underline{\beta})\right)^{1-y_i},$$

simply the joint probability mass function for $y_1, \ldots, y_n$, viewed as a function of β (with the sampled $y_1, \ldots, y_n$ viewed as fixed). Maximum likelihood theory says that the maximum likelihood estimator (MLE), $\underline{\beta}$, is chosen so that $L(\underline{\beta}^*, \underline{x}_i)$ is maximized over all possible values of $\underline{\beta}^*$. Or, equivalently, choose $\hat{\underline{\beta}}$ so that $\ln L(\underline{\beta}^*, \underline{x}_i)$ is maximized, where *ln* refers to the natural logarithm. Now, it is easy to see that

$$\ln L(\underline{\beta}, \underline{x}_i) = \sum_{i=1}^{n} \left(y_i \ln\left(F(\underline{x}_i' \underline{\beta})\right) + (1 - y_i) \ln\left(1 - F(\underline{x}_i' \underline{\beta})\right)\right).$$

The MLE is found by computing $\dfrac{\partial \ln L(\underline{\beta}^*, \underline{x}_i)}{\partial \underline{\beta}^*}$ and setting the results equal to $\underline{0}$, the $px1$ vector of zeros, and solving for $\hat{\beta}$. After some algebra, we obtain

$$\frac{\partial \ln L(\underline{\beta}^*, \underline{x}_i)}{\partial \underline{\beta}^*} = \sum_{i=1}^{n} \left[\frac{y_i - F(\underline{x}_i' \underline{\beta}^*)}{F(\underline{x}_i' \underline{\beta}^*)\left(1 - F(\underline{x}_i' \underline{\beta}^*)\right)} \right] f(\underline{x}_i' \underline{\beta}^*) \underline{x}_i,$$

where $f(u) = \dfrac{dF(u)}{du}$ is the probability density function (PDF) associated with the CDF, $F(u)$. In the above equation, we note that the expected value of y_i is subtracted from y_i in the numerator of the terms within the square brackets and that the variance of y_i is the denominator. Also, note that the derivative of $F(u)$ is one of the multipliers of the term in square brackets. This form is important not only for the quantal regression problem considered in this section but for all other examples within the GLM – a concept that will be illustrated and in the next two sections as well. In fact, the above system of equations represents a very general form of equations used to obtain the MLEs in a broad variety of situations.

Although the above system of equations appears to be very complicated, it is simply a more general form of equations previously observed. Consider first the usual form for the MLR model, discussed in Section 7.1. Recall in the MLR model that $y_i = \underline{x}_i'\underline{\beta} + \varepsilon_i$ with the assumption that $\underline{\varepsilon} \sim N(\underline{0}, \sigma^2 I)$. Then the likelihood function of the random sample $y_1, \ldots, y_n$ is

$$L(\underline{\beta}^*, \underline{x}_i) = \prod_{i=1}^{n} (2\pi\sigma^2)^{\frac{1}{2}} \exp\left(-\frac{(y_i - \underline{x}_i' \underline{\beta}^*)^2}{2\sigma^2} \right),$$

and after a little algebra, we find that

$$\frac{\partial \ln L(\underline{\beta}^*, \underline{x}_i)}{\partial \underline{\beta}^*} = \sum_{i=1}^{n} \left(\frac{y_i - \underline{x}_i' \, \underline{\beta}^*}{\sigma^2} \right) (1) \underline{x}_i,$$

or in matrix notation, we have

$$\frac{\partial \ln \ L(\underline{\beta}^*, \underline{x}_i)}{\partial \underline{\beta}^*} = \sigma^{-2} X'(\underline{y} - X\underline{\beta}^*).$$

We should recognize the above results. When this derivative is set equal to $\underline{0}$, we have the "normal equations" used in Section 7.1 to obtain the least squares estimates. We note that the form of the above equations matches exactly the form of $\frac{\partial \ln L(\underline{\beta}^*, \underline{\mathbf{x}}_i)}{\partial \underline{\beta}^*}$ for the MLEs in the quantal regression setting. That is, the expected value of y_i is subtracted from y_i in the numerator of the term in square brackets and the variance of y_i is the term in the denominator. We note that the function $F(u)$ that relates the expected value of y_i to $\underline{x}_i' \, \underline{\beta}^*$ is the identity function – that is, $E(y_i) = F(\underline{x}_i' \, \underline{\beta}^*) = \underline{x}_i' \, \underline{\beta}^*$. If we let $u = \underline{x}_i' \, \underline{\beta}^*$, then $\frac{\mathrm{dF(u)}}{\mathrm{du}} = 1$. We note that one is one of the multipliers of the term in square brackets. So, we see that the usual normal equations are set up in the same way as the $\frac{\partial \ln L(\underline{\beta}^*, \underline{x}_i)}{\partial \underline{\beta}^*}$ in the quantal regression setting.

8.1.3 The Numerical Method

We note that the nonlinear nature of $\frac{\partial \ln L(\underline{\beta}^*, \underline{x}_i)}{\partial \underline{\beta}^*}$ in the quantal regression setting does not lead to a closed-form solution to the MLEs as in the normal MLR case where F is the identity function and $\mathrm{Var}(\mathrm{y}_i)$ is constant. Solving the nonlinear system of equations for the quantal regression setting must be achieved using some appropriate numerical method.

The most common numerical method used to obtain the solution to $\frac{\partial \ln L(\underline{\beta}^*, \underline{\mathbf{x}}_i)}{\partial \underline{\beta}^*} = \underline{0}$ is the Gauss-Newton method. The method works by first expanding the function $F(\underline{x}_i{'}\underline{\beta}^*)$ about some initial estimate of $\underline{\beta}$, denoted by $\hat{\beta}_0$, using a first-order multivariate Taylor series expansion. (The reader will recall the use of the Taylor series in the delta method of Section 2.4). That is, we obtain the linear approximation to $F(\underline{x}_i \underline{\beta}^*)$ as

$$\mathrm{F}(\underline{x}_\mathrm{i}' \, \underline{\beta}^*) \approx \mathrm{F}(\underline{\mathrm{x}}_\mathrm{i}' \, \hat{\underline{\beta}}_0) + \mathrm{f}(\underline{\mathrm{x}}_\mathrm{i}' \, \hat{\underline{\beta}}_0) \underline{\mathrm{x}}_\mathrm{i}' \, (\underline{\beta}^* - \hat{\underline{\beta}}_0).$$

Substitution of the right-hand side of the above expression for $F\left(\underline{x}_i{}'\underline{\beta}^*\right)$ in the numerator within the square brackets of the $\frac{\partial \ln L\left(\underline{\beta}^*, \underline{x}_i\right)}{\partial \underline{\beta}^*} = \underline{0}$ expression for quantal regression and replacing $\underline{\beta}^*$ by $\hat{\beta}_0$ gives us, after a little algebra,

$$\frac{\partial \ln L\left(\beta^*, \underline{x}_i\right)}{\partial \underline{\beta}^*} \approx \sum_{i=1}^{n}\left[y_i^{**} - \underline{x}_i'\,\underline{\beta}^*\right]\left[\frac{f^2\left(\underline{x}_i'\,\hat{\underline{\beta}}_0\right)}{F\left(\underline{x}_i'\,\hat{\underline{\beta}}_0\right)\left(1 - F\left(\underline{x}_i'\,\hat{\underline{\beta}}_0\right)\right)}\right]\underline{x}_i = \underline{0},$$

where

$$y_i^{**} = \frac{y_i - F\left(\underline{x}_i'\,\underline{\beta}_0\right) + f\left(\underline{x}_i'\,\hat{\underline{\beta}}_0\right)\underline{x}_i'\,\underline{\beta}_0}{f\left(\underline{x}_i'\,\hat{\underline{\beta}}_0\right)}.$$

If we let $w_i = \frac{f^2\left(\underline{x}_i\hat{\underline{\beta}}_0\right)}{F\left(\underline{x}_i{}'\hat{\underline{\beta}}_0\right)\left(1 - F\left(\underline{x}_i\,\hat{\underline{\beta}}_0\right)\right)}$, we can rewrite the above expression as

$$\frac{\partial \ln L\left(\underline{\beta}^*, \underline{x}_i\right)}{\partial \underline{\beta}^*} \approx \sum_{i=1}^{n}\left[y_i^{**} - \underline{x}_i'\,\underline{\beta}^*\right]w_i\underline{x}_i = \underline{0},$$

a set of weighted least squares normal equations. It is easy to solve for the value of $\underline{\beta}^*$, which achieves the solution as

$$\hat{\beta}_1 = \left(X'WX\right)^{-1}X'Wy^{**},$$

where W is an nxn diagonal matrix, called the weight matrix, with $w_1, \ldots, w_n$ as the diagonal elements. A little algebra reveals that we can write the value for $\hat{\beta}_1$ as

$$\hat{\beta}_1 = \hat{\beta}_0 + \left(X'WX\right)^{-1}X'Wy^*,$$

where the ith element of $\underline{y}^*$ is $y_i^* = \frac{y_i - F\left(\underline{x}_i\,\underline{\beta}_0\right)}{f\left(\underline{x}_i'\,\hat{\beta}_0\right)}$. Unfortunately, $\hat{\beta}_1$ is seldom equal to the MLE because the closeness of $\frac{\partial \ln\ L\left(\underline{\beta}^*, \underline{x}_i\right)}{\partial \underline{\beta}^*}$ to its linearized version depends a great deal on the closeness of $\hat{\beta}_0$ to the MLE.

We should think of $\hat{\beta}_1$, expressed above, as an "improved" estimator of $\underline{\beta}$ over the initial estimator $\hat{\beta}_0$. However, we can now use $\hat{\beta}_1$ as the new initial estimator and obtain an improved version of it using the same process as described above. We can continue to obtain improved versions of the estimator by iterating the process many times until convergence is achieved. We employ the iteratively reweighted least squares (IRLS) algorithm of Chapters 6 and 7.

The IRLS algorithm, adjusted for quantal regression, is as follows:

Step 1: Obtain the initial estimator, $\hat{\beta}_0$. More will be said about how this is accomplished in the comments to follow the algorithm.

Step 2: Compute the y_i^* as $y_i^* = \dfrac{y_i - F\left(\underline{x}_i\underline{\hat{\beta}}_0\right)}{f\left(\underline{x}_i'\underline{\hat{\beta}}_0\right)}$ and the w_i as

$$w_i = \frac{f^2\left(\underline{x}_i'\underline{\hat{\beta}}_0\right)}{F\left(\underline{x}'_i\underline{\hat{\beta}}_0\right)\left(1-F\left(\underline{x}'\underline{i}_0\underline{\hat{\beta}}_0\right)\right)}, \text{ for i} = 1,\ldots,\text{n}.$$

Step 3: Compute $\hat{\beta}_1 = \hat{\beta}_0 + \left(\mathrm{X'WX}\right)^{-1}\mathrm{X'Wy}^*$.

Step 4: If $\hat{\beta}_1 \approx \hat{\beta}_0$, quit. Declare $\hat{\beta} = \hat{\beta}_1$. Otherwise, replace $\hat{\beta}_0$ by $\hat{\beta}_1$, and return to step 2.

The most common way to start the algorithm is to apply the MLR model with y as the response and obtain $\hat{\beta}_0$ as the least squares estimate.

It should be noted at this time that finding $\underline{\hat{\beta}}$ so that $\ln L\left(\underline{\beta}, \underline{x}_i\right)$ is maximized is equivalent to minimizing $-2\ln L(\beta, \underline{x}_i)$, a quantity that we will use throughout this chapter. It is easy to see that $L(\beta, \underline{x}_i)$, the product of n probabilities must be a number between zero and one, and that $\ln L\left(\underline{\beta}, \underline{x}_i\right)$ must be between $-\infty$ and zero. Consequently, $-2\ln L(\beta, \underline{x}_i)$, a quantity we want to make as small as possible, will be a number between zero and $+\infty$.

8.1.4 Specific Forms of $F\left(\underline{x}_i'\underline{\beta}\right)$

The discussion to this point has been developed for any arbitrary CDF, F. However, several forms of this function are used far more than others. If $F\left(\underline{x}_i'\underline{\beta}\right)$ is the function

$$F(x_i'\beta) = \frac{1}{1+\exp(-x_i'\beta)'}$$

the CDF for the standard logistic random variable (a logistic with a mean of zero and a scale of one), then quantal regression is referred to as "logistic regression." This is the most common contemporary form of F but not the only one.

If $F\left(\underline{x}_i\underline{\beta}'\right)$ is the function

$$F\left(\underline{x}_i{}'\underline{\beta}\right) = \frac{1}{\sqrt{2\pi}}\int_{-\infty}^{\underline{x}_i'\underline{\beta}} \exp\left(\frac{-z^2}{2}\right) dz = \Phi\left(\underline{x}_i{}'\underline{\beta}\right),$$

the CDF of a standard normal random variable, then quantal regression is called "Probit regression" or "Probit analysis." See the books by Finney (1971,1978) for more on this topic. Other forms of F that have been used are the Weibull and the Gompertz. Due to its diversity and current popularity, only the logistic will be illustrated here.

8.1.5 Model Evaluation and Hypothesis Testing

Once $\hat{\beta}$ is obtained by MLE, the model can be applied to the data. We obtain

$$\hat{P}(\underline{x}_i) = F(\underline{x}_i\underline{\hat{\beta}}) = \hat{y}_i$$

as the estimate of the mean response of $y, E(y_i) = P(\underline{x}_i)$. This is the estimated probability that the ith subject will respond. A good model is one where the estimated responses, the $\hat{P}(\underline{x}_i), i = 1,\ldots,n$, are in "agreement" with the observed responses. By this agreement, we mean that if $y_i = 0$, then we would hope that $\hat{P}(\underline{x}_i)$ is small, implying that the model says the subject is unlikely to respond. And if $y_i = 1$, then we would hope $\hat{P}(\underline{x}_i)$ is large, implying that the model says that the subject is likely to respond. Methods to quantify this agreement will be discussed shortly.

Hypothesis tests can also be made to evaluate the statistical contribution of each regressor to the quantal regression model. This is measured in terms of the contribution of each regressor to the likelihood function or the −2log likelihood function in that the stronger the variable in the model, the greater the increase in likelihood or, equivalently, the greater the decrease in 2log likelihood.

The following fact summarizes a key hypothesis testing concept to keep in mind when using the theory of MLEs. As variables are added to the model, the likelihood function cannot decrease, or, equivalently, −2 log-likelihood cannot increase. Subsequent material will demonstrate applications of this fact.

8.1.6 Generalized Likelihood Ratio Tests

In a manner similar to the concepts presented in Section 7.1 for MLR, partition $\underline{\beta}$, a $px1$ vector, and X as follows. Let $\underline{\beta} = \begin{bmatrix} \underline{\beta}_1 \\ \underline{\beta}_2 \end{bmatrix}$, where $\underline{\beta}_1$ is $p_1x1, \underline{\beta}_2$ is p_2x1 with $p_1 + p_2 = p$. Furthermore, write $X = [X_1 \quad X_2]$, where X_1 is nxp_1 and X_2 is nxp_2. Suppose we wish to test the following hypotheses:

$H_0 : \underline{\beta}_2 = \underline{0}$ versus $H_1 : \underline{\beta}_2 \neq \underline{0}$, in the presence of X_1.

Of course, this is equivalent to testing the null hypothesis that the model using only the X_1 regressors, the "reduced model," is appropriate for the data versus the alternative hypothesis that the model using the regressors in both X_1 and X_2, the "full model," is appropriate. In Chapter 7, the general F-test was used to make this test. However, this chapter does not deal with normal data with constant variance. Just as the estimation technique differs, so does the testing method.

We define the generalized likelihood ratio test (GLRT) as

$$\text{GLRT} = -2\ln\left[\frac{L\left(\hat{\underline{\beta}}_1, \underline{x}_i'\right)}{L\left(\hat{\underline{\beta}}, \underline{x}_i'\right)}\right],$$

where $\hat{\beta}_1$ and $\hat{\beta}$ are the MLEs for the reduced and full models, respectively. We see that the GLRT is the ratio of $-2\log$ likelihood for the reduced to the full model.

How should we interpret the GLRT? The above fact tells us that $L\left(\hat{\underline{\beta}}, \underline{x}_i'\right) \geq L\left(\underline{\beta}_1, \underline{x}_i'\right)$, since the full model contains p_2 more regressors than the reduced model. Thus, the likelihood ratio, $\dfrac{L(\hat{\underline{\beta}}_1, \underline{x}_i')}{L(\hat{\underline{\beta}}, \underline{x}_i')}$, must be greater than or equal to zero but less than or equal to one (recall that the likelihood must be between zero and one for quantal regression). If H_0 is true, the regressors associated with $\underline{\beta}_2$ contribute little to the model in terms of likelihood. Therefore, $L\left(\hat{\underline{\beta}}_1, \underline{x}_i'\right) \approx L\left(\hat{\underline{\beta}}, \underline{x}_i'\right)$, the likelihood ratio is close to one, and the GLRT is close to zero. Thus, small values of GLRT indicate that the null hypothesis is true. On the other hand, if H_1 is true, the regressors associated with $\underline{\beta}_2$ contribute significantly to the likelihood. From the above fact, it follows that $L\left(\hat{\underline{\beta}}_1, \underline{x}_i'\right)$ may be substantially smaller than $L\left(\hat{\underline{\beta}}, \underline{x}_i'\right)$, resulting in the likelihood ratio close to zero and the GLRT being large. Thus, large values of GLRT indicate that the alternative hypothesis is true. In order to quantify the size of the GLRT, we need its sampling distribution. Likelihood theory states that the asymptotic distribution of the GLRT is that of a chi-square with p_2 degrees of freedom.

It will be convenient for us later to express the GLRT as a difference in −2 log-likelihood rather than the −2log of the likelihood ratio.

That is, we can write

$$\text{GLRT} = -2\ln \text{L}\left(\hat{\underline{\beta}}_1, \underline{\text{x}}_i{}'\right) - \left[-2\ln \text{L}\left(\hat{\underline{\beta}}, \underline{\text{x}}_i{}'\right)\right].$$

We note that, in general, $-2\ln L\left(\hat{\underline{\beta}}, \underline{\text{x}}_i\right) \leq -2\ln \text{L}\left(\hat{\underline{\beta}}, \underline{\text{x}}_i{}'\right)$ (this statement follows from the previous fact). If H_0 is true, $-2\ln \text{L}\left(\hat{\underline{\beta}}, \underline{\text{x}}_i{}'\right) \approx -2\ln \text{L}\left(\hat{\underline{\beta}}_1, \underline{\text{x}}_i{}'\right)$, and the GLRT is approximately zero. And, if H_1 is true, $-2\text{In } L\,(\hat{\underline{\beta}}, \underline{\text{x}}_i{}') < -2\text{In } L\,(\hat{\underline{\beta}}, \underline{\text{x}}_i{}')$ and the GLRT is greater than zero. Thus, the GLRT measures the amount of reduction in $-2\ln L\left(\underline{\beta}_1, \underline{x}_i{}'\right)$ due to the contribution of adding the regressors in X_2 to the model. The greater the contribution, the more significant the variables, where the benchmark for significance is the chi-square distribution with p_2 degrees of freedom.

The analogous concept in the MLR model from Chapter 7 is measured by the error sum of squares for the full and reduced models. That is, for the normal data case with known variance, it may be shown that the GLRT would be

$$\text{GLRT} = \frac{\left(\text{SS}_{\text{error, reduced}} - \text{SS}_{\text{error, full}}\right)}{\sigma^2}.$$

There are two specific examples of using the GLRT applicable to this section's quantal regression setting and the GLM of the next section. Consider testing the hypotheses:

$\text{H}_0 : \beta_1 = \beta_2 = \ldots = \beta_k = 0$ versus H_1 : not H_0, (in the presence of the intercept).

In this example, $\underline{\beta} = \begin{bmatrix} \underline{\beta}_0 \\ \underline{\beta}_2 \end{bmatrix}$, where $\underline{\beta}_2 = \begin{bmatrix} \beta_1 \\ \vdots \\ \beta_k \end{bmatrix}$ and the GLRT test and its asymptotic distribution become

$$\text{GLRT} = -2\ln \text{L}\left(\hat{\underline{\beta}}_0, \underline{\text{x}}_i{}'\right) - \left[-2\ln \text{L}\left(\hat{\underline{\beta}}, \underline{\text{x}}_i{}'\right)\right] \sim \chi^2_{\text{k}},$$

an approximate chi-square random variable with *k* degrees of freedom. This particular form of the GLRT is referred to as the "model chi-square or model χ^2" test since it is testing all the regressors in the model simultaneously. It is analogous to the model or overall *F*-test in the MLR setting of Chapter 7.

The quantal regression model based only on the intercept, the equivalent of the location model in the usual linear model framework, provides us with some interesting insights. The model states that the probability of response can be written as

$$P(\underline{x}_i) = F(\beta_0) = P,$$

a constant independent of the values of the regressors. It is easy to show from the ML theory, presented in Section 2.2, that the MLE of P is $\hat{P} = \bar{y} = \frac{r}{n}$, the sample proportion, where r is the total number responding in the sample, and n is the total sample size.

By the invariance property of the MLE (which states that if $\hat{P}$ is the MLE of P, then $g(\hat{P})$ is the MLE of $g(P)$, where $g()$ is some function), it follows that the MLE of β_0 is $\underline{\beta}_0 = F^{-1}(\hat{P})$. In the case of logistic regression, where F is the logistic CDF, the inverse CDF is the "logit," where the logit of P is defined as $\text{logit}(P) = \ln\left(\frac{P}{1-P}\right)$.

Thus, the MLE of β_0 is $\hat{\beta}_0 = \text{logit}(\hat{P}) = \ln\left(\frac{\bar{y}}{1-\bar{y}}\right)$. For this simple example, the IRLS algorithm is not required.

It is also easy to show for the intercept-only quantal model, the −2log likelihood reduces to the following form:

$$-2\ln L(\hat{\beta}_0, \underline{x}_i') = -2[r\ln(y) + (n-r)\ln(1-\bar{y}).$$

The importance of $-2\ln L(\underline{\hat{\beta}}_0, \underline{\underline{x}}_i{}')$ is that since there are no regressors in the model, this is the largest value that −2log likelihood can achieve, as it can only get smaller (or stay the same size) as regressors are added to the model. On the other hand, if there is a perfect agreement between the model fits and the observed data (for $y_i = 1$, the model gives $\hat{P}(\underline{x}_i)$ and for $y_i = 0$, the model gives $\hat{P}(\underline{x}_i) = 0$), it is easily shown that the minimum value that can be achieved by −2log likelihood is zero. Thus, zero and $-2\ln L(\underline{\hat{\beta}}_0, \underline{x}_i{}')$ provide the minimum and maximum bounds that any model can achieve by −2 log-likelihood.

Again, there are obvious analogies with the MLR model. We can think of $-2\ln L(\underline{\underline{\beta}}_0, \underline{x}_i{}')$ in quantal regression as playing the role of SS_{total} from MLR, the model χ^2 from quantal regression as SS_{model} from MLR and $-2\ln L(\underline{\hat{\beta}}, \underline{x}_i')$ from quantal regression as SS_{error} from MLR. The concept of the partitioning of the sum of squares in the MLR model is replaced by the partitioning of the −2log likelihood as

$$-2\ln \mathrm{L}\left(\hat{\underline{\beta}}_0, \underline{\mathrm{x}}_i'\right) = -2\ln \mathrm{L}\left(\hat{\underline{\beta}}, \underline{\mathrm{x}}_i'\right) + \mathrm{model}\chi^2.$$

One goal in quantal regression is to find the proper regressors for the model to drive −2log likelihood down from $-2\ln L\left(\hat{\underline{\beta}}_0, \underline{x}'\right)$ toward zero by a significant amount.

A second illustration of the GLRT is the testing of the hypotheses:

$\mathrm{H}_0 : \beta_j = 0$ versus $\mathrm{H}_1 : \beta_j \neq 0$, in the presence of the other $k-1$ regressors.

Let $X = \left[X_{-j} \;\; \underline{x}_j\right]$, where X_{-j} is the *nxk* matrix of all regressors (including the intercept) included in the model except the regressor $\underline{x}_j$, the regressor being tested for possible inclusion into the model. Then $\underline{\beta} = \begin{bmatrix} \underline{\beta}_{-j} \\ \beta_j \end{bmatrix}$, where $\underline{\beta}_{-j}$ is the $kx1$ vector of all coefficients for the model except β_j. The GLRT test statistic appropriate for testing H_0 and its distribution are

$$\mathrm{GLRT} = -2\ln L\left(\hat{\underline{\beta}}_{-j}, \underline{\mathrm{x}}_i'\right) - \left[-2\ln L\left(\hat{\underline{\beta}}, \underline{\mathrm{x}}_i'\right)\right] \sim \chi_1^2,$$

where the single degree of freedom results from the fact that only one regressor $\left(\underline{x}_j\right)$ is being tested by the GLRT. This is called the partial chi-square (or partial χ^2) test and is analogous to the partial F-test in the MLR model. Using the GLRT allows us to make all tests of interest: on all variables, any single variable, or any combination of variables.

8.1.7 Some Additional MLE Theory

There are other approaches to testing hypotheses involving regressors when the estimates are obtained by the maximum likelihood method, as in our quantal regression case. One approach is based on the asymptotic distribution of the MLEs. Let $\hat{\beta}$ denote the MLE of $\underline{\beta}$. Then it can be shown that the asymptotic distribution of $\hat{\beta}$ is that of a multivariate normal with mean $\underline{\beta}$ and variance Σ. Furthermore, it can be shown that $\Sigma = \mathrm{I}^{-1}\left(\hat{\underline{\beta}}\right)$, where $I\left(\hat{\beta}\right) = -E\left(\dfrac{\partial^2 \ln L\left(\underline{\beta}, \underline{x}_i\right)}{\partial \underline{\beta} \partial \underline{\beta}'}\right)$, a *pxp* matrix of second derivatives of the log-likelihood function. This matrix is called the Fisher information matrix, the multivariate counterpart of the univariate Fisher information number from Section 2.2.

Thus, we can write

$$\hat{\hat{\beta}} \sim \mathrm{N}\left(\underline{\beta}, \mathrm{I}^{-1}\left(\underline{\hat{\beta}}\right)\right).$$

The information matrix $I\left(\underline{\hat{\beta}}\right)$, will typically involve the unknown parameters, $\underline{\beta}$, and, therefore, must be estimated. One way to accomplish this is to compute

$$\hat{\mathrm{I}}\left(\underline{\hat{\beta}}\right) = -\left(\frac{\partial^2 \ln L\left(\underline{\beta}_{\beta}, \underline{x}_i\right)}{\partial \underline{\beta} \partial \underline{\beta}}\right)_{\underline{\beta}=\underline{\hat{\beta}}}$$

the negative second derivative of the log-likelihood, with all parameters replaced by their MLE. Consequently, the estimated variance-covarinace matrix of the MLE is $\hat{\mathrm{V}}\mathrm{ar}\left(\underline{\hat{\beta}}\right) = \hat{\mathrm{I}}(\underline{\hat{\beta}})^{-1}$. It turns out that this estimated variance-covarinace matrix, $\hat{\mathrm{I}}(\underline{\hat{\beta}})^{-1}$ can also be written as $\left(\mathrm{X}'\mathrm{W}\mathrm{X}\right)^{-1}$, the same quantity needed in the IRLS algorithm used to obtain the MLE. The diagonal elements of W, the n weights, are functions of the MLEs. We see that quantal regression, in particular, or the GLMs in general, is yet another illustration of the use of weighted least squares. Previous examples occurred in computing the M- or BI-estimates of Chapter 7 and in computing the LLR estimates of Chapter 6. With this information, we can test hypotheses of the form

$\mathrm{H}_0 : \beta_j = 0$ versus $\mathrm{H}_1 : \beta_j \neq 0$, in the presence of the other $k-1$ regressors.

Using the test statistic

$$z_j = \frac{\hat{\beta}_j}{\mathrm{se}\left(\hat{\beta}_j\right)},$$

where $\mathrm{se}\left(\hat{\beta}_j\right)$ is the jth diagonal element of $\mathrm{Var}\left(\underline{\hat{\beta}}\right)$. This statistic, commonly called a "Wald's z-test" has an asymptotic standard normal distribution. It is readily seen that z_j is analogous to the t-test in the usual MLR model.

Alternatively, we could compute $z_j^2 = X^2$, a test statistic with an asymptotic chi-square distribution with one degree of freedom. This version of the test statistic is called "Wald's chi-square" test and is analogous to the partial F-test in MLR.

We now have two approaches for testing the significance of a single regressor in a k-variable model, either by using the GLRT or by using either of the Wald tests. Actually, several other possibilities exist as well, and all are asymptotically equivalent but may differ for any finite sample size problem. And, for any given dataset, one cannot say which procedure is the best. In the examples to follow, we will compare the GLRT and the Wald chi-square test.

By the way, the reader is probably already familiar with one application of the MLE theory presented above. Consider again the usual MLR model written as $\underline{y} = X\underline{\beta} + \underline{\varepsilon}$ with $\underline{\varepsilon} \sim N\left(\underline{0}, \sigma^2 I\right)$. We have seen earlier in this section the MLE for $\underline{\beta}$ is just the least squares estimator (LS).

What about the variance-covariance matrix of $\hat{\beta}$?

For the given model, we have

$$L\left(\underline{\beta}, \underline{x}_i\right) = \prod_{i=1}^{n} \left(2\pi\sigma^2\right)^{\frac{1}{2}} \exp\left(-\frac{\left(y_i - \underline{x}_i{}'\underline{\beta}\right)^2}{2\sigma^2}\right)$$

and

$$\ln L\left(\underline{\beta}, \underline{x}_i\right) = -\frac{n}{2}\ln(2\pi) - \frac{\text{n}}{2}\ln\left(\sigma^2\right) - \frac{\left(\underline{y} - X\underline{\beta}\right)'\left(\underline{y} - X\underline{\beta}\right)}{2\sigma^2}.$$

It follows that

$$\frac{\partial \ln L\left(\underline{\beta}, \underline{x}_i\right)}{\partial \underline{\beta}} = \frac{X'\left(\underline{y} - X\underline{\beta}\right)}{\sigma^2}$$

and

$$-\left(\frac{\partial^2 \ln L\left(\underline{\beta}, \underline{x}_i\right)}{\partial \underline{\beta} \partial \underline{\beta}'}\right) = \frac{X'X}{\sigma^2} = I\left(\underline{\beta}\right).$$

And we then have

$$\text{I}(\underline{\beta})^{-1} = \sigma^2 (X'X)^{-1},$$

which we know is not only the asymptotic variance but the exact variance-covariance of the LS estimator $(=\text{mle})$ of β. That is, this variance-covariance expression is correct regardless of sample size. Of course, it would be estimated by $\hat{\text{I}}(\beta)^{-1} = \hat{\sigma}2\left(X'X\right)^{-1} = MS_{\text{error}}\left(X'X\right)^{-1}$, as indicated in Chapter 6.

8.1.8 Measures of Model Performance

While tests of hypotheses are useful in building quantal regression models, or any GLM for that matter, the significance of variables in the model does not indicate whether the model is providing a useful description of the data. We saw this point illustrated in Chapter 6, where many other measures are available to quantify model performance. One such measure is the R^2, the ratio of the sum of squares for the model divided by the sum of squares total (both sums of squares are corrected for the mean).

In the quantal regression model, the −2 log-likelihood terms are the counterparts to the MLR sum of squares. Therefore, a logical " R^2-type" measure, a pseudo R^2, for quantal regression would be

$$R^2 = \frac{-2\ln L\left(\hat{\beta}_0, \underline{x}_i\right) - \left[-2\ln L\left(\hat{\underline{\beta}}, \underline{x}_i\right)\right]}{-2\ln L\left(\hat{\beta}_0, \underline{x}_i\right)} = \frac{\text{model}\chi^2}{-2\ln L\left(\hat{\beta}_0, \underline{x}_i\right)}.$$

From earlier, it is clear that this R^2 must satisfy $0 \le R^2 \le 1$. One must use caution when using this R^2 in quantal regression, however. A good quantal regression model may have smaller R^2 than we would expect for a good MLR model. The justification for this statement is illustrated by examining a plot of the model's estimated responses versus the observed responses. A plot of $\hat{y}_i$ versus y_i is presented in Figure 8.2 for the MLR model using the breadwrapper data as an example. Recall from Chapter 7 that the final MLR equation modeled the breaking strength of the breadwrapper seal (*s*) as a function of sealing temperature (*st*) and percent polyethylene (*pe*). The model had an R^2 of 0.986, obviously an excellent fit. The R^2 for the MLR model may be viewed as the square of the correlation between y_i and $\hat{y}_i$, the multiple correlation coefficient. That is, $R^2 = \text{corr}(\hat{y}, y)^2$. Thus, the R^2 is a measure of the linear association between y and $\hat{y}$. In Figure 8.2, a straight line, representing the LS fit of $\hat{y}$ to y via a simple linear regression is superimposed on the scatterplot, indicating that y and $\hat{y}$ are nearly perfectly linearly related. In fact, corr $(\hat{s}, s) = .993$, resulting in $R^2 = .993^2 = .986$. The R^2 is a valid summary measure of linearity, appropriate for a linear model.

On the other hand, the quantal regression model, and other GLMs, are nonlinear models, and R^2 is no longer an appropriate measure of model performance. This is seen in Figure 8.3, where a plot of $\hat{y} = \hat{P}(\underline{x}_i)$ versus y is given for an artificial dataset with five responses equal to zero and five responses equal to one. It is noted that y can take on only the values zero and one, but that $\hat{P}(\underline{x}_i)$ can take on any value between zero and one. Thus, it is impossible for the continuous $\hat{P}(\underline{x}_i)$ to be strongly linearly associated with the dichotomous y. This fact is revealed by noticing that the $\hat{P}(\underline{x}_i)$ in the plot Figure 8.3 do not lie on the straight line, representing the best linear fit to this data. In this example, corr $\left(\hat{P}(\underline{x}), y\right) = .917$, and $R^2 = .917^2 = .841$, indicating a moderately good fit for the data.

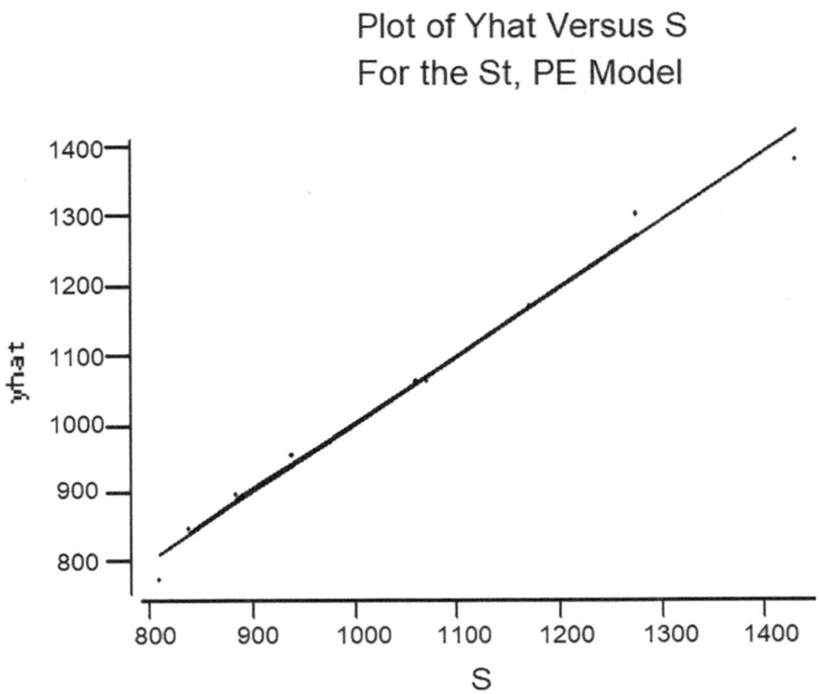

FIGURE 8.2
The plot of *yhat* versus S for the ST, PE models.

One may argue, however, that the quantal regression model is much better than "moderately good" for this example. We note that whenever $y = 0$, the estimated probability of response is smaller than the estimated probability of response whenever $y = 1$. In fact, one could argue that there is a near-perfect agreement between the true response and the estimated probabilities in that when $y = 0$, all estimated probabilities are "small," and when $y = 1$, all estimated probabilities are "large" – a fact inconsistent with the small R^2 value.

The problem arises because the response, y, is a $0-1$ variable and $\hat{P}(\underline{x}) = \hat{y}$ is a continuous variable on the $0-1$ scale. Since we cannot make y continuous, one way to compare them is to make $\hat{P}(\underline{x})$ discrete. This is accomplished by picking an arbitrary cutoff on the $\hat{P}(\underline{x})$ axis – typically, 0.5 is chosen – and using the following rule:

$$\text{if } \hat{P}(\underline{x}_i) \geq \text{cutoff, predict a 1 ("response")}$$

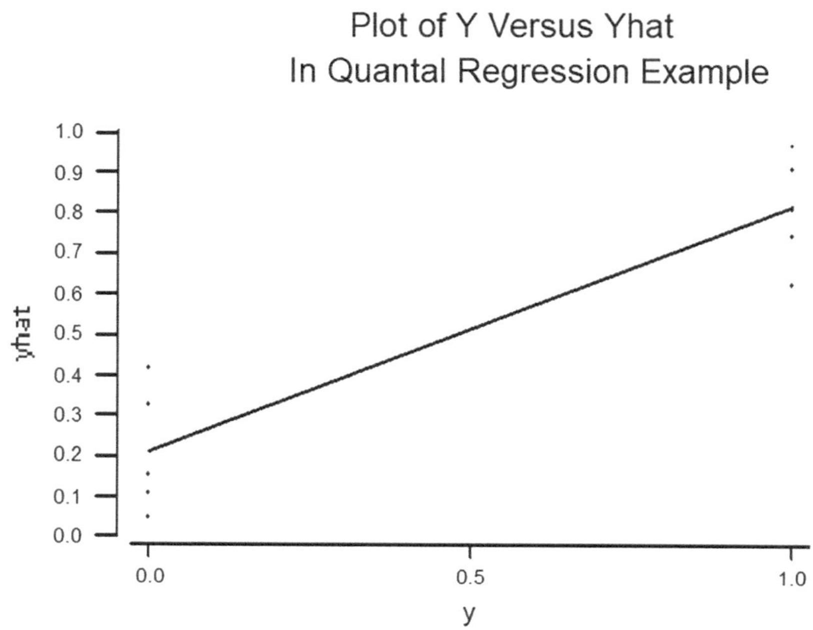

FIGURE 8.3
The plot of y versus *yhat* in quantal regression.

TABLE 8.1
Classification Matrix

		Predicted (based on rule)		
		1	0	
Actual (y)	1	A	B	$A + B = n_1$
	0	C	D	$C + D = n_0$
		A + C	B + D	n

and

$$\text{if } \hat{P}(\underline{x}_i) < \text{ cutoff, predict a 0 ("nonresponse").}$$

These new predicted responses can be compared to the actual responses by setting up Table 8.1.

For example, A is the number of subjects with $y = 1$ and $\hat{y} = 1$, where $\hat{y}$ is the predicted response based on the above rule, B is the number with $y = 1$ and $\hat{y} = 0$, and so on. The total sample size is n, with n_1 and n_0 representing the actual number of respondents who responded and those who did not respond, respectively. Now, the quantity of interest is the fraction of subjects that are correctly classified, FCC, computed as $\text{FCC} = (\text{A} + \text{D}) / \text{n}$.

Obviously, $0 \le \text{FCC} \le 1$, and we would like FCC to be as close to one as possible. One drawback with computing FCC is that it requires a "cutoff." A reasonable one is 0.5, since if $\hat{\text{P}}(\underline{\text{x}}_\text{i}) > 0.5$, the odds are in favor of a "response."

A second drawback with using FCC is that it is really a measure of the quality of fit of the model to the existing data and not a measure of the model's ability to predict accurate response/nonresponse to a new subject. The FCC tends to overestimate the model's ability to predict since each $\hat{\text{P}}(\underline{x}_i)$ is based on the n responses, $\text{y}_1,\ldots,\text{y}_\text{n}$, used to calculate $\hat{\underline{\beta}}$. Thus, an FCC of 0.9 does not imply that the model will predict correct response/nonresponse in 90% of newly available subjects where their regressor information is known but their response is unknown.

One way to counter this objection is to compute the prediction of response for the ith subject as $\hat{P}(x_i) = F(\underline{x}_i'\hat{\underline{\beta}}_{-i})$ (instead of $\hat{P}(\underline{x}_i) = F\left(\underline{x}_i \, {}^{\hat{\beta}}_{\hat{\underline{\beta}}}\right)$), where $\hat{\underline{\beta}}_{-\text{i}}$ is the MLE of $\underline{\beta}$ based on the $n-1$ observations obtained by deleting the ith subject's data, $(y_i, x_i), i = 1,\ldots,n$. This "single point deletion scheme" allows $\hat{\text{P}}(\underline{\text{x}}_\text{i})$ to be computed for all n subjects with each $\hat{\beta}_{-i}$ obtained independently of $(y_i, \underline{x}_i)$. These $\hat{P}(\underline{x}_i)$ are then used in the rule stated above to compute the FCC as a measure of the model's ability to predict. At first glance, it appears that computing the n values of $\hat{\text{P}}(\underline{\text{x}}_\text{i}) = \text{F}\left(\underline{\text{x}}_\text{i} \hat{\underline{\beta}}_{-i}\right)$ would require n separate logistic regressions, each based on $\text{n}-1$ observations.

This is not the case, however, as the $\hat{\beta}_{-\text{i}}$ may be approximated quite accurately by using an equation based on $\hat{\beta}$ (not given here). Consequently, only one logistic regression is needed based on the full n-point dataset. To avoid confusion, we will label the FCC computed in this way as FCCP, representing the fraction correctly classified for prediction.

All the ideas presented in this section will now be illustrated using an example.

8.1.9 The Teaching Method Application

The study aimed to determine the effect of a special teaching method called personalized system of instruction (PSI) on the grades of students in intermediate statistics. Did students exposed to PSI increase their chances of receiving an A in the course?

Traditional methods of teaching intermediate statistics were employed in two separate classes, but the students were also exposed to PSI in one class. The data from the study is reproduced in Table 8.2. Note that there are 18 students in the non-PSI class and 14 students in the PSI class, which

TABLE 8.2

Data on the Effect of "Personalized System of Instruction" (PSI) on Course Grades

	GPA	PRESTAT			Letter		GPA	TUCE			Letter
Obs.	Grade	Grade	PSI	Grade	Grade	Obs.	Grade	Grade	PSI	Grade	Grade
1	2.66	20	0	0	C	17	2.75	25	0	0	C
2	2.89	22	0	0	B	18	2.83	19	0	0	C
3	3.28	24	0	0	B	19	3.12	23	1	0	B
4	2.92	12	0	0	B	20	3.16	25	1	1	A
5	4.00	21	0	1	A	21	2.06	22	1	0	C
6	2.86	17	0	0	B	22	3.62	28	1	1	A
7	2.76	17	0	0	B	23	2.89	14	1	0	C
8	2.87	21	0	0	B	24	3.51	26	1	0	B
9	3.03	25	0	0	C	25	3.54	24	1	1	A
10	3.92	29	0	1	A	26	2.83	27	1	1	A
11	2.63	20	0	0	C	27	3.39	17	1	1	A
12	3.32	23	0	0	B	28	2.67	24	1	0	B
13	3.57	23	0	0	B	28	3.65	21	1	1	A
14	3.26	25	0	1	A	30	4.00	23	1	1	A
15	3.53	26	0	0	B	31	3.10	21	1	0	C
16	2.74	19	0	0	B	32	2.39	19	1	1	A

is a total of 32 students. Also recorded were other information related to each student, such as the letter grade received by each student, whether the student received an A or not (GRADE), the entering grade point average of the student (GPA), the score on a preliminary statistics exam given at the beginning of the course (PRESTAT), and whether or not the student received PSI (a 0 for no and a 1 for yes), the example introduce by Aldrich and Nelson (1984).

It is seen that 8 of the 14 PSI students received an A for the course, but only 3 of 18 non-PSI students received an A. Thus, it appears that PSI does increase the likelihood of receiving an A in the course. It is easy to show that the students in the two classes were very similar at the beginning of the semester, with an average GPA of 3.10 and 3.14 and an average PRESTAT score of 21.6 and 22.4 for the non-PSI and PSI classes, respectively.

On the other hand, students who received an A had an average GPA of 3.43 and an average PRESTAT score of 23.6 compared to the non-A students with averages of 2.95 and 21.1 for GPA and PRESTAT, respectively. Thus, it appears that the A students were already more likely to learn intermediate statistics than the students who did not receive an A. A multivariate quantal regression analysis is needed to evaluate the impact of the regressors GPA, PRESTAT, and PSI on the quantal dependent variable, GRADE.

The data will be analyzed in several different ways. First, the quantal aspect of GRADE will be ignored and an MLR model will be obtained using all three regressor variables. The limitations of this approach will be very apparent. Second, different quantal regression models using different combinations of regressors will be compared. Finally, several graphical displays will be introduced to enhance the interpretation of the performance of the quantal regression model. Using the coming R code, let's look at a sample scatterplot, as shown in Figure 8.4.

```
plot(data$GPA, data$GRADE,
    xlab = "GPA", ylab = "GRADE", main = "GRADE Versus GPA",
    pch = 20,# Plot character type, making the points small dots
    col = "black",    # Color of the points
    xlim = c(2, 4),   # Set limits for GPA axis based on the data
    ylim = c(0, 1),   # Set limits for GRADE axis as it is
                      binary (0, 1)
    las = 1)          # Rotate axis labels for easier reading
```

We see that students with A (the 1s) seem to have higher values of GPA than students with grades of 0. We note that scatterplots with a quantal variable do not have the "linear" appearance that we see when using a measurement response variable. The MLR model is based on all three regressors.

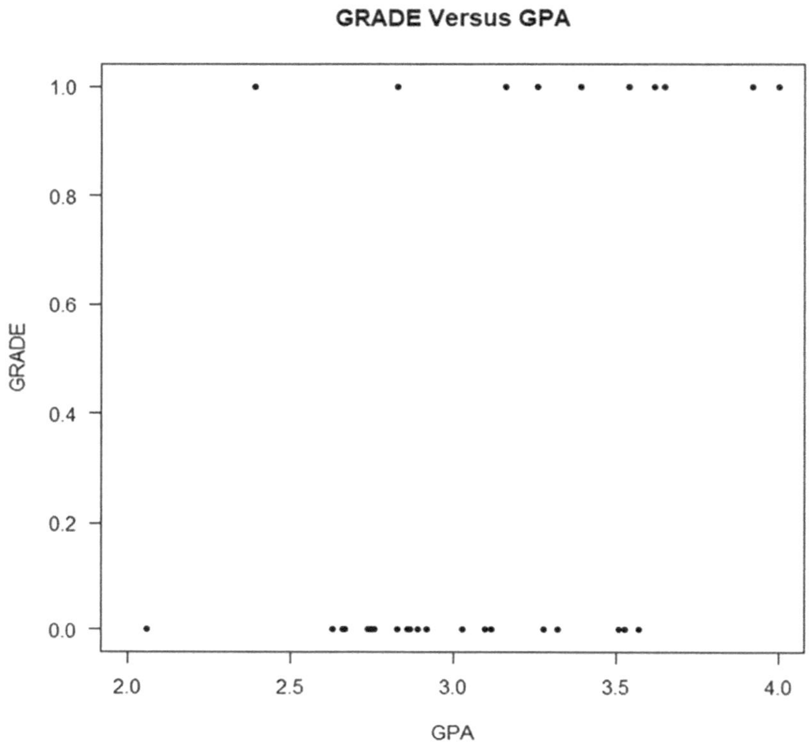

FIGURE 8.4
The grade versus GPA.

```
> # Fit the multiple linear regression model
> model <- lm(GRADE ~ GPA + PSI + PRESTAT, data = data)
> # Display the summary of the regression model
> summary(model)
Call:
lm(formula = GRADE ~ GPA + PSI + PRESTAT, data = data)
Coefficients:
            Estimate Std. Error t value Pr(>|t|)
(Intercept) -1.49802    0.52389  -2.859  0.00793 **
GPA          0.46385    0.16196   2.864  0.00784 **
PSI          0.37855    0.13917   2.720  0.01109 *
PRESTAT      0.01050    0.01948   0.539  0.59436
---
Signif. codes: 0 '***' 0.001 '**' 0.01 '*' 0.05 '.' 0.1 ' ' 1
Residual standard error: 0.3881 on 28 degrees of freedom
Multiple R-squared: 0.4159,     Adjusted R-squared: 0.3533
F-statistic: 6.646 on 3 and 28 DF, p-value: 0.001571
```

The regression equation is

```
GRADE = - 1.49802 + 0.464 GPA + 0.379 PSI + 0.0105 PRESTAT
> # Calculate residuals and fitted values
> data$residuals <- residuals(model)
> data$fitted <- fitted(model)
> # Identify unusual observations based on standardized residuals
> data$std_residuals <- rstandard(model)
> # Find unusual observations with standardized residuals
beyond +/- 2
> unusual_obs <- data[abs(data$std_residuals) > 2, ]
> # Display unusual observations
> print(unusual_obs)
OBS GPA PRESTAT PSI GRADE Letter_Grade residuals fitted
  std_residuals
24 24 3.51 26    1   0   B   -0.7815302   0.7815302   -2.130299
32 32 2.39 19    1   1   A    0.8114495   0.1885505    2.277573
Analysis of Variance
SOURCE         DF        SS       MS         F        p
Regression      3    3.0023   1.0008      6.65    0.002
Error          28    4.2165   0.1506
Total          31    7.2187
```

We note that this three-variable model has an R^2 of only 41.6% and that PRESTAT is highly insignificant. In addition, two observations (24 and 32) are detected as unusual data points. The fits based on the MLR model are displayed in the index plot in Figure 8.5 (left) by the number of fits by observation.

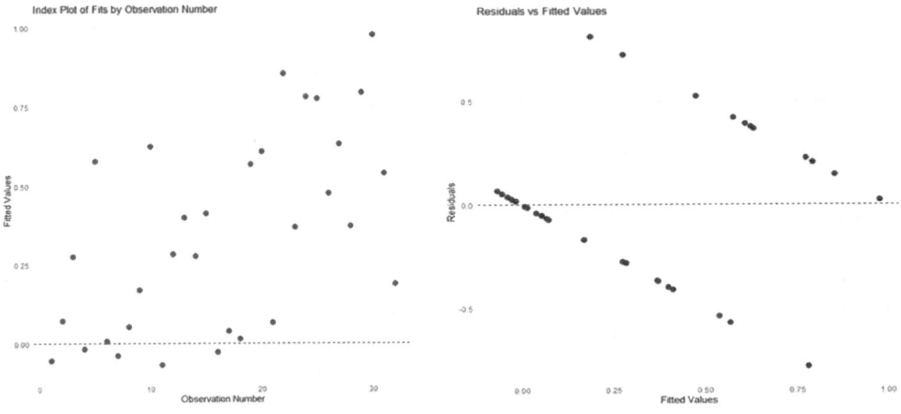

FIGURE 8.5
The index plot on the left and the residuals plot on the right.

Note that five estimated probabilities are under the line drawn at fits = 0, indicating five negative probabilities. This, of course, is nonsense and is the result of using a model that cannot constrain the estimated probabilities to be between zero and one. A plot of the residuals versus fits is shown in Figure 8.5 (right).

Note the parallel lines representing the two groups, those receiving an A and those not receiving an A. This model has an equal ability to fit the A group as it does the non-A group. The reader should also note that the pattern of the residuals, Figure 8.5 (right), differs greatly from the random, patternless scatter of points we would expect with measurement data. We can obtain a quantal regression model, specifically logistic or probit regression, as shown here.

```
> # Logistic regression using the glm function
> logistic_model <- glm(GRADE ~ GPA + PSI + PRESTAT, data =
data, family = binomial(link = "logit"))
> # Display the summary of the logistic regression model
> summary(logistic_model)
Call:
glm(formula = GRADE ~ GPA + PSI + PRESTAT, family = binomial
  (link = "logit"),
    data = data)
Deviance Residuals:
    Min        1Q     Median        3Q       Max
-1.9551  -0.6453   -0.2570   0.5888   2.0966
Coefficients:
             Estimate Std. Error z value Pr(>|z|)
(Intercept) -13.02135   4.93127   -2.641   0.00828 **
GPA           2.82611   1.26293    2.238   0.02524 *
PSI           2.37869   1.06456    2.234   0.02545 *
PRESTAT       0.09516   0.14155    0.672   0.50143
---
Signif. codes: 0 '***' 0.001 '**' 0.01 '*' 0.05 '.' 0.1 ' ' 1
(Dispersion parameter for binomial family taken to be 1)
    Null deviance: 41.183 on 31 degrees of freedom
Residual deviance: 25.779 on 28 degrees of freedom
AIC: 33.779
Number of Fisher Scoring iterations: 5
```

We see that the IRLS algorithm has converged in five iterations. The MLEs are given above along with their standard errors, the Wald z-test, and the p-value. It is noted that the coefficient on the third regressor (the fourth coefficient), prestat, is highly insignificant in the presence of the other two regressors. For this reason, it is probably best to ignore the rest of this analysis and run the GPA, PSI model. We will go to this two-variable model after we look at the GLRT statistic, as shown hereafter.

```
> # Calculate and display -2 log likelihood for intercept-only
model
> logistic_model_null <- glm(GRADE ~ 1, data = data, family =
binomial(link = "logit"))
> null_log_likelihood <- logLik(logistic_model_null)
> full_log_likelihood <- logLik(logistic_model)
> model_chi_square <- -2 * (as.numeric(null_log_likelihood)−
as.numeric(full_log_likelihood))
> # Output -2 log likelihood results
>   cat("Intercept-only model: -2 log likelihood =", -2 * as.
numeric(null_log_likelihood), "\n")
Intercept-only model: -2 log likelihood = 41.18346
> cat("Full model: -2 log likelihood =", -2 * as.numeric(full_
log_likelihood), "\n")
Full model: -2 log likelihood = 25.77927
> cat("Model Chi-square =", model_chi_square, "\n")
Model Chi-square = 15.40419
> cat("P-value =", pchisq(model_chi_square, df = 3, lower.
tail = FALSE), "\n")
       P-value = 0.001501879
```

We note that the –2 log-likelihood for the intercept-only model is 41.1835 and that for the three-variable model is 25.7793, resulting in a model chi-square of 15.4043 and a p-value of 0.0015. Thus, we conclude that there is at least one useful variable in the model. Let's evaluate the two-variable regressor model, GPA and PSI.

```
> # Create a reduced logistic model excluding PRESTAT (optional
based on summary output)
> reduced_logistic_model <- glm(GRADE ~ GPA + PSI, data = data,
family = binomial(link = "logit"))
> summary(reduced_logistic_model)
Call:
glm(formula = GRADE ~ GPA + PSI, family = binomial(link = "logit"),
    data = data)
Coefficients:
             Estimate Std. Error z value Pr(>|z|)
(Intercept)   -11.602      4.213  -2.754 0.00589 **
GPA             3.063      1.223   2.505 0.01224 *
PSI             2.338      1.041   2.246 0.02470 *
---
Signif. codes: 0 '***' 0.001 '**' 0.01 '*' 0.05 '.' 0.1 ' ' 1
(Dispersion parameter for binomial family taken to be 1)
Null deviance: 41.183 on 31 degrees of freedom
Residual deviance: 26.253 on 29 degrees of freedom
AIC: 32.253
Number of Fisher Scoring iterations: 5
```

All coefficients are significant at the 0.025 level. Consequently, we will evaluate this model in greater detail than the last one. Our estimated probabilities are determined by the formula:

```
phat=F(-11.6016 + 3.0634gpa + 2.3378 psi),
```

where F(u)=1/(1+exp(-u)), the standard logistic cdf. It is readily seen that as gpa increases or as psi changes from 0 to 1 the phat increases.

```
> # Output -2 log likelihood results
> cat("Intercept-only model: -2 log likelihood =", -2 *
as.numeric(null_log_likelihood), "\n")
Intercept-only model: -2 log likelihood = 41.18346
> cat("Full model: -2 log likelihood =", -2 * as.numeric(full_
log_likelihood), "\n")
Full model: -2 log likelihood = 26.25315
> cat("Model Chi-square =", model_chi_square, "\n")
Model Chi-square = 14.93031
> cat("P-value =", pchisq(model_chi_square, df = 2, lower.
tail = FALSE), "\n") # df = 2 for GPA and PSI
P-value = 0.0005726957
> # Calculate Pseudo R-squared
> pseudo_r2 <- 1 - (as.numeric(full_log_likelihood) /
as.numeric(null_log_likelihood))
> cat("Pseudo R-squared =", pseudo_r2, "\n")
Pseudo R-squared = 0.3625318
```

The model chi-square for this two-variable model is 14.93, highly significant. One measure of model performance is the quantal regression R^2, which is 36.25% for our model. Remember that this statistic tends to underestimate the performance of the model in quantal regression.

```
> # Calculate fraction of fits correctly classified (FCC)
> fitted_probs <- ifelse(fitted(logistic_model) > 0.5, 1, 0)
> fcc <- mean(fitted_probs == data$GRADE)
> cat("Fraction of Fits Correctly Classified (FCC) =", fcc,
"\n")
Fraction of Fits Correctly Classified (FCC) = 0.8125
```

If 0.5 is used as a cutoff value for "fitting A" versus "fitting non-A," we find that the model correctly classifies 26 of the 32 students, giving us FCC = 26/32 = 81.25%, the fraction correctly classified. Thus, apparently, this is a pretty good model, much better than the R^2 of 36.3% suggests. Figure 8.6 shows the estimated probabilities along with their 95% confidence intervals.

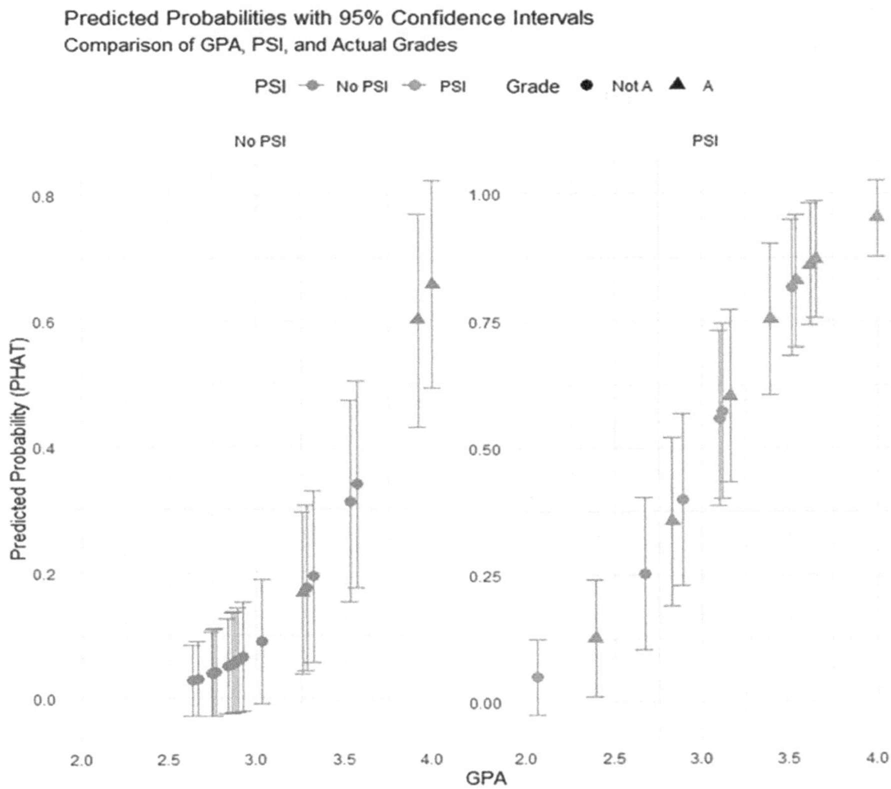

FIGURE 8.6
The estimated probabilities with their 95% confidence intervals.

A quick glance tells us that the model is working reasonably well. Note, for example, observation 30, where the student had a GPA of 4.00 and had the PSI course. This model estimates that the probability that this student will receive an A is 0.95. The student got an A, so the model works for this student.

On the other hand, student 32 had a GPA of 2.39 and had the PSI course. The model estimates the probability of A for this student of only 0.12, but the student did receive an A. The logistic model has failed to capture the behavior of this student. One simple graph that is very informative is to plot the estimated probabilities (*phat*) versus the response (y) in Figure 8.7.

The model's correct classifications are labeled above by the letters "A" and "B." Note that the student IDs are separated for ease of reading in the lower-left portion of the plot. The model's mistakes are labeled in the above

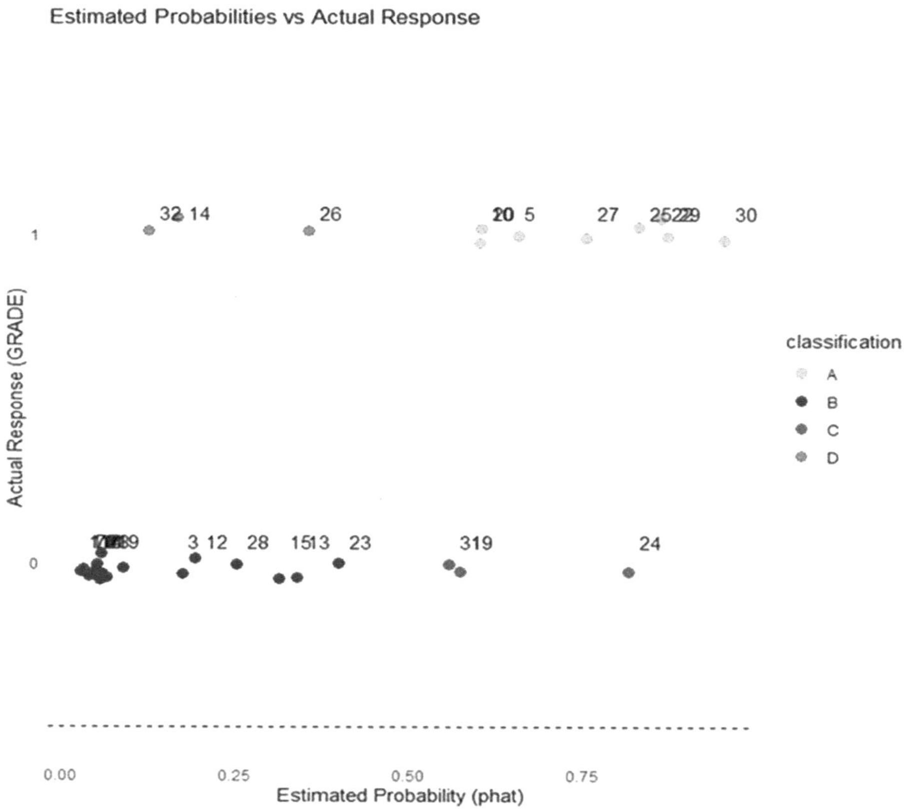

FIGURE 8.7
The estimated probabilities versus the response.

plot by the letters "D" and "C." Three students (students 19, 24, and 31) did not receive an A, but the model gives them a high probability, at least above the cutoff of 0.5 (the horizontal line). Three students (students 14, 26, and 32) received an A but were given low probabilities by the model. We will evaluate these mistakes in greater detail in subsequent material.

8.1.10 Interpretation of Coefficients in Quantal Regression

The user of quantal regression often desires to apply an interpretation to the estimated coefficients in a manner similar to the interpretation given to coefficients in the MLR model. Recall that in the MLR model, we can write $y_i = x_i \underline{\beta} + \varepsilon_i$. Taking the derivative with respect to the jth regressor on both sides of the model, we obtain $\frac{\partial y_i}{\partial x_{ji}} = \beta_j$. Thus, β_j measures the slope of the fitted surface in the direction of the *j*th regressor. Or, using another tactic, consider the change in expected response as the *j*th regressor is increased by one unit. It is easy to show that this change is equal to β_j. Hence, the interpretation of $\hat{\beta}_j$ is that it represents the estimated change in the expected response per unit change of the *j*th regressor. For example, if $\hat{\beta}_j = 2.33$, we conclude that if x_j is increased by one unit, then the estimated mean response will increase by approximately 2.33 units.

The issue is not as clear in the quantal regression setting. Here, $P(\underline{x}_i) = F(\underline{x}_i{}'\underline{\beta})$, where F is generally a nonlinear function. Taking the derivative with respect to the *j*th regressor on both sides of the model, we obtain $\frac{\partial y_i}{\partial x_{ji}} = f(\underline{x}_i{}'\underline{\beta})\beta_j$, where *f* is the derivative of F(*u*) with respect to *u*. We see that the derivative is a function of β_j and the values of x_j through the function *f*. Thus, the effect of changes in the *j*th regressor on the estimated probability of response will reflect not only the size of β_j but the values of the *k* regressors through the function *f* as well.

Using the "teaching method" example to illustrate, let us examine the effect of PSI on the estimated probabilities. Consider the GPA, PSI model and the estimated probabilities from the logistic regression model as a function of GPA for both values of PSI: zero and one. Since the MLE equation for this model is

$$F(\underline{x}_i{}'\hat{\underline{\beta}}) = \frac{1}{1+\exp(-(-11.6016+3.0634GPA+2.3378PSI))},$$

it is easy to show that the conditional probability of response at PSI = 0 is

$$F(\underline{x}_i{}'\hat{\underline{\beta}} \mid PSI = 0) = \frac{1}{1+\exp(-(-11.6016+3.0634GPA))}$$

and the conditional probability of response at PSI = 1 is

$$F(\underline{x}_i{}'\hat{\underline{\beta}} \mid PSI = 1) = \frac{1}{1+exp(-(-9.2638+3.0634GPA))}.$$

Using three different but representative values of GPA results in the estimated probabilities given in Table 8.3.

It is clear from the probabilities above that the "effect" of PSI is to increase the likelihood of receiving an A in the class, but this effect is different for different GPA values. A graphical display of this effect is presented in Section 8.1.11 (Figures 8.8 and 8.9).

TABLE 8.3

Estimated Probabilities of "A" as a Function of GPA and PSI

Probability of "A"	GPA = 2.67	GPA = 3.53	GPA = 4.00
PSI = 0	0.031	0.313	0.658
PSI = 1	0.253	0.829	0.952
Difference	0.222	0.516	0.294

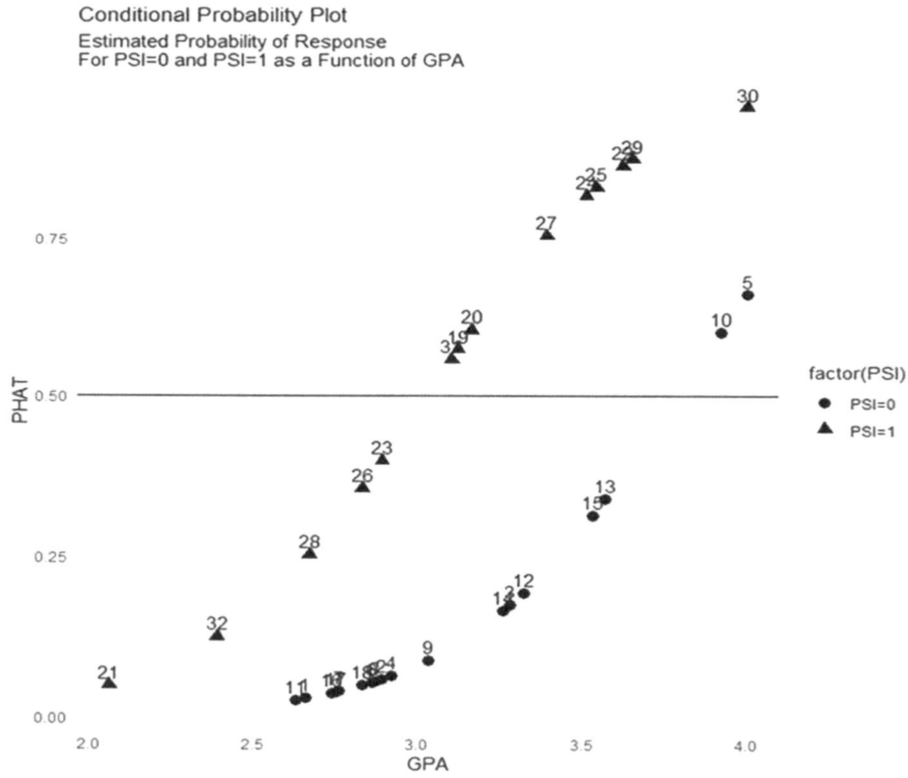

FIGURE 8.8
The conditional probability plot.

8.1.11 Graphical Display: The Conditional Probability Plots

One useful graphical display of the quantal regression fits, the estimated probabilities, is the conditional probability plots, sometimes referred to as the conditional effect plot (Hamilton, 1992). Here, the fits are plotted against one of the measurement variable regressors, while holding the other $k - 1$

regressors constant at some reasonable values. In the "teaching method" example, the measurement variable is GPA. The conditional probability plot for this example would be a plot of the estimated probabilities versus GPA at $PSI = 0$ and at $PSI = 1$. Thus, two probability curves would be superimposed on the same plot. The curves plotted would be $F\left(\underline{x}_i{}'\underline{\beta}|PSI = 0\right)$ and $F\left(\underline{x}_i{}'\underline{\beta}|PSI = 1\right)$ as in Figure 8.8.

On R, type as follows:

```
# Plot phat against GPA with different groups for PSI
ggplot(data, aes(x = GPA, y = phat, label = OBS)) +
  geom_point(aes(shape = factor(PSI)), size = 3) +
  geom_text(aes(label = OBS), vjust = -0.5, hjust = 0.5) +
  geom_hline(yintercept = 0.5, linetype = "solid", color =
    "black") +
  labs(title = "Conditional Probability Plot",
       subtitle = "Estimated Probability of Response\nFor
PSI=0 and PSI=1 as a Function of GPA", x = "GPA", y = "PHAT") +
  scale_shape_manual(values = c(16, 17), labels = c("PSI=0",
  "PSI=1")) + theme_minimal()
```

In Figure 8.8, we see the effect of PSI in the logistic regression equation. We note, for example, that logistic-shaped probability curves are not parallel, indicating that the effect of PSI changing from zero to one is not constant. In fact, the model estimates that the greatest benefit of taking the PSI course occurs for those students with GPA in the 3.5 range and the least benefit to students with GPA around 2.6. Another useful plot (Figure 8.9) that should accompany the one above is a plot of the conditional probability using the response variable as the label for each point rather than the observation number. This will enable us to identify those subjects whose estimated probability does not agree with the observed responses.

The Figures 8.7 and 8.8 show how well the model agrees with the data graphically. Consider the data under the horizontal cutoff line at 0.5. Three students with an A (labeled with 1) received low estimated probabilities. These errors are called "false negatives" representing those declared as "non-A" by the model when they really are A students. Consider the data above the cutoff line. Three non-A students received high estimated probabilities by the model, another type of error, called "false positive," declares a nonresponder as a responder. We note that our model classified 26 out of 32 correctly and misclassified three as false positives and three as false negatives.

Information can be gathered from the top two graphs to shed light on the model's mistakes. The students who were classified can be identified by matching the mistakes in the bottom plot with the observation numbers in the top plot. Thus, the three false negatives were students 14, 26, and 32. Students 26 and 32 were given low estimated probabilities because of low GPAs, while student 14 was assigned a low probability because of not taking PSI.

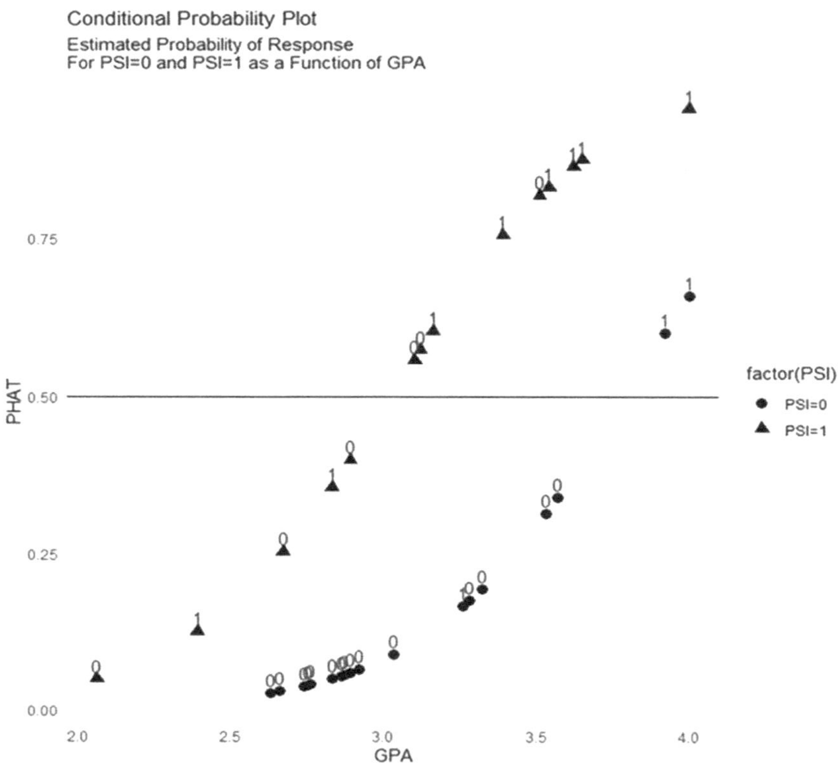

FIGURE 8.9
The conditional probability plot using the response variable as a label.

The three students who received false positives were students 19, 24, and 31. All three were assigned a high probability because they had relatively high GPAs and received PSI.

8.1.12 Another Interpretation of the Coefficients

Another interpretation of the estimated coefficients in the quantal regression setting is obtained by using the inverse transformation F^{-1}. It follows that if $\hat{y}_i = \hat{P}(\underline{x}_i) = F(\underline{x}_i \hat{\underline{\beta}})$, then $F^{-1}(\hat{P}(\underline{x}_i)) = \underline{x}_i{}'\hat{\underline{\beta}}$, a linear equation in $\hat{\underline{\beta}}$. Thus, F^{-1} may be viewed as a "linearizing transformation." In the logistic regression setting, it is easy to show that F^{-1} may be written as

$$F^{-1}(\hat{P}(\underline{x}_i)) = \ln\left(\frac{\hat{P}(\underline{x}_i)}{1-\hat{P}(\underline{x}_i)}\right) = \underline{x}_i'\hat{\underline{\beta}}\,'$$

where the transformation $\ln\left(\frac{\hat{P}(\underline{x}_i)}{1-\hat{P}(\underline{x}_i)}\right)$ is called the "logit" transformation, denoted by logit $\left(\hat{P}(\underline{x}_i)\right)$. The ratio $\frac{\hat{P}(\underline{x}_i)}{1-\hat{P}(\underline{x}_i)}$ is of interest as well. It is the ratio of the probability that a subject will respond to the probability that the subject will not respond – that is, the "odds of responding," and denoted by O_i.

For example, if $\hat{P}(\underline{x}_i)=.667$, then $1-\hat{P}(\underline{x}_i)=.333$, and the odds of responding are $\frac{.667}{.333}=\frac{2}{1}$. We simply say "The odds are 2 to 1, in favor of response." Obviously, the greater the odds (compared to 1), the greater the probability of response (closer to 1), and the smaller the odds (compared to 1), the lesser the probability of response (closer to 0).

In quantal regression, the estimated coefficient for the *j*th regressor may be isolated by considering the change in the transformed estimated response, given a unit change in the *j*th regressor. That is, if we let $\underline{x}_i' = \left(1x_{1i}\cdots x_{ji}\cdots x_{ki}\right)$ and $\underline{x}_i^{*\prime} = (1x_{1i}\cdots x_{ji}\cdots x_{ki})$, it is easy to show that

$$F^{-1}\left(\hat{P}\left(\underline{x}_i^*\right)\right)-F^{-1}\left(\hat{P}\left(\underline{x}_i\right)\right)=\hat{\beta}_j\cdot$$

The logistic regression setting again provides a nice illustration of the above result. We obtain

$$\log it\left(\hat{P}\left(\underline{x}_i^*\right)\right)-\log it\left(\hat{P}\left(\underline{x}_i\right)\right)=\hat{\beta}_j.$$

But, $\log it\left(\hat{P}(\underline{x}_i)\right)=\ln\left(\frac{\hat{P}(\underline{x}_i)}{1-\hat{P}(\underline{x}_i)}\right)=\ln(O_i)$, so that the above equation implies that

$$\ln\left(\frac{\hat{P}\left(\underline{x}_i^*\right)}{1-\hat{P}\left(\underline{x}_i^*\right)}\right)-\ln\left(\frac{\hat{P}\left(\underline{x}_i\right)}{1-\hat{P}\left(\underline{x}_i\right)}\right)=ln\left(\frac{\frac{\hat{P}(\underline{x}_i^*)}{1-\hat{P}(\underline{x}_i^*)}}{\frac{\hat{P}(\underline{x}_i)}{1-\hat{P}(\underline{x}_i)}}\right)=ln\left(\frac{o_i^*}{o_i}\right)=\hat{\beta}_j\cdot$$

Thus, the estimated coefficient in a logistic regression is the log of the ratio of the odds at x_j +1 to the odds at x_j, with the other $k-1$ regressors held constant, a quantity called the "odds ratio" and denoted by OR_j. Thus, ln $(OR_j)=\hat{\beta}_j$. It follows that $OR_j=\frac{O_i^*}{O_i}=e^{\hat{\beta}_j}$, or writing our result in terms of odds, we have

$O_i^* = OR_j O_i = e^{\hat{\beta}_j} O_i$. We see that the odds ratio for the jth regressor, OR_j, represents the multiplicative increase in odds when the regressor x_j has been increased by one unit. Odds and odds ratios are important quantities in many areas of biostatistics.

A numerical example of the use of these concepts can be found by considering the teaching method example and the GPA, PSI model. Since $\hat{\beta}_1 = 2.33$ for GPA, it follows that $OR_1 = e^{2.33} = 10.385$. In Table 8.3, we note, for example, that if GPA = 4.00, then the estimated probability of A for a non-PSI student is 0.658, for odds of $.658/(1-.658) = 1.924$. The estimated probability of A for a PSI student is 0.952, for odds of 19.833.

The odds ratio is $19.833/1.924 = 10.3085$. Thus, the odds of an A increase by a factor of 10.3085 for a student with a GPA of 4.00 when they take the PSI. The reader should be able to verify that the odds ratio is 10.385 for the other two values of GPA given in this table. In fact, the odds ratio is 10.385 for all values of GPA, since the odd ratio depends only on $\hat{\beta}_j$ and not on $\underline{x}_i$, the values of the regressors.

8.1.13 The –2 Log-Likelihood Components

The reader will recall from our discussions of the MLR model that the residuals played a major role in determining the appropriateness of a model to the data. One goal in building an MLR model is to minimize the MSE, the sum of squared residuals divided by the error degrees of freedom. In addition, the residuals are found in many key diagnostic formulas. The quantal regression problem is complicated by the fact that the residuals for the quantal regression model, due to the 0-1 nature of the response variable, are not as informative as they are in the MLR model. We have already seen an example of this for the teaching method data when we plotted the residuals against the *phat*s. The elements corresponding to the residuals in the quantal regression model come directly from the –2 log-likelihood formula.

The reader will recall the log-likelihood formula given previously as

$$\ln L\left(\underline{\beta}^*, \underline{x}_i\right) = \sum_{i=1}^{n} \left(y_i \ln\left(F\left(\underline{x}_i' \underline{\beta}^*\right)\right) + (1-y_i)\ln\left(1-F\left(\underline{x}_i' \underline{\beta}^*\right)\right)\right).$$

The MLEs are chosen to maximize $\ln L\left(\underline{\beta}^*, \underline{x}_i\right)$ or, equivalently, to minimize $-2\ln L\left(\underline{\beta}^*, \underline{x}_i\right)$. We can think of $-2\ln L\left(\underline{\beta}^*, \underline{x}_i\right)$ as being composed of the sum of n values, the $-2\log$ likelihood components (LLC), defined as

$$llc_i = -2(y_i \ln(F(\underline{x}_i' \underline{\beta}^*)) + (1-y_i) \ln(1-F(\underline{x}_i' \underline{\beta}^*))),$$

where llc_i denotes the −2log-likelihood component for the *i*th subject using coefficients $\underline{\beta}^*$. Clearly, the value of −2 log-likelihood, based on the MLEs, for any given model is the sum of these *n* log-likelihood components. As mentioned previously, −2 log-likelihood in quantal regression is analogous to the sum of squared residuals in the MLR model. And just as a large residual indicates that the MLR model provides a poor fit to a particular subject, a large value of LLC indicates that the quantal regression model provides a poor fit.

To see this, recall that a quantal regression model is in "agreement" with the data if the model estimates a large probability for a subject with y = 1 and a small estimated probability if the subject has $y = 0$. Indeed, it is easy to see that if y = 1 and $F(\underline{x}_i'\hat{\underline{\beta}}) = 1$, then $llc_i = 0$. Similarly, if $y = 0$ and $F(\underline{x}_i'\hat{\underline{\beta}}) = 0$, then llc = 0 as well. On the other hand, if y = 1 and $F(\underline{x}_i'\hat{\underline{\beta}}) = .01$, then LLC = 9.21. The plot in Figure 8.10 shows how LLC varies with $F(\underline{x}_i\,{}_i\hat{\underline{\beta}})$, for y = 1 and 0.

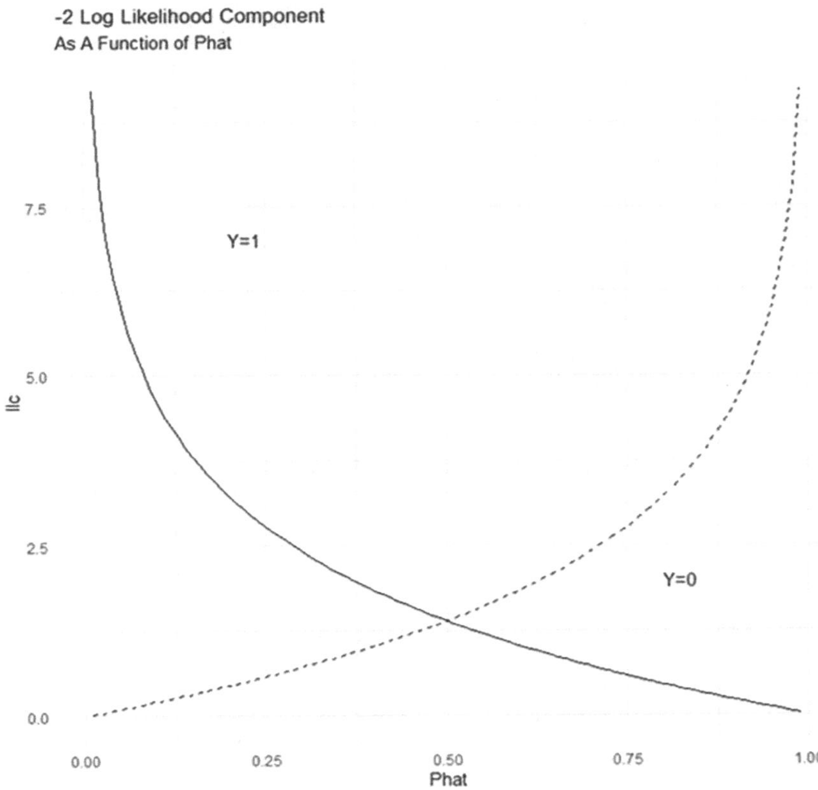

FIGURE 8.10
The LLC as a function of *phat*.

The plot in Figure 8.10 clearly shows that a subject's contribution to −2 log-likelihood will be small if $y = 1$ and the estimated probabilities are large or if $y = 0$ and the estimated probabilities are small.

A useful way to identify the subjects whose responses do not agree with the model is to plot each LLC value versus the subject number. Such a plot might be called the "−2 log-likelihood component index plot." If LLC for any subject is very much larger than the other LLC values, this subject might be labeled as an "outlier." Plotting the LLC values versus the other regressors will help to identify why the point is an outlier.

To demonstrate these ideas, the teaching method application should be considered again. We have already identified six points (three false positives and three false negatives) where model disagreement occurs. The R code will also return the LLC values as illustrated in the coming:

```
> # Calculate predicted probabilities (phat)
> data$phat <- predict(logistic_model, type = "response")
> # Calculate the -2 log likelihood component (llc) for each
  observation
> data$llc <- -2 * (data$GRADE * log(data$phat) + (1 - data$GRADE)
  * log(1 - data$phat))
> # Print the required columns to identify potential outliers
> print(data[, c("GPA", "PSI", "GRADE", "phat", "llc")])
    GPA PSI GRADE       phat        llc
1  2.66   0     0 0.03067928 0.06231949
2  2.89   0     0 0.06017518 0.12412357
3  3.28   0     0 0.17454959 0.38365218
4  2.92   0     0 0.06558752 0.13567463
5  4.00   0     1 0.65743940 0.83880536
6  2.86   0     0 0.05518309 0.11352824
7  2.76   0     0 0.04122264 0.08419278
8  2.87   0     0 0.05680221 0.11695854
9  3.03   0     0 0.08951574 0.18755734
10 3.92   0     1 0.60032865 1.02055605
11 2.63   0     0 0.02806109 0.05692466
12 3.32   0     0 0.19291402 0.42865014
13 3.57   0     0 0.33954022 0.82963809
14 3.26   0     1 0.16589729 3.59277278
15 3.53   0     0 0.31262437 0.74974874
16 2.74   0     0 0.03886807 0.07928719
17 2.75   0     0 0.04002877 0.08170392
18 2.83   0     0 0.05058286 0.10381403
19 3.12   1     0 0.57295469 1.70173032
20 3.16   1     1 0.60263374 1.01289132
21 2.06   1     0 0.04958301 0.10170890
22 3.62   1     1 0.86123674 0.29877171
23 2.89   1     0 0.39875437 1.01750346
24 3.51   1     0 0.81587251 3.38425376
25 3.54   1     1 0.82927967 0.37439565
```

```
26 2.83    1      1 0.35561219 2.06782900
27 3.39    1      1 0.75417891 0.56425131
28 2.67    1      0 0.25263674 0.58240785
29 3.65    1      1 0.87185952 0.27425395
30 4.00    1      1 0.95210570 0.09815844
31 3.10    1      0 0.55790137 1.63244454
32 2.39    1      1 0.12539084 4.15263935
```

Identify potential outliers based on high LLC values. A quick scanning of the LLC values results in three subjects, 14, 24, and 32 (the boldfaced lines above), being considered as possible outliers.

Three points stand out clearly above the others: subjects 14, 24, and 32. A reference line is drawn at the average LLC value of the other 29 points of 0.52. The average LLC value for these three points is 3.71. The standard deviation of LLC for the other 29 points is 0.55. Thus, the LLC values for the three "outliers" are over 5 standard deviations away from the average. In addition, these three subjects, representing, 9% of the sample, have LLC values that contribute over 42% of the –2 log-likelihood for this model.

To see why these are usual students, we can examine their GPA, PSI, and grade values to see that student 14 had an average GPA and no PSI, so the model estimates a low probability of A, but the student received an A—an overachiever. Student 24 had a high GPA and had PSI, so the model estimates a high probability of A, but the student did not get an A—an underachiever. Student 32 had a low GPA and had PSI, so the model estimates a low probability of A, but the student received an A—an overachiever.

Both SPSS and R have built-in functions to perform logistic regression. Using R performs logistic regression for the ungrouped data case illustrated in this chapter. For example, consider the teaching method example again and the model containing all three regressors.

To examine this model using R, type as follows:

```
> response_info <- table(data$GRADE)
Response Information
> cat("Variable Value       Count\n")
Variable  Value       Count
GRADE     1            11
          0            21
          Total        32

> logistic_summary <- summary(logistic_model)
Logistic Regression Table
Predictor     Coef      StDev    Z    P Odds Ratio    Lower  Upper
Constant  -13.021     4.9313 -2.64 0.008
GPA          2.826     1.2629  2.24 0.025  16.88     1.9    316.48
PRESTAT     2.3787     1.0646  2.23 0.025  10.79   1.61    122.76
PSI          0.095     0.1416  0.67 0.501    1.1   0.84    1.5
> # Log-Likelihood, Model Chi-square (G), and P-Value
> log_likelihood <- logLik(logistic_model)
```

```
> G_statistic <- -2 * (logLik(update(logistic_model, ~ 1))-
  log_likelihood)
Log-Likelihood = -12.89
Test that all slopes are zero: G = 15.404, DF = 3, P-Value = 0.002
> # Goodness-of-Fit Tests
Goodness-of-Fit Tests
Method           Chi-Square          DF          P
Pearson          27.257              28          0.504
Deviance         25.779              28          0.585
Hosmer-Lemeshow  6.095               8           0.637
```

In the above output, we note many of the statistics discussed and illustrated in this section. The statistic G = 15.404 is the model chi-square, and the deviance is the –2 log for the model. We note that R prints out the odds ratio as well as 95% CI for the odds ratio. R code also supplies other information regarding model performance and diagnostics for detecting the influence of individual observations. This information will not be given here, but the interested reader is encouraged to examine the output independently using the previous codes as guidelines.

In the next section, the quantal regression methodology (the GLM) will be generalized to analyze non-normal data. The method will be demonstrated a second time for Poisson data.

8.2 The Generalized Linear Model (GLM)

The quantal regression material discussed in the last section is an example of the GLM. Many of the issues that were presented in Section 8.1 are also issues for GLMs. For example, estimating parameters and testing hypotheses are accomplished using the ML theory. Essentially, GLM is a technique for obtaining regression models for non-normal data, appropriate when the response variable of interest follows a distribution that is a member of the exponential family. Recall that the exponential family includes a wide variety of distributions, including the Bernoulli, binomial, Poisson, gamma (which includes the exponential and chi-square), negative binomial, and normal. While there is a large body of literature on the theoretical development of the exponential family, our purpose is not to present the GLMs in this general way. The reader is referred to McCullagh and Nelder (1994) for more theoretical details. Rather, we will give many of the important results pertinent to the practical application of GLM and illustrate these ideas with one further example. The quantal regression setting of the last section already serves to illustrate most of the important concepts.

Suppose that the random variable y_i is a member of the exponential family. We can write $E(y_i)$ as $E(y_i) = F(\underline{x}_i'\underline{\beta}) = \mu_i$, where F is some arbitrary function and $\underline{x}_i'\underline{\beta}$ is the "linear predictor" based on k regressors and the unknown coefficients, β. In the GLM setting, some of our goals are to estimate β, make inferences on β, and assess the quality of our model.

The problem of estimating $\underline{\beta}$ is complicated by the fact that the response, y, often has a nonconstant variance that is a function of the mean response, σ_i. That is, $\text{Var}(y_i) = G(\mu_i) = \sigma_i^2$, where G is some function that depends on the distribution of y through the mean of y. The variance of y is a function of the mean of y, which in turn is a function of the regressors. Thus, subjects with different regressor values will also have different variances. Some examples will help to clarify this point.

Consider first the usual MLR model where it is assumed that $y_i \sim N(\mu_i, \sigma^2)$. Here, $\mu_i = \text{F}(\underline{\text{x}}_i'\underline{\beta}) = \underline{\text{x}}_i'\underline{\beta}$ and $\sigma_i^2 = \text{G}(\mu_i) = \sigma^2$. In this model, the variance of each response is not a function of the mean response but a constant, σ^2. This represents the "constant variance" assumption.

As a second example, consider the quantal regression model where it is assumed that the Bernoulli random variable y has a probability of response P. Then if P is also a function of the k regressors, we can show that $\text{E}(\text{y}_i) = \mu_i = \text{P}(\underline{\text{x}}_i) = \text{F}(\underline{\text{x}}_i\underline{\beta})$ and

$$\text{Var}(\text{y}_i) = \sigma_i^2 = \text{P}(\underline{\text{x}}_i)(1-\text{P}(\underline{\text{x}}_i)) = \text{F}(\underline{\text{x}}_i'\underline{\beta})(1-\text{F}(\underline{\text{x}}_i'\underline{\beta})) = \mu_i(1-\mu_i).$$

The notation above is used to show that the variance of y is $\text{P}(1-\text{P})$, the usual variance of the Bernoulli random variable, that this variance is a function of regressors, and that the variance is a function of the mean. Here, $\text{G}(\mu_i) = \mu_i(1-\mu_i)$. All these facts were exploited in the previous section.

For a third example, consider the Poisson random variable y_i, a member of the exponential family. The Poisson distribution is often used to model count data, such as the number of bubbles in a glass pane or red blood cells on a petri dish. We know that $\text{E}(\text{y}_i) = \lambda$, the Poisson density parameter. Suppose that λ is viewed as a function of regressor information through the linear score $\underline{x}_i'\beta$. In the number of bubbles example, the regressors might represent the amounts of certain chemicals used in the composition of the glass and/or represent certain properties of the pane of the glassmaking process such as the speed of the rollers and the temperature of the composition, among other characteristics. If this is the case, then λ is not a constant but will vary with varying values of the k regressors. Thus, we can write $E(y_i) = \lambda_i = F(\underline{x}_i'\underline{\beta})$, for some function F. In addition, we know that $\text{Var}(y_i) = E(y_i) = \lambda_i = F(\underline{x}_i\underline{\beta})$. It follows that $G(\mu_i) = \mu_i$. So as the regressors change values, the mean changes according to the function $F(\underline{x}'\underline{\beta})$, with the variance changing by exactly the same amount.

As a final example, consider the exponential distribution. The exponential distribution is often used to model the waiting time required to observe the first occurrence of an event of a specified type when events of this type are occurring randomly at a mean rate λ per unit of time. Or, as another example, the lifetimes of certain electrical components, such as light bulbs or batteries, are modeled with the exponential distribution. The PDF for the exponential distribution is defined as

$$\text{pdf}(y) = \lambda \exp(-\lambda y) \text{ for } y > 0.$$

It is easy to show that $E(y) = \lambda^{-1}$ and $\text{Var}(y) = \lambda^{-2}$. As in the other examples presented above, this rate parameter λ may be dependent on regressor information through the linear predictor. Thus, we can write $\text{E}(y_i) = \mu_i = \lambda_i^{-1} = \text{F}(\underline{x}_i'\underline{\beta})$ and $\text{Var}(y_i) = \lambda_i^{-2} = \text{F}(\underline{x}_i'\underline{\beta})^{-2}$. It follows that $\text{G}(\mu_i) = \mu_i^2$.

8.2.1 Estimation of $\underline{\beta}$

In each of the last three examples, the estimation of $\underline{\beta}$ will enable us to estimate both the mean and the variance of the response. As in the quantal regression section, the estimator of $\underline{\beta}$, $\hat{\underline{\beta}}$, is found by the method of ML estimation. The method is exactly the same as illustrated in Section 8.1. That is, once we find the log-likelihood function, we compute $\dfrac{\partial \ln L(\underline{\beta}^*, \underline{x}_i)}{\partial \underline{\beta}^*}$, set the resulting p equations equal to the $p \times 1$ vector of zeros, and solve for the MLE, $\underline{\beta}$.

It turns out that as long as the distribution of y is a member of the exponential family, the resulting form for $\dfrac{\partial \ln L(\underline{\beta}^*, \underline{x}_i)}{\partial \underline{\beta}^*}$ may be written as

$$\frac{\partial \ln L(\underline{\beta}^*, \underline{x}_i)}{\partial \underline{\beta}^*} = \sum_{i=1}^{n} \left[\frac{y_i - F(\underline{x}_i'\underline{\beta}^*)}{\text{Var}(y_i)} \right] f(\underline{x}_i'\underline{\beta}^*)\underline{x}_i,$$

where $f(\underline{x}_i'\underline{\beta}^*) = \dfrac{\partial F(\underline{x}_i'\underline{\beta}^*)}{\partial \underline{x}_i'\underline{\beta}^*}$.

The above form is a generalization of $\dfrac{\partial \ln L(\underline{\beta}^*, \underline{x}_i)}{\partial \underline{\beta}^*}$, already presented in Section 6.1 for the usual MLR model and in Section 8.1 for the quantal regression model. If we assume $y_i \sim N(\mu_i, \sigma^2)$, then for the MLR model, we have $F(\underline{x}_i'\underline{\beta}^*) = \mu_i = \underline{x}_i'\underline{\beta}$, $\text{Var}(y_i) = \sigma^2$, and $\dfrac{\partial \underline{x}_i\underline{\beta}'}{\partial \underline{x}_i\underline{\beta}^*} = 1$. Thus,

$$\frac{\partial \ln L(\underline{\beta}^*, \underline{x}_i)}{\partial \underline{\beta}^*} = \sum_{i=1}^{n} \left[\frac{y_i - \underline{x}_i'\underline{\beta}^*}{\sigma^2} \right] \underline{x}_i$$

and when set equal to $\underline{0}$, gives us the usual "normal equations."

The resulting MLE of $\underline{\beta}$ is the usual LS estimator. For quantal regression,

$F(\underline{x}_i\underline{\beta}) = P(y_i = 1 \mid \underline{x}_i{}'\underline{\beta})$, $\text{Var}(y_i) = F(\underline{x}_i\underline{\beta})(1 - F(\underline{x}_i\underline{\beta}))$ and
$\text{f}(\underline{x}'_i\underline{\beta}^*) = \dfrac{\partial \text{F}(\underline{x}_i{}'\underline{\beta}^*)}{\partial \underline{x}_i{}'\underline{\beta}^*}$,

the PDF associated with the CDF F. The resulting equation for $\dfrac{\partial \ln L(\underline{\beta}^*, \underline{x}_i)}{\partial \underline{\beta}^*}$ matches exactly the equation derived in Section 8.1.

To demonstrate the technique one more time, consider the situation where the distribution of response is assumed to be a Poisson with $E(y_i) = \lambda_i = F(\underline{x}_i\underline{\beta}')$ and $\text{Var}(y_i) = E(y_i) = \lambda_i = F(\underline{x}_i\underline{\beta})$. The PMF for y_i is

$$\text{pmf}(\text{y}_\text{i}) = \frac{\exp(-\text{F}(\underline{\text{x}}'_\text{i}\underline{\beta}))(\text{F}(\underline{\text{x}}'_\text{i}\underline{\beta}))^{\text{y}_\text{i}}}{\text{y}_\text{i}!}.$$

It follows that the likelihood function for $\text{y}_1, \ldots, \text{y}_\text{n}$ is

$$L(\underline{\beta}, \underline{x}_i) = \prod_{i=1}^{n} \frac{\exp(-F(\underline{x}'_i\underline{\beta}))(F(\underline{x}'_i\underline{\beta}))^{y_i}}{y_i!}$$

with the log-likelihood function of

$$\ln L(\underline{\beta}, \underline{x}_i) = \sum_{\text{i}=1}^{\text{n}} \text{y}_\text{i} \ln(\text{F}(\underline{\text{x}}'_\text{i}\underline{\beta})) - \sum_{\text{i}=1}^{\text{n}} \text{F}(\underline{\text{x}}'_\text{i}\underline{\beta}) - \sum_{\text{i}=1}^{\text{n}} \ln(\text{y}_\text{i}!).$$

Viewing the above as a function of β^*, where β^* represents an arbitrary value of the parameter vector, the MLE is found by taking the derivative of $\ln L(\underline{\beta}^*, \underline{x}_i)$ with respect to β^*.

We obtain, after a little algebra, the expression

$$\frac{\partial \ln L(\underline{\beta}^*, \underline{x}_i)}{\partial \underline{\beta}^*} = \sum_{i=1}^{n} \left[\frac{y_i - F(\underline{x}'_i\underline{\beta}^*)}{F(\underline{x}'_i\underline{\beta}^*)}\right] f(\underline{x}'_i\underline{\beta}^*)\underline{x}_i$$

in agreement with the general form given earlier.

8.2.2 The Numerical Method

As in the quantal regression case, setting $\dfrac{\partial \ln L(\underline{\beta}^*, \underline{x}_i)}{\partial \underline{\beta}^*}$ equal to the $p \times 1$ vector of zeros and solving for $\underline{\beta}$ requires the use of some appropriate numerical method. We employ the Gauss-Newton method illustrated in detail in Section 8.1. The reader is encouraged to look again at the steps outlined in the GLM as well. Essentially, these steps include approximating $F(\underline{x}_i\underline{\beta}')$ with a linear version obtained via a first-order multivariate Taylor series expansion about some initial estimator, $\underset{=0}{\beta}$, and rewriting the resulting equation as "weighted normal" equations. The IRLS algorithm is used to solve these equations. The IRLS algorithm, appropriate for any GLM problem, is listed in subsection 8.2.3.

8.2.3 The IRLS Algorithm Used to Obtain the MLE in GLM

Step 1: Obtain the initial estimator, $\hat{\beta}_0$. More will be given concerning the initial estimator at the end of the algorithm.

Step 2: Compute $y_i^* = \dfrac{y_i - F\left(\underline{x}_i' \underline{\hat{\beta}}_0\right)}{\mathrm{f}\left(\underline{x}_i' \hat{\beta}_0\right)}$ and $w_i = \dfrac{f^2\left(\underline{x}_i' \underline{\hat{\beta}}_0\right)}{\mathrm{Var}(y_i)}$, for $i = 1,..,n$,

where $\mathrm{Var}(y_i) = G\left(F\left(\underline{x}_i \underline{\beta}'\right)\right)$.

Step 3: Compute $\hat{\beta}_1 = \hat{\beta}_0 + \left(\mathrm{X'WX}\right)^{-1} \mathrm{X'Wy}^{**}$, where W is the diagonal matrix with $\mathrm{w}_1,\ldots,\mathrm{w}_\mathrm{n}$ as the diagonal elements.

Step 4: If $\hat{\beta}_1 \approx \hat{\beta}_0$, quit. Declare $\hat{\beta} = \hat{\beta}_1$. Otherwise, replace $\hat{\beta}_0$ by $\hat{\beta}_1$, and return to step 2.

Since $E(y_i) = F\left(\underline{x}_i' \underline{\beta}\right)$, it follows that $F^{-1}\left(E(y_i)\right) = \underline{x}_i' \underline{\beta}$, where F^{-1} is the inverse function of F. Thus, the function, F^{-1}, results in a linearization of the expected response. In the GLM framework, F^{-1} is referred to as the "link" function.

The link function is the transformation required to linearize the response. An application of the use of this link function is in obtaining the initial estimates for the IRLS algorithm, usually obtained as $\hat{\beta}_0 = \left(\mathrm{X'X}\right)^{-1} \mathrm{X'}\underline{F}^{-1}$, where $\underline{F}^{-1}$ is the $n \times 1$ vector with elements $F_i^{-1} = F^{-1}(y_i), i = 1,\ldots,n$. For example, in quantal regression, the link function is the logit of $\hat{\mathrm{P}}(\underline{x}_i)$, where $\mathrm{logit}\left(\hat{\mathrm{P}}(\underline{\mathrm{x}}_\mathrm{i})\right) = \ln\left(\dfrac{\hat{\mathrm{P}}(\underline{\mathrm{x}}_\mathrm{i})}{1-\hat{\mathrm{P}}(\underline{\mathrm{x}}_\mathrm{i})}\right)$. In Poisson regression, if the function relating the mean response to the linear predictor is the exponential function, $\lambda_i = F\left(\underline{x}_i \underline{\beta}\right) = \exp\left(\underline{x}' \underline{\beta}\right)$, then the link function is the natural log since $\ln(\lambda_i) = \underline{x}_i{}' \beta$.

8.2.4 Hypothesis Testing for GLM

As in Section 8.1, the GLRT can be used to test the contribution of any single regressor or any collection of regressors to the model. That is, we can compare any reduced model to a full model, so long as the reduced model is "nested" within the full model by using the GLRT. A nested model is one where the regressors in the reduced model are also regressors within the full model. For example, the reduced model containing the regressor x_1 is nested within the model containing the three regressors x_1, x_2, and x_3 since the reduced model is contained within the full model. On the other hand, the reduced model containing x_1 and x_4 is not nested within the full model containing x_1, x_2, and x_3 since x_4 is not contained within the full model. The GLRT may be applied to the former example but not the latter. The general form for the GLRT may be written as

$$\text{GLRT} = -2\ln\left[\frac{\left(\hat{\beta}_{\text{reduced}}, \underline{x}_{\text{i}}\right)}{\text{L}\left(\hat{\beta}_{\text{full}}, \underline{x}_{\text{i}}\right)}\right],$$

where $\hat{\beta}_{\text{reduced}}$ and $\hat{\beta}_{\text{full}}$ refer to the MLEs for the reduced and full models, respectively. Using the same notation as in Section 8.1, the full model has p parameters, while the reduced model has p_1 parameters, where $p = p_1 + p_2$. The GLRT has an asymptotic chi-square distribution with p_2 degrees of freedom, where p_2 is the difference in the number of parameters estimated between the full and reduced models.

According to the ML theory, the asymptotic distribution of the MLE, $\hat{\beta}$, is a multivariate normal distribution with a mean vector $\underline{\beta}$ and variance-covariance matrix $I^{-1}\left(\underline{\hat{\beta}}\right)$, where $I\left(\underline{\beta}\right)$ is the Fisher information matrix, $I\left(\underline{\hat{\beta}}\right) = -E\left(\frac{\partial^2 \ln L\left(\underline{\beta}, \underline{x}_i\right)}{\partial \underline{\beta} \partial \underline{\beta}}\right)$. In practice, $I^{-1}\left(\underline{\hat{\beta}}\right)$ must be estimated as $\text{Var}\left(\underline{\hat{\beta}}\right) = I^{-1}\left(\underline{\hat{\beta}}\right) = \left(X'WX\right)^{-1}$, where the matrix W is the same weight matrix used at the final iteration of the IRLS algorithm. With this information, we can perform tests on any single regressor coefficient by using the Wald *z*-test, introduced in the previous section. That is, to test hypotheses of the form

$H_0: \beta_j = 0$ versus $H_1: \beta_j \neq 0$, in the presence of the other $k-1$ regressors, for $j = 1, \ldots, k$, use the test statistic

$$z_j = \frac{\hat{\beta}_j}{\text{se}\left(\hat{\beta}_j\right)},$$

where $\text{se}\left(\hat{\beta}_j\right)$ is the square root of the *j*th diagonal element of the $\left(X'WX\right)^{-1}$ matrix. The asymptotic distribution of z_j is the standard normal. Of course, we could also use $X_j^2 = z_j^2$, the Wald chi-square test, with one degree of freedom. This is an equivalent test.

8.2.5 Measure of Model Performance for GLM

The ability of the GLM to adequately describe the behavior of the mean response can be measured with several statistics that are extensions of measures used in the quantal regression section. There, one value of interest was an "R^2-type" measure. An adjustment must be made for the GLM, however. Recall that for quantal regression, the value of $-2\ln L\left(\hat{\beta}_0, \underline{x}_i\right)$ provided the maximum value that could be achieved by −2 log-likelihood. The smallest value possible for −2 log-likelihood was zero. While $-2\ln L\left(\hat{\beta}_0, \underline{x}_i\right)$ still provides the maximum possible value for −2 log-likelihood in the GLM case, zero is no longer the minimum possible value.

For the ungrouped data case such as we are considering, this minimum value is achieved by an unusual "model" where we have a perfect fit at each of the n observations. That is, $\hat{y}_i = \hat{E}(y_i) = y_i$ for all $i = 1,\ldots,n$. In general, this model is not a function of regressors and coefficients through the linear predictor but rather a model where n "parameters" are used to obtain the n fits. These n parameters are the n response observations, $y_1,\ldots,y_n$ themselves. Clearly, this represents an "oversaturated" model with possibly many more parameters than we would like. It is also a model that ignores the information contained in the k regressors. Nonetheless, the value of −2 log-likelihood obtained from this "idealized model" provides the lower bound that we can achieve with our GLM. The reader will recall that a fundamental principle of using −2 log-likelihood is that its value cannot increase as regressors are added to the model or, stated another way, as parameters are added to the model. This perfect fit model represents the most parameters that could be used in a model.

To illustrate this concept, consider the Poisson regression case. As indicated above, the log-likelihood function for a sample of n responses from a Poisson distribution can be written as

$$\ln L\left(\underline{\beta}, \underline{x}_i\right) = \sum_{i=1}^{n} y_i \ln\left(F\left(\underline{x}_i' \underline{\beta}\right)\right) - \sum_{i=1}^{n} F\left(\underline{x}_i' \underline{\beta}\right) - \sum_{i=1}^{n} \ln\left(y_i!\right).$$

Consequently, −2 log-likelihood for this perfect fit model, where $\hat{E}(y_i) = y_i$, is

$$-2\ln L\left(\underline{y}\right) = -2\left[\sum_{i=1}^{n} y_i \ln(y_i) - \sum_{i=1}^{n} y_i - \sum_{i=1}^{n} \ln(y_i!)\right],$$

where the notation $\ln L\left(\underline{y}\right)$ is used to denote that the log-likelihood is a function of y and no longer a function of $\underline{\beta}$ and regressor information. It is clear that $-2\ln L\left(\underline{y}\right)$ will not be zero unless $\sum_{i=1}^{n} y_i \ln(y_i) = \sum_{i=1}^{n} y_i + \sum_{i=1}^{n} \ln(y_i!)$.

If, on the other hand, we have the "grouped" data case where there are s distinct settings of the k regressors, $\underline{x}_1{}', \ldots, \underline{x}_s{}'$, the sample means of the responses at each set represent the "best" fit to the data that any model could achieve. These s sample means are collectively denoted by $\bar{y}' = (\mathrm{y}_1 \quad \bar{\mathrm{y}}_2 \quad \cdots \quad \bar{\mathrm{y}}_s)$. So if y_{ij} represents the ith response at the jth setting of the regressors, $\underline{x}_j{}'$, we have, for this best model, $\hat{\mathrm{E}}(\mathrm{y}_{ij}) = \bar{y}_j$, for $\mathrm{j} = 1,..,\mathrm{s}$. As in the ungrouped data situation, this model represents the saturated model in that s is the most parameters that could be fit, and consequently, this best model achieves the smallest possible value of –2 log-likelihood.

Poisson regression can be used to provide an example of the grouped data case. Suppose that we have s distinct combinations of k regressors and n_j response observations at each combination. If y_{ij} denotes the ith response at the jth combination, then the minimum possible value of –2 log-likelihood is

$$-2\ln L(\underline{\mathrm{y}}) = -2\left[\sum_{j=1}^{s}\sum_{i=1}^{n_j}\mathrm{y}_{ij}\ln(\bar{\mathrm{y}}_j) - \sum_{j=1}^{s}\mathrm{n}_j\bar{\mathrm{y}}_j - \sum_{j=1}^{s}\sum_{i=1}^{n}\ln(\mathrm{y}_{ij}!)\right],$$

where the notation $\ln L(\bar{y})$ is used to signify that this -2 log-likelihood is a function of the s sample means and not a function of coefficients and regressors.

Other GLMs can now be compared to the "best model" by comparing their –2 log-likelihood values. One way to accomplish this is to compute the GLRT, where the "reduced model," required by the GLRT, is the current model being considered and the "full model" is the idealized "best model." In this case, the GLRT becomes for the ungrouped data situation

$$\mathrm{GLRT} = -2\ln\left[L(\hat{\beta}, \underline{x}_i)/L(\underline{y})\right],$$

a statistic with an asymptotic chi-square distribution with $n - p$ degrees of freedom. For the grouped data case, L(y) would be used instead of the ungrouped $L(y)$, and the degrees of freedom would be $s - p$ instead of $n - p$.

In the GLM literature, this GLRT statistic is referred to as the "deviance" of the current model. We will denote this deviance as $\mathrm{D}(\underline{\hat{\beta}})$. The term deviance is appropriate in that $\mathrm{D}(\underline{\hat{\beta}})$ can be expressed as

$$\mathrm{D}(\underline{\hat{\beta}}) = -2\ln\mathrm{L}\left(\hat{\underline{\beta}}, \underline{\mathrm{x}}_i\right) - \left(-2\ln\mathrm{L}(\underline{\mathrm{y}})\right) \sim \chi^2_{n-p}$$

where $D(\underline{\beta})$ represents that deviation (or distance) of the model $E(y_i) = F(\underline{x}_i{}'\underline{\beta})$ from the best fit obtainable (with the idealized model) in terms of –2 log-likelihood units.

To extend the analogies that were used in the quantal regression section, the worst possible model, with the largest value of –2 log-likelihood, is the model with no regressors (the "intercept-only" model); the smallest value is achieved by the idealized model; and somewhere in between is the value of –2 log-likelihood obtained by the current model. Thus, the deviance (or distance) between the best and the worst, denoted by $D(\hat{\beta}_0)$, is analogous to the SS_{total} in the classic MLR model; $D(\hat{\beta}_0)$ is analogous to SS_{error}; and $D(\hat{\beta}_0) - D(\hat{\underline{\beta}})$, the model χ^2, is analogous to the SS_{model}. Our partitioning of deviance becomes

$$D(\hat{\beta}_0) = D(\hat{\underline{\beta}}) + \text{ model } \chi^2.$$

Graphically, this partitioning is represented in Figure 8.11.

We will consider four primary uses of deviance. First, the deviance can be used as a type of "goodness of fit" test for the current model under consideration. Since $D(\hat{\underline{\beta}})$ represents the deviation of the current model to the idealized model, large values of $D(\hat{\underline{\beta}})$, compared to a chi-square distribution with $n - p$ degrees of freedom, indicate that the current model is further from the best model by more than chance would dictate. The null hypothesis is that the current model provides an adequate description of the data, while

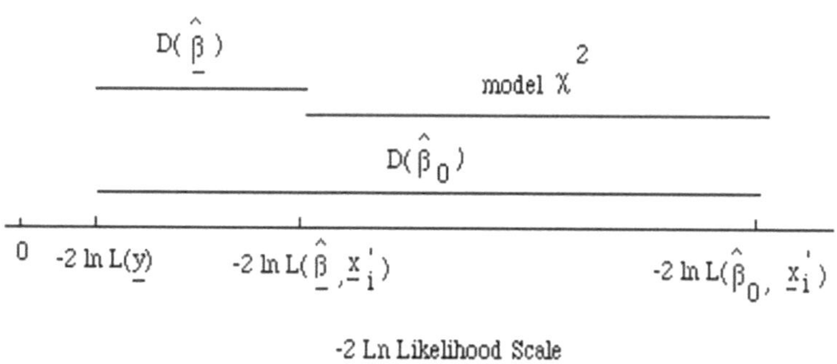

FIGURE 8.11
Deviance partitioning.

the alternative hypothesis is that it does not. Rejection of the null hypothesis, through large values of $D(\hat{\underline{\beta}})$, indicates that the model needs to be improved in some significant way.

Second, simple algebra reveals that the GLRT may be expressed in terms of the deviance for the reduced and full models as

$$\text{GLRT} = -2\ln\left[\frac{L\left(\hat{\underline{\beta}}_{\text{reduced}}, \underline{x}_i\right)}{L\left(\hat{\underline{\beta}}_{\text{full}}, \underline{x}_i\right)}\right] = D\left(\hat{\underline{\beta}}_{\text{reduced}}\right) - D\left(\hat{\underline{\beta}}_{\text{full}}\right).$$

Since the asymptotic distribution of $D\left(\hat{\underline{\beta}}_{\text{reduced}}\right)$ is a chi-square with $n - p_1$ degrees of freedom and that of $D\left(\hat{\underline{\beta}}_{\text{full}}\right)$ is also a chi-square with $n - p$ degrees of freedom, it follows that the difference between them is a chi-square with $p - p_1 = p_2$ degrees of freedom.

Third, using the analogies between the deviance and the sum of squares in the MLR model, we can define a "R^2"-type statistic as

$$R^2 = \frac{\text{model } \chi^2}{D\left(\hat{\beta}_0\right)}.$$

From the above display of the deviance, it is clear that $0 \leq R^2 \leq 1$. A word of caution is in order. As in the quantal regression case, a very good GLM may have an R^2 smaller than what one would consider "very good" for the MLR model. That is, the R^2 tends to underrepresent the GLM's ability to describe the data. Our next example will demonstrate this point.

The fourth use of the deviance is to help identify the unusual observations in the dataset whose responses or regressors are out of line with the other observations. In Section 8.1, we considered the n individual –2 log-likelihood components, the sum of which resulted in the 2 log-likelihood. Indeed, in the GLM, the deviance is also composed of the sum of n individual components, now called the "deviance components" and denoted by $\text{dc}\left(\hat{\underline{\beta}}\right)_i$ for $i = 1, \ldots, n$, where $\hat{\beta}$ is used in the notation to remind us that the deviance and the deviance components are particular to each model under consideration. Each $\text{dc}\left(\hat{\beta}\right)_i$ is computed (for the ungrouped data case) as

$$\text{dc}\left(\hat{\underline{\beta}}\right)_i = \text{llc}\left(\hat{\underline{\beta}}\right)_i - \text{llc}(\underline{y})_i,$$

where $\text{llc}\left(\underline{\hat{\beta}}\right)_i$ and $\text{llc}(\underline{y})_i$ represent the ith elements of $-2\ln L\left(\underline{\hat{\beta}}, \underline{x}_i\right)$ and $-2\ln L\left(\underline{y}\right)$, respectively. Thus, it is easy to show that

$$D(\hat{\beta}) = \sum_{i=1}^{n} \text{dc}\left(\underline{\beta}_i\right)_{i=1} = \sum_{i}^{n} \text{llc}(\underline{\hat{\beta}})_i - \sum_{i=1}^{n} \text{llc}(\underline{y})_i = -2\ln L\left(\underline{\hat{\beta}}, \underline{x}_i\right) - \left(-2\ln L\left(\underline{y}\right)\right).$$

Disproportionably large values of $\text{dc}(\hat{\beta})_i$ are indicative that the ith subject is anomalous in either the response or the regressors or both. For the grouped data situation, we simply replace $\underline{y}$ in the above notation with $\bar{y}$.

As a final note, in the quantal regression setting, the idealized model would result in $\hat{P}(\underline{x}_i) = 1$ whenever $y_i = 1$ and $\hat{P}(\underline{x}_i) = 0$ whenever $y_i = 0$, resulting in

$$\text{llc}(\underline{y})_i = -2\ln\left(\hat{P}(\underline{x}_i)^{y_i}\left(1 - \hat{P}(\underline{x}_i)\right)^{1-y_i}\right) = -2\ln\left(1^1 0^0\right) = -2\ln(1) = 0,$$

and, it follows that $-2\ln L(y) = \sum_{i=1}^{n} 11c(y) = 0$. Consequently,

$$D\left(\hat{\beta}\right) = -2\ln L\left(\hat{\beta}, \underline{x}_i\right) - 0 = -2\ln L\left(\hat{\beta}, \underline{x}_i\right).$$

That is, the deviance is just $-2\ln L\left(\underline{\hat{\beta}}, \underline{x}_i\right)$ for the ungrouped data case. For this reason, the concept of deviance was not introduced for the quantal regression model.

For any given distribution within the exponential family, it is easy to give the formula for the deviance. For example, consider again the Poisson regression situation. Expressions for $-2\ln L\left(\underline{\hat{\beta}}, \underline{x}_i\right)$ and $-2\ln L\left(\underline{y}\right)$ have been given above. Using them, it is easy to see that

$$\begin{aligned}\text{dc}\left(\underline{\hat{\beta}}\right)_i = \text{llc}\left(\underline{\hat{\beta}}\right)_i - \text{llc}\left(\underline{y}_i\right)_i &= -2\left(\left(y_i - F\left(\underline{x}_i' \underline{\hat{\beta}}\right)\right) + y_i \ln\left(\frac{F\left(\underline{x}_i' \underline{\hat{\beta}}\right)}{y_i}\right)\right) \\ &= -2\left(r_i + y_i \ln\left(\frac{F\left(\underline{x}_i' \underline{\hat{\beta}}\right)}{y_i}\right)\right),\end{aligned}$$

where $r_i = y_i - F\left(\underline{x}_i \underline{\hat{\beta}}\right)$ is the ith residual. As in MLR, a good model will have small residuals. If the GLM results in $F\left(\underline{x}_i{}' \underline{\hat{\beta}}\right) \approx y_i$, then $\text{dc}\left(\underline{\beta}_i\right)_i$ will be close to zero. If this is true for all n observations, then $D(\underline{\hat{\beta}})$ will be small. On the other hand, it would be interesting to know whether, if the size of $D(\underline{\hat{\beta}})$ is due to a small percent of the data, those observations where the model fit may be inadequate. We will demonstrate the use of the deviance components in the example.

8.2.6 An Example of Poisson Regression

All the features of GLM that have been discussed above will now be demonstrated with Poisson regression for the ungrouped data situation. Recall that for the Poisson random variable, y_i, that $E(y_i) = \lambda_i = F(\underline{x}'\underline{\beta}) = \text{Var}(y_i)$. One popular choice of the function F is the exponential function where $F(\underline{x}_i' \underline{\beta}) = \exp(\underline{x}_i\underline{\beta})$. This implies that $\ln(F(\underline{x}_i\underline{\beta}')) = \underline{x}_i{}'\underline{\beta}$, so in GLM terminology, we are using the log link function. We desire to estimate the mean response of y_i. This is accomplished by estimating $\underline{\beta}$ and computing $\hat{\lambda}_i = F(\underline{x}_i{}'\underline{\hat{\beta}})$. We need the MLE of $\underline{\beta}$. Since the Poisson distribution is a member of the exponential family, we can apply the GLM results. We know that $\text{Var}(y_i) = \lambda_i = F(\underline{x}_i{}'\underline{\beta})$ and if $F(\underline{x}_i{}'\underline{\beta}) = \exp(\underline{x}_i{}'\underline{\beta})$, for example, and noting that $\frac{d\exp(u)}{du} = f(u) = \exp(u)$, we find the *mle* of $\underline{\beta}$ by solving

$$\frac{\partial \ln L(\underline{\beta}^*, \underline{x}_i)}{\partial \underline{\beta}^*} = \sum_{i=1}^{n} \left[\frac{y_i - \exp(\underline{x}_i' \underline{\beta}^*)}{\exp(\underline{x}_i' \underline{\beta}^*)} \right] \exp(\underline{x}_i' \underline{\beta}^*) \underline{x}_i = \sum_{i=1}^{n} \left[y_i - \exp(\underline{x}_i' \underline{\beta}^*) \right] \underline{x}_i .$$

Again, we note that the above, when set equal to $\underline{0}$, represents a nonlinear system of equations that must be solved via appropriate numerical methods. The IRLS algorithm is used to accomplish this with the elements of the weights matrix set at

$$w_i = \frac{f^2(\underline{x}_i' \underline{\hat{\beta}}_0)}{\text{Var}(y_i)} = \frac{\exp(\underline{x}_i' \underline{\hat{\beta}}_0)^2}{\exp(\underline{x}_i' \underline{\hat{\beta}}_0)} = \exp(\underline{x}_i' \underline{\hat{\beta}}_0).$$

Simple algebra using the above expressions gives the deviance for the model with estimated parameters, $\hat{\beta}$, as

$$D(\underline{\hat{B}}) = -2\left(\sum_{i=1}^{n} r_i - \sum_{i=1}^{n} y_i \ln\left(\frac{y_i}{F(\underline{x}_i \underline{\hat{\beta}}} \right) \right),$$

where $r_i = y_i - F(\underline{x}_i' \hat{\underline{\beta}})$. Thus, if the model is the "intercept-only" model with $E(y_i) = F(\beta_o) = \lambda$, we know from Section 2.3 that $\hat{\lambda} = F(\hat{\beta}_0) = \bar{y}$, and consequently, $r_i = y_i - \bar{y}$, and the deviance for this model may be written as

$$D(\hat{\beta}_0) = -2\left(\left(\sum_{i=1}^{n} y_i\right)\left(\ln\left(-\bar{y}\right)\right) - \left(\sum_{i=1}^{n} y_i \ln(y_i)\right)\right).$$

This, of course, is the maximum value of the deviance over all possible values of the linear predictor.

8.2.7 The Mining Application

In the mining application (Myers, 1990), the response of interest is the number of fractures that occur in the upper seam of mines in the coalfields of the Appalachian region in Western Virginia. Forty-four observations were obtained on mines in this area. The response is thought to be a function of the material in the land and the mining area. Thus, the four regressors are x_1 = inner burden thickness (feet), which is the shortest distance between the seam floor and the lower seam; x_2 = percent of extraction of the lower previously mined seam; x_3= lower seam height (in feet); and x_4= time (years) that the mine has been opened. The data are available on the website in Mining_Data.xlsx, and a sample is given in Table 8.4.

The R codes were written to analyze this data using Poisson regression. The analysis of this data, including the inappropriate MLR analysis, variable selection using Poisson regression, and evaluation of the final selected model, is provided in the subsequent discussion.

TABLE 8.4

The Mining Data

Obs.	y	x1	x2	x3	x4
1	2	50	70	52	1.0
2	1	230	65	42	6.0
3	0	125	70	45	1.0
4	4	75	65	68	0.5
5	1	70	65	53	0.5
6	2	65	70	46	3.0
7	0	65	60	62	1.0
8	0	350	60	54	0.5
9	4	350	90	54	0.5
10	4	160	80	38	0.0
		....			
43	0	100	65	60	25.0
44	3	50	88	60	20.0

The scatterplots in Figure 8.12 show that the plot is unusual due to the discrete values taken by the response (y) variable. Also, note that observation 30 clearly has an unusually large value of $x1$. The trend appears to be weakly negative. From the scatterplot of y versus $X2$, there is a weak positive trend and unusual values for observations 27 and 34. Note that there is very little trend to this plot for y versus $x3$. Also, observation 25 has an unusual value for $x3$. Note the weak negative trend and the unusual behavior of observation 30 from the scatterplot of y versus $x4$. Let's look at the correlations (inappropriate since y is not a continuous measurement variable but helpful nonetheless).

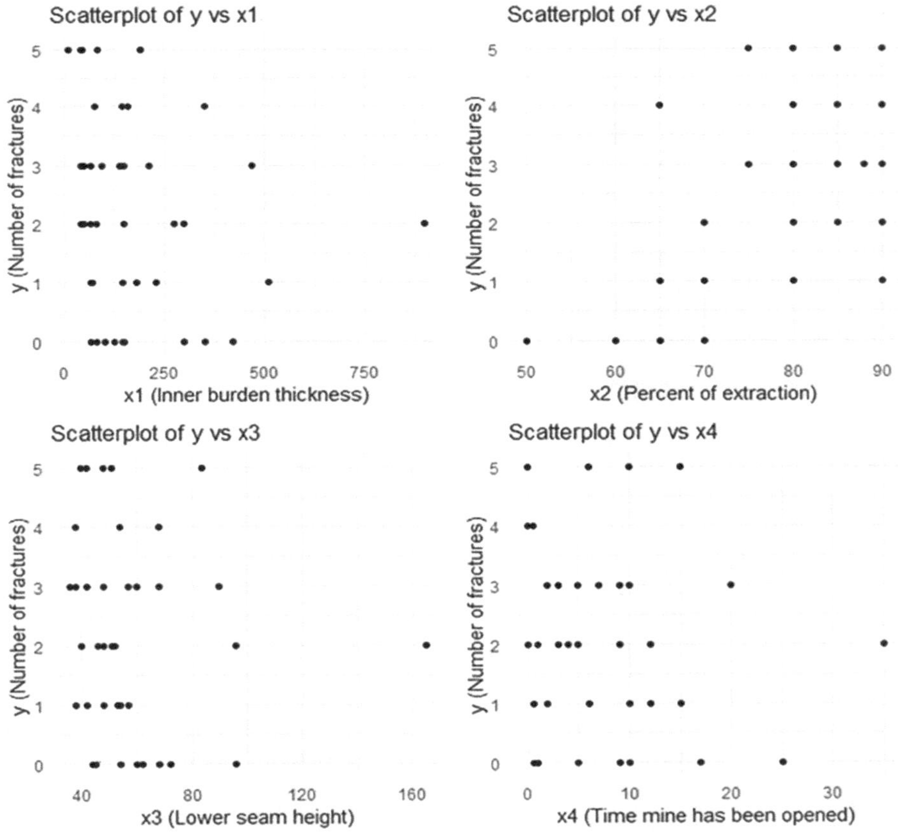

FIGURE 8.12
The scatterplot matrix.

```
> # 2. Correlation matrix (note: y is not continuous, but we
calculate correlations anyway)
> cor_matrix <- cor(mining_data[, c("y", "x1", "x2", "x3", "x4")])
> print(cor_matrix)
             y         x1          x2           x3          x4
y    1.0000000 -0.2139824 0.632108950 -0.131946101 -0.21046858
x1  -0.2139824  1.0000000 0.101589387  0.105018774  0.42333096
x2   0.6321090  0.1015894 1.000000000  0.005481737  0.13398320
x3  -0.1319461  0.1050188 0.005481737  1.000000000  0.06019417
x4  -0.2104686  0.4233310 0.133983205  0.060194173  1.00000000
```

The correlations between *y* and the regressor are quite small, with *x*2 being the largest. Let's do the MLR model. This is the wrong analysis, but let's see what happens.

```
> #Fit the multiple linear regression (MLR) model
> mlr_model <- lm(y ~ x1 + x2 + x3 + x4, data = mining_data)
> #Display the summary of the MLR model
> summary(mlr_model)
Call:
lm(formula = y ~ x1 + x2 + x3 + x4, data = mining_data)
Coefficients:
             Estimate Std. Error t value Pr(>|t|)
(Intercept) -4.579885   1.381809  -3.314  0.00199 **
x1          -0.001896   0.001290  -1.469  0.14975
x2           0.104574   0.017082   6.122 3.51e-07 ***
x3          -0.007993   0.008531  -0.937  0.35457
x4          -0.050250   0.027942  -1.798  0.07986.
---
Signif. codes: 0 '***' 0.001 '**' 0.01 '*' 0.05 '.' 0.1 ' ' 1
Residual standard error: 1.262 on 39 degrees of freedom
Multiple R-squared: 0.5283, Adjusted R-squared: 0.4799
F-statistic: 10.92 on 4 and 39 DF, p-value: 4.902e-06
The regression equation is
   y = - 4.58 - 0.00190 x1 + 0.105 x2 - 0.00799 x3 - 0.0503 x4
```

The MLR model provides a weak fit with an R^2 of 52.8; X3, and possibly x1, may be removed from the model. There are five observations that we should keep an eye on. In addition, though not shown here, there are two negative fitted values. This is nonsense since the counts must be positive values. Stepwise regression gives the following results.

```
> Perform stepwise selection using AIC (both directions)
> stepwise_model <- stepAIC(full_model, direction = "both")
Start: AIC=25.19
y ~ x1 + x2 + x3 + x4
       Df Sum of Sq     RSS    AIC
- x3    1     1.399  63.540 24.169
```

```
<none>                    62.142 25.190
- x1    1      3.440     65.582 25.561
- x4    1      5.153     67.295 26.695
- x2    1     59.717    121.859 52.822
Step: AIC=24.17
y ~ x1 + x2 + x4
      Df Sum of Sq        RSS    AIC
<none>                    63.540 24.169
- x1    1      3.868     67.408 24.769
+ x3    1      1.399     62.142 25.190
- x4    1      5.253     68.793 25.664
- x2    1     59.849    123.389 51.371
> # Display the summary of the stepwise-selected model
> summary(stepwise_model)
Call:
lm(formula = y ~ x1 + x2 + x4, data = mining_data)
Coefficients:
             Estimate Std. Error  t value   Pr(>|t|)
(Intercept) -5.019685   1.297646   -3.868   0.000394 ***
x1          -0.002003   0.001283   -1.560   0.126528
x2           0.104686   0.017055    6.138   3.02e-07 ***
x4          -0.050724   0.027894   -1.818   0.076492.
---
Residual standard error: 1.26 on 40 degrees of freedom
Multiple R-squared: 0.5176, Adjusted R-squared: 0.4815
F-statistic: 14.31 on 3 and 40 DF, p-value: 1.757e-06
> selected_model <- lm(y ~ x2 + x4, data = mining_data)
> Display the summary of the selected MLR model
> summary(selected_model)
Call:
lm(formula = y ~ x2 + x4, data = mining_data)
Coefficients:
            Estimate Std. Error t value Pr(>|t|)
(Intercept) -5.12812   1.31827   -3.890   0.00036 ***
x2           0.10336   0.01733    5.964   4.86e-07 ***
x4          -0.06881   0.02581   -2.666   0.01093 *
---
Residual standard error: 1.282 on 41 degrees of freedom
Multiple R-squared: 0.4883, Adjusted R-squared: 0.4633
F-statistic: 19.56 on 2 and 41 DF, p-value: 1.085e-06
```

The results indicate that we have a model that does not significantly lack of fit (p-value of lof test is 0.478), and there is at least one significant regressor (p-value for the model chi-square is 0.000). The R^2 is modest in size, indicating that the fit is not very good. The Wald z-tests indicate that $x3$ is a very weak variable in the presence of the other three regressors. We will drop $x3$ from the model and not consider this model further. One can see that only $x2$ and $x4$ are selected to be included in the MLR model. We can obtain the

Poisson regression results by using the R program. The Poisson model with $x1$, $x2$, and $x4$ is as follows:

```
> Poisson regression model
> poisson_model <- glm(y ~ x1 + x2 + x3 + x4, data = mining_
data, family = poisson())
> Summary of the Poisson regression model
> summary(poisson_model)
Call:
glm(formula = y ~ x1 + x2 + x3 + x4, family = poisson(),
data = mining_data)
Coefficients:
              Estimate Std. Error z value Pr(>|z|)
(Intercept) -3.5930896  1.0256803  -3.503  0.00046 ***
x1          -0.0014066  0.0008358  -1.683  0.09240 .
x2           0.0623458  0.0122862   5.074 3.89e-07 ***
x3          -0.0020803  0.0050661  -0.411  0.68134
x4          -0.0308135  0.0162648  -1.894  0.05816 .
---
Signif. codes: 0 '***' 0.001 '**' 0.01 '*' 0.05 '.' 0.1 ' ' 1
(Dispersion parameter for poisson family taken to be 1)
   Null deviance: 74.984 on 43 degrees of freedom
Residual deviance: 37.856 on 39 degrees of freedom
AIC: 144.13
Number of Fisher Scoring iterations: 5
> Variable selection using stepwise AIC (optional for final
model selection)
> stepwise_model <- stepAIC(poisson_model, direction = "both")
Start: AIC=144.13
y ~ x1 + x2 + x3 + x4
       Df Deviance    AIC
- x3    1   38.031 142.30
<none>      37.856 144.13
- x1    1   41.023 145.29
- x4    1   41.750 146.02
- x2    1   69.807 174.08
Step: AIC=142.3
y ~ x1 + x2 + x4
       Df Deviance    AIC
<none>      38.031 142.30
- x1    1   41.626 143.90
+ x3    1   37.856 144.13
- x4    1   42.094 144.37
- x2    1   70.426 172.70
> summary(stepwise_model)
Call:
glm(formula = y ~ x1 + x2 + x4, family = poisson(), data =
mining_data)
```

```
Coefficients:
              Estimate Std. Error z value Pr(>|z|)
(Intercept) -3.7206821  0.9788770  -3.801 0.000144 ***
x1          -0.0014793  0.0008244  -1.794 0.072757 .
x2           0.0627011  0.0122711   5.110 3.23e-07 ***
x4          -0.0316514  0.0163095  -1.941 0.052298 .
---
Signif. codes: 0 '***' 0.001 '**' 0.01 '*' 0.05 '.' 0.1 ' ' 1
(Dispersion parameter for poisson family taken to be 1)
    Null deviance: 74.984 on 43 degrees of freedom
Residual deviance: 38.031 on 40 degrees of freedom
AIC: 142.3
Number of Fisher Scoring iterations: 5
Deviance explained (Poisson R-sq): 0.4928
```

First, we note that there is little difference in the performance criteria between this three-variable model and the four variables considered above. The three-variable model has a nonsignificant lof test, the model chi-square is highly significant, and the R^2 is nearly equal to the one for the four-variable model. By the way, we can easily compute the GLRT for comparing the reduced model with $x1$, $x2$, and $x4$ to the full model with $x1$, $x2$, $x3$, and $x4$ by subtracting their deviances as GLRT = 38.0307 – 37.8560 = 0.1747, a chi-square with one degree of freedom, with a p-value of 0.675968. Clearly, $x4$ should be dropped from the full model. Note also that the GLRT is similar to the Wald z-test result (with a p-value of 0.6813 for $x4$ in the full model). Recall these two tests are asymptotically equivalent. Should $x1$ be removed from the model?

Let's see.

```
> # Fit the Poisson regression model with x2 and x4
> poisson_model_x2_x4 <- glm(y ~ x2 + x4, data = mining_data,
family = poisson(link = "log"))
> # Summary of the Poisson model
> summary(poisson_model_x2_x4)
Call:
glm(formula = y ~ x2 + x4, family = poisson(link = "log"),
data = mining_data)
Coefficients:
            Estimate Std. Error z value Pr(>|z|)
(Intercept) -3.58943    0.94471  -3.800 0.000145 ***
x2           0.05875    0.01169   5.027 4.98e-07 ***
x4          -0.03802    0.01545  -2.460 0.013888 *
---
(Dispersion parameter for poisson family taken to be 1)
    Null deviance: 74.984 on 43 degrees of freedom
Residual deviance: 41.626 on 41 degrees of freedom
AIC: 143.9
Number of Fisher Scoring iterations: 5
```

We note that all remaining regressors in the model ($x2$ and $x4$) are highly significant. The model does not significantly lack fit. The GLRT for $x1$ given $x2$ and $x4$ is: GLRT = 41.6261 – 38.0307 = 3.5954 with a p-value of 0.05457, just barely insignificant of the 0.05 level. As expected, the GLRT is slightly different in p-value from the Wald's z-test (0.0737). Also, note that the R^2 dropped by about 5%. Consequently, we will keep $x1$ in the model with $x2$ and $x4$. We now proceed with a more detailed evaluation of the model including $x1$, $x2$, $x4$. Rerun the model, printing out the fits, and their 95% confidence limits as in Figure 8.13.

Using the MLE, the model for the mean response is

```
Lambda=exp(-3.72 - 0.0015*x1 + 0.0627*x2 - 0.0316*x4).
```

Note that the mean response will decrease as $x1$ and $x4$ increase, and as $x2$ increases, so does the mean response. Note the fits. One advantage of the GLM over MLR is that the fits are all positive, as theory says they should be.

We note that several observations do not fit particularly well, notably observations 4, 21, 29, and 42. There does appear to be a strong positive linear trend to the data.

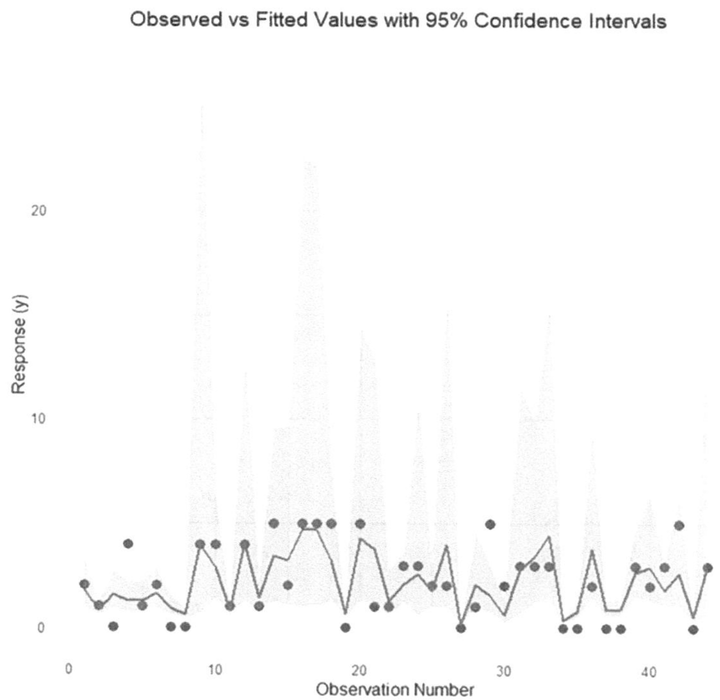

FIGURE 8.13
Observed versus fitted values with 95% CI.

```
> # Calculate the correlation between y and estmean
> correlation <- cor(mining_data$y, fitted_values)
> # Print the correlation value
Correlation of y and estmean = 0.712
```

The correlation of 0.71 supports this observation. Remember, however, that this is not the proper statistic when evaluating GLMs other than the normal. Plotting the response (using the observation number as the label) versus the linear predictor, xbeta, with the estimated fits, represented as the curved exponential function, is informative, as in Figure 8.14.

The vertical lines connecting several points to the curved line (the fits) reveal which observations are outliers with respect to our model. Clearly, observations 3, 4, 18, 29, 30, and 42 are among the poor fits. The deviance component index figure also reveals the same results.

Several observations have very large deviance components, especially observations 29, 4, 8, and 21. Observation 29 is clearly the "sore thumb." It will be up to the reader to verify that if the above-mentioned four observations are removed from the dataset and the three-variable Poisson regression

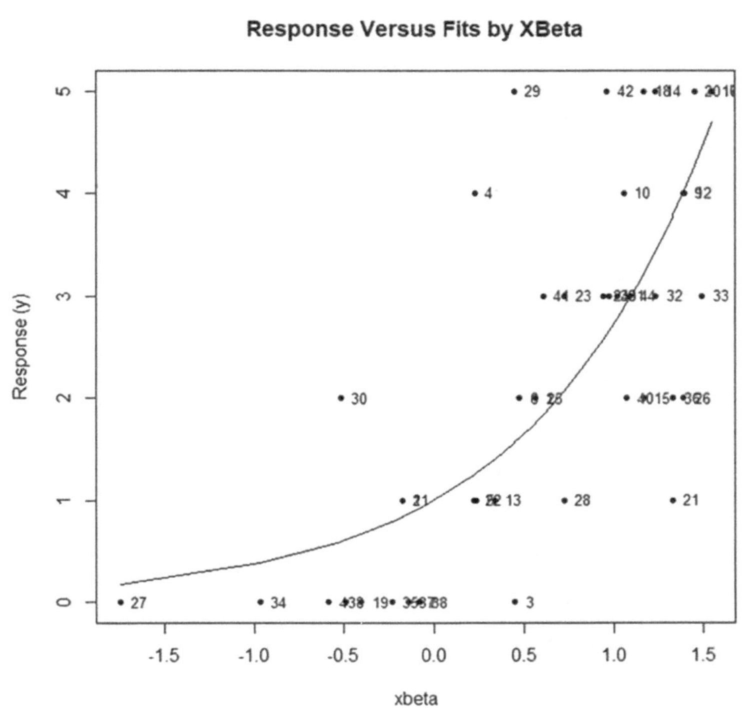

FIGURE 8.14
The response versus fitted values.

model rerun, the R^2 increases to 64%. Apparently, the four unusual observations are hurting the fit to the remaining 40 observations to some extent.

Of course, other aspects of GLM could be discussed. For example, diagnostic information such as outlier and leverage diagnostics have been developed for GLM as they have been for the MLR model. Pregibon (1981) is a good source for this material as is McCullagh and Nelder (1994). Rather, we will illustrate some practical applications of GLM in the next section. For example, we will consider again, as in Chapter 7, the two-sample and *k*-sample problems, where we have quantal data and count data instead of normal data.

8.2.8 GLM versus MLR

We have discussed several ways to model regression data. Each method depends on the distribution of the response variable and the corresponding form of the means function. If we can assume that the response variable follows a normal distribution with the mean function written as $E(y_i|\underline{x}_i) = \mu_i = F(\underline{x}_i'\ \underline{\beta}) = \underline{x}_i'\ \underline{\beta}$, the identity link, then we use the standard MLR model and the corresponding least squares analysis.

On the other hand, if the distribution of the response variable is another member of the exponential family, such as the Poisson or the exponential, with $E(y_i|\underline{x}_i) = \mu_i = F(\underline{x}_i{}'\underline{\beta})$, we apply the GLM and the corresponding ML analysis. As we have noted several times in this chapter, the MLR model and the LS analysis are just a special case of the more general GLM ML approach.

Of course, if the response data follows a Poisson distribution with a log link – that is, $E(y_i|\underline{x}_i) = \mu_i = F(\underline{x}_i\underline{\beta}_i) = \exp(\underline{x}_i'\ \underline{\beta})$, then the optimal analysis is the GLM using $F(\underline{x}_i{}'\underline{\beta}) = \exp(\underline{x}_i{}'\underline{\beta})$. Use of any other form for $F(\underline{x}_i{}'\underline{\beta})$ or assuming any other distribution would give results that are less than optimal.

It is easy to believe these statements because every situation has an optimal technique appropriate for that situation. The only trouble with the theory is that it requires knowledge of both the distribution of the response and the form of the mean function, both of which are unknown to the researcher. This, of course, is the same dilemma faced by the researcher in earlier chapters: the need to make certain assumptions about the model.

The choice of distribution of the response variable can often be justified on the basis of a combination of the underlying theory of the subject matter and a graphical study of the data. For example, the obvious model for quantal response data is quantal regression. The choice of logistic regression (the logit link) or probit regression (the inverse standard normal CDF link) or some other means model is often a matter of convenience—logistic regression is well known, often works very well in many diverse situations, and software packages have logistic regression programs readily available.

This fact does not imply that the logistic model is the best quantal regression model for a given set of data but only that logistic regression can work well on many sets of quantal regression data.

Count data is often modeled with GLM, assuming a Poisson distribution with a log link. Continuous measurement data is often modeled as normally distributed. Graphical analysis of the residuals is used to justify the constant variance assumption and appropriateness of the linear means model.

Mistakes in the choice of the error distribution and/or form of the means models (the proper form of the regression function) are a fact of life in the area of statistical model building and evaluation. While model misspecification may be common, the researcher hopes that reasonable, nearly correct models are used, giving near-optimal results.

A reasonable question is what penalty is paid if one uses the wrong error distribution and/or wrong regression function? While a detailed study of the properties of GLM under various forms of model misspecification are beyond the scope of this book, we can get an idea of the importance of proper choice of model through a simple example.

Let us consider the situation where the true error distribution is Poisson, where the means function is based on the identity link – that is, $E(y_i|\underline{x}_i) = F(\underline{x}_i' \underline{\beta}) = \underline{x}_i'\underline{\beta}$. The researcher, unaware of this fact, chooses to use the MLR model, assuming that the response variable is normal with $E(y_i|\underline{x}_i) = \underline{x}_i'\underline{\beta}$.

We recall from early in this section that the MLE of $\underline{\beta}$, for the general GLM, can be written at the convergence of the IRLS algorithm, as

$$\hat{\beta}_1 = \hat{\beta}_0 + (X'WX)^{-1}X'Wy^*,$$

where $y_i^* = \dfrac{y_i - F(\underline{x}_i' \underline{\beta}_0)}{f(\underline{x}'_i\hat{\underline{\beta}}_0)}$, $w_i = \dfrac{f^2(\underline{x}'_i\underline{\beta}_0)}{\hat{\text{Var}}(y_i)}$, and $\text{Var}(y_i) = G(F(\underline{x}_i\hat{\underline{\beta}}_0))$.

Under the conditions of a Poisson distribution-identity link, the MLE simplifies to (using the fact that

$$F(\underline{x}'\hat{\underline{\beta}}_0) = \underline{x}_i' \hat{\underline{\beta}}_0, f(\underline{x}'_i\hat{\underline{\beta}}_0) = 1, \text{ and } G(F(\underline{x}_i' \hat{\underline{\beta}}_0)) = F(\underline{x}'\hat{\underline{\beta}}_0) = \underline{x}_i' \underline{\beta}_{=0},$$

$$\hat{\beta}_{\text{MLE}} = (X'\hat{V}^{-1}X)^{-1}X'\hat{V}^{-1}\underline{y}$$

where $\hat{V}$ is an *nxn* diagonal matrix with the estimated variances of response on the diagonal, $\text{Var}(y_1) = \underline{x}_1'\hat{\beta}_{\text{MLE}}, \ldots, \text{Var}(y_n) = \underline{x}_n'\hat{\beta}_{\text{MLE}}$. This is the optimal estimator of $\underline{\beta}$.

The researcher, on the other hand, uses the MLR model and obtains the usual LS estimator of $\underline{\beta}$ as

$$\hat{\beta}_{\text{LS}} = \left(X'X\right)^{-1} X'y.$$

The LS estimator is less than optimal in this situation.

A comparison of the two estimators reveals that the difference between them lies in the use of the $\hat{\text{V}}^{-1}$ matrix. The optimal estimator gives more weight to those observations with a smaller estimated variance of response and less weight to those observations with a larger estimated variance of response. The closer the matrix $\hat{\text{V}}^{-1}$ is to the identity matrix, the closer the properties of $\hat{\beta}_{\text{LS}}$ are to the optimal properties of $\hat{\beta}_{\text{MLE}}$.

On the other hand, if the variance estimates in $\hat{V}$ differ considerably, resulting in $\hat{V}^{-1}$ being very different from the identity matrix, one would expect the properties of $\hat{\underline{\beta}}_{\text{MLE}}$ to be much better than those of $\underline{\beta}_{\text{LS}}$. This is indeed the case. The two procedures may be compared theoretically by looking at their asymptotic expectation and variance. The asymptotic expectations of $\hat{\beta}_{\text{MLE}}$ and $\hat{\beta}_{\text{LS}}$ are both equal to β, as they are both asymptotically unbiased estimators.

The asymptotic variance of $\hat{\beta}_{\text{MLE}}$ can be shown to be $\text{Var}\left(\underline{\beta}_{\text{MLE}}\right) = \left(\text{X}'\text{V}^{-1}\text{X}\right)^{-1}$, where V is the true variance-covariance matrix of the n responses. For our example, $\text{Var}\left(y_i \mid \underline{x}_i\right) = \underline{x}_i\beta$ is the ith element of V.

The variance of $\hat{\beta}_{\text{LS}}$ can be shown to be $\text{Var}\left(\hat{\beta}_{\text{LS}}\right) = \left(X'X\right)^{-1} X'VX\left(X'X\right)^{-1}$. These two variance-covariance matrices are equal only if $V = I$; otherwise, it can be shown that

$$\text{Var}\left(\hat{\beta}_{LSj}\right) \geq \text{Var}\left(\hat{\beta}_{MLEj}\right)$$

for all $j = 0,\ldots,p$. The MLE is a more precise estimator of each unknown coefficient.

8.3 Some Applications of GLM

Many practical applications of GLM can be expressed in terms of very simple models. Two examples we will consider in this section are the two-sample problem and the k-sample problem. In Section 6.4, both these models were illustrated as examples of the GLM. That is, the models could be expressed as $y_i = \underline{x}'_i\underline{\beta} + \varepsilon_i$ where the regressors took the form of indicator variables. The error term, ε_i, was assumed to follow either a normal distribution or some symmetric, heavy-tailed distribution. In either case, the response, y_i, is a continuous measurement variable.

The two-sample analysis of means and the *k*-sample analysis of means could be expressed in the GLM framework with the resulting analysis equivalent to the classical two-sample *t*-test and the one-way analysis of variance (ANOVA), respectively.

Similar results are available for the GLM, as will be illustrated in this section. The major difference now is that the response variable's distribution is considered a member of the exponential family, specifically distributions other than the normal. And, just as in Section 6.4 where several methods were available to analyze each model, several techniques will be available to analyze these simple non-normal data situations.

For example, as discussed in Sections 8.1 and 8.2, one complication in estimating GLM parameters is that the variance of the response is nonconstant, depending on the mean response. The techniques of those sections dealt with the nonconstant variance problem. However, in earlier sections of this book, specifically Sections 1.8, 2.4, 5.2, and 5.3, transformations were discussed that would result in transformed data with constant variance of response. Thus, data transformations may be viewed as an alternative to GLM for analyzing data of the type considered here. In the applications to follow, these two methods, the GLM approach and the transformation approach, will be compared.

8.3.1 The Two-Sample Problem

If we desire to compare two means, μ_1 and μ_2, and we have the assumption of normal distributions with common variance, the optimal method is the two-sample t-test, as illustrated in Section 6.4. If outliers are present in the data, perhaps resulting from contaminated normal distributions, alternative methods are available. Two methods, the two-sample *t*-test based on ranks and the two-sample *t*-test based on robust weights, were introduced and illustrated in Section 6.4 as well. Suppose, however, that the response measured in the two-sample problem is quantal data (0–1 data) or count data or highly skewed continuous data.

Clearly, the symmetric continuous measure variable assumptions made in Section 6.4 are no longer valid. However, the distributions that apply to these situations, the Bernoulli, Poisson, and exponential, all are examples of the exponential family, and consequently, the GLM analysis is appropriate. That is, parameters will be estimated using ML and tests of the hypothesis made using either the GLRT or the Wald *z* (or chi-square) test.

We could simply use the techniques of Section 8.1 for quantal regression or Section 8.2 for Poisson regression. But, for these simple models, the MLEs may be obtained very easily without the need for the IRLS algorithm. In addition, the GLRT also has a very simple form.

8.3.2 Quantal Response

One application of the two-sample problem for quantal data is to compare population proportions, P_1 and P_2, across the two populations. These proportions represent the proportion of "successes" in each population. Random samples of size n_1 and n_2 are obtained from population 1 and population 2, respectively. In each sample, the quantal response variable, y, is observed where $y = 0$ if the response is a "failure" and $y = 1$ if the response is a "success." If r_1 and r_2 represent the number of successes in the samples from populations 1 and 2, respectively, then the MLE of P_1 is $\hat{P}_1 = \frac{r_1}{n_1}$, the sample proportion of successes from population 1, and the MLE of P_2 is $\hat{P}_2 = \frac{r_2}{n_2}$, the sample proportion of successes from population 2.

It is a simple exercise to show that each sample proportion is an unbiased estimator of the population proportion and that $\mathrm{Var}(P_i) = \frac{P_i(1-P_i)}{n_i}$, for $i = 1$ and 2.

If we desire to test the hypotheses $H_0 : P_1 = P_2 = P$ versus $H_1 : P_1 > ($ or $<$ or $\neq) P_2$, we could use the two-sample test of proportions, presented in most elementary statistics texts (see Moore and McCabe, 1998, for example) as

$$z = \frac{\hat{P}_2 - \hat{P}_1}{\mathrm{se}(\hat{P}_2 - \hat{P}_1)},$$

where $\mathrm{se}(\hat{P}_2 - \hat{P}_1) = \sqrt{\hat{P}(1-\hat{P})\left(\frac{1}{n_1} + \frac{1}{n_2}\right)}$ with $\hat{P} = \frac{r_1 + r_2}{n_1 + n_2}$, the pooled estimator of the common P.

The test statistic z has approximate $N(0,1)$ distribution, a result based on the central limit theorem, provided that the sample sizes are large enough. The traditional recommendation is that "large enough" is achieved if $n_1\hat{P}_1, n_1(1-\hat{P}_1), n_2\hat{P}_2$, and $n_2(1-\hat{P}_2)$ all are five or more.

The above hypotheses could also be tested using the GLRT. The full model would be that there are two separate population proportions, P_1 and P_2, with the MLEs given above. The reduced model resulting from the null hypothesis is that there is only the common proportion, P, whose MLE is $\hat{P} = \frac{r_1 + r_2}{n_1 + n_2}$. It is easy to show using the results of Section 8.1 that the GLRT may be written as

$$\text{GLRT} = -2\left[\ln L\left(\hat{P}\right) - \ln L\left(\hat{P}_1, \hat{P}_2\right)\right]$$

$$= -2\left[\left[r \ln\left(\frac{\hat{P}}{1-\hat{P}}\right) + n \ln\left(1-\hat{P}\right)\right] - \sum_{i=1}^{2}\left[r_i \ln\left(\frac{\hat{P}_i}{1-\hat{P}_i}\right) + n_i \ln\left(1-\hat{P}_i\right)\right]\right],$$

a test statistic with an approximate Chi-squared distribution with one degree of freedom.

Of course, the GLRT actually tests the two-sided alternative between P_1 and P_2. But, if either the upper one-sided alternative or the lower one-sided alternative is desired, it is easy to compute the *p*-value for the one-sided test as half the chi-square *p*-value and conclude that $P_1 > P_2$ if the GLRT is significant and $\hat{P}_1 > \hat{P}_2$ or conclude that $P_1 < P_2$ if the GLRT is significant and $\hat{P}_1 < \hat{P}_2$. Of course, these two tests are asymptotically equivalent but may differ for finite sample samples. Both are easily computed on R.

8.3.3 The Smoking Habits Application

Students in an introductory statistics class at a large land-grant university in the south, VT, conducted a study of the smoking habits of students attending the university. Random samples of 100 male and 100 female students revealed 32 male and 26 female smokers. Based on this information, what can be concluded concerning the proportion of males at the university who smoke compared to the proportion of females at the university who smoke?

To use R to compute either test, type as follows:

```
> z_test_statistic <- (p1_hat-p2_hat) / sqrt(p_hat * (1-p_hat) *
(2/100))
> print(z_test_statistic)
[1] 0.9349924 #Clearly, the test statistic is nonsignificant.
> # Compute the GLRT
> glrt <- -2 * (log_likelihood_reduced-log_likelihood_full)
> print(glrt)
[1] 41.21526
> # Comparing the two test statistics. Remember that
z**2=Chi-squared.
> z_squared <- z_test_statistic^2
> print(glrt)
[1] 41.21526
> print(z_squared)
[1] 0.8742108
```

For this example, we note that there is very little difference between the two tests, which both have asymptotic chi-squared distributions with one degree of freedom.

We notice that the GLRT is far more complicated than the standard two-sample z-test for proportions, and since they typically give very similar results, it is no wonder that the two-sample z-test for proportions is the method presented in most textbooks. One advantage of the GLRT is that we already have a built-in function to perform this analysis for any GLM based on the logistic distribution. In the spirit of Section 6.4, we can rewrite the two-sample problem as a GLM with the aid of indicator variables.

Let the indicator variable z be zero if the subject receives treatment 1 and one if the subject receives treatment 2. Then the two-sample problems for quantal data may be written as

$$E(y_{ij}) = P_i = F(\beta_0 + \beta_1 z_{ij}).$$

It follows from this model that $\mathrm{E}(\mathrm{y}_{1j}) = \mathrm{P}_1 = \mathrm{F}(\beta_0)$ for an observation from treatment 1 and $E(y_{2j}) = P_2 = F(\beta_0 + \beta_1)$ if an observation is from treatment 2.

We recall from Section 8.1 that the coefficients may be isolated by applying the inverse function, F^{-1}, the logit in the case of logistic regression. In particular, β_1 is expressed as the log of the "odds ratio," –that is, $\beta_1 = \ln\left(\dfrac{\frac{P_2}{1-P_2}}{\frac{P_1}{1-P_1}}\right)$.

Consequently, an equivalent test of $\mathrm{H}_0 : \mathrm{P}_1 = \mathrm{P}_2 = \mathrm{P}$ is $\mathrm{H}_0 : \beta_1 = 0$ in the GLM.

Of course, the test can be computed by using the logistic regression model with the regressor z and using the Wald z-test on β_1.

To perform the analysis on R, type as follows:

```
> # Creating the data frame
> data <- data.frame(response = responses, group = groups)
> # Fit the logistic regression model
> model <- glm(response ~ group, family = binomial(link =
"logit"), data = data)
> # Summarize the model
> summary(model)
Call:
glm(formula = response ~ group, family = binomial(link =
  "logit"), data = data)
Coefficients:
             Estimate Std. Error z value Pr(>|z|)
(Intercept)   -1.0460     0.2280  -4.588 4.48e-06 ***
group          0.2922     0.3129   0.934     0.35
---
```

```
Dispersion parameter for binomial family taken to be 1)
    Null deviance: 240.86 on 199 degrees of freedom
Residual deviance: 239.99 on 198 degrees of freedom
AIC: 243.99
Number of Fisher Scoring iterations: 4 #The number of itera-
  tions required was 4.
> # Get the odds ratio for the group effect (males vs.
  females)
> exp(coef(model)["group"])
   group
1.339367
> # Chi-square test statistic and p-value
> chi_square <- logLik_null-logLik_model
> p_value <- pchisq(chi_square, df = 1, lower.tail = FALSE)
> # Print the results
Analysis of Maximum Likelihood Estimates
> print(coef(summary(model)))
              Estimate Std. Error    z value     Pr(>|z|)
(Intercept) -1.0459686  0.2279802 -4.5879801 4.475552e-06
group        0.2921968  0.3129390  0.9337179 3.504494e-01
-2 Log Likelihood information
Intercept only: 240.8607 Full model: 239.9853 Model Chi-
Square: 0.8754 P-value: 0.3495
The Pseudo R Square is
0.00363446
> cat(sprintf("\nThe Fraction of Fits Correctly Classified
is\n%.3f\n",
+ mean((predict(model, type = "response") >
  0.5) == data$response)))
The Fraction of Fits Correctly Classified is
0.710
```

The results are in complete agreement with those achieved earlier. The Wald z-test is 0.9337, which is very close to the two-sample z-test of proportions seen above, which is 0.934. The model chi-square is the same as the GLRT computed earlier using the special formula.

8.3.4 Count Data: The Two-Sample Problem for Poisson Data

Suppose instead of observing quantal data for each treatment the response is count data. The Poisson distribution is often the appropriate model for this situation. Recall that if the response, y, follows a Poisson distribution with parameter λ –that is, $y \sim P(\lambda)$, that $E(y) = Var(y) = \lambda$. Our model for the two-sample problem would be $y_{ij} \sim P(\lambda_i), i = 1,2$ and $j = 1,\ldots,n_i$. The hypotheses of interest are $H_0 : \lambda_1 = \lambda_2$ versus $H_1 : \lambda_1 \neq \lambda_2$ (or $\lambda_1 > \lambda_2$ or $\lambda_1 < \lambda_2$).

Several approaches are available to deal with this problem. We will list four methods. First, since the MLEs of λ_1 and λ_2 are $\bar{y}_1$ and $\bar{y}_2$, respectively, the point estimator of $\lambda_2 - \lambda_1$ is $\bar{y}_2 - \bar{y}_1$.

Using the variance operator, we know that $\text{Var}(\bar{y}_2 - \bar{y}_2) = \frac{\lambda_2}{n_2} - \frac{\lambda_1}{n_1}$. By the central limit theorem, we know that the large sample distribution of this point estimator is $\bar{y}_2 - \bar{y}_1 \overset{d}{\sim} N\left(\lambda_2 - \lambda_1, \frac{\lambda_2}{n_2} + \frac{\lambda_1}{n_1}\right)$.

Consequently, a reasonable test statistic is

$$z = \frac{\bar{y}_2 - \bar{y}_1}{\sqrt{\frac{\bar{y}_2}{n_2} + \frac{\bar{y}_1}{n_1}}},$$

a statistic with an approximate $N(0,1)$ distribution. This is the two-sample *z*-test for Poisson data.

The second approach is to apply the delta method (see Section 2.4) and the theory of variance stabilizing transformations (see Section 5.2) to suggest that we convert from the original count response, y_{ij}, to the transformed response, y^*_{ij}, by the square-root transformation. That is, we define the transformed response as y^*_{ij}, where $y^*_{ij} = \sqrt{y_{ij}}$.

The theory suggests that $\text{Var}(y^*_{ij}) = \frac{1}{4}$, regardless of the value of λ. In Section 2.4, we argued that $E(y^*_{ij}) = E(\sqrt{y_{ij}}) \approx \sqrt{E(y_{ij})} = \sqrt{\lambda_i}$, at least for λ_i greater than about six. Thus, $\bar{y}^*_2 - \bar{y}^*_1$ is an approximate estimate of $\sqrt{\lambda_2} - \sqrt{\lambda_1}$ with $\text{Var}(\bar{y}^*_2 - \bar{y}^*_1) \approx \frac{1}{4n_1} + \frac{1}{4n_2}$.

Thus, to test $H_0 : \lambda_1 = \lambda_2$ (which is equivalent to testing $H_0 : \sqrt{\lambda_1} = \sqrt{\lambda_2}$), we can use the test

$$z = \frac{\bar{y}^*_2 - \bar{y}^*_1}{\sqrt{\frac{1}{4}\left(\frac{1}{n_1} + \frac{1}{n_2}\right)}},$$

a statistic with an asymptotic $N(0,1)$ distribution.

A third test is based on the GLRT. Using the formulas for –2 log-likelihood in Section 8.2 and a little algebra, we can derive the expression for the GLRT for this problem as

$$GLRT = -2lnL(\lambda) - \left[-2lnL(\hat{\lambda}_1, \hat{\lambda}_2)\right] = 2\left[n_1\bar{y}_1 \ln\left(\frac{\bar{y}}{\bar{y}_1}\right) + n_2\bar{y}_2 \ln\left(\frac{\bar{y}}{\bar{y}_2}\right)\right],$$

a test with an asymptotic chi-square distribution with one degree of freedom.

In the above notation, $\hat{\lambda}$ is the MLE of λ, the common value of λ_1 and λ_2 under the null hypothesis. The MLE of λ is $\bar{y}$, the grand mean.

The fourth approach is to rewrite the two-sample problem as a GLM with $E(y_{ij}) = \lambda_i = F(\beta_0 + \beta_1 z_i)$, where z is an indicator variable set to zero if the observation receives treatment 1 and set to one if the observation receives treatment 2. As in the last example for the two-sample problem for quantal data, we have $E(y_{1j}) = \lambda_1 = F(\beta_0)$ for treatment 1 and $E(y_{2j}) = \lambda_2 = F(\beta_0 + \beta_1)$.

Thus, the parameter β_1 characterizes the difference between λ_1 and λ_2. Consequently, an equivalent test of interest is $H_0 : \beta_1 = 0$, which can be tested with either the Wald *z*-test or the GLRT.

We can easily perform all these tests on R to compute the Wald *z*-test and the GLRT.

8.3.5 Epileptic Seizures Application

In a study of the anti-epileptic drug progabide, 59 patients suffering from epileptic seizures were randomly assigned two treatments. One group with 28 patients received the anti-epileptic drug progabide, while the second group with 31 patients received a placebo. The data are displayed in Table 8.5 (Data from Leppick et al., "A Double-Blind Crossover Evaluation of Progabide in Partial Seizures," *Neurology* 35 (1985): 285.) The variables are the treatment (0 for the placebo and 1 for progabide); the age in years of the patient; the number of epileptic seizures in the eight weeks before administration of the treatment; and the number of seizures in the eight weeks after start of treatment. The data is available on the website in Epileptic_Seizures_Data.xlsx.

One question of interest to address with this data is to compare the two treatment groups.

We would hope that the two groups are nearly alike before the administration of the treatment. To compare them in terms of the mean number of seizures occurring during the eight weeks before the application of the treatment, we assume that this variable is a Poisson variable with a mean λ_i, for $i = 1$ (the placebo group) and $i = 2$ (the progabide group). We want to test the hypotheses $H_0 : \lambda_1 = \lambda_2$ versus $H_1 : \lambda_1 \neq \lambda_2$.

We can use any of the four procedures given above. On R, to obtain the four tests mentioned, type as follows:

```
> # Calculate the mean pre-treatment seizure counts for each
group
> means  <- aggregate(cbind(Placebo_Pre_Count, Drug_Pre_Count)
~ Placebo_Treatment + Drug_Treatment, data = data, FUN = mean,
na.rm = TRUE)
> print(means)
  Placebo_Treatment Drug_Treatment Placebo_Pre_Count Drug_Pre_Count
1         0              1          30.78571       33.21429
> # Z-Test for Difference in Means for the Original Data
```

TABLE 8.5

Epileptic Seizures Data

Subject	Placebo	Age	Pre treat Count	Post treat Count	Drug	Age	Pre treat Count	Post treat Count
1	0	31	11	14	1	18	76	42
2	0	30	11	14	1	32	38	28
3	0	25	6	11	1	20	19	7
4	0	36	8	13	1	30	10	13
5	0	22	66	55	1	18	19	19
6	0	29	27	22	1	24	24	11
7	0	31	12	12	1	30	31	74
8	0	42	52	95	1	35	14	20
9	0	37	23	22	1	27	11	10
10	0	28	10	33	1	20	67	24
11	0	36	52	66	1	22	41	29
12	0	24	33	30	1	28	7	4
13	0	23	18	16	1	23	22	6
14	0	36	42	42	1	40	13	12
15	0	26	87	59	1	33	46	65
16	0	26	50	16	1	21	36	26
17	0	28	18	6	1	35	38	39
18	0	31	11	123	1	25	7	7
19	0	32	118	15	1	26	36	32
20	0	21	20	16	1	25	11	3
21	0	29	12	14	1	22	151	302
22	0	21	9	14	1	32	22	13
23	0	32	17	13	1	25	41	26
24	0	25	28	30	1	35	32	10
25	0	30	55	143	1	21	56	70
26	0	40	9	6	1	41	24	13
27	0	19	10	10	1	32	16	15
28	0	22	47	53	1	26	22	51
29					1	21	25	6
30					1	36	13	0
31					1	37	12	10

```
> mean_placebo <- mean(data$Placebo_Pre_Count, na.rm = TRUE)
> mean_drug <- mean(data$Drug_Pre_Count, na.rm = TRUE)
> se_placebo <- sd(data$Placebo_Pre_Count, na.rm = TRUE) /
sqrt(n_placebo)
> se_drug <- sd(data$Drug_Pre_Count, na.rm = TRUE) /
sqrt(n_drug)
> se_diff <- sqrt(se_placebo^2 + se_drug^2)
> z_value <- (mean_drug-mean_placebo) / se_diff
```

Z-test statistic for original counts: 0.1153263, P-value: 0.9081865

```
> # Z-Test for Difference in Means for the Square Root-
Transformed Data
```

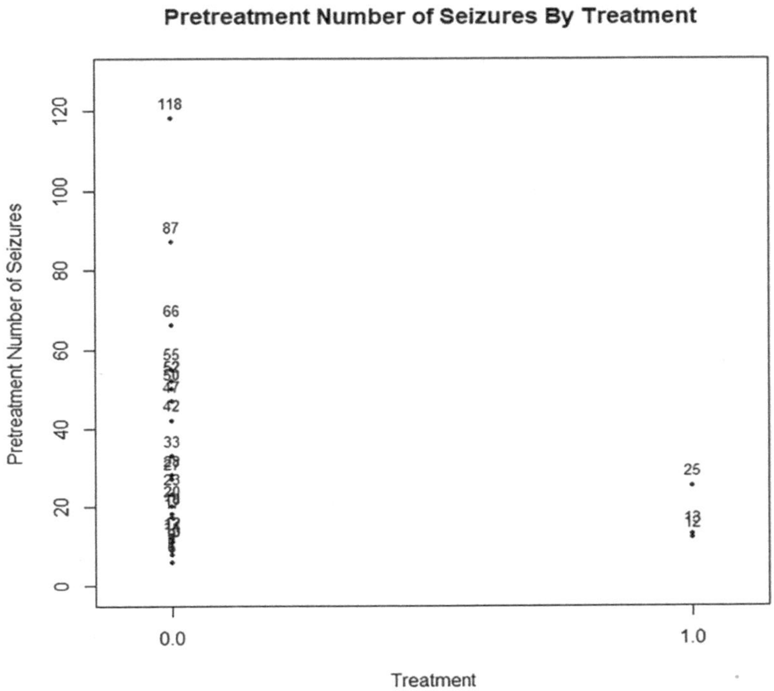

FIGURE 8.15
The scatterplot for the pretreatment number by treatment.

```
> data$Sqrt_Placebo_Pre_Count <- sqrt(data$Placebo_Pre_Count)
> data$Sqrt_Drug_Pre_Count <- sqrt(data$Drug_Pre_Count)
> mean_placebo_sqrt <- mean(data$Sqrt_Placebo_Pre_Count,
na.rm = TRUE)
> mean_drug_sqrt <- mean(data$Sqrt_Drug_Pre_Count, na.rm =
TRUE)
> se_placebo_sqrt <- sd(data$Sqrt_Placebo_Pre_Count, na.rm =
TRUE) / sqrt(n_placebo)
> se_drug_sqrt <- sd(data$Sqrt_Drug_Pre_Count, na.rm = TRUE) /
sqrt(n_drug)
> se_diff_sqrt <- sqrt(se_placebo_sqrt^2 + se_drug_sqrt^2)
> z_value_sqrt <- (mean_drug_sqrt—mean_placebo_sqrt) /
se_diff_sqrt
Z-test statistic for square root transformed counts: 0.258064,
P-value: 0.7963575
```

Figure 8.15 gives the scatterplot of the pretreatment number of seizures versus treatment. There appears to be little difference between the two treatment groups, with both groups having similar means and standard deviations. Hypothesis testing will confirm this observation. Also, we saw from the scatterplot for the square-root transformation that there is very little difference between the two groups.

The third test is based on the GLRT, and the fourth is the Wald z-test as follows:

```
> # Fit the models
> full_model <- glm(Pre_treat_Count ~ factor(Treatment), family =
poisson(link = "log"), data = analysis_data)
> null_model <- glm(Pre_treat_Count ~ 1, family = poisson(link =
"log"), data = analysis_data)
> # Perform GLRT
> glrt_result <- anova(null_model, full_model, test = "Chisq")
> print(glrt_result)
Analysis of Deviance Table
Model 1: Pre_treat_Count ~ 1
Model 2: Pre_treat_Count ~ factor(Treatment)
  Resid. Df Resid. Dev Df Deviance Pr(>Chi)
1        58     1088.4
2        57     1088.0  1   0.32259   0.5701
> # Wald Test from Poisson Regression
>   wald_stat   <-   coef(summary(full_model))["factor(Treatment)1","z
value"]^2
>  wald_p_value     <-    coef(summary(full_model))["factor
(Treatment)1", "Pr(>|z|)"]
Wald test statistic: 0.3224143, P-value: 0.5701604
```

The Wald z-test and the GLRT are in bold above. We note that the two GRLT statistics are equal except for the round-off error. We note that the Wald z (when squared) is nearly equal to the GLRT as expected. The Wald z is nearly equal to the first z-test. The second z-test is quite different from the other tests.

All tests, however, lead to the same conclusion: they fail to reject the null hypothesis and conclude that the average number of seizures during the eight weeks prior to the application of the treatment is the same for both groups.

8.3.6 The *K*-Sample Problem

Suppose now that we have data from k populations, $k \geq 2$, where the data is from any distribution in the exponential family. This situation represents the exponential family counterpart to the one-way ANOVA model for the normal data case that we addressed in Chapter 4. The analysis is based on GLM for exponential family data. One specific example will be illustrated here: the quantal data case.

8.3.7 The *K*-Sample Problem for Quantal Data

To perform the analysis for the k-sample problem for quantal data, simply extend the GLRT from the $k = 2$ situation discussed above to the general

$k \geq 2$ case. That is, let y_{ij} be the 0-1 response variable with y_{ij} equal to one if the jth subject in the sample from the ith population is a "success" and a zero otherwise. The hypotheses of interest are H_0:P_1=P_2=L=P_k=P versus H_1: not H_0, where P_i is the proportion of "successes" in population $i, i = 1,\ldots,k$. The full model has k parameters, and the MLEs are the k individual sample proportions $\hat{P}_i = \bar{y}_{i'}$, the average number of successes in the ith population. The reduced model, the null model, has only one parameter, P, with the MLE $\hat{P}=\bar{y}$, the grand average number of successes over all k populations. If the sample size for each independent sample is n_is and r_i is the number of successes in i^{th} sample, then the total sample size is $n = \sum_{i=1}^{n} n_i$ and the total number of successes is $r = \sum_{i=1}^{n} r_i$.

The GLRT is easy to derive for this case and can be expressed as

$$GLRT = -2\left[\ln L\left(\hat{P}\right) - \ln L\left(\hat{P}_{1,}, \hat{P}_2, \ldots, \hat{P}_k\right)\right] = -2\left[r \ln\left(\frac{\hat{P}}{1-\hat{P}}\right) + n \ln\left(1-\hat{P}\right)\right.$$

$$\left.- \sum_{i=1}^{k}\left[r_i \ln\left(\frac{\hat{P}_i}{1-\hat{P}_i}\right) + n_i \ln\left(1-\hat{P}_i\right)\right]\right].$$

The GLRT can be computed using R. Perhaps the better approach is to use indicator variables to write the k-sample problem as a GLM. This is the approach taken in Section 6.4 for the normal data case and used for the quantal data case for the $K = 2$ sample problem, illustrated earlier in this section.

Following the setup of Section 6.4, if we have k populations, we need $k-1$ indicator variables, denoted by z_j, where j represents the jth indicated population.

To set a more specific example, suppose $k = 3$ and the researcher has determined that population 3 will be the reference population, indicated when both z_1 and z_2 are zero.

Let z_1 indicate population 1 and z_2 indicate population 2. Then our GLM is $E\left(y_{ij}\right) = P_i = F\left(\beta_0 + \beta_1 z_1 + \beta_2 z_2\right)$. When $z_1 = z_1 = 0$, our model states that $E\left(y_{3j}\right) = P_3 = F\left(\beta_0\right)$. When $z_1 = 1$ and $z_1 = 0$, the GLM states that $E\left(y_{1j}\right) = P_1 = F\left(\beta_0 + \beta_1\right)$.

Implying that the difference between P_1 and P_3 is represented by β_1.

We know from previous discussions that β_1 is the log odds ratio, $\beta_1 = \ln\left(\dfrac{\frac{P_1}{1-P_1}}{\frac{P_3}{1-P_3}}\right)$. We can see that if $P_3 = P_1$, then $\beta_1 = 0$. Similarly, when $z_1 = 0$ and $z_1 = 1$, the GLM states that $E\left(y_{2j}\right) = P_2 = F\left(\beta_0 + \beta_2\right)$, implying that the difference between P_2 and P_3 is β_2. Since $\beta_2 = \ln\left(\frac{\frac{P_2}{1-P_2}}{\frac{P_3}{1-P_3}}\right)$, it follows that if $P_2 = P_3$, then $\beta_2 = 0$.

Thus, an equivalent test of the null hypothesis $H_0: P_1 = P_2 = P_3 = P$ is $H_0: \beta_1 = \beta_2 = 0$. The GLRT is just the model chi-square test for this GLM.

The Wald z-tests can be used to make the comparisons between pairs of population proportions. For example, the Wald z-test for β_1 is testing the null hypotheses $H_0: P_1 = P_3$.

8.3.8 The Smoking Habits Example (Revisited)

Recall the smoking habit example presented earlier in this section. As part of the study, the students also sampled 100 male and 100 female students at two other universities. At universities RU and QU, it was found that 24% and 16% of the sample of males smoked, respectively.

Does this information suggest that there is a difference in the proportion of males that smoke at these three universities?

Setting up the problem as a GLM using VT as the reference university, we have $\hat{P}_3 = .32$ for VT, $\hat{P}_1 = .24$, for RU, and $\hat{P}_2 = .16$, for QU.

The indicator z_1 indicates RU, and z_2 indicates QU. On R, type as follows:

```
> # Create the dataset
> # VT: 32 smokers out of 100, RU: 24 smokers out of 100, QU:
16 smokers out of 100
> smoking_data <- data.frame(
+ smoke = c(rep(1, 32), rep(0, 68), rep(1, 24), rep(0, 76),
rep(1, 16), rep(0, 84)), + university = factor(c(rep("VT",
100), rep("RU", 100), rep("QU", 100)), + levels = c("VT",
"RU", "QU"))+ )
> # Create indicator variables
> smoking_data <- smoking_data %>%
+   mutate(RU = if_else(university == "RU", 1, 0),
+          QU = if_else(university == "QU", 1, 0))
> logistic_model <- glm(smoke ~ RU + QU, family =
binomial(link = "logit"), data = smoking_data)
> model_summary <- summary(logistic_model)
> print(model_summary)
Call:
glm(formula = smoke ~ RU + QU, family = binomial(link =
  "logit"),
data = smoking_data)
Coefficients:
            Estimate Std. Error z value Pr(>|z|)
(Intercept)  -0.7538     0.2144  -3.516 0.000438 ***
RU           -0.3989     0.3175  -1.257 0.208912
QU           -0.9045     0.3469  -2.607 0.009133 **
```

```
---
(Dispersion parameter for binomial family taken to be 1)
    Null deviance: 330.65 on 299 degrees of freedom
Residual deviance: 323.52 on 297 degrees of freedom
AIC: 329.52
Number of Fisher Scoring iterations: 4
> # Calculate the p-value for the model using anova (GLRT)
> glrt_result <- anova(logistic_model, test = "Chisq")
> print(glrt_result)
Analysis of Deviance Table
Model: binomial, link: logit
Response: smoke
Terms added sequentially (first to last)
     Df Deviance Resid. Df Resid. Dev Pr(>Chi)
NULL                  299     330.65
RU    1   0.0000      298     330.65 1.000000
QU    1   7.1241      297     323.52 0.007605 **
---
Signif. codes: 0 '***' 0.001 '**' 0.01 '*' 0.05 '.' 0.1 ' ' 1
intercept only: 330.6480 full model: 323.5239 model chi-
square: 7.1241 p-value: 0.0284 dfmodel: 2
> cat("Coefficients' p-values:\n")
Coefficients' p-values:
> print(coefficients_p_values)
(Intercept)          RU          QU
0.000437826 0.208911628 0.009133241
The Fraction of Fits Correctly Classified is
fcc 0.760
```

We find that the GLRT = model chi-square is significant at p-value = 0.0284. So at least one Pi is different. The Wald test reveals the P2 is significantly less than P3 (less smoking by males at QU than at VT) and that P1 = P3 (no difference in smoking by males at RU and VT). If we want to compare RU to UV, we must rerun the GLM with either RU or QU (instead of VT) as the reference. This is a simple matter of redefining the indicator variables. When this was run, (the results not shown here), it was shown that P1 = P2 (no difference in smoking by males at RU and QU). Thus, the ordering of the population proportions is (QU = RU) < (RU = VT).

We have illustrated several applications of GLM using quantal data and count data. The interested reader should refer to Kleinbaum, Kupper, and Muller (1988) and Ramsey and Schafer (2002) among others for applications of more complex logistic and Poisson regression models.

Exercises

1) In a random sample of students at Qatar University (QU), 32 out of 100 students said that they would attend the FIFA 2022 in Doha. Let the random variable X be one if the student will attend the game and zero otherwise. Let P represent the population proportion of

students at QU who will attend this football game. Answer the following questions.

For all parts of this problem, do all calculations by hand.

a) i) Derive the MLE of P. ii) What is the $E(\hat{P})$? iii) What is the $Var(\hat{P})$?.

b) What test statistic would you use to test $H_0 : P = P_0$ versus $H_1 : P \neq P_0$, provided that the sample size, n, is large? What is its sampling distribution? Exclude the GLRT, as it will be used in part vi).

c) Using the information given in this problem, what is the numerical value of your test in part b) for testing $H_0 : P = 0.5$ versus $H_0 : P \neq 0.5$. Base your conclusion on the p-value (bounds on the p-value may be necessary).

d) Now test the same hypotheses as in part c) above using the GLRT test. Base your conclusion on the p-value (bounds on the p-value may be necessary).

e) Compare your results in parts c) and d) above. Should they be the same or different?

Explain briefly.

2) Consider the following situation where we have a 0-1 response variable y, one regressor x, and a 0-1 variable. The data is summarized in Table 8.6.

TABLE 8.6

Cross-Table for 50 Observations

x				
		0	**1**	**Total**
y		12	13	25
		18	7	25
	Total	30	20	50

Let $P_0 = \text{Prob}\,(y = 1 \mid x = 0)$ and $P_1 = \text{Prob}(y = 1 \mid x = 1)$. For all parts of this problem, do all calculations by hand.

a) What are the numerical values of the MLEs of P_0 and P_1 $(\hat{P}_0$ and $\hat{P}_1)$?

b) Test the hypothesis: $H_0 : P_0 = P_1$ and $H_1 : P_0 \neq P_1$ using the two-sample tests of proportions. Base your conclusion on the p-values (bounds may be necessary).

Now consider the logistic regression model $P(\underline{x}_i) = \text{Prob}(y_i = 1 \mid \underline{x}_i) = F(\beta_0 + \beta_1 x_i)$.

Answer the following questions.

c) Find the value of $-2\ln L\left(\hat{\beta}_0, \hat{\beta}_1, \underline{x}_i\right)$ for the logistic model.

d) Use the GLRT to test the same hypotheses as in part b) above. State the test statistic and its distribution. Evaluate the test statistic, and state your conclusions, basing your result on the *p*-value.

e) Compare your answers from parts b) and d). Should they be equal? Explain?

3) Consider the linear model $\underline{y} = X\underline{\beta} + \underline{\varepsilon}$, where $\underline{\varepsilon} \sim N\left(\underline{0}, \sigma^2 I\right)$ with σ^2 known. Recall that the normal density for ε_i is written as

$$f(\varepsilon_i) = \frac{1}{\sqrt{2\pi\sigma^2}} \exp\left[-\frac{1}{2}\left(\frac{y_i - \underline{x}_i{}'\underline{\beta}}{\sigma}\right)^2\right].$$

a) Find an expression for $-2\ln L\left(\hat{\underline{\beta}}, \underline{x}_i\right)$, where $\hat{\underline{\beta}}$ is the MLE of $\underline{\beta}$.

b) Test the hypotheses: $H_0 : \beta_1 = \ldots = \beta_k$ versus H_1: not H_0 using the GLRT. Be sure to state the complete test statistic and its distribution, including all parameters. State how you would obtain the *p*-value for this test.

4) Consider the following situation where we have a 0-1 response variable *y*, one regressor *x*, and a 0-1 variable. The data is summarized in Table 8.7.

TABLE 8.7

Cross-Table for 90 Observations

x				
		0	**1**	
y	0	15	12	40
	1	15	48	50
		30	60	90

Let $P_0 = \text{Prob}\,(y = 1 \mid x = 0)$ and $P_1 = \text{Prob}\left(y = 1 \mid x = 1\right)$. Do all calculations by hand.

a) What are the numerical values of the MLEs of P_0 and P_1 $\left(\hat{P}_0 \text{ and } \hat{P}\right)$?

Consider the logistic regression model $P(\underline{x}_i) = \text{Prob}\left(y_i = 1 \mid \underline{x}_i\right) = F(\beta_0 + \beta_1 x_i)$. Answer the following questions.

b) Write P_0 and P_1 in terms of $P(\underline{x}_i)$.

c) Write β_0 and β_1 in terms of some functions of P_0 and P_1.

d) Find the value of $-2\ln L\left(\hat{\beta}_0, \hat{\beta}_1, \underline{x}_i\right)$ for the logistic model.

e) Argue that testing the hypotheses $H_0 : P_0 = P_1$ and $H_1 : P_0 \neq P_1$ is equivalent to testing the hypotheses $H_0 : \beta_1 = 0$ and $H_1 : \beta_1 \neq 0$.

f) Use the GLRT to test $H_0 : \beta_1 = 0$ and $H_1 : \beta_1 \neq 0$. State the test statistic and its distribution. Evaluate the test statistic, and state your conclusions, basing your result on the p-value.

g) It is a fact that the $\text{Var}\left(\text{log}it\left(\hat{P}_i\right)\right) = \dfrac{1}{n_i P_i (1 - P_i)}$, where $F(u)$ is the standard logistic CDF. Use this fact to find an expression for $\text{Var}\left(\hat{\beta}_1\right)$. Use this expression to compute the Wald chi-squared test for $H0{:}\beta_1 = 0$ and $H_1 : \beta_1 \neq 0$. What is the p-value, and state your conclusion.

5) Consider Poisson regression as an example of the GLM. Specifically, suppose that the random variable y_i follows a Poisson distribution with parameter $\lambda_i = F\left(\underline{x}'_i\, \underline{\beta}\right)$, where $F(u)$ is some arbitrary function of regressors, $\underline{x}'_i = (1 \quad x_{1i} \quad \ldots \quad x_{ki})$. Suppose we have a random sample, $y_1, \ldots, y_n$. Answer the following questions. Do all calculations by hand.

a) Give an expression for the log-likelihood, $\ln L\left(\underline{\beta}; \underline{x}_i\right)$.

b) Using part a) above, derive an expression for $\dfrac{\partial \ln L(\beta; x_i)}{\partial \underline{\beta}}$ that you would use to solve the MLE of $\underline{\beta}$. Be sure that your expression is in the form we used in class.

c) Recall the IRLS algorithm used to obtain the MLE of β. In the algorithm, how would you compute each w_i, for $i = 1, \ldots n$?

d) Suppose that you are using $F\left(\underline{x}_i \underline{\beta}\right) = e^{\underline{x}'_i \underline{\beta}}$, a commonly used function to model λ_i. Consider the model $\lambda_i = F(\beta_0) = e^{\beta_0}$. Write out the formula for the MLE of β_0. Be sure to justify that your formula is truly the MLE.

e) Consider the following Table 8.8 data, where y represents the number of fractures per subregion of a mine and x represents the age of the mine, where 0 indicates an "old" mine and 1 indicates a "new" mine.

TABLE 8.8

The Number of Fractures per Subregion

Obs.	1	2	3	4	5	6	7	8	9	10	11	12	13	14	15	16	17	18	19	20
y	2	1	0	4	1	2	0	0	4	4	1	4	1	5	2	5	5	2	0	2
x	0	1	0	0	0	1	0	0	0	0	1	0	0	0	1	0	0	1	1	1

Summary information: $\bar{y}_0 = \frac{35}{13}, \bar{y}_1 = \frac{10}{7}$, and $\bar{y} = \frac{45}{20}$.

i. Using the x variable as a regressor and the Poisson regression model with $\lambda_x = F(\beta_0 + \beta_1 x) = e^{\beta_0 = \beta_1 x}$. Find the numerical values of the MLEs of β_0 and β_1.

ii. Derive the formula (in a nice, simplified form) for the appropriate GLRT for testing $H_0 : \lambda_0 = \lambda_1$ and $H_1 : \lambda_0 \neq \lambda_1$. Evaluate this test, and state your decision based on a guess of the p-value.

6) Consider the logistic regression model $P_i = F(\underline{x}_i \beta)$, and define the log-likelihood for y_i as $L_i(P_i) = y_i \ln(P_i) + (1 - y_i)\ln(1 - P_i)$, where y_i is either zero or one. Further, define the "deviance component," d_i, for observation y_i as $d_i = 2\,L(y_i) - 2\,L(\hat{P}_i)$. Finally, define the "deviance" as $deviance = \sum_{i=1}^{n} d_i$, the sum of the deviance components. Show that the $deviance = -2lnL(\hat{\beta}, \underline{x}_i)$, where $L(\hat{\beta}, x_i)$ is the likelihood function for $y_1, \ldots, y_n$, as defined before.

7) In studying a production process, three regressor variables are observed and a response variable. The response is the status of a component produced by the process: either defective ("1") or non-defective ("0"). Suppose that in 50 observations, we note that 32 are defective. Let Prob(response $= 1) = P(\underline{x}_i) = F(\underline{x}_i \beta')$ denote our quantal regression model.

a) Consider the model $P(\underline{x}_i) = F(\beta_0)$, the "intercept-only model."

i. What is the numerical value of the MLE of $\beta_0, \hat{\beta}_0$, if $F(u)$ is the logistic CDF?

ii. What is the numerical value of the MLE of $\beta_0, \hat{\beta}_0$, if $F(u)$ is the normal CDF?

iii. What is the numerical value of the MLE of $\beta_0, \hat{\beta}_0$, if $F(u)$ is the uniform CDF on the interval from 0 to 1?

iv. For this model and $F(u)$ equal to the logistic CDF, find the fraction correctly classified if $\hat{P}(\underline{x}_i) = F(\hat{\beta}_0)$ (Use $F(\hat{\beta}_0) \geq .5$ as the cutoff for a "defective").

For the remaining parts of this problem, assume that $F(u)$ is the logistic CDF.

b) The following lists values of $-2\ln L(\hat{\beta}; \underline{\underline{x}})$ for a number of different models using combinations of the three regressors:

$-2\ln L\left(\hat{\beta}_0\right) = 65.34$, $-2\ln L\left(\hat{\beta}_0, \hat{\beta}_1; x_1\right) = 61.8$, $-2\ln L\left(\hat{\beta}_0, \hat{\beta}_2; x_2\right) = 63.4$,
$-2\ln L\left(\hat{\beta}_0, \hat{\beta}_3; x_3\right) = 59.4$, $-2\ln L\left(\hat{\beta}_0, \hat{\beta}_2, \hat{\beta}_3; x_2, x_3\right) = 55.4$,
$-2\ln L\left(\hat{\beta}_0, \hat{\beta}_1, \hat{\beta}_2, \hat{\beta}_3; x_1, x_2, x_3\right) = 46.8$

i. Test $H_0 : \beta_1 = \beta_2 = \beta_3 = 0$ versus H_1: not H_0. State the distribution of your test statistic. Evaluate it, and give the p-value.

ii. What is the best single-variable model? Justify your answer.

iii. Test the hypotheses:$H_0 : \beta_2 = \beta_3 = 0$ versus H_1: not H_0, in the presence of x_1.

Be sure to state the test statistic and its distribution. Find the p-value for the test.

8) A geneticist is interested in the proportion of males and females in a population that has a certain minor blood disorder. In a random sample of 100 males, 31 were found to be afflicted, whereas only 24 of 100 females tested appeared to have the disorder. Can it be concluded that the proportion of men in the population afflicted with this blood disorder is significantly greater than the proportion of women afflicted?

a) Use the z-test to test the above hypothesis. Find the p-value.

b) Use the GLRT to test the above hypothesis. Calculate the GLRT by hand and by using the R program. Compare your results to those obtained in part a).

9) A study was made to determine whether more Italians than Americans prefer white champagne to pink champagne at weddings. Of the 300 Italians selected at random, 72 preferred white champagne, and of the 400 Americans selected, 70 preferred white champagne rather than pink. Can we conclude that a high proportion of Italians rather than Americans prefer white champagne at weddings?

a) Use the z-test to test the above hypothesis. Find the p-value.

b) Use the GLRT to test the above hypothesis. Calculate the GLRT by hand and by using the R program. Compare your results to those obtained in a).

10) The data (Table 8.9) lists the number of suicides in a large city by month for the 12 months of a recent year. Also listed are the average temperature and percent of the time it was cloudy during the day for the same 12 months. (Data from Dr. George Terrell, Virginia Tech).

TABLE 8.9

The Number of Suicides in a Large City by Month

month	number of suicides	ave. temp	percent cloudy
1	7	55	48
2	12	43	75
3	11	35	56
4	15	41	63
5	9	47	80
6	16	59	75
7	6	68	33
8	4	71	25
9	9	74	38
10	8	72	21
11	18	65	74
12	16	59	65

Use your software package and Poisson regression (with the log link) to predict the suicide rate. Evaluate your model using the methods discussed in this chapter. Discuss your results.

Multiple-Choice Questions

1. What type of distribution is often used to model a binary response variable (e.g., success or failure)?
   a) Normal distribution
   b) Poisson distribution
   c) Bernoulli distribution
   d) Exponential distribution

   Answer: c)

2. Which of the following is not a characteristic of the generalized linear model (GLM)?
   a) The response variable is normally distributed.
   b) The response variable can follow an exponential family of distributions.
   c) The link function connects the linear predictor to the mean of the distribution function.
   d) It includes both linear regression and logistic regression as special cases.

   Answer: a)

3. In the context of logistic regression, which function is commonly used as the link function?
   a) Identity function
   b) Logarithm function
   c) Logit function
   d) Exponential function

   Answer: c)
4. What is the main issue with using the least squares method in the context of binary response variables?
   a) It assumes homoscedasticity.
   b) It can predict probabilities outside the [0, 1] range.
   c) It does not account for multicollinearity.
   d) It cannot be used for continuous response variables.

   Answer: b)
5. Which of the following models is suitable for counting data?
   a) Logistic regression
   b) Linear regression
   c) Poisson regression
   d) Probit regression

   Answer: c)
6. What is the purpose of the link function in a GLM?
   a) To transform the predictor variables
   b) To linearize the relationship between the predictors and response variable
   c) To ensure the response variable is normally distributed
   d) To standardize the predictor variables

   Answer: b)
7. Which of the following is a common method used to estimate the parameters in a logistic regression model?
   a) Ordinary least squares (OLS)
   b) Ridge regression
   c) Maximum likelihood estimation (MLE)
   d) Bayesian estimation

   Answer: c)
8. In a logistic regression model, if the coefficient of a predictor variable is positive, what does this indicate about the relationship between the predictor and the probability of success?

a) The predictor decreases the probability of success.
b) The predictor has no effect on the probability of success.
c) The predictor increases the probability of success.
d) The predictor is not significant.

Answer: c)

9. The deviance in a GLM is used to measure
   a) The correlation between predictor variables.
   b) The fit of the model compared to a saturated model.
   c) The standard deviation of residuals.
   d) The variance is explained by the model.

   Answer: b)

10. What is pseudo R^2 in the context of GLM?
    a) A measure of the correlation between predicted and observed values
    b) A measure used for model comparison in GLMs
    c) An exact measure of the model's fit similar to the R^2 in linear regression
    d) A measure of the model's predictive power for continuous variables

    Answer: b)

References

Abdel-Salam, A. S. G., Birch, J. B., & Jensen, W. A. (2013). A semiparametric mixed model approach to phase I profile monitoring. *Quality and Reliability Engineering International, 29*(4), 555–569.

Agard, D. B. (1991). *Robust inferential procedures applied to regression* [Unpublished PhD dissertation. Department of Statistics, Virginia Polytechnic Institute and State University].

Aldrich, J. H., & Nelson, F. D. (1984). *Linear probability, logit, and probit models. Sage university papers: Quantitative applications in the social sciences, 07-045*. Sage.

Alquier, P., & Gerber, M. (2024). Universal robust regression via maximum mean discrepancy. *Biometrika, 111*(1), 71–92.

Andrews, D. R. (1974). A robust method for multiple linear regression. *Technometrics, 16*, 523–531.

Andrews, D. R., Bickel, P. J., Hampel, F. R., Huber, P. J., Rogers, W. H., & Tukey, J. W. (1972). *Robust estimates of location: Survey and advances*. Princeton University Press.

Assaid, C. A., & Birch, J. B. (1998). *Outlier resistant model robust regression. Technical report number 78*, Virginia Polytechnic Institute and State University.

Assaid, C. A., & Birch, J. B. (2000). Automatic bandwidth selection in robust nonparametric regression. *Journal of Statistical Computation and Simulation, 66*, 259–272.

Bartlett, M. S. (1947). The use of transformations. *Biometrics, 3*, 39–52.

Beaton, A. E., & Tukey, J. W. (1974). The fitting of power series, meaning polynomials, illustrated on band spectroscopic data. *Technometrics, 16*, 147–185.

Bickel, P. J. (1973). On some analogies to linear combination of order statistics in the linear model. *Annals of Statistics, 1*, 597–616.

Bickel, P. J. (1975). One-step Huber estimates in the linear model. *Journal of the American Statistical Association, 70*, 428–433.

Bickel, P. J., & Doksum, K. (1977). *Mathematical statistics: Basic ideas and selected topics*. Holden-Day.

Birch, J. B. (1980). Some convergence properties of iterated reweighted least squares in the location model. *Communications in Statistics B, 9*(4), 359–369.

Birch, J. B. (1982). Robust analysis of covariance. *Biometrics, 38*, 699–713.

Birch, J. B. (1983). On the power of robust tests in analysis of covariance. *Communications in Statistics B, 12*, 59–82.

Birch, J. B. (1992). *Estimation and inference in multiple linear regression using robust weights. Technical report number 92-11*. Virginia Tech, Department of Statistics.

Birch, J. B., & Agard, D. B. (1993). Robust inference in multiple regression: A comparative study. *Communications in Statistics, 22*(1), 217–244.

Birch, J. B., & Fleischer, S. (1984). Diagnostic methods for detecting outliers in regression analysis. *Environmental Entomology, 13*(1), 19–25.

Blom,G. (1958). *Statistical estimates and transformed beta-variables*. John Wiley & Sons.

Box, G. E. P., & Cox, D. R. (1964). An analysis of transformations (with discussion). *Journal of the Royal Statistical Society, Series B, 26*, 211–246.

Brownlee, K. A. (1965). *Statistical theory and methodology in science and engineering* (2nd ed.). John Wiley & Sons.

Burkhart, Harold & Parker, R. & Oderwald, R.. (1972). *Yields for natural stands of loblolly pine. Div. For. Wildl. Resour.* Va. Polytech. Inst. State Univ. Publ. FWS-2-72.
Cleveland, W. S. (1979). Robust locally weighted regression and smoothing scatterplots. *Journal of the American Statistical Association, 74*, 829–836.
Conover, W. J. (1971). *Practical Nonparametric Statistics.* John Wiley & Sons.
Conover, W. J., & Iman, R. L. (1981). Rank transformations as a bridge between parametric and nonparametric statistics. *The American Statistician, 35*(3), 124–129.
Cook, R. D., & Weisberg, S. (1982). *Residuals and Influence in Regression.* Chapman and Hall.
Crawley, M. J. (2012). *The R book.* John Wiley & Sons.
Dallal, G. E., & Wilkinson, L. (1986). An analytic approximation to the distribution of Lilliefor's test statistics for normality. *The American Statistician, 40*, 294–296.
Daniel, C., & Wood, F. S. (1980). *Fitting equations to data* (2nd ed.). John Wiley & Sons.
Deaton, M. L., Reynolds, M. R., Jr., & Myers, R. H. (1983). Estimation and hypothesis testing in regression in the presence of nonhomogeneous error variance. *Communications in Statistics B, 12*(1), 45–66.
Efron, B., & Tibshirani, R. J. (1993). *An introduction to the bootstrap.* Chapman and Hall.
Fan, J., & Gijbels, I. (1996). *Local polynomial modeling and its applications.* Chapman and Hall.
Filliben, J. J. (1975). The probability plot correlation coefficient test for normality. *Technometrics, 17*, 111–118.
Finney, D. J. (1978). *Probit anaysis* (3rd ed.). The University Press.
Finney, D. J. (1978). *The statistics of biological assays* (3rd ed.). MacMillan Publishing Company.
Gareth, J., Daniela, W., Trevor, H., & Robert, T. (2013). *An introduction to statistical learning: With applications in R.* Springer.
Gelman, A., & Hill, J. (2006). *Data analysis using regression and multilevel/hierarchical models.* Cambridge University Press.
Gomaa, A. S., & Birch, J. B. (2019). A semiparametric nonlinear mixed model approach to phase I profile monitoring. *Communications in Statistics-Simulation and Computation, 48*(6), 1677–1693.
Hamilton, L. C. (1992). *Regression with graphics: A second course in applied statistics.* Brooks/Cole.
Hampel, F. R. (1971). A general qualitative definition of robustness. *Annals of Mathematical Statistics, 42*, 1887–1896.
Hampel, F. R. (1974). The influence curve and its role in robust estimation. *Journal of the American Statistical Association, 69*, 383–393.
Hampel, F. R., Ronchetti, E. M., Rousseeuw, P. J., & Stahel, W. J. (1986). *Robust statistics, the approach based on influence functions.* John Wiley & Sons.
Hardle, W. (1991). *Applied nonparametric regression.* Cambridge University Press.
Hastie, T., & Tibshirani, R. (1990). *Generalized additive models.* Chapman and Hall.
Hastie, T., Tibshirani, R., Friedman, J. H., & Friedman, J. H. (2009). *The elements of statistical learning: data mining, inference, and prediction* (Vol. 2, pp. 1–758). Springer.
Hettmansperger, T. P., & Sheather, S. J. (1992). A cautionary note on the method of least median squares. *The American Statistician, 46*(2), 79–84.
Hoaglin, D. C., Mosteller, F., & Tukey, J. W. (1983). *Understanding robust and exploratory data analysis.* John Wiley & Sons.
Holland, P. W., & Welsch, R. E. (1977). Robust regression using iteratively reweighted least squares. *Communications in Statistics (Theory and Methods), 6*, 813–828.

Hong, S., & Reed, W. R. (2021). Using Monte Carlo experiments to select meta-analytic estimators. *Research Synthesis Methods, 12*(2), 192–215.

Hosmer, D. W., Jr., & Lemeshow, S. (1989). *Applied logistic regression*. John Wiley and Sons.

Huber, P. J. (1964). Robust estimation of a location parameter. *Annals of Mathematical Statistics, 35*, 73–101.

Huber, P. J. (1972). Robust statistics: A review. *Annals of Mathematical Statistics, 43*, 1041–1067.

Huber, P. J. (1973). Robust regression: Asymptotics, conjectures, and Monte Carlo. *Annals of Statistics, 1*, 799–821.

Huber, P. J. (1981). *Robust statistics*. John Wiley and Sons.

Iman, R. L. (1994). *A data-based approach to statistics*. Duxbury Press.

Johnson, R. A., & Wichern, D. W. (2007). *Applied Multivariate Statistical analysis*. Prentice Hall.

Jones, E., Harden, S., & Crawley, M. J. (2022). *The R book*. John Wiley & Sons.

Kleinbaum, D. G., Kupper, L. L., & Muller, K. E. (1988). *Applied regression analysis and other multivariable methods*. PWS-Kent.

Koopmans, L. H. (1987). *Introduction to contemporary statistical methods* (2nd ed.). PWS Publishers.

Krasker, W. S., & Welsch, R. E. (1982). Efficient bounded-influence regression estimation. *Journal of the American Statistical Association, 77*, 595–604.

Larsen. R. J., & Marx, M. L. (1981). *An introduction to mathematical statistics and its applications*. Prentice-Hall.

Lilliefors, H. W. (1967). On the Kolmogorov-Smirnov test for normality with mean and variance unknown. *Journal of the American Statistical Association, 62*, 399–402.

Mallows, C. L. (1973). Some comments on. *Technometrics, 15*, 661–675.

Mallows, C. L. (1975). *On some topics in robustness. Unpublished Bell Telephone laboratories report*. Murray Hill.

Mandansky, A. (1988). *Prescriptions for working statisticians*. Springer-Verlag.

Markatou, M., & Hettmansperger, T. P. (1990). Robust bounded influence tests in linear models. *Journal of the American Statistical Association, 85*, 187–190.

Maronna, R. A., Martin, R. D., Yohai, V. J., & Salibián-Barrera, M. (2019). *Robust statistics: Theory and methods (with R)*. John Wiley & Sons.

Mason, A. L., & Bell, C. B. (1986). New Lilliefors and Srinivasan tables with applications. *Communications in Statistics: Simulation and Computation, 15*(2), 451–477.

Mays, J. E., Birch, J. B., & Einsporn, R. L. (2000). An overview of model robust regression. *Journal of Statistical Computation and Simulation, 66*, 79–100.

Mays, J. E., Birch, J. B., & Starnes, B. A. (2001). Model robust regression: Combining parametric, nonparametric, and semiparametric methods. *Journal of Nonparametric Statistics, 12*(2), 245–277.

McCullagh, P., & Nelder, J. A. (1989). *Generalized linear models* (2nd ed.). Chapman and Hall.

MINITAB User's Guide: Release 10 Xtra for Windows and Macintosh. (1995).

Mood, A. M., Graybill, F. A., & Boes, D. C. (1974). *Introduction to the theory of statistics* (3rd ed.). McGraw-Hill.

Moore, D. S. (1995). *The basic practice of statistics*. W. H. Freeman and Company.

Moore, D. S., & McCabe, G. P. (1998). *Introduction to the practice of statistics*. W. H. Freeman and Company.

Mosteller, F., & Tukey, J. W. (1977). *Data analysis and regression*. Addison-Wesley.

Mount, J., & Zumel, N. (2019). *Practical data science with R*. Simon and Schuster.

Myers, R. H. (1976). *Response surface methodology*. Published by author.

Myers, R. H. (1990). *Classical and modern regression with applications* (2nd ed.). PWS-Kent.

Neter, J., Natcheshim, C., Wasserman, W., & Kutner, M. H. (1996). *Applied linear statistical models: Regression, analysis of variance, and experimental designs* (4th ed.). Irwin.

Nottingham, Q. J., & Birch, J. B. (2000). A semiparametric approach to analyzing dose-response data. *Statistics in Medicine, 19,* 389–404.

Ott, R. L., & Longnecker, M. (2001). *An introduction to statistical methods and data analysis* (5th ed.). Duxbury Press.

Pearson, E. S., & Hartley, H. O. (1958). *Biometrika tables for statisticians* (Vol. I). Cambridge University Press.

Pearson, K. (1900). On the criterion that a given system of deviations from the probable in the case of a correlated system of variables is such that it can be reasonably supposed to have arisen from random sampling. *Philosophical Magazine, 5*(50), 157–172.

Pendelton, O. J., & Hocking, R. R. (1981). Diagnostics techniques in multiple linear regression using proc matrix. *SUGI,* 195–201.

Pregibon, D. (1981). Logistic regression diagnostics. *Annals of Statistics, 9*(4), 705–724.

Ramsey, F. L., & Shafer, D. W. (2002). *The statistical Sleuth-A course in methods of data analysis* (2nd ed.). Duxbury Press.

Rao, C. R. (1965). *Linear statistical inference and its applications.* John Wiley & Sons.

Ripley, B. D. (2002). *Modern applied statistics with S.* Springer.

Robinson, T. J., & Birch, J. B. (1998). *Dual model robust regression. Technical report number 78.* Virginia Polytechnic Institute and State University.

Ronchetti, E. (1982). *Robust testing in linear models: The infinitesimal approach* [Unpublished PhD dissertation, ET Zurich].

Rousseeuw, P. J., & Leroy, A. M. (1987). *Robust regression and outlier detection.* John Wiley & Sons.

Ryan, T. P. (1997). *Modern regression methods.* John Wiley and Sons.

Salibian-Barrera, M. (2023). Robust nonparametric regression: Review and practical considerations. *Econometrics and Statistics.* https://doi.org/10.1016/j.ecosta.2023.04.004.

Schabenberger, O., & Pierce, F. J. (2002). *Contemporary statistical models for the plant and soil sciences.* CRC Press.

Schrader, R. M., & Hettmansperger, T. P. (1980). Robust analysis of variance based upon a likelihood ratio criterion. *Biometrika, 67,* 93–101.

Schrader, R. M., & McKean, J. W. (1977). Robust analysis of variance. *Communications in Statistics A, 6,* 879–894.

Sen, P. M. (1982). On M tests in linear models. *Biometrika, 69,* 245–248.

Shao, J., & Tu, D. (1995). *The jackknife and bootstrap.* Springer-Verlag.

Shapiro, S. S., & Wilk, M. B. (1965). An analysis of variance test for normality (complete samples). *Biometrika, 52,* 591–611.

Shapiro, S. S., Wilk, M. B., & Chen, H. J. (1968). A comparative study of various tests for normality. *Journal of the American Statistical Association, 63,* 1343–1372.

Simpson, D. G., Ruppert, D., & Carroll, R. J. (1992). On one-step GM-estimates and stability of inference in linear regression. *Journal of the American Statistical Association, 87,* 439–450.

Speakman, P. (1988). Kernel smoothing in partial linear models. *Journal of the Royal Statistical Society, Series B, 50,* 413–436.

Staudte, R. G., & Sheather, S. J. (1990). *Robust estimation and testing.* John Wiley & Sons.

Stigler, S. M. (1977). Do robust estimators work with real data? *Annuals of Statistics, 5*, 1055–1098.
Tukey, J. W. (1970). *Exploratory data analysis* (Limited preliminary ed.). Addison-Wesley.
Tukey, J. W. (1977). *Exploratory and data analysis*. Addison-Wesley.
Vahdat, K., & Shashaani, S. (2022). Robust prediction error estimation with Monte-Carlo methodology. *arXiv preprint arXiv:2207.13612*.
Verma, J. P., & Abdel-Salam, A. S. G. (2019). *Testing statistical assumptions in research*. John Wiley & Sons.
Waterman, M. J., Birch, J. B., & Abdel-Salam, A. S. G. (2015). Several nonparametric and semiparametric approaches to linear mixed model regression. *Journal of Statistical Computation and Simulation, 85*(5), 956–977.
Wilk, M. B., & Gnanadesikan, R. (1968). Probability plotting methods for the analysis of data. *Biometrika, 55*, 1–17.
Winsor, C. P., & Clarke, C. L. (1940). *Journal of Marine Research, 3*, 1.
Zhao, P., & Wan, Z. (2024). Robust Nonparametric Regression under Poisoning Attack. Proceedings of the AAAI Conference on Artificial Intelligence, 38(15), 17007–17015. https://doi.org/10.1609/aaai.v38i15.29644.

Index

Page numbers in italic indicate a figure and page numbers in bold indicate a table on the corresponding page.

H

I

K

L

M

S

T

FSC
www.fsc.org
MIX
Papier | Fördert gute Waldnutzung
FSC® C083411

Kolibri 360 GmbH
Ferdinand-Jühlke-Straße 7
99095 Erfurt, Deutschland
produktsicherheit@kolibri360.de